function spaces

PURE AND APPLIED MATHEMATICS

A Program of Monographs, Textbooks, and Lecture Notes

LECTURE NOTES IN PURE AND APPLIED MATHEMATICS

1. *N. Jacobson,* Exceptional Lie Algebras
2. *L.-Å. Lindahl and F. Poulsen,* Thin Sets in Harmonic Analysis
3. *I. Satake,* Classification Theory of Semi-Simple Algebraic Groups
4. *F. Hirzebruch, W. D. Newmann, and S. S. Koh, Differentiable Manifolds and Quadratic Forms*
5. *I. Chavel,* Riemannian Symmetric Spaces of Rank One
6. *R. B. Burckel,* Characterization of C(X) Among Its Subalgebras
7. *B. R. McDonald, A. R. Magid, and K. C. Smith,* Ring Theory: Proceedings of the Oklahoma Conference
8. *Y.-T. Siu,* Techniques of Extension on Analytic Objects
9. *S. R. Caradus, W. E. Pfaffenberger, and B. Yood,* Calkin Algebras and Algebras of Operators on Banach Spaces
10. *E. O. Roxin, P.-T. Liu, and R. L. Sternberg,* Differential Games and Control Theory
11. *M. Orzech and C. Small,* The Brauer Group of Commutative Rings
12. *S. Thomier,* Topology and Its Applications
13. *J. M. Lopez and K. A. Ross,* Sidon Sets
14. *W. W. Comfort and S. Negrepontis,* Continuous Pseudometrics
15. *K. McKennon and J. M. Robertson,* Locally Convex Spaces
16. *M. Carmeli and S. Malin,* Representations of the Rotation and Lorentz Groups: An Introduction
17. *G. B. Seligman,* Rational Methods in Lie Algebras
18. *D. G. de Figueiredo,* Functional Analysis: Proceedings of the Brazilian Mathematical Society Symposium
19. *L. Cesari, R. Kannan, and J. D. Schuur,* Nonlinear Functional Analysis and Differential Equations: Proceedings of the Michigan State University Conference
20. *J. J. Schäffer,* Geometry of Spheres in Normed Spaces
21. *K. Yano and M. Kon,* Anti-Invariant Submanifolds
22. *W. V. Vasconcelos,* The Rings of Dimension Two
23. *R. E. Chandler,* Hausdorff Compactifications
24. *S. P. Franklin and B. V. S. Thomas,* Topology: Proceedings of the Memphis State University Conference
25. *S. K. Jain,* Ring Theory: Proceedings of the Ohio University Conference
26. *B. R. McDonald and R. A. Morris,* Ring Theory II: Proceedings of the Second Oklahoma Conference
27. *R. B. Mura and A. Rhemtulla,* Orderable Groups
28. *J. R. Graef,* Stability of Dynamical Systems: Theory and Applications
29. *H.-C. Wang,* Homogeneous Branch Algebras
30. *E. O. Roxin, P.-T. Liu, and R. L. Sternberg,* Differential Games and Control Theory II
31. *R. D. Porter,* Introduction to Fibre Bundles
32. *M. Altman,* Contractors and Contractor Directions Theory and Applications
33. *J. S. Golan,* Decomposition and Dimension in Module Categories
34. *G. Fairweather,* Finite Element Galerkin Methods for Differential Equations
35. *J. D. Sally,* Numbers of Generators of Ideals in Local Rings
36. *S. S. Miller,* Complex Analysis: Proceedings of the S.U.N.Y. Brockport Conference
37. *R. Gordon,* Representation Theory of Algebras: Proceedings of the Philadelphia Conference
38. *M. Goto and F. D. Grosshans,* Semisimple Lie Algebras
39. *A. I. Arruda, N. C. A. da Costa, and R. Chuaqui,* Mathematical Logic: Proceedings of the First Brazilian Conference
40. *F. Van Oystaeyen,* Ring Theory: Proceedings of the 1977 Antwerp Conference
41. *F. Van Oystaeyen and A. Verschoren,* Reflectors and Localization: Application to Sheaf Theory
42. *M. Satyanarayana,* Positively Ordered Semigroups
43. *D. L Russell,* Mathematics of Finite-Dimensional Control Systems
44. *P.-T. Liu and E. Roxin,* Differential Games and Control Theory III: Proceedings of the Third Kingston Conference, Part A
45. *A. Geramita and J. Seberry,* Orthogonal Designs: Quadratic Forms and Hadamard Matrices
46. *J. Cigler, V. Losert, and P. Michor,* Banach Modules and Functors on Categories of Banach Spaces

47. *P.-T. Liu and J. G. Sutinen,* Control Theory in Mathematical Economics: Proceedings of the Third Kingston Conference, Part B
48. *C. Byrnes,* Partial Differential Equations and Geometry
49. *G. Klambauer,* Problems and Propositions in Analysis
50. *J. Knopfmacher,* Analytic Arithmetic of Algebraic Function Fields
51. *F. Van Oystaeyen,* Ring Theory: Proceedings of the 1978 Antwerp Conference
52. *B. Kadem,* Binary Time Series
53. *J. Barros-Neto and R. A. Artino,* Hypoelliptic Boundary-Value Problems
54. *R. L. Sternberg, A. J. Kalinowski, and J. S. Papadakis,* Nonlinear Partial Differential Equations in Engineering and Applied Science
55. *B. R. McDonald,* Ring Theory and Algebra III: Proceedngs of the Third Oklahoma Conference
56. *J. S. Golan,* Structure Sheaves Over a Noncommutative Ring
57. *T. V. Narayana, J. G. Williams, and R. M. Mathsen,* Combinatorics, Representation Theory and Statistical Methods in Groups: YOUNG DAY Proceedings
58. *T. A. Burton,* Modeling and Differential Equations in Biology
59. *K. H. Kim and F. W. Roush,* Introduction to Mathematical Consensus Theory
60. *J. Banas and K. Goebel,* Measures of Noncompactness in Banach Spaces
61. *O. A. Nielson,* Direct Integral Theory
62. *J. E. Smith, G. O. Kenny, and R. N. Ball,* Ordered Groups: Proceedings of the Boise State Conference
63. *J. Cronin,* Mathematics of Cell Electrophysiology
64. *J. W. Brewer,* Power Series Over Commutative Rings
65. *P. K. Kamthan and M. Gupta,* Sequence Spaces and Series
66. *T. G. McLaughlin,* Regressive Sets and the Theory of Isols
67. *T. L. Herdman, S. M. Rankin III, and H. W. Stech,* Integral and Functional Differential Equations
68. *R. Draper,* Commutative Algebra: Analytic Methods
69. *W. G. McKay and J. Patera,* Tables of Dimensions, Indices, and Branching Rules for Representations of Simple Lie Algebras
70. *R. L. Devaney and Z. H. Nitecki,* Classical Mechanics and Dynamical Systems
71. *J. Van Geel,* Places and Valuations in Noncommutative Ring Theory
72. *C. Faith,* Injective Modules and Injective Quotient Rings
73. *A. Fiacco,* Mathematical Programming with Data Perturbations I
74. *P. Schultz, C. Praeger, and R. Sullivan,* Algebraic Structures and Applications: Proceedings of the First Western Australian Conference on Algebra
75. *L Bican, T. Kepka, and P. Nemec,* Rings, Modules, and Preradicals
76. *D. C. Kay and M. Breen,* Convexity and Related Combinatorial Geometry: Proceedings of the Second University of Oklahoma Conference
77. *P. Fletcher and W. F. Lindgren,* Quasi-Uniform Spaces
78. *C.-C. Yang,* Factorization Theory of Meromorphic Functions
79. *O. Taussky,* Ternary Quadratic Forms and Norms
80. *S. P. Singh and J. H. Burry,* Nonlinear Analysis and Applications
81. *K. B. Hannsgen, T. L. Herdman, H. W. Stech, and R. L. Wheeler,* Volterra and Functional Differential Equations
82. *N. L. Johnson, M. J. Kallaher, and C. T. Long,* Finite Geometries: Proceedings of a Conference in Honor of T. G. Ostrom
83. *G. I. Zapata,* Functional Analysis, Holomorphy, and Approximation Theory
84. *S. Greco and G. Valla,* Commutative Algebra: Proceedings of the Trento Conference
85. *A. V. Fiacco,* Mathematical Programming with Data Perturbations II
86. *J.-B. Hiriart-Urruty, W. Oettli, and J. Stoer,* Optimization: Theory and Algorithms
87. *A. Figa Talamanca and M. A. Picardello,* Harmonic Analysis on Free Groups
88. *M. Harada,* Factor Categories with Applications to Direct Decomposition of Modules
89. *V. I. Istrăţescu,* Strict Convexity and Complex Strict Convexity
90. *V. Lakshmikantham,* Trends in Theory and Practice of Nonlinear Differential Equations
91. *H. L. Manocha and J. B. Srivastava,* Algebra and Its Applications
92. *D. V. Chudnovsky and G. V. Chudnovsky,* Classical and Quantum Models and Arithmetic Problems
93. *J. W. Longley,* Least Squares Computations Using Orthogonalization Methods
94. *L. P. de Alcantara,* Mathematical Logic and Formal Systems
95. *C. E. Aull,* Rings of Continuous Functions
96. *R. Chuaqui,* Analysis, Geometry, and Probability
97. *L. Fuchs and L. Salce,* Modules Over Valuation Domains

98. *P. Fischer and W. R. Smith,* Chaos, Fractals, and Dynamics
99. *W. B. Powell and C. Tsinakis,* Ordered Algebraic Structures
100. *G. M. Rassias and T. M. Rassias,* Differential Geometry, Calculus of Variations, and Their Applications
101. *R.-E. Hoffmann and K. H. Hofmann,* Continuous Lattices and Their Applications
102. *J. H. Lightbourne III and S. M. Rankin III,* Physical Mathematics and Nonlinear Partial Differential Equations
103. *C. A. Baker and L, M. Batten,* Finite Geometrics
104. *J. W. Brewer, J. W. Bunce, and F. S. Van Vleck,* Linear Systems Over Commutative Rings
105. *C. McCrory and T. Shifrin,* Geometry and Topology: Manifolds, Varieties, and Knots
106. *D. W. Kueker, E. G. K. Lopez-Escobar, and C. H. Smith,* Mathematical Logic and Theoretical Computer Science
107. *B.-L. Lin and S. Simons,* Nonlinear and Convex Analysis: Proceedings in Honor of Ky Fan
108. *S. J. Lee,* Operator Methods for Optimal Control Problems
109. *V. Lakshmikantham,* Nonlinear Analysis and Applications
110. *S. F. McCormick,* Multigrid Methods: Theory, Applications, and Supercomputing
111. *M. C. Tangora,* Computers in Algebra
112. *D. V. Chudnovsky and G. V. Chudnovsky,* Search Theory: Some Recent Developments
113. *D. V. Chudnovsky and R. D. Jenks,* Computer Algebra
114. *M. C. Tangora,* Computers in Geometry and Topology
115. *P. Nelson, V. Faber, T. A. Manteuffel, D. L. Seth, and A. B. White, Jr.,* Transport Theory, Invariant Imbedding, and Integral Equations: Proceedings in Honor of G. M. Wing's 65th Birthday
116. *P. Clément, S. Invernizzi, E. Mitidieri, and I. I. Vrabie,* Semigroup Theory and Applications
117. *J. Vinuesa,* Orthogonal Polynomials and Their Applications: Proceedings of the International Congress
118. *C. M. Dafermos, G. Ladas, and G. Papanicolaou,* Differential Equations: Proceedings of the EQUADIFF Conference
119. *E. O. Roxin,* Modern Optimal Control: A Conference in Honor of Solomon Lefschetz and Joseph P. Lasalle
120. *J. C. Díaz,* Mathematics for Large Scale Computing
121. *P. S. Milojević,* Nonlinear Functional Analysis
122. *C. Sadosky,* Analysis and Partial Differential Equations: A Collection of Papers Dedicated to Mischa Cotlar
123. *R. M. Shortt,* General Topology and Applications: Proceedings of the 1988 Northeast Conference
124. *R. Wong,* Asymptotic and Computational Analysis: Conference in Honor of Frank W. J. Olver's 65th Birthday
125. *D. V. Chudnovsky and R. D. Jenks,* Computers in Mathematics
126. *W. D. Wallis, H. Shen, W. Wei, and L. Zhu,* Combinatorial Designs and Applications
127. *S. Elaydi,* Differential Equations: Stability and Control
128. *G. Chen, E. B. Lee, W. Littman, and L. Markus,* Distributed Parameter Control Systems: New Trends and Applications
129. *W. N. Everitt,* Inequalities: Fifty Years On from Hardy, Littlewood and Pólya
130. *H. G. Kaper and M. Garbey,* Asymptotic Analysis and the Numerical Solution of Partial Differential Equations
131. *O. Arino, D. E. Axelrod, and M. Kimmel,* Mathematical Population Dynamics: Proceedings of the Second International Conference
132. *S. Coen,* Geometry and Complex Variables
133. *J. A. Goldstein, F. Kappel, and W. Schappacher,* Differential Equations with Applications in Biology, Physics, and Engineering
134. *S. J. Andima, R. Kopperman, P. R. Misra, J. Z. Reichman, and A. R. Todd,* General Topology and Applications
135. *P Clément, E. Mitidieri, B. de Pagter,* Semigroup Theory and Evolution Equations: The Second International Conference
136. *K. Jarosz*, Function Spaces
137. *J. M. Bayod, N. De Grande-De Kimpe, and J. Martínez-Maurica*, *p*-adic Functional Analysis
138. *G. A. Anastassiou*, Approximation Theory: Proceedings of the Sixth Southeastern Approximation Theorists Annual Conference
139. *R. S. Rees*, Graphs, Matrices, and Designs: Festschrift in Honor of Norman J. Pullman
140. *G. Abrams, J. Haefner, and K. M. Rangaswamy*, Methods in Module Theory

141. *G. L. Mullen and P. J.-S. Shiue*, Finite Fields, Coding Theory, and Advances in Communications and Computing
142. *M. C. Joshi and A. V. Balakrishnan*, Mathematical Theory of Control: Proceedings of the International Conference
143. *G. Komatsu and Y. Sakane*, Complex Geometry: Proceedings of the Osaka International Conference
144. *I. J. Bakelman*, Geometric Analysis and Nonlinear Partial Differential Equations
145. *T. Mabuchi and S. Mukai*, Einstein Metrics and Yang–Mills Connections: Proceedings of the 27th Taniguchi International Symposium
146. *L. Fuchs and R. Göbel*, Abelian Groups: Proceedings of the 1991 Curaçao Conference
147. *A. D. Pollington and W. Moran*, Number Theory with an Emphasis on the Markoff Spectrum
148. *G. Dore, A. Favini, E. Obrecht, and A. Venni*, Differential Equations in Banach Spaces
149. *T. West*, Continuum Theory and Dynamical Systems
150. *K. D. Bierstedt, A. Pietsch, W. Ruess, and D. Vogt*, Functional Analysis
151. *K. G. Fischer, P. Loustaunau, J. Shapiro, E. L. Green, and D. Farkas*, Computational Algebra
152. *K. D. Elworthy, W. N. Everitt, and E. B. Lee*, Differential Equations, Dynamical Systems, and Control Science
153. *P.-J. Cahen, D. L. Costa, M. Fontana, and S.-E. Kabbaj*, Commutative Ring Theory
154. *S. C. Cooper and W. J. Thron*, Continued Fractions and Orthogonal Functions: Theory and Applications
155. *P. Clément and G. Lumer*, Evolution Equations, Control Theory, and Biomathematics
156. *M. Gyllenberg and L. Persson*, Analysis, Algebra, and Computers in Mathematical Research: Proceedings of the Twenty-First Nordic Congress of Mathematicians
157. *W. O. Bray, P. S. Milojević, and Č. V. Stanojević*, Fourier Analysis: Analytic and Geometric Aspects
158. *J. Bergen and S. Montgomery*, Advances in Hopf Algebras
159. *A. R. Magid*, Rings, Extensions, and Cohomology
160. *N. H. Pavel*, Optimal Control of Differential Equations
161. *M. Ikawa*, Spectral and Scattering Theory: Proceedings of the Taniguchi International Workshop
162. *X. Liu and D. Siegel*, Comparison Methods and Stability Theory
163. *J.-P. Zolésio*, Boundary Control and Variation
164. *M. Křížek, P. Neittaanmäki, and R. Stenberg*, Finite Element Methods: Fifty Years of the Courant Element
165. *G. Da Prato and L. Tubaro*, Control of Partial Differential Equations
166. *E. Ballico*, Projective Geometry with Applications
167. *M. Costabel, M. Dauge, and S. Nicaise*, Boundary Value Problems and Integral Equations in Nonsmooth Domains
168. *G. Ferreyra, G. R. Goldstein, and F. Neubrander*, Evolution Equations
169. *S. Huggett*, Twistor Theory
170. *H. Cook, W. T. Ingram, K. T. Kuperberg, A. Lelek, and P. Minc*, Continua: With the Houston Problem Book
171. *D. F. Anderson and D. E. Dobbs*, Zero-Dimensional Commutative Rings
172. *K. Jarosz,* Function Spaces: The Second Conference

Additional Volumes in Preparation

function spaces

the second conference

proceedings of the conference at Edwardsville

edited by
Krzysztof Jarosz

Southern Illinois University at Edwardsville
Edwardsville, Illinois

Marcel Dekker, Inc. **New York • Basel • Hong Kong**

Library of Congress Cataloging-in-Publication

Function spaces, the second conference : proceedings of the conference at Edwardsville / edited by Krzysztof Jarosz.
p. cm. — (Lecture notes in pure and applied mathematics ; 172)
"The Second Conference on Function Spaces was held at Southern Illinois University at Edwardsville, from May 24 through May 28, 1994"—Pref.
Includes bibliographic references and index.
ISBN 0-8247-9665-9 (pbk. : acid-free)
1. Function spaces—Congresses. I. Jarosz, Krzysztof II. Conference on Function Spaces (2nd : 1994 : Southern Illinois University) III. Series: Lecture notes in pure and applied mathematics : v. 172.
QA323.F875 1995
515'.7—dc20 95-9482
CIP

The publisher offers discounts on this book when ordered in bulk quantities. For more information, write to Special Sales/Professional Marketing at the address below.

This book is printed on acid-free paper.

MARCEL DEKKER, INC.
270 Madison Avenue, New York, New York 10016

Current printing (last digit):
10 9 8 7 6 5 4 3 2 1

PRINTED IN THE UNITED STATES OF AMERICA

Preface

The *Second Conference on Function Spaces* was held at Southern Illinois University at Edwardsville. It attracted approximately 120 participants representing 20 countries.

Most of the papers in this volume are expanded texts of lectures delivered at the conference, while some others have been included by invitation. Some of the articles contain expositions of known results; some of them present fresh discoveries, and still others contain both ingredients. The papers cover a wide range of topics, including spaces and algebras of analytic functions of one and of many variables, L_p-spaces, spaces of Banach-valued functions, isometries of function spaces, geometry of Banach spaces, Banach algebras, and others.

The editor would like to thank everyone who contributed to the Proceedings: the authors, the referees, and Marcel Dekker, Inc.

Krzysztof Jarosz

Contents

Preface iii

Contributors ix

Two Constant Theorems for Hardy Spaces in Several Variables 1
John T. Anderson and Joseph A. Cima

On Locally Convex Algebras with Cyclic Bases 11
Hugo Arizmendi and Angel Carrillo

Some Remarks on Norm-Attaining n-Linear Forms 19
R. M. Aron, C. Finet, and E. Werner

On Uniformly Closed Ideals in f-Algebras 29
Mohamed Basly and Abdelmajid Triki

A Strong Superdensity Property for Some Subspaces of $C(X)$ 35
Alain Bernard

Ultraseparating Function Spaces, Operating Functions, and Spaces of Continuous Functions Whose nth Powers Span a Dense Subspace of $C(X)$ 43
Eggert Briem

Small-Bound Isomorphisms of Function Spaces 51
C.-H. Chu and H. B. Cohen

Amenability, Weak Amenability, and the Close Homomorphism Property for Commutative Banach Algebras 59
P. C. Curtis, Jr.

Kadec–Klee Properties for $L(l_p, l_q)$ 71
S. J. Dilworth and Denka Kutzarova

Regular Retractions onto Finite Dimensional Convex Sets 85
Tadeusz Dobrowolski and Jerzy Mogilski

Contractive Projections on Lebesgue–Bochner Spaces 101
Ian Doust

Some Measures of Convexity in Banach Spaces 111
Patrick N. Dowling, Zhibao Hu, and Douglas Mupasiri

Regularity Conditions for Banach Function Algebras 117
J. F. Feinstein

Some Bargmann Spaces of Analytic Functions 123
D. J. H. Garling and P. Wojtaszczyk

Characters of Function Algebras on Banach Spaces 139
M. I. Garrido, J. Gómez Gil, and J. A. Jaramillo

On the Corona Theorem 147
Pamela Gorkin and Raymond Mortini

Spectral Synthesis of Ideals in Classical Algebras of Smooth Functions 167
Leonid G. Hanin

A Characterization of Lacunary Sets and Spectral Properties of Fourier Multipliers 183
Osamu Hatori

On the Extremal Structure of the Unit Ball of the Space $C(K,X)^*$ 205
Zhibao Hu and Mark A. Smith

On Certain Banach Algebras of Vector-Valued Functions 223
Robert Kantrowitz and Michael M. Neumann

Weighted Spaces of Holomorphic Functions and Analytic Functionals 243
Le Hai Khoi

Isometries of L_p-Spaces of Solutions of Homogeneous Partial Differential Equations 251
Alexander Koldobsky

L-Projections on Banach Lattices 265
Pei-Kee Lin

A Survey of Mean Convergence of Orthogonal Polynomial Expansions 281
D. S. Lubinsky

Copies of Classical Sequence Spaces in Vector-Valued-Function Banach Spaces 311
José Mendoza

Boyd Indices of Orlicz–Lorentz Spaces 321
Stephen J. Montgomery-Smith

A Note on the Looman–Menchoff Theorem 335
N. V. Rao

On the Extreme Point Intersection Property 339
T. S. S. R. K. Rao

A Singular Integral Operator with a Large Eigenvector 347
Richard Rochberg

Some Very Dense Subspaces of $C(X)$ 353
S. J. Sidney

Antosik's Interchange Theorem 361
Charles Swartz

On the Continuity of Random Derivations 371
M. V. Velasco and A. R. Villena

Subdifferentiability and the Noncommutative Banach–Stone Theorem 377
W. Werner

Index 387

Contributors

JOHN T. ANDERSON College of the Holy Cross, Worcester, Massachusetts

HUGO ARIZMENDI Universidad Nacional Autónoma, Mexico City, Mexico

R. M. ARON Kent State University, Kent, Ohio

M. BASLY University of Tunis, Tunis, Tunisia

ALAIN BERNARD Fourier Institute, University of Grenoble, St. Martin d'Hères, France

EGGERT BRIEM Science Institute, University of Iceland, Reykjavik, Iceland

ANGEL CARILLO Universidad Nacional Autónoma, Mexico City, Mexico

C.-H. CHU Goldsmiths College, University of London, London, England

JOSEPH A. CIMA University of North Carolina, Chapel Hill, North Carolina

H. B. COHEN University of Pittsburgh, Pittsburgh, Pennsylvania

P. C. CURTIS, JR. University of California, Los Angeles, California

S. J. DILWORTH University of South Carolina, Columbia, South Carolina

TADEUSZ DOBROWOLSKI Pittsburg State University, Pittsburg, Kansas

IAN DOUST University of New South Wales, Sydney, Australia

PATRICK N. DOWLING Miami University, Oxford, Ohio

J. F. FEINSTEIN University of Nottingham, Nottingham, England

C. FINET Université de Mons-Hainaut, Mons, Belgium

D. J. H. GARLING University of Cambridge, Cambridge, England

M. I. GARRIDO Universidad de Extremadura, Badajoz, Spain

J. GÓMEZ GIL Universidad Complutense de Madrid, Madrid, Spain

PAMELA GORKIN Bucknell University, Lewisburg, Pennsylvania

LEONID G. HANIN Michigan Technological University, Houghton, Michigan

OSAMU HATORI Tokyo Medical College, Tokyo, Japan

ZHIBAO HU Miami University, Oxford, Ohio

J. A. JARAMILLO Universidad Complutense de Madrid, Madrid, Spain

ROBERT KANTROWITZ Hamilton College, Clinton, New York

LE HAI KHOI Uppsala University, Uppsala, Sweden

ALEXANDER KOLDOBSKY University of Texas at San Antonio, San Antonio, Texas

DENKA KUTZAROVA Institute of Mathematics, Bulgarian Academy of Sciences, Sofia, Bulgaria

PEI-KEE LIN University of Memphis, Memphis, Tennessee

D. S. LUBINSKY University of the Witwatersrand, Wits, Republic of South Africa

JOSÉ MENDOZA Universidad Complutense de Madrid, Madrid, Spain

JERZY MOGILSKI Universidad Autonoma Metropolitana, Mexico City, Mexico

STEPHEN J. MONTGOMERY-SMITH University of Missouri at Columbia, Columbia, Missouri

RAYMOND MORTINI Mathematical Institute, Universität Karlsruhe, Karlsruhe, Germany

DOUGLAS MUPASIRI University of Northern Iowa, Cedar Falls, Iowa

MICHAEL M. NEUMANN Mississippi State University, Mississippi State, Mississippi

N. V. RAO The University of Toledo, Toledo, Ohio

T. S. S. R. K. RAO Indian Statistical Institute, Bangalore, India

RICHARD ROCHBERG Washington University, St. Louis, Missouri

S. J. SIDNEY The University of Connecticut, Storrs, Connecticut

MARK A. SMITH Miami University, Oxford, Ohio

CHARLES SWARTZ New Mexico State University, Las Cruces, New Mexico

ABDELMAJID TRIKI University of Tunis, Tunis, Tunisia

M. V. VELASCO Universidad de Granada, Granada, Spain

A. R. VILLENA Universidad de Granada, Granada, Spain

E. WERNER Case Western Reserve University, Cleveland, Ohio, and Université de Lille 1, Lille, France

W. WERNER Universität-GH-Paderborn, Paderborn, Germany

P. WOJTASZCZYK Institute of Mathematics, Polish Academy of Sciences, Warsaw, Poland

Two Constant Theorems for Hardy Spaces in Several Variables

JOHN T. ANDERSON Department of Mathematics, College of the Holy Cross, Worcester, Massachusetts 01610-2395

JOSEPH A. CIMA Department of Mathematics, University of North Carolina, Chapel Hill, North Carolina, 27599-3250

1 INTRODUCTION

Let Ω be a domain in the complex plane bounded by a simple closed rectifiable curve Γ. The classical maximum principle states that any function f holomorphic in Ω and continuous on the closure of Ω assumes its maximum modulus on Γ. Among the many refinements of the maximum principle is the "Two-Constant Theorem" discovered by the Nevanlinna brothers and A. Ostrowski in 1922: (see [3], chapter VI) suppose that Γ is divided into two subarcs Γ_+ and Γ_-, that f is holomorphic and bounded in Ω, and that

$$\limsup_{z\in D\to\zeta} |f(z)| \leq \begin{cases} M & \text{for } \zeta \in \Gamma_+ \\ m & \text{for } \zeta \in \Gamma_- \end{cases}$$

Then for $z \in \Omega$,

$$|f(z)| \leq m^{1-\omega(z,\Gamma_+,\Omega)}\ M^{\omega(z,\Gamma_+,\Omega)}$$

where $\omega(z, \Gamma_+, \Omega)$ is harmonic measure for Γ_+ with respect to Ω at $z \in \Omega$ (i.e., the solution of the Dirichlet problem on Ω with boundary data 1 on Γ_+, 0 on Γ_-). For domains such as the unit disk where harmonic measure can be calculated explicitly, the two constant theorem gives precise estimates on $|f|$ in terms of the distance of z to Γ_+. Fuchs' book [3] contains several applications of the two-constant theorem, including the Phragmén-Lindelöf theorem for half-planes.

There are many well-known classes of functions which are not bounded but still admit boundary values in some restricted sense. For example, consider on the unit disk $D = \{z \in \mathbf{C} : |z| < 1\}$ the classical Hardy space $H^p(D)$ $(0 < p < \infty)$ consisting of all functions f holomorphic in D such that the integral means

$$M_f(r) = \int_0^{2\pi} |f(re^{i\theta})|^p \, d\theta$$

are uniformly bounded for $0 < r < 1$. If $f \in H^p(D)$ then the radial limits

$$f^*(e^{i\theta}) = \lim_{r \to 1^-} f(re^{i\theta})$$

exist for almost all $\theta \in [0, 2\pi)$. On the other hand, $\limsup_{z \in D \to \zeta} |f(z)|$ may be infinite on any subarc of $\Gamma = \{\zeta : |\zeta| = 1\}$, so there is no hope of establishing a pointwise estimate such as that in the classical two-constant theorem. However, there is a possible substitute for the maximum principle in terms of the integral means $M_f(r)$. In fact, $M_f(r)$ is increasing for $0 < r < 1$ and

$$\lim_{r \to 1^-} M_f(r) = \int_0^{2\pi} |f^*(e^{i\theta})|^p \, d\theta$$

This suggests the possibility of a two-constant theorem for *integrals* of functions in the Hardy classes. A result of this type was proved by Akopian [1] (stated by him for the upper half plane, but easily transferred to the unit disk): let Γ_+ be the upper semicircle $\{z \in \Gamma : \operatorname{Im}(z) > 0\}$, Γ_- the lower semicircle $\Gamma \setminus \Gamma_+$, and Γ_θ the circular arc in D joining $z = 1$ and $z = -1$, making an angle θ $(0 < \theta < \pi)$ with Γ at $z = 1$. If $f \in H^p(D)$, Akopian's inequality states that

$$\int_{\Gamma_\theta} |f(z)|^p \, dz \le \left(\int_{\Gamma_+} |f^*(e^{i\psi})|^p \, d\psi \right)^{1 - \frac{\theta}{\pi}} \left(\int_{\Gamma_-} |f^*(e^{i\psi})|^p \, d\psi \right)^{\frac{\theta}{\pi}} \tag{1}$$

A further analogy with the pointwise two-constant theorem is provided by the fact that the curves Γ_θ are in fact level sets for harmonic measure (for Γ_+) on D. We mention one application of Akopian's theorem: if we take $\theta = \pi/2$ in (1) ($\Gamma_{\pi/2}$ is just the segment [-1,1]) and apply the arithmetic-geometric mean inequality to the right-hand side, we obtain a classical inequality of Fejer and Reisz:

$$\int_{-1}^{1} |f(x)|^p \, dx \le \frac{1}{2} \int_0^{2\pi} |f^*(e^{i\theta})|^p \, d\theta$$

This inequality has a well-known geometric consequence: if ψ maps D conformally to a domain Ω with rectifiable boundary Γ, then $\psi' \in H^1(D)$. Taking $f = \psi'$ and $p = 1$ in we find that the length of the image under ψ of any diameter of D is no more than one-half the length of Γ.

Our main goal in this paper is to extend Akopian's result to the setting of the unit ball and the Siegel upper half-space (a natural analogue of the upper half-plane) in $\mathbf{C}^n, n > 1$, which we do in section 3. In the next section we comment in more detail on Akopian's theorem for the upper half-plane and some of its consequences.

2 AKOPIAN'S THEOREM

Let $U = \{z \in \mathbf{C} : \mathrm{Im}(z) > 0\}$ be the upper half-plane. For $0 < p < \infty$, $H^p(U)$ will denote the classical Hardy space of functions holomorphic in U such that

$$\|f\|_p = \sup_{y>0} \left[\int_{-\infty}^{\infty} |f(x+iy)|^p dx \right]^{1/p} < \infty$$

For $f \in H^p(U)$, the boundary values $f^*(x) = \lim_{y \to 0} f(x+iy)$ exist for almost all x, and $f^* \in L^p(-\infty, \infty)$ with respect to Lebesgue measure.

We will consider integral means on the rays $\gamma_\theta = \{z \in U : \mathrm{Arg(z)} = \theta\}$. For $0 < \theta < \pi$ and a measurable function g defined on U set

$$M_\theta(g) = \int_0^\infty |g(re^{i\theta})|^p dr = e^{-i\theta} \int_{\gamma_\theta} |g|^p dz$$

It is not clear *a priori* that $M_\theta(f)$ is finite for $f \in H^p(U)$, although this can be established, for example, by a simple maximal function argument (a similar argument in several variables is given in detail in section 2). If $f \in H^p(U)$, we can define

$$M_+(f) = \int_0^\infty |f^*(r)|^p dr \quad \text{and} \quad M_-(f) = \int_0^\infty |f^*(-r)|^p dr$$

and so extend M_θ to $[0, \pi]$ (with $M_0 \equiv M_+, M_\pi \equiv M_-$). Then we have the following "two-constant" theorem due to Akopian [1]:

THEOREM 1. *If* $0 < p < \infty$, *and* $f \in H^p(U)$, *then* $\log M_\theta(f)$ *is a convex function of* θ *in* $[0, \pi]$. *In particular,*

$$M_\theta(f) \le M_+(f)^{1-\frac{\theta}{\pi}} M_-(f)^{\frac{\theta}{\pi}} \tag{2}$$

It is interesting that the uniform boundedness of the means $M_\theta(f)$ actually characterizes $H^p(U)$, as was shown by Sedletckii.

THEOREM 2. (Sedletckii [10]) *If* $0 < p < \infty$, *and* $H^p_*(U)$ *denotes the class of functions holomorphic in* U *such that*

$$\|f\|_{*,p} = \sup_{\theta \in (0,\pi)} [M_\theta(f)]^{1/p} < \infty$$

then $H^p_*(U) = H^p(U)$. *Moreover there exist positive constants* C, C' *so that* $C\|f\|_p \le \|f\|_{*,p} \le C'\|f\|_p$.

As noted in the introduction, it is straightforward to transfer Akopian's theorem to H^p spaces on the unit disk D, using the fact (see [5], p.128ff.) that for $1 \leq p \leq \infty$,

$$f \in H^p(D) \Longleftrightarrow \frac{f \circ \phi^{-1}(z)}{(1-iz)^{2/p}} \in H^p(U) \tag{3}$$

where $\phi(z) = i\left(\frac{1-z}{1+z}\right)$ maps D conformally to U. More generally, we can exploit the fact that the curves Γ_θ appearing in (1) are level sets for harmonic measure of Γ_+ to establish the following fact: Let Ω be a simply connected plane domain bounded by a simple closed rectifiable curve Γ. Define $H^p(\Omega)$ as follows: choose a sequence of closed rectifiable curves C_n lying in Ω, such that any compact set $K \subset \Omega$ lies in the region bounded by C_n for all large n. Then $H^p(\Omega)$ is the class of functions holomorphic in Ω with $\sup_n \int_{C_n} |f(z)|^p |dz| < \infty$. It is shown in [2], chp.10 (where this class is denoted $E^p(\Omega)$) that the resulting space is independent of the choice of the curves C_n and that $f \in H^p(\Omega)$ if and only if $(f \circ \psi)(\psi')^{-1/p} \in H^p(D)$ for any conformal map $\psi : D \mapsto \Omega$. Let Γ_+ be a subarc of $\Gamma, \Gamma_- = \Gamma \setminus \Gamma_+$, and $\omega(z, \Gamma_+, \Omega)$ the harmonic measure of Γ_+ with respect to Ω at $z \in \Omega$. For $0 < t < 1$ set $\Gamma_t = \{z \in \Omega \, : \, \omega(z, \Gamma_+, \Omega) = t\}$. Then for $f \in H^p(\Omega)$,

$$\int_{\Gamma_t} |f(z)|^p |dz| \leq \left(\int_{\Gamma_+} |f^*(z)|^p |dz|\right)^{1-t} \left(\int_{\Gamma_-} |f^*(z)|^p |dz|\right)^t \tag{4}$$

In the next section we turn to the generalization of Akopian's theorem to several variables. It turns out to be easiest to start with Akopian's theorem on the upper half-plane, lift this to the Siegel upper-half space in $\mathbf{C}^n$, then transfer the result to the unit ball.

3 AKOPIAN'S THEOREM IN $C^n, n > 1$

Throughout this section we denote coordinates in $\mathbf{C}^n$ as either $\zeta = (\zeta_1, \zeta')$ or $z = (z_1, z')$ where $z_1, \zeta_1 \in \mathbf{C}$ and $\zeta', z' \in \mathbf{C}^{n-1}$. We let B_n denote the unit ball in $\mathbf{C}^n, U_n$ the Siegel upper-half space

$$U_n = \{z \in \mathbf{C}^n : \mathrm{Im}(z_1) > |z'|^2\}$$

The Cayley map

$$\phi(z) = (i\frac{1-z_1}{1+z_1}, \frac{z'}{1+z_1})$$

maps B_n biholomorphically to U_n. We recall below some basic facts about Hardy spaces on U_n and B_n; see [12], chapter XII and [9] as general references on $H^p(U_n)$ and $H^p(B_n)$ respectively.

The classical Hardy space $H^p(U_n)$ consists of those functions holomorphic in U_n satisfying

$$\|f\|_{H^p(U_n)} = \sup_{y>0} \left[\int_{\partial U_n} |f(\zeta_1 + iy, \zeta')|^p d\mu(\zeta)\right]^{1/p} < \infty$$

where $d\mu(\zeta) = d\mathrm{Re}(\zeta_1) d\lambda(\zeta'), d\lambda$ being $2n-2$-dimensional Lebesgue measure on $\mathbf{C}^{n-1}$ For $f \in H^p(U_n), f^*(\zeta) = \lim_{y \to 0} f(\zeta_1 + iy, \zeta')$ exists for $d\mu$ - a.a. $\zeta \in \partial U_n$, $f^* \in L^p(\partial U_n, d\mu)$, and the norm in $L^p(\partial U_n, d\mu)$ is equivalent to the $H^p(U_n)$ on boundary values of H^p functions. $H^p(B_n)$ consists of those functions holomorphic in B_n satisfying

$$\|f\|_{H^p(B_n)} = \sup_{r<1}\left[\int_{\partial B_n} |f(r\zeta)|^p d\sigma(\zeta)\right]^{1/p} \leq \infty$$

where $d\sigma$ is normalized Euclidean measure on ∂B_n. For $f \in H^p(B_n), \lim_{r\to 1} f(r\zeta) = f^*(\zeta)$ exists σ - a.e. on $\partial B_n, f^* \in L^p(\partial B_n, d\sigma)$, and, as for U_n ,the boundary norm is equivalent to the H^p norm.

It is useful to have a description of the relationship between $H^p(U_n)$ and $H^p(B_n)$ analogous to (2). For $p = 2$ this correspondence is given in [7]; see also [12], p.575. The following proposition, which gives the correspondence for general p, is perhaps well-known, but we have been unable to find a reference. We give a proof along the lines of the one-variable proof in [5]. Set $q(z) = 1 - iz_1$. q is non-vanishing on U_n, so we may fix a branch of $\log(q)$ on U_n (and the corresponding branch of the logarithm of $q \circ \phi(z) = 2(1+z_1)^{-1}$ on B_n).

PROPOSITION 1. *For* $1 < p < \infty$,

(a) *If* $f \in H^p(U_n)$*then* $f \circ \phi \in H^p(B_n)$;

(b) *If* f *is holomorphic in* B_n, *then* $f \in H^p(B_n) \Longleftrightarrow F = (f \circ \phi^{-1})q^{-2n/p} \in H^p(U_n)$

Proof: (a) For g nonnegative and plurisubharmonic in U_n with

$$\sup_{y>0} \int_{\partial U_n} g(\zeta_1 + iy, \zeta')\, d\mu(\zeta) = M < \infty$$

it can be shown, using arguments similar to those in ([2], chp.11.1,Lemma 1) that for $z \in U_n$,

$$g(z) \leq C\frac{M}{(\mathrm{Im}(z_1) - i|z'|^2)^n}$$

for some positive constant C independent of g. This implies that $g \circ \phi$ is bounded, for $t > 0$ fixed, on the ellipsoid

$$\phi^{-1}(\{(\mathrm{Im}(z_1) - i|z'|^2 > t\}) = \{|z_1 + \frac{t}{1+t}|^2 + \frac{1}{1+t}|z'|^2 < \frac{1}{(1+t)^2}\}$$

Using this fact it is not hard to show (cf. [2], chp.11.1, Lemma 2) that $g \circ \phi$ has a harmonic majorant in B_n. Taking $g = |f|^p$ where $f \in H^p(U_n)$, we find that $|f \circ \phi|^p$ has a harmonic majorant in B_n, which implies (see [11], Theorem 1) that $f \circ \phi \in H^p(B_n)$.

(b) We let $\mathcal{P}_{B_n}$ and $\mathcal{P}_{U_n}$ denote the Poisson-Szegö kernels for B_n and U_n respectively (see [6]):

$$\mathcal{P}_{B_n}(z, \zeta) = \frac{1 - |z|^2}{|1 - \langle z, \zeta\rangle|^{2n}}$$

$$\mathcal{P}_{U_n}(z,\zeta) = c_n \frac{(\mathrm{Im}(z_1) - |z'|^2)^n}{|i(\bar{z_1} - \zeta_1) - 2\langle z', \zeta' \rangle|^{2n}}, \quad c_n = \frac{4^{n-1}(n-1)!}{\pi^n}$$

If $(\mathcal{D}, d\nu)$ denotes either $(B_n, d\sigma)$ or $(U_n, d\mu)$ then we set for $g \in L^p(\partial\mathcal{D}, d\nu)$

$$\mathcal{P}_{\mathcal{D}}[g](z) = \int_{\partial\mathcal{D}} g(\zeta)\mathcal{P}_{\mathcal{D}}(z,\zeta)d\nu(\zeta)$$

It is easy to check that

$$d\sigma(\phi^{-1}(z)) = c_n|q(z)|^{-2n}d\mu(z), \tag{5}$$

$$\mathcal{P}_{B_n}[\phi^{-1}(z), \phi^{-1}(\zeta)] = c_n^{-1}|q(\zeta)|^{2n}\mathcal{P}_{U_n}(z,\zeta) \tag{6}$$

and hence that

$$\mathcal{P}_{B_n}[g \circ \phi] = \mathcal{P}_{U_n}[g] \circ \phi \tag{7}$$

The following facts can be found in [6]:

$$\sup_{y>0} \int_{\partial U_n} |\mathcal{P}_{U_n}[g](\zeta_1 + iy, \zeta')|^p \, d\mu(\zeta) < \infty \qquad \text{for } g \in L^p(U_n, d\mu) \tag{8}$$

$$\sup_{r<1} \int_{\partial B_n} |\mathcal{P}_{B_n}[g](r\zeta)|^p \, d\sigma(\zeta) < \infty \qquad \text{for } g \in L^p(B_n, d\sigma) \tag{9}$$

Moreover,

$$\mathcal{P}_{\mathcal{D}}[g^*] = g \ \text{ if } \ g \in H^p(\mathcal{D})$$

Now if $f \in H^p(B_n)$ and $(f \circ \phi^{-1})q^{-2n/p} = F$, then F is holomorphic in U_n. Set $F^* = (f^* \circ \phi^{-1})q^{-2n/p}$ on ∂U_n; then it follows from (5) that $F^* \in L^p(\partial U_n, d\mu)$. Moreover, by (5) and (7),

$$\mathcal{P}_{U_n}[F^*] = \mathcal{P}_{B_n}[F^* \circ \phi] \circ \phi^{-1} = \mathcal{P}_{B_n}[f^*(q^{-2n/p} \circ \phi)] \circ \phi^{-1} = [f(q^{-2n/p} \circ \phi)] \circ \phi^{-1} = F$$

so by (8), $F \in H^p(U_n)$. Conversely, if f is holomorphic in B_n and $F \in H^p(U_n)$, then by (a), $F \circ \phi = f(q^{-2n/p} \circ \phi) \in H^p(B_n)$. To show $f \in H^p(B_n)$ we argue as in ([5], Lemma, p.129). Using the plurisubharmonicity of $\log|F \circ \phi|$ and the fact that $\log|G^*| \in L^1(B_n)$ for $G \in H^p(B_n)$ (see [9], Theorem 5.6.4) it is not hard to show that

$$\log|F \circ \phi| \leq \mathcal{P}_{B_n}[\ \log|F^* \circ \phi|\ ] \tag{10}$$

while

$$\log|q^{-2n/p} \circ \phi| = \mathcal{P}_{B_n}[\ \log|q^{-2n/p} \circ \phi|\ ] \tag{11}$$

on B_n, since $\log(q \circ \phi) \in H^p(B_n)$ and $\mathcal{P}_{B_n}$ is real. Subtracting (11) from (10) we obtain

$$\log|f| \leq \mathcal{P}_{B_n}[\ \log|f^*|\ ] \tag{12}$$

where $f^* = (F^* q^{2n/p}) \circ \phi$ on ∂B_n. Using (5) we see that $f^* \in L^p(\partial B_n, d\sigma)$. Exponentiating both sides of (12) and applying Jensen's inequality gives $|f| \leq \mathcal{P}_{B_n}[f^*]$ in B_n; by (9), $f \in H^p(B_n)$. □

REMARK: The equivalence of the H^p norms with the L^p boundary norms together with the transformation formula (5) shows that $\|F\|_{H^p(U_n)}$ is equivalent to $\|f\|_{H^p(B_n)}$.

Now we state our version of Akopian's Theorem for $H^p(U_n)$. Define the hypersurfaces $\Sigma_\theta, \Sigma_+, \Sigma_-$ as follows: set $h(z) = z_1 - i|z'|^2$, then

$$\Sigma_\theta = \{z \in U_n : \operatorname{Arg} h(z) = \theta\}$$

$$\Sigma_+ = \{z \in \partial U_n : \operatorname{Re}(z_1) > 0\}, \quad \Sigma_- = \{z \in \partial U_n : \operatorname{Re}(z_1) < 0\}$$

and let $d\mu_\theta(z)$ be the restriction of $d|h(z)|d\lambda(z')$ to Σ_θ. Note that $d\mu_\theta$ agrees with (the previously defined) $d\mu$ on $\Sigma_\pm$ (i.e., for $\theta = 0, \pi$).

THEOREM 3. *For $f \in H^p(U_n), 1 < p < \infty, \theta \in (0, \pi)$, we have*

$$\int_{\Sigma_\theta} |f|^p d\mu_\theta \leq \left(\int_{\Sigma_+} |f|^p d\mu\right)^{1-\frac{\theta}{\pi}} \left(\int_{\Sigma_-} |f|^p d\mu\right)^{\frac{\theta}{\pi}} \tag{13}$$

Proof. For f holomorphic in U_n and $z' \in \mathbf{C}^{n-1}$ set

$$f_{z'}(z_1) = f(z_1 + i|z'|^2, z')$$

Then $f_{z'}$ is holomorphic as a function of z_1 in $U_1 \equiv U$. Now suppose $f \in H^p(U_n)$ and furthermore assume that $f_{z'} \in H^p(U)$ for each $z' \in \mathbf{C}^{n-1}$. Then by Akopian's Theorem,

$$\int_0^\infty |f_{z'}(re^{i\theta})|^p dr \leq \left(\int_0^\infty |f_{z'}(r)|^p dr\right)^{1-\frac{\theta}{\pi}} \left(\int_0^\infty |f_{z'}(-r)|^p dr\right)^{\frac{\theta}{\pi}} \tag{14}$$

Integrating both sides of (14) with respect to $d\lambda(z')$ over $\mathbf{C}^{n-1}$ and applying Hölder's inequality to the right-hand side gives (13), provided that each slice function $f_{z'}$ belongs to $H^p(U)$. To remove this restriction, we use the following two lemmas. Theorem 3 is an immediate consequence of these.

LEMMA 1. *The set of all functions $f \in H^p(U_n)$ with the property that $f_{z'} \in H^p(U)$ for all $z' \in \mathbf{C}^{n-1}$ is dense in $H^p(U_n)$.*

Proof: Since polynomials are dense in $H^p(B_n)$, by Proposition 1 (and the remark following its proof) the set of all functions of the form

$$\tilde{P} = (P \circ \phi^{-1}) q^{-2n/p}$$

where P is a polynomial, is dense in $H^p(U_n)$. It easy to check that $\tilde{P}_{z'} \in H^p(U)$ for all $z' \in \mathbf{C}^{n-1}$.

LEMMA 2. *For $1 < p < \infty$, there exists $C_p > 0$ such that for all $f \in H^p(U_n)$,*

$$\int_{\Sigma_\theta} |f|^p d\mu \leq C_p \|f\|^p_{H^p(U_n)} \tag{15}$$

Proof: For $\zeta \in \partial U_n, \alpha > 0$, let $K_\alpha(\zeta)$ be the Korányi region:

$$K_\alpha(\zeta) = \{z \in U_n : \frac{|i(\bar{z}_1 - \zeta_1) - 2\langle \zeta', z' \rangle|}{\text{Im}(z_1) - |z'|^2} \leq 1 + \alpha\}$$

and let M_α be the associated maximal function; for a function g defined on U_n and $\zeta \in \partial U_n$ set

$$M_\alpha(g)(\zeta) = \sup\{|g(z)| : z \in K_\alpha(\zeta)\}$$

Korányi [6] showed that for all $\alpha > 0, p > 1$ there exists $C_{\alpha,p} > 0$ so that for all $f \in H^p(U_n)$,

$$\int_{\partial U_n} M^p_\alpha(p) d\mu \ \leq \ C_{\alpha,p} \|f\|^p_{H^p(U_n)} \tag{16}$$

To estimate f on Σ_θ, take $\alpha = 1$. An easy computation shows that

$$(re^{i\theta} + i|\zeta'|^2, \zeta') \in K_1((r\cos(\theta) + r\sin(\theta) + i|\zeta'|^2, \zeta'))$$

for all $r \in (0, \infty), \zeta' \in \mathbf{C}^{n-1}, \theta \in (0, \pi)$. Thus for $f \in H^p(U_n)$,

$$\begin{aligned}
\int_{\Sigma_\theta} |f|^p d\mu &= \int_{\mathbf{C}^{n-1}} \int_0^\infty |f(re^{i\theta} + i|\zeta'|^2, \zeta')|^p \, dr d\lambda(\zeta') \\
&\leq \int_{\mathbf{C}^{n-1}} \int_0^\infty M^p_1(f)(r\cos(\theta) + r\sin(\theta) + i|\zeta'|^2, \zeta') \, dr d\lambda(\zeta') \\
&= \frac{1}{\cos(\theta) + \sin(\theta)} \int_{\mathbf{C}^{n-1}} \int_0^\infty M^p_1(f)(x + i|\zeta'|^2, \zeta') \, dx d\lambda(\zeta'))
\end{aligned}$$

If $\theta \in (0, \frac{\pi}{2}], \cos(\theta) + \sin(\theta) \geq 1$, hence

$$\int_{\Sigma_\theta} |f|^p \leq \int_{\partial U_n} M^p_1(f) d\mu \tag{17}$$

Combining (16) and (17) gives the desired inequality (15) for $\theta \in (0, \pi/2]$. For $\theta \in (\pi/2, \pi)$, use the fact that

$$(re^{i\theta} + i|\zeta'|^2, \zeta') \in K_1(r\cos(\theta) - r\sin(\theta) + i|\zeta'|^2, \zeta'))$$

to obtain

$$\int_{\Sigma_\theta} |f|^p d\mu \leq \frac{1}{|\cos(\theta) - \sin(\theta)|} \int_{\partial U_n} M^p_1(f) d\mu \leq \int_{\partial U_n} M^p_1(f) d\mu$$

for $\theta \in (\pi/2, \pi)$ and conclude that (15) holds for all $\theta \in (0, \pi)$. This completes the proof of the lemma and Theorem 3. □

One can also prove a version of Theorem 3 for the ball. For $\theta \in [0,\pi), \theta \neq \pi/2$ set

$$S_\theta = \{z \in B_n : |z_1 + i\tan(\theta)|^2 + |z'|^2 = \sec^2(\theta)\}$$

Let

$$S_{\frac{\pi}{2}} = B_n \cap \{\mathrm{Im}(z_1) = 0\}, S_+ = \partial B_n \cap \{\mathrm{Im}(z_1) > 0\}, S_- = \partial B_n \setminus S_+,$$

and let $\tilde{\sigma}_\theta$ be $2n-1$ dimensional Euclidean volume measure on S_θ, *not* normalized. On $S_0 = \partial B_n$, let $\tilde{\sigma} \equiv \tilde{\sigma}_0$; then $\tilde{\sigma} = \lambda_n \sigma$, where $\lambda_n = 2\pi^n/(n-1)!$ is the Euclidean $2n-1$-dimensional volume of ∂B_n.

THEOREM 3$'$. *For $f \in H^p(B_n), 1 < p < \infty$,*

$$\int_{S_\theta} |f|^p d\tilde{\sigma}_\theta \leq \left(\int_{S_+} |f|^p d\tilde{\sigma}\right)^{1-\frac{\theta}{\pi}} \left(\int_{S_+} |f|^p d\tilde{\sigma}\right)^{\frac{\theta}{\pi}}$$

The proof consists of applying Proposition 1 and checking that $\phi^{-1}(\Sigma_\theta) = S_\theta, \phi^{-1}(\Sigma_\pm) = S_\pm$, and $d\tilde{\sigma}_\theta(\phi^{-1}(z)) = 2^{2n-1}|q(z)|^{-2n} d\mu_\theta(z)$.

As in the one-dimensional case, Theorems 3 and 3$'$ lead to Fejér-Riesz-type inequalities:

$$\int_{U_n \cap \{\mathrm{Re}(z_1)=0\}} |f|^p \, d\mathrm{Im}(z_1) d\lambda(z') \;\leq\; \frac{1}{2} \int_{\partial U_n} |f|^p d\mu \tag{18}$$

$$\int_{B_n \cap \{\mathrm{Im}(z_1)=0\}} |f|^p \, d\mathrm{Re}(z_1) d\lambda(z') \;\leq\; \frac{1}{2} \int_{\partial B_n} |f|^p d\tilde{\sigma} \tag{19}$$

for $f \in H^p(U_n)$ and $f \in H^p(B_n)$ respectively. (18) and (19) have been established (for $0 < p < \infty$), using different techniques, by Hasumi and Mochizuki in [8],[4]. It is tempting to try to use (18) to draw some conclusions about volumes of images of the hypersurfaces Σ_θ under holomorphic mappings ψ of B_n into $\mathbf{C}^n$ (taking $f = |J_{\mathbf{C}}(\psi)|$, where $J_{\mathbf{C}}(\psi)$ is the complex Jacobian of ψ), analogous to the one dimensional result discussed in the introduction in connection with the Fejér - Riesz inequality. However, it is not easy to relate $\int_{\Sigma_\theta} |J_{\mathbf{C}}(\psi)| d\sigma_\theta$ to the volume of $\psi(\Sigma_\theta)$. Some estimates along this line are given in [4] under the assumption that ψ is quasiconformal on the closed ball, but they are not simple to state.

We conclude with two questions:

1) Is there a more general result, analogous to (4), which holds for a larger class of domains in $\mathbf{C}^n$? One would first like to understand the role of the sets Σ_θ, S_θ - are they level sets of some natural function (as in the one-dimensional case) on these domains?

2) Does Sedletckii's result (Theorem 2) hold for U_n? That is, if f is holomorphic in U_n and

$$\sup_{\theta \in (0,\pi)} \int_{\Sigma_\theta} |f|^p d\mu_\theta < \infty$$

is $f \in H^p(U_n)$?

ACKNOWLEDGEMENTS: Part of this research was carried out while the first author was a visiting scholar at Brown University. He thanks the Brown mathematics department for their hospitality.

References

[1] S.A. Akopian, Two Constants Theorems for Functions Belonging to the Class H_p, *Izv. Akad. Nauk Armenskoi* **2**, no.2, 1967, 123-127 (in Russian).

[2] P. Duren, *Theory of H^p Spaces*, Academic Press, New York and London, 1970.

[3] W. H. J. Fuchs, *Topics in the Theory of Functions of One Complex Variable*, D. Van Nostrand, Priceton, N.J. 1967

[4] M. Hasumi and N. Mochizuki, Fejer-Riesz Inequality for Holomorphic Functions of Several Complex Variables, *Tôhuku Math. J.* **33** (1981) 493-501.

[5] K. Hoffman, *Banach Spaces of Analytic Functions*, Prentice-Hall Inc.,Englewood Cliffs, N.J. 1962.

[6] A. Korányi, Harmonic Functions on Hermitian Hyperbolic Space, *Trans. of the A.M.S.* **135** (1969) 507-516.

[7] A. Korányi and S. Vági, Singular Integrals in Homogeneous Spaces and Some Problems of Classical Analysis, *Ann. Scuola Norm. Sup. Pisa* **25** (1971) 575-648.

[8] N. Mochizuki, The Fejér-Riesz Inequality for Siegel Domains, *Tôhuku Math. J.* **36** (1984) 581-590.

[9] W. Rudin, *Function Theory on the Unit Ball of* $\mathbf{C}^n$, Springer, Berlin 1980

[10] A.M. Sedletckii, An Equivalent Definition of H^p spaces in the Half-Plane and some Applications, *Math. USSR Sbornik* **25**, no.1 (1975), 69-76.

[11] E.M. Stein, *Boundary Behavior of Holomorphic Functions of Several Complex Variables*, Princeton University Press, Princeton, N.J 1972.

[12] E.M. Stein, *Harmonic Analysis: Real-Variable Methods, Orthogonality, and Oscillatory Integrals*, Princeton University Press, Princeton, N.J. 1993.

On Locally Convex Algebras with Cyclic Bases

HUGO ARIZMENDI AND ANGEL CARRILLO Instituto de Matematicás, Universidad Autónoma de Mexico, México City, México

ABSTRACT

We study the locally convex algebras with cyclic bases, establish some properties of the extended spectrum of the generator of such an algebra and give a characterization of the continuous homomorphisms of such an algebra to an arbitrary B_0-algebra.

1 INTRODUCTION

The m-convex B_0-algebras with cyclic bases were introduced by T. Husain and S. Watson. As a matter of fact, S. Watson proved in [5] that if A is a m-convex B_0-algebra with *cyclic* basis $\{z^n\}$ (i.e. a Schauder basis of the form $\{z^n,\ n = 0, 1, \ldots\}$, being z a fixed element in A), and the spectrum $\sigma(z)$ is an open set Ω, then A is isomorphic to the algebra $H(\Omega)$ of all holomorphic functions on Ω.

Here we shall consider B_0-algebras, not necessarily m-convex, with cyclic bases $\{z^n\}$.

We prove, among other properties of such algebras, that if for the generator of the algebra z we have that $R_7(z) = r > 0$, where $R_7(z)$ is the extended spectral radius of z, defined by W. Zelazko in [6], then

$$D_r(0) \subseteq \sigma(z) \subseteq \overline{D_r(0)}$$

where $D_r(0)$ is the complex open disk of radius r centred at 0, if $r < \infty$, and $D_r(0) = \mathbf{C}$ if $r = \infty$.

In the other hand, we give a sufficient condition for the existence of a continuous homomorphism (i.e. linear and multiplicative map) from a commutative m-convex B_0-algebra A, with unit e and with a basis cyclic, into an arbitrary B_0-algebra B. Explicitely, we shall prove that if A and B are as above, and x is an element of B such that

$$R_7(x) < R_6(z),$$

(the definition of $R_6(x)$ is recalled below), then there exists a continuous homomorphism $T: A \to B$ such that

$$T(z) = x$$

Actually,

$$T(y) = \sum_{k=0}^{\infty} a_k x^k,$$

for $y = \sum_{k=0}^{\infty} a_k x^k \in A$.

2 DEFINITIONS AND NOTATION

A *locally convex* algebra A is a topological Hausdorff algebra which is a locally convex space. The topology of such an algebra can be introduced by a family of seminorms $\{\| \quad \|_\alpha\}$, $a \in \Lambda$, such that for any α there is a β such that

$$\|xy\|_\alpha \leq \|x\|_\beta \cdot \|y\|_\beta \tag{1}$$

for all $x, y \in A$.

A locally convex metrizable and complete algebra is called a B_0-*algebra.* In a B_0-algebra A the topology can be introduced by a sequence $(\| \quad \|_i)$, $i = 1, 2, \ldots$, of seminorms satisfying

$$\|x\|_i \leq \|x\|_{i+1} \tag{2}$$

and

$$\|xy\|_i \leq \|x\|_{i+1} \cdot \|y\|_{i+1} \tag{3}$$

for $i = 1, 2, \ldots$ and all $x, y \in A$.

If for a locally convex algebra A condition (1) can be replaced by

$$\|xy\|_\alpha \leq \|x\|_\alpha \cdot \|y\|_\alpha \tag{4}$$

for all $x, y \in A$, then we call A *locally multiplicatively convex* (shortly m-*convex*).

DEFINITION 2.1 [6] Let A be a complete complex locally convex algebra with unit e. For an $x \in A$ the extended spectrum of x is defined as

$$\Sigma(x) = \sigma(x) \cup \sigma_d(x) \cup \sigma_\infty(x),$$

where

$$\sigma(x) = \{\lambda \in \mathbf{C}: x - \lambda e \quad \text{is not invertible in} \quad A\},$$

$\sigma_d(x) = \{\lambda \in \mathbf{C}: R(t,x) = (te - x)^{-1}$ is discontinuous at $t = \lambda\}$, and

$$\sigma_\infty(x) = \begin{cases} \emptyset & \text{if } t \to R(1, tx) \text{ is continuous at } t = 0 \\ \infty & \text{otherwise.} \end{cases}$$

The extended spectral radius of x is defined as

$$R(x) = \sup \{|\lambda|: \lambda \in \Sigma(x)\}$$

If A is a commutative complete m-convex algebra, then $\Sigma(x) = \sigma(x)$.

DEFINITION 2.2 Let A be a complete locally convex unital algebra, and let Λ be a family of seminorms satisfying (1) and defining the topology of A. For any $x \in A$, we define

$$R_1(x) = \sup_{\alpha \in \Lambda} \limsup_n \sqrt[n]{\|x^n\|_\alpha}.$$

$$R_2(x) = \sup \left\{ \limsup_n \sqrt[n]{\|x^n\|}: \| \ \| \text{ is a continuous seminorm on } A \right\}.$$

$$R_3(x) = \sup_{f \in A^*} \limsup_n \sqrt[n]{|f(X^n)|},$$

where A^* is the set of all continuous linear functionals on A.

$$R_4(x) = \sup_{f \in \mathcal{M}(A)} |f(x)|$$

where $\mathcal{M}(A)$ is the set of all continuous complex homomorphisms on A.

$$R_5(x) = \inf \{r: x - \lambda e \in G(A) \text{ for all } |\lambda| > r\}$$

$$R_6(x) = \inf \{0 < r \leq \infty: \quad \text{there is a sequence of complex}$$

numbers (α_i) such that the radius of convergence of $\sum_{i=0}^{\infty} \alpha_i \lambda^i$ is r and $\sum_{i=0}^{\infty} \alpha_i x^i$ converges in $A\}$.

$$R_7(x) = \inf \{0 < r \leq \infty: \text{ for any sequence of complex numbers } (\alpha_i) \text{ such that}$$

the radius of convergence of $\sum_{i=0}^{\infty} \alpha_i \lambda^i$ is r, $\sum_{i=0}^{\infty} \alpha_i x^i$ converges in $A\}$.

$$R_*(x) = \sup_{\alpha \in \Lambda} \limsup_n \sqrt[n]{\|x^n\|_\alpha}.$$

In [6] W. Zelazko proposed the above definitions for $R_i(x)$, $1 \leq i \leq 7$, In [3] it was introduced $R_*(x)$ and it was proved the following

THEOREM 2.3 Let A be a complex, unital, commutative complete m-convex algebra. Then for any $x \in A$, we have

$$\begin{aligned} R(x) &= R_1(x) = R_2(x) = R_3(x) = R_4(x) = \\ &= R_5(x) = R_6(x) = R_7(x) = R_*(x). \end{aligned}$$

THEOREM 2.4 Let A be a complex unital B_0-algebra. Then, for any $x \in A$ we have

$$R(x) = R_1(x) = R_2(x) = R_3(x) = R_7(x) \geq R_6(x) = R_*(x) \geq R_4(x).$$

THEOREM 2.5 Let A be a complex unital, commutative, complete locallly convex algebra. Then, for any $x \in A$, we have

$$R(x) = R_1(x) = R_2(x) = R_3(x) = R_7(x) \geq R_6(x) \geq R_*(x) \geq R_4(x).$$

Throughout this paper A will denote a commutative complex B_0-algebra with unit e.

DEFINITION 2.6 [5] Let z be a fixed element of A. The sequence $\{z^n: n \geq 0\}$ is called a *cyclic basis* of A, if it is a Schauder basis for A. So for any x in A there is a unique sequence $(\lambda_i(x))$ of complex numbers such that $x = \sum_{i=0}^{\infty} \lambda_i(x) z^i$

EXAMPLE 2.7 Let Ω denote a simply connected domain in the complex plane and let $H(\Omega)$ denote the algebra of all holomorphic functions on Ω, endowed with the compact-open topology. $H(\Omega)$ is a commutative m-convex B_0-algebra with unit e. If $\Omega = \mathbf{C}$, then $H(\Omega) = \mathcal{E}$ the algebra of entire functions and the sequence $\{z^n\}$, where z is the identity function, is a cyclic basis in $\mathcal{E}$. In the case when Ω is a proper subset of $\mathbf{C}$, let $\gamma = \Psi(\xi)$ be a conformal map from Ω onto the unit disk D of $\mathbf{C}$. For $f \in H(\Omega)$ we have

$$f(\xi) = f(\Psi^{-1}(\gamma)) = \sum_{n=0}^{\infty} \alpha_n \gamma^n = \sum_{n=0}^{\infty} \alpha_n (\Psi(\xi))^n$$

and the series converges in the topology of $H(\Omega)$. This series representation is unique and thus $\{\Psi^n\}$ is a cyclic basis for $H(\Omega)$.

EXAMPLE 2.8 For the algebra $H(D_1(0))$, where $D_1(0)$ is the complex open disc with radius 1 centred at 0, the open compact topology can be given by the seminorms $(\| \quad \|_n)$, $n = 1, 2, \ldots$, where

$$\|x\|_n \| \sum_{i=0}^{\infty} \lambda_i(x) z^i \|_n = \sum_{i=0}^{\infty} |\lambda_i(x)| r_n^i,$$

for all $x = \sum_{i=0}^{\infty} \lambda_i(x) z^i \in H(D_1(0))$, where (r_n), $n = 1, 2, \ldots$, is an increasing sequence of positive numbers r_n converging to 1.

EXAMPLE 2.9 Let $\bar{\alpha} = (\alpha_j)$ be a finite or infinite sequence of complex unitary numbers, and let $H_{\bar{\alpha}}(D_1(0))$ be the algebra of all functions $x = \sum_{i=0}^{\infty} \lambda_i(x) z^i \in H(D_1(0))$ such that $\sum_{i=0}^{\infty} \lambda_i(x) \alpha_j^i$ converges in $\mathbf{C}$, for every $\alpha_j \in \bar{\alpha}$. We endow the algebra $H_{\bar{\alpha}}(D_1(0))$ with the topology given by the sequence of seminorms $(\| \quad \|_n) \cup (\|x\|_{\alpha_j})$, where $(\| \quad \|_n)$ are the seminorms defined in example 2.8 and

$$\|x\|_{\alpha_j} = |\sum_{i=0}^{\infty} \lambda_i(x) \alpha_j^i|$$

for each $\alpha_j \in \bar{\alpha}$, and for each $x = \sum_{i=0}^{\infty} \lambda_i(x) z^i \in A$. It is easy to see that $H_{\bar{\alpha}}(D_1(0))$ is a m-convex, metrizable algebra with that sequence of seminorms. It is not complete, what we show by assuming, by simplicity, that $\bar{\alpha} = \{1\}$. In this case the $H_{\bar{\alpha}}(D_1(0))$ is not complete because, for example, $\left(\sum_{i=0}^{m} z^i - \sum_{i=m+1}^{2m+1} z^i\right)_{m=1}^{\infty}$ is a Cauchy sequence and it does not converges in $H_{\bar{\alpha}}(D_1(0))$.

The completion $\widetilde{H}_{\bar{\alpha}}(D_1(0))$ of $H_{\bar{\alpha}}(D_1(0))$ is $H(D_1(0)) \times \mathbf{C}$ with pointwise operations and the topology generated by the following seminorms: $\|(x, z_0)\|_n = \|x\|_n$, $n \geq 1$, $\|(x, z_0)\|_\alpha = |z_0|$.

This can be seen noticing that $H(D_1(0)) \times \mathbf{C}$ is complete, the mapping $x \to (x, x(1))$, from $H_{\bar{\alpha}}(D_1(0))$ to $H(D_1(0)) \times \mathbf{C}$ is a seminorm preserving homomorphism and the image of $H_{\bar{\alpha}}(D_1(0))$ is dense in $H(D_1(0)) \times \mathbf{C}$, actually given (x, z_0) with $x = \sum_{i=0}^{\infty} \lambda_i z^i \in H(D_1(0))$ and $z_0 \in \mathbf{C}$, the sequence

$$\left(\left(\sum_{i=0}^{m} \lambda_i z^i, \sum_{i=0}^{n} \lambda_i\right) - \left(\sum_{i=m+1}^{2m+1} \lambda_{i-m} z^i, \sum_{i=m+1}^{2m+1} \lambda_{i-m}\right) - \left(z_0 z^{2m+2}, z_0\right)\right)_{m=1}^{\infty}$$

converges to (x, z_0).

$\widetilde{H}_{\bar{\alpha}}(D_1(0))$ is commutative m-convex B_0-algebra wich has the same multiplicative functionals as $H_{\bar{\alpha}}(D_1(0))$, but it has not a cyclic basis.

DEFINITION 2.10 Let A be a commutative complex unital B_0-algebra with basis cyclic $\{z^n\}$. We define the following spectrum of the generator z

$$\sigma_1(z) = \left\{\lambda: \sum_{i=0}^{\infty} \lambda_i(x)\lambda^i \text{ converges in } \mathbf{C}, \text{ for all } x \text{ in } A\right\}$$

PROPOSITION 2.11 Let A be with cyclic basis $\{z^n\}$. Then

$$\sigma_1(z) = \{f(z): f \in \mathcal{M}(A)\}$$

Proof: It is clear that $f(z) \in \sigma_1(z)$ if $f \in \mathcal{M}(A)$. So to get the above equality we shall prove that if $\lambda \in \sigma_1(z)$, then the map $f_\lambda: A \to \mathbf{C}^n$ defined as

$$f_\lambda(x) = \sum_{i=0}^{\infty} \lambda_i(x)\lambda^i,$$

for each $x = \sum_{i=0}^{\infty} \lambda_i(x) z^i \in A$, is a continuous homomorphism.

It is easy to see that this map f_λ is an homomorphism because A is an algebra and $\sum_{i=0}^{\infty} \lambda_i(x)\lambda^i$ converges for all $x = \sum_{i=0}^{\infty} \lambda_i(x) z^i$ in A.

The continuity of f_λ follows from the following theorem of R. Arsove

Theorem [4]: *Let $\{z_n\}$ be a basis in a Fréchet space $\mathcal{U}$. If $\{x_n\}$ is a sequence in another Frechet space $\mathcal{V}$ for which the condition: for all sequences $\{a_n\}$ of scalars*

$$(A) \qquad \sum_{k=0}^{\infty} a_k z_k \textit{ convergent} \implies \sum_{k=0}^{\infty} a_k x_k \textit{ convergent},$$

holds, then there exists a continuous linear mapping T of $\mathcal{U}$ into $\mathcal{V}$ such that $Tz_n = x_n$.

COROLLARY 2.12 Let A be as before. Then

$$\sigma_1(z) \subseteq \sigma(z)\,.$$

PROPOSITION 2.13 Let A be with cyclic basis $\{z^n\}$. Let $R_6(z) = r > 0$, then

$$D_r(0) \subseteq \sigma_1(z) \subseteq \overline{D_r(0)}\,,$$

where $D_r(0)$ is the complex open disc of radius r centred at 0 and $\overline{D}_r(0)$ its closure if $r > 0$, and $D_r(0) = \overline{D}_r(0) = \mathbf{C}$ if $r = \infty$.

Proof: Let λ be in $\sigma_1(z)$, so $\lambda \in \sigma(z)$ then $\sum_{k=0}^{\infty} \lambda_k(y)\lambda^k$ converges for all $y = \sum_{k=0}^{\infty} \lambda_k(y)z^k$ in A, so $|\lambda| \leq R_4(z) \leq R_6(z) = r$.

Suppose now that $\lambda \notin \sigma_1(z)$, then there exists $y = \sum_{k=0}^{\infty} \lambda_k(y)z^k$ in A, such that $\sum_{k=0}^{\infty} \lambda_k(y)\lambda^k$ does not converges in $\mathbf{C}$. Let $r(y)$ the radius of convergence of the series $\sum_{k=0}^{\infty} \lambda_k(y)\lambda^k$, then we have that $r(y) \leq |\lambda|$. On the other hand, we have, because of the definition of $R_6(z)$, that

$$R_6(z) \leq r(y) \leq |\lambda|\,.$$

So $\lambda \notin D_r(0)$, and we we have the conclusion.

COROLLARY 2.14 Let A be with cyclic basis $\{z^n\}$ with $R_6(z) = r$, $0 < r \leq \infty$. Then we have

$$R(z) = R_1(z) = R_2(z) = R_3(z) = R_7(z) \geq R_6(z) = R_*(z) = R_4(z)\,.$$

PROPOSITION 2.15 Let A be with cyclic basis $\{z^n\}$ such that $R_7(z) = r > 0$. Then

$$D_r(0) \subseteq \sigma(z) \subseteq \overline{D_r(0)},$$

Proof: Let $\lambda \in \mathbf{C}$, $|\lambda| < r = R_7(z)$ and assume that $\lambda \notin \sigma(z)$, i.e. that $(z - \lambda e)^{-1} \in A$, then since $(z - \lambda e)^{-1}$ has a unique representation as power series in A, we have

$$(z - \lambda e)^{-1} = -\sum_{n=0}^{\infty} z^n/\lambda^{n+1}\,.$$

Let $(\|\ \ \|_i)$, $i = 1, 2, \ldots$, be sequence of seminorms defining the topology of A and satisfying (1). Then by the definition of $R_7(z)$ there exists a seminorm $\|\ \ \|_i$ such that

$$\limsup_n \sqrt[n]{\|z^n\|_i} > |\lambda|$$

then for this seminorm $\|\ \ \|_i$ we have that

$$\limsup_n \sqrt[n]{\|z^n/\lambda^{n+1}\|_i} > 1\,,$$

which contradicts the assumption that the series $\sum_{n=0}^{\infty} z^n/\lambda^{n+1}$ converges in A, thus if $|\lambda| < R_7(z)$, then $\lambda \in \sigma(z)$.

3 HOMOMORPHISM FROM B_0-ALGEBRAS WITH CYCLIC BASES.

Now, we want to give the following result about the continuous homomorphisms from a complex unital B_0-algebra with cyclic basis into an arbitrary B_0-algebra.

THEOREM 3.1 Let B be a B_0-algebra with cyclic basis $\{z^n\}$ with $R(z) = r$, $0 < r \leq \infty$, and let x in aB_0-algebra B_1 such that

$$R_7(x) < R_6(z),$$

then there exists a continuous homomorphism $T: B \to B_1$ such that

$$T_x(z) = x.$$

Proof: Let $x \in B_1$ such that $R_7(x) < R_6(z)$, and let $y \in B$ such that $y = \sum_{k=0}^{\infty} \lambda_k(y) z^k$, then we have that the radius of convergence of the series $\sum_{k=0}^{\infty} \lambda_k(y)\lambda^k$ is $r \geq R_6(z) > R_7(x)$, then the series $\sum_{k=0}^{\infty} \lambda_K(y)x^K$ converges in B, and so $T_x(y) = \sum_{k=0}^{\infty} \lambda_k(y)x^k$ belongs to A, and this formula defines a map $T: B \to B_1$, which is clearly an homomorphism.

Using, once again, the theorem of R. Arsove the continuity of T_x follows.

The condition $R_7(x) < R_6(z)$ in the above Theorem 1.16 is essencial. In [1] a m-convex B_0-algebra A with cyclic basis $\{z^n\}$ is constructed, such that $z^3 \in A$, is such that

$$R(z^3) = R_6(z^3) = R_7(z^3) = R_6(z) = R_7(z) = R(z),$$

and nevertheless there does not exist an endomorphism ϕ of A (continuous or not) such that $\phi(z) = z^3$. This algebra A is such that contains elements $y = \sum_{k=0}^{\infty} a_k z^k \in A$, such that $w = \sum_{k=0}^{\infty} a_k z^{3k}$ does not converges in A.

We do not know if all the endomomorphisms of a B_0-algebra with cyclic basis $\{z^n\}$ are continuous.

REFERENCES

1. H.Arizmendi, A. Carrillo and L. Palacios, *On the continuous homomorphism of m-convex algebras.* Manuscript.
2. H. Arizmendi, *On the spectral radius of a matrix algebra*, Funct. Approx. Comment. Math. **19** (1990), 167–176.
3. H. Arizmendi and K. Jarosz, *Extended spectral radius in topological algebras*, Rocky Mount. Jour.Math. Vol. **23**, No. 4, 1993, 1179–1195.
4. R. Arsove, *Similar bases and isomorphisms in Frechet spaces*, Math. Ann. **135** (1958), 283–293.
5. S. Watson, *F-algebras with cyclic bases*, Comm. Math. **23** (1982), 141-146.
6. W. Zelazko, *Selected topics in topological algebras,* Aarhus University Lecture Notes Series **31**, 1971.

Some Remarks on Norm-Attaining n-Linear Forms

R. M. ARON* Kent State University, Kent, Ohio 44242, U.S.A.

C. FINET† Université de Mons-Hainaut, Belgique

E. WERNER‡ Case Western Reserve University, Cleveland, Ohio 44106-4901, U.S.A. and Université de Lille 1, Lille, France

Abstract

We show the denseness of the set of norm-attaining n-linear forms on a Banach space with the Radon-Nikodym property. We also get some renorming results.

Introduction

This work is motivated by the Bishop-Phelps Theorem. That is, if X is a Banach space, then the collection of norm-attaining functionals of $X^\star$ (the dual space of X) is dense in $X^\star$ [5] (see also [18] and [9]). Let us recall that a functional $x^\star$ in $X^\star$ is called *norm-attaining* if there is an $x \in S(X)$, the unit sphere of X, such that $|x^\star(x)| = \|x^\star\|$. This result leads to a natural question : What about denseness of norm-attaining bilinear forms, and more generally, n-linear forms ? We let $F_n(X)$ denote the space of continuous n-linear forms on X, endowed with the norm $||A|| = \sup\{|A(x_1, ..., x_n)| : ||x_1|| \leq 1, ..., ||x_n|| \leq 1\}$. The subspace consisting

*Research supported in part by NSF Grant Int-9023951

†Research supported by Human Capital and Mobility grant and FNRS grant, Programme Tournesol

‡Research supported in part by NSF Grant DMS-9108003

of norm-attaining n-linear forms is denoted $AF_n(X)$. Unlike the linear case, it is not true in general that $\overline{AF_n(X)} = F_n(X)$. Indeed, M.D. Acosta, F.J. Aguirre, and R. Payá [3] have recently proved that $AF_2(G)$ is not dense in $F_2(G)$ when G is the Banach sequence space used by W.T. Gowers to show that ℓ_p $(1 < p < \infty)$ fails property B [12].

In this note, we prove that if X has the Radon-Nikodym property then $AF_n(X)$ is dense in $F_n(X)$ for every n. The proof is based on an "optimization" principle due to C. Stegall [20] (see also [4] and [6]). Section 2 is devoted to the role which property (α), introduced by W. Schachermayer [19], plays in the study of norm-attaining operators (see also [6], [13], [15]). We prove that if X has property (α) then the space $AF_n(X)$ is dense in $F_n(X)$ for every n. We then get some renorming results.

Before proceeding, several remarks are in order. First, each element $A \in F_n(X)$ can be associated to a unique functional ϕ in the dual of the n-fold projective tensor product $X\hat{\bigotimes}_\pi \ldots \hat{\bigotimes}_\pi X$, and vice versa. Consequently, there is an element $\tilde{\phi} \in (X\hat{\bigotimes}_\pi \ldots \hat{\bigotimes}_\pi X)^*$ which is close to ϕ and which attains its norm at a point $z \in S(X\hat{\bigotimes}_\pi \ldots \hat{\bigotimes}_\pi X)$. If z were of the form $z = x_1 \bigotimes \ldots \bigotimes x_n$ where each $x_i \in S(X)$, then it would follow that the n-linear mapping $\tilde{A}$ associated to $\tilde{\phi}$ would belong to $AF_n(X)$, and of course $\tilde{A}$ would be close to A. However, there is no reason for z to be of this form. We recall that every strongly exposed point of $B(X\hat{\bigotimes}_\pi X)$ is of the form $x_1 \bigotimes x_2$, where x_1 and x_2 are strongly exposed points of $B(X)$ [22] (Here, $B(Y)$ denotes the closed unit ball of Banach space Y.) When $\tilde{\phi}$ attains its norm at a strongly exposed point, then we are done. Thus our result is obvious for those Banach spaces X for which the Radon-Nikodym property is preserved under completed projective tensor products. However, it may happen that X has the Radon-Nikodym property while $X\hat{\bigotimes}_\pi X$ fails to have it [7].

Let us mention the following result of J. Bourgain: X has the Radon-Nikodym property if and only if for every equivalent norm on X, the space $NA(X, Y)$ of norm-attaining linear operators is dense in $\mathcal{L}(X, Y)$ [6]. Note also that every $A \in F_2(X)$ can be associated to an element $L_A \in \mathcal{L}(X, X^*)$, by $L_A(x_1)(x_2) = A(x_1, x_2)$. However, the above result of J. Bourgain [6] cannot be directly applied to our situation, since it can happen that L_A attains its norm although $A \notin AF_2(X)$. For example, the bilinear form $A : \ell_1 \times \ell_1 \to C$ given by $A(x, y) = x_1 \sum_{n=1}^{\infty} \frac{n}{n+1} y_n$ is not in $AF_2(\ell_1)$, although $1 = ||L_A|| = ||L_A(e_1)||$. On the other hand, by using Theorem 1 of [15] and the fact that continuous linear forms on a reflexive space attain their norms, it is easy to see that $AF_2(X)$ is dense in $F_2(X)$ if X is reflexive. In fact, we do not know an example of a Banach space X such that $\overline{NA(X, X^*)} = \mathcal{L}(X, X^*)$ but such that $\overline{AF_2(X)} \neq F_2(X)$. In addition, if $X = G \bigoplus_1 \ell_p \ \ (1 < p < \infty)$, then $\overline{NA(X, X^*)} \neq \mathcal{L}(X, X^*)$ [3].

We remark that closely related problems involving polynomials rather than multilinear forms have recently been investigated by Y. S. Choi and S. G. Kim [8]. In addition, parallel to the study of norm-attaining operators is the study of numerical radius attaining operators. The space of numerical radius attaining operators on X is not dense in general in $\mathcal{L}(X)$ [17], [2], although it is dense when X has the Radon-Nikodym property [4]. In this setting, there are connections with property (β) (introduced by [15] and also studied by [10], [11], [16], [19]) and the corresponding renorming result in [1].

Finally, the authors acknowledge with thanks several conversations with Charles Stegall during the period when this manuscript was being prepared [21]. The second author thanks Professor D. J. H. Garling for the fruitful conversations they had on this subject while he was visiting the Université de Mons-Hainaut.

1 Norm-attaining n-linear forms on Banach spaces with Radon-Nikodym property

As in [4], our proof is based on "a non linear optimization principle" due to C. Stegall [20], see also [6]. Recall that a closed bounded convex set B in a real Banach space X is called a *Radon-Nikodym set* if, for any closed subset $\mathcal{C}$ of B, the set of functionals that strongly expose $\mathcal{C}$ is norm dense in the dual space X^*. A bounded above function $\Phi : \mathcal{C} \to \mathbb{R}$ *strongly exposes* $\mathcal{C}$ if Φ attains its maximum at a point x_0 of $\mathcal{C}$, and any sequence (x_n) of points in $\mathcal{C}$ satisfying $(\Phi(x_n)) \to \Phi(x_0)$ converges to x_0.

Let us recall the following result of Stegall [20].

Theorem : Let Φ be an upper semicontinuous, real-valued, bounded above function on a Radon-Nikodym set B of a real Banach space Y. Then the set

$$\{f \in Y^\star : \Phi + f \text{ strongly exposes } B\}$$

is a G_δ-dense subset of $Y^\star$.

We are now ready to prove :

Theorem 1 Let X be a Banach space having Radon-Nikodym property. Then the space $AF_n(X)$ is dense in $F_n(X)$ for every $n \geq 1$.

Proof. Let $A : X \times X \times \cdots \times X \to C$ be in $F_n(X)$. We will assume that $n = 3$. The general case is done in the same way. Suppose $\|A\| = 1$.

If X is a complex Banach space then we consider its real restriction. Let Y be the space $X \oplus_\infty X \oplus_\infty X$ with norm given by $\|(x, y, z)\| = \max\{\|x\|, \|y\|, \|z\|\}$ for $x, y, z \in X$. Since X has the Radon-Nikodym property, we can easily see that Y also has the Radon-Nikodym property. Let $\Phi : B(Y) \to \mathbb{R}$ be the function defined by $\Phi(x, y, z) = |A(x, y, z)|$, $(x, y, z) \in B(Y)$. Then Φ is a bounded, upper semicontinuous function on $B(Y)$. Let $0 < \epsilon < 1$. By the preceding theorem, there exists $(x_0^\star, y_0^\star, z_0^\star) \in Y^\star$ such that $0 < \|(x_0^\star, y_0^\star, z_0^\star)\|_1 < \epsilon$, and there is $(x_0, y_0, z_0) \in B(Y)$ such that :

$$\begin{aligned}\Phi(x_0, y_0, z_0) &+ Re[(x_0^\star, y_0^\star, z_0^\star)(x_0, y_0, z_0)] \\ &\geq \Phi(x, y, z) + Re[(x_0^\star, y_0^\star, z_0^\star)(x, y, z)] \qquad (*)\end{aligned}$$

for all $(x, y, z) \in B(Y)$.

By rotating x, y and z, we get that :

$$\begin{aligned}\Phi(x_0, y_0, z_0) &+ Re[(x_0^\star, y_0^\star, z_0^\star)(x_0, y_0, z_0)] \\ &\geq \Phi(x, y, z) + |x_0^\star(x)| + |y_0^\star(y)| + |z_0^\star(z)|\end{aligned}$$

for all $(x, y, z) \in B(Y)$.

In particular if $x = x_0$, $y = y_0$ and $z = z_0$, then :

$$(Re[(x_0^\star, y_0^\star, z_0^\star)(x_0, y_0, z_0)] \geq |x_0^\star(x_0)| + |y_0^\star(y_0)| + |z_0^\star(z_0)|)$$

implies that :

$$((x_0^\star, y_0^\star, z_0^\star)(x_0, y_0, z_0) \geq 0) \quad .$$

It follows by (*) that:

$$\Phi(x_0, y_0, z_0) + x_0^\star(x_0) + y_0^\star(y_0) + z_0^\star(z_0)$$
$$\geq \Phi(x, y, z) + |x_0^\star(x)| + |y_0^\star(y)| + |z_0^\star(z)| \qquad (**)$$

for all $(x, y, z) \in B(Y)$.

Moreover, $x_0 \neq 0$, $y_0 \neq 0$ and $z_0 \neq 0$. Indeed, suppose for example that $y_0 = 0$. Then it follows by (**) that $x_0^\star(x_0) + z_0^\star(z_0) \geq \Phi(x, y, z) + |x_0^\star(x)| + |z_0^\star(z)|$, for all $(x, y, z) \in B(Y)$. Since $\|(x_0^\star, y_0^\star, z_0^\star)\|_1 = \|(x_0^\star, z_0^\star)\|_1 < \epsilon$, one has $\epsilon > |A(x, y, z)| + |x_0^\star(x)| + |z_0^\star(z)|$, for all $(x, y, z) \in B(Y)$. It follows that $\|A\| < \epsilon < 1$, contradicting $\|A\| = 1$. In fact, we have $\|x_0\| = 1 = \|y_0\| = \|z_0\|$ (take first $x = \frac{x_0}{\|x_0\|}$ in (**) to get $\|x_0\| = 1$, and then take $y = \frac{y_0}{\|y_0\|}$ and $z = \frac{z_0}{\|z_0\|}$ in (**)). By the Hahn-Banach theorem, there exist $a_0^\star$, $b_0^\star, c_0^\star \in X^\star$, such that :

$$\begin{cases} \|a_0^\star\| = 1 = a_0^\star(x_0) \\ \|b_0^\star\| = 1 = b_0^\star(y_0) \\ \|c_0^\star\| = 1 = c_0^\star(z_0) \end{cases}$$

For $x, y, z \in X$, define

$$S(x, y, z) = \lambda[x_0^\star(x)b_0^\star(y)c_0^\star(z) + a_0^\star(x)y_0^\star(y)c_0^\star(z) + a_0^\star(x)b_0^\star(y)z_0^\star(z)]$$

with $\lambda \in \mathbb{C}$ chosen so that $|\lambda| = 1$ and $A(x_0, y_0, z_0) = \lambda|A(x_0, y_0, z_0)|$.
Then $S \in F_3(X)$ and $\|S\| \leq \|x_0^\star\| + \|y_0^\star\| + \|z_0^\star\| = \|(x_0^\star, y_0^\star, z_0^\star)\|_1 \quad < \epsilon$.
Moreover,

$$\begin{aligned} |(A + S)(x_0, y_0, z_0)| &= |A(x_0, y_0, z_0) + \lambda[x_0^\star(x_0) + y_0^\star(y_0) + z_0^\star(z_0)]| \\ &= \left| |A(x_0, y_0, z_0)| + \underbrace{x_0^\star(x_0) + y_0^\star(y_0) + z_0^\star(z_0)}_{\geq 0} \right| \\ &= |A(x_0, y_0, z_0)| + x_0^\star(x_0) + y_0^\star(y_0) + z_0^\star(z_0) \\ &= \Phi(x_0, y_0, z_0) + (x_0^\star, y_0^\star, z_0^\star)(x_0, y_0, z_0) \quad . \end{aligned}$$

On the other hand,

$$|(A + S)(x, y, z)|$$

$$= |A(x,y,z) + \lambda[x_0^\star(x)b_0^\star(y)c_0^\star(z) + a_0^\star(x)y_0^\star(y)c_0^\star(z) + a_0^\star(x)b_0^\star(y)z_0^\star(z)]|$$
$$\leq \Phi(x,y,z) + |x_0^\star(x)| + |y_0^\star(y)| + |z_0^\star(z)|$$
$$\leq |(A+S)(x_0,y_0,z_0)|, \qquad \text{for all} \quad (x,y,z) \in B(Y).$$

Q.E.D.

Remarks (1) More generally, following the proof of Theorem 1, we also have that if $X_1, ..., X_n$ are Banach spaces having the Radon-Nikodym property and if Y is any Banach space, then the space of norm-attaining n-linear maps from $X_1 \times \cdots \times X_n \to Y$ is dense in the space of n-linear maps.

(2) The converse of Theorem 1 is not true : For example, $X = c_0$ does not have the Radon-Nikodym property although $AF_2(c_0)$ is dense in $F_2(c_0)$. This follows from the facts that every operator from c_0 to ℓ_1 is compact and that ℓ_1 has the approximation property.

2 Renorming results

In the study of norm-attaining operators, W. Schachermayer introduced the following "property (α)":

Definition 2 [19] A Banach space X has property (α) if there exist $\lambda \in [0,1)$ and a family $\{(x_\alpha, x_\alpha^\star)\}_{\alpha\in I}$ of $S(X) \times S(X^\star)$ satisfying the following assertions :

1) $x_\alpha^\star(x_\alpha) = 1$,

2) $|x_\alpha^\star(x_\beta)| \leq \lambda$, for $\alpha \neq \beta$,

3) The unit ball of X is the closed convex circled hull of $(x_\alpha)_{\alpha\in I}$.

A typical example of a Banach space with property (α) is ℓ^1. W. Schachermayer proved that if X has property (α) then for every Banach space Y, the norm-attaining operators from X to Y is a dense set in $\mathcal{L}(X,Y)$ [19]. We will prove the analogous result for n-linear forms.

Theorem 3 Let X be a Banach space with property (α). Then $AF_n(X)$ is dense in $F_n(X)$, for every $n \geq 1$.

Proof. Let $A : X \times X \times \cdots \times X \to C$ be in $F_n(X)$. As before, it is enough to consider the case $n = 3$. By [19], we may suppose that the corresponding operator $T : X \to F_2(X)$ attains its norm and that $\|T\| = 1$. Further, as in [19] there exists $\alpha \in I$ such that $\|Tx_\alpha\| = 1$. Let $\epsilon_0 > 0$ and $\delta > 0$ be given. δ will be fixed later. Then there exist $\gamma, \mu \in I : |Tx_\alpha(x_\gamma, x_\mu)| > 1-\delta$. Choose ϵ and a such that $\epsilon + a < \epsilon_0$ and such that $1 + \epsilon\lambda + a < 1 + \epsilon - a$. We now define $\tilde{T} : X \to F_2(X)$ by :

$$\tilde{T}x(y,z) = Tx(y,z) + \epsilon x_\alpha^\star(x)Tx_\alpha(y,z) + ax_\gamma^\star(y)x_\mu^\star(z)Tx(x_\gamma, x_\mu) \quad .$$

We have

$$\tilde{T}x_\alpha(y,z) = (1+\epsilon)Tx_\alpha(y,z) + ax_\gamma^\star(y)x_\mu^\star(z)Tx_\alpha(x_\gamma, x_\mu),$$

from which it follows that :

$$\left\|\tilde{T}x_\alpha\right\| \geq 1 + \epsilon - a \quad .$$

On the other hand, for $\beta \neq \alpha$, $\left\|\tilde{T}x_\beta\right\| \leq 1 + \epsilon\lambda + a$. By the choice of ϵ and a, $\left\|\tilde{T}\right\| = \left\|\tilde{T}x_\alpha\right\|$. Furthermore, $\left\|\tilde{T}x_\alpha\right\| = \left|\tilde{T}x_\alpha(x_\gamma, x_\mu)\right|$, since on the one hand $\left|\tilde{T}x_\alpha(x_\gamma, x_\mu)\right| = (1+\epsilon+a)\left|Tx_\alpha(x_\gamma, x_\mu)\right| \geq (1-\delta)(1+\epsilon+a)$ and on the other hand, for $\beta \neq \gamma$, $\eta \neq \mu$,

$$\tilde{T}x_\alpha(x_\beta, x_\eta) = Tx_\alpha(x_\beta, x_\eta) + \epsilon Tx_\alpha(x_\beta, x_\eta) + ax_\gamma^\star(x_\beta)\,x_\mu^\star(x_\eta)\,Tx_\alpha(x_\gamma, x_\mu) \quad .$$

It follows that :

$$\left|\tilde{T}x_\alpha(x_\beta, x_\eta)\right| \leq 1 + \epsilon + a\lambda^2.$$

Now choose δ such that $1 + \epsilon + a\lambda^2 < (1-\delta)(1+\epsilon+a)$. Moreover $\tilde{T}$ is close to T. Q.E.D.

This result can also be obtained as a corollary of the following observation due to D. J. H. Garling:

Proposition 4 Suppose that X has property (α) with constant λ and that Y has property (α) with constant μ. Then the projective tensor product $X\hat{\bigotimes}_\pi Y$ has property (α) with constant $\max(\lambda, \mu)$.

Proof. Let $\{(x_\alpha, x_\alpha^\star)\}_{\alpha\in A}$ and $\{(y_\beta, y_\beta^\star)\}_{\beta\in B}$ satisfy the required properties. Then $B(X\hat{\bigotimes}_\pi Y)$ is the closed convex circled hull of the set $\{x_\alpha \bigotimes y_\beta\}_{\alpha\in A,\ \beta\in B}$. Also,

$(x_\alpha^\star \bigotimes y_\beta^\star)(x_\alpha \bigotimes y_\beta) = 1$ and $|(x_\alpha^\star \bigotimes y_\beta^\star)(x_{\alpha'} \bigotimes y_{\beta'}| \leq \max(\lambda, \mu)$, for $(\alpha, \beta) \neq (\alpha', \beta')$. Q.E.D.

We now recall some renorming results. W. Schachermayer proved that if X is weakly compactly generated then X has a renorming with property (α) [19]. This result was extended by B. Godun and S. Troyanski [13] to Banach spaces having a biorthogonal system with cardinality equal to dens X. Therefore, the following is immediate :

Corollary 5 Suppose X has a biorthogonal system with cardinality equal to dens X. Then for every n, X can be renormed in such a way that the space $AF_n(X)$ is dense in $F_n(X)$.

Let us mention that not every Banach space can be renormed to have property (α). One counterexample is the space $X = \mathcal{C}(K)$ constructed by K. Kunen, [14], which has the property that for any uncountable set $A \subset X$, there exists a point $a \in A$ such that $a \in \overline{\text{conv}}(A \setminus \{a\})$ Finally, we do not know if every Banach space X can be renormed in such a way that the space $AF_n(X)$ is dense in $F_n(X)$.

References

[1] M. D. Acosta, Every real Banach space can be renormed to satisfy the denseness of numerical radius attaining operators. Israel. J. Math. **81** (1993), 273-280.

[2] M. D. Acosta, F.J. Aguirre, and R. Payá, A space by W. Gowers and new results on norm and numerical radius attaining operators. Acta Univ. Carolinae Math. & Phys. **33** (1992), 5-13.

[3] M. D. Acosta, F.J. Aguirre, and R. Payá, There is no bilinear Bishop-Phelps theorem, Isr. J. Math (to appear).

[4] M. D. Acosta and R. Payá, Numerical radius attaining operators and the Radon-Nikodym property. Bull. London Math. Soc. **25** (1993), 67-73.

[5] E. Bishop and R.R. Phelps, The support functionals of a convex set, Proc. Sympos. Pure Math VII, Convexity, A.M.S., Providence, R.I. (1963).

[6] J. Bourgain, La propriété de Radon-Nikodym, cours de 3ème cycle, (1979). Publications mathématiques de l'Université Pierre et Marie Curie.

[7] J. Bourgain and G. Pisier, A construction of $\mathcal{L}_\infty$−spaces and related Banach spaces, Bol. Soc. Bras. Mat. **14** (1983), 109-123.

[8] Y. S. Choi and S. G. Kim, Norm or numerical radius attaining multilinear mappings and polynomials, J. London Math. Soc. (to appear).

[9] J. Diestel, Geometry of Banach spaces. Selected Topics. Lecture Notes in Math. **485**, Springer-Verlag, (1975).

[10] C. Finet, Renorming Banach spaces with many projections and smoothness properties. Math. Ann. **284** (1989), 675-679.

[11] C. Finet, W. Schachermayer, Equivalent norms on separable Asplund spaces. Studia Math., **92.3** (1989), 275-283.

[12] T. W. Gowers, Symmetric block bases of sequences with large average growth, Israel J. Math. **69** (1990), 129-149.

[13] B. Godun and S. Troyanski, Renorming Banach spaces with fundamental biorthogonal systems. Contemporary Math. **144** (1993), 119-126.

[14] K. Kunen and J. Vaughan, Handbook of set-theoretic topology, North Holland, (1988).

[15] J. Lindenstrauss, On operators which attain their norm. Israel. J. Math. **1** (1963), 139-148.

[16] J. Partington, Norm attaining operators. Israel J. Math. **43** (1982), 273-276.

[17] R. Payá, A counterexample on numerical radius attaining operators, Israel J. Math., **79** (1992), 83-101.

[18] R.R. Phelps, The Bishop-Phelps theorem in complex spaces: an open problem, in Function spaces (K. Jarosz, editor), Lecture Notes in Pure and Applied Math., M. Dekker (N.Y.) **136** (1992), 337-340.

[19] W. Schachermayer, Norm attaining operators and renormings of Banach spaces. Israel. J. Math. **44** (1983), 201-212.

[20] C. Stegall, Optimization and differentiation in Banach spaces. Linear Algebra Appl. **84** (1986), 191-211.

[21] C. Stegall, Some remarks on norm-attaining functions, preprint.

[22] D. Werner, Denting points in tensor products of Banach spaces. Proc. A.M.S. **101**, 1 (1987), 122-126.

On Uniformly Closed Ideals in *f*-Algebras

MOHAMED BASLY and ABDELMAJID TRIKI Département de Mathématiques, Faculté des Sciences de Tunis, Université de Tunis, 1060 Tunis, Tunisie

ABSTRACT

In this note we give necessary and sufficient conditions in order that in a given uniformly complete Archimedian *f*-algebra, every maximal modular ring ideal is relatively uniformly closed.

1. PRELIMINARIES

For the concepts not explained in this paper we refer to [4, 5, 7, 8]. All Riesz spaces and Riesz algebras under consideration are supposed to be Archimedean.

Let A be a Riesz space with a positive cone $A^+ = \{a \in A : \ a \geq 0\}$. Given the element $b \in A^+$, the sequence $\{a_n : n = 1, 2, ...\}$ in A, is said to converge b-uniformly to the element a, whenever for $\epsilon > 0$ there exist a natural number N_ϵ such that $|a - a_n| \leq \epsilon b$ for all $n > N_\epsilon$. The sequence $\{a_n \ : n = 1, 2, ...\}$ in A is said to converge relatively uniformly to a if it converge b-uniformly to a for some $b \in A^+$. This is denoted by $a_n \to a \ (ru)$. The Riesz space is Archimedean if and only if every uniformly convergent sequence in A has a unique limit. In the obvious way, the notions of b-uniformly Cauchy sequences and relatively uniformly Cauchy sequences are defined. A is called b-uniformly complete whenever every b-uniformly Cauchy sequence in A has a unique limit. A is called uniformly complete if A is b-uniformly complete for every $b \in A^+$.

For any non empty subset D of A we define the pseudo-closure $D'$of D to be the set of all $a \in A$ for which there exist a sequence $(a_n)_{n \in N}$, $a_n \in D$, such that $a_n \to a \ (ru)$. Furthermore, D is called uniformly closed, whenever $D = D'$.

The real algebra A is called a Riesz algebra if A is simultaneously a Riesz space such that the ordering and the multiplication are compatible (i.e. : $a, b \in A^+$ implies $ab \in A^+$). A is said to be an *f*-algebra if A satisfies the condition that $a \wedge b = 0$ implies $ca \wedge b = ac \wedge b = 0$ for $c \in A^+$. A is an almost *f*-algebra if A satisfies the condition that $a \wedge b = 0$ implies $ab = 0$.

The ring ideal J is said to be modular if there exists u in A such that $(x - xu) \in J$ for every $x \in A$. An order bounded operator π from A into itself is called an orthomorphism if it follows from $|a| \wedge |b| = 0$ that $|\pi(a)| \wedge |b| = 0$. The ordered vector space $Orth(A)$ of all orthomorphisms in the *f*-algebra A is itself an *f*-algebra with respect to the usual vector space operations and with the composition of applications as multiplication. The principal order ideal generated by the identity I_A is called the center of A and is denoted by $Z(A)$. Let ρ be the mapping from A into $Orth(A)$ defined by $\rho(a) = \pi_a$ where $\pi_a(b) = ab$ for all $b \in A$.

2 THE MAIN RESULTS

We recall that a ring ideal J is said to be semiprime whenever

$$J = \{a \in A : a^n \in J \text{ for some } n \in N.\}$$

Lemma 1. *Every semiprime ideal in an uniformly complete almost f-algebra is an order ideal.*

Proof. It follows from the identities $f^2 = |f^2| = |f|^2$ that $f \in I \Longrightarrow |f| \in I$. Let $0 \leq u \leq v$ and $v \in I$, we have to prove that $u \in I$. It is easy to see that $N(A) \subset I$. Then $\bar{I} := \{\bar{x} \in A/N(A) : x \in I\}$ is a semiprime ideal of $A/N(A)$ which is an uniform complete semiprime *f*-algebra. Then ([3]; Theorem 4.7) $\bar{I}$ is an order ideal i.e. : $\bar{0} \leq \bar{u} \leq \bar{v}$ and $\bar{v} \in \bar{I} \Rightarrow \bar{u} \in \bar{I}$. That's means that there exist x in I and a in $N(A)$ such that $u - x = a \in N(A) \subset I$. This proves that $u \in I$ and that I is an order ideal.

Proposition 2. *Let A be a relatively uniformly complete Archimedean almost f-algebra. The following conditions are equivalent:*
(i) *J is a relatively uniformly closed maximal modular ring ideal in A.*
(ii) *J is the kernel of a non zero multiplicative positive functional on A.*

Proof. (i)$\Rightarrow$(ii). Let J be a relatively uniformly maximal modular ring ideal in A. By Lemma 1, J is an order ideal. It follows from ([4] ; 33.2 and 60.2) that A/J is a linearly ordered Archimedean field, hence A/J is Riesz and algebra isomorphic to R. Let i denotes this isomorphism. If q is the Riesz and algebra canonical homomorphism from A into A/J, then $f = i \circ q$ is a positive multiplicative linear functional on A for which we have $J = \ker f$.

(ii)$\Rightarrow$(i) Suppose that f is non zero positive multiplicative functional on A. Then $\ker(f)$ is a maximal modular ring ideal in A and it follows that $\ker(f)$ is a prime ring ideal. Hence $\ker(f)$ is a maximal order ideal, so $\ker(f)$ is a relatively uniformly closed ideal in A.

Remark 3. It is well known that in a Banach algebra every maximal modular ring ideal is closed. But even if A is an uniformly complete *f*-algebra it may happen that A contains maximal modular ring ideals which are not uniformly closed. By way of example, consider the *f*-algebra $m([0,1])$ of all real Lebesgue measurable functions on $[0,1]$ with the usual identification of almost equal functions. $m([0,1])$ is a semiprime (i.e. $N(A) = \{0\}$) and a semisimple (i.e. $Rad(A) = \{0\}$) *f*-algebra, but as it is well known, the only linear functional on $m([0,1])$ is the null functional and hence, by Proposition 2, $m([0,1])$ doesn't contain any relatively uniformly closed maximal modular ring ideal.

Let A be an uniformly complete *f*-algebra , we give now necessary and sufficient conditions in order that any maximal modular ring ideal in A is relatively uniformly closed in A.

Theorem 4. *If A is an uniformly complete f- algebra , the following properties are equivalent:*
(i) *Any maximal modular ring ideal in A is relatively uniformly closed in A.*
(ii) $\rho(A) \subset Z(A)$.
(iii) *Any order ideal in A is a ring ideal .*
(iv) *Any maximal modular ring ideal* is a maximal order ideal.
(v) *Any maximal modular ring ideal is the kernel of a positive multiplicative linear functional on A.*

Proof. Since by the preceding proposition (i) is equivalent to (v). We shall prove the following implications : (ii) $\Rightarrow$ (iii) $\Rightarrow$ (iv) $\Rightarrow$ (i) $\Rightarrow$ (ii).

The implication (ii) $\Rightarrow$ (iii) follows from ([1];1.3).

(iii) $\Rightarrow$ (iv). Let J be a maximal modular ring ideal in A, J is a prime ring ideal then an order ideal in A. Assume that J is not a maximal order ideal, then there exist a proper ideal I in A strictly containing the order ideal J. By (iii) I is a modular ring ideal which contains strictly J. This contradicts the maximality of J as a ring ideal , then J is necessarily a maximal order ideal.

(iv)$\Rightarrow$(i) Follows from the fact any maximal order ideal is relatively uniformly closed.

(i)$\Rightarrow$(ii) Assume that there exists $x \in \rho(A)^+$such that $x \notin Z(A)$. $\rho(A)$ is relatively uniformly complete since A is so and ρ is a Riesz homomorphism. Then $\rho(A)$ satisfies the Stone condition, in particular $u = x \wedge I_A \in \rho(A)$.
For a fixed integer n, we define x_n and y_n as follows :$x_n = x \vee nu - x$ and $y_n = x - x \wedge nu$
It is easy to show that :
$x_n y_n = x^2 \vee nux - x^2 - (x \vee nu)(x \wedge nu) + x^2 \wedge nux$
Using the formula $ab = (a \wedge b)(a \vee b)$ which holds in every Archimedean *f*-algebra, we get

$x_n y_n = 0$ for all $n \geq 1$. Hence x_n is an element of the band $J_n = \{y_n\}^d$ with $\{y_n\}^d = \{z \in \rho(A) : |z| \wedge y_n = 0\}$. Since $x \notin Z(A)$ and $x \wedge nu \in Z(A)$ we have that $y_n = x - x \wedge nu \neq 0$. Consequently we have that $\rho(A) \neq J_n$, since J_n is relatively uniformly closed in $\rho(A)$ then ([3]; proposition 3.1) J_n is a proper ring ideal in $\rho(A)$. The inequality $y_n \geq y_{n+1}$ implies $J_n \subset J_{n+1}$ for all $n \geq 1$. Hence $J = \bigcup_{n\geq 1} J_n$ is a proper ideal. Moreover we have

$$(I_A - u) \wedge (x - x \wedge I_A) = (I_A - x \wedge I_A) \wedge (x - x \wedge I_A) = 0 \text{ in } Orth(A)$$

Hence $(I_A - x \wedge I_A)(x - x \wedge I_A) = 0$ in $Orth(A)$ and then for all $z \in \rho(A)$, we have $z(I_A - u)(x - x \wedge I_A) = 0$. If we note $y_1 = x - x \wedge I_A$ it follows that $(z - uz)y_1 = 0$. This implies that $| z - uz | \wedge y_1 = 0$ since $\rho(A)$ is semiprime. Hence, $(z - uz) \in \{y_1\}^d = J_1 \subset J$, it is shown that u is an identity modulo the proper ring ideal J in $\rho(A)$. Then there exists in $\rho(A)$ a maximal modular ring ideal M containing J. Let v be an element of A^+ such that $\rho(v) = u$. It is not too hard to verify that $\rho^{-1}(M)$ is a maximal modular ring ideal in A, and that v is an identity modulo $\rho^{-1}(M)$ in A. Consequently $\rho^{-1}(M)$ is relatively uniformly closed and by ([4]; Theorem 59.3) $M = \rho(\rho^{-1}(M))$ is relatively uniformly closed in $\rho(A)$. Then $\rho(A)/M$ is an Archimedean field ([4] Theorem 60.2). For every $n \geq 1$, we have $x_n = (x \vee nu - x) \in J_n \subset M$ and if $\bar{x}$ is the class of x modulo M then we have $\bar{x}_n = \bar{x} \vee n\,\bar{u} - \bar{x} = \bar{0}$ for all $n \geq 1$.
In other words we have $\bar{x} \vee n\,\bar{u} = \bar{x}$ and hence $\bar{0} \leq n\,\bar{u} \leq \bar{x}$ for all $n \geq 1$.
Since $\rho(A)/M$ is Archimedean, it follows that $\bar{u} = \bar{0}$ which is contradictory with the fact $\bar{u}$ is the unit element of the field of $\rho(A)/M$. This completes the proof.

For an uniformly complete semiprime f-algebra , we can deduce immediately from ([1]; Theorem 5) that:

Corollary 5. *Let A be an uniformly complete semiprime f-algebra, then the following conditions are equivalent :*
(i) Every maximal modular ring ideal in A is uniformly closed.
(ii) There exists an M-norm in A.
(iii) There exists a Riesz norm in A.

REFERENCES

1. M. Basly and A. Triki , *f*-algebras in which order ideals are ring ideals. *Indagation Math. V* 50, *fasc 3* (1988) 231-234.
2. G. Birkhoff , Lattice theory, *3rd Amer. Math Soc. colloquium publication vol. 25*(1973).
3. C.B. Huijsmans and B. de Pagter , Ideal theory in *f*-algebra , *Trans. Amer. Soc.* (1982). 225 – 245.
4. W.A.J. Luxemburg and A.C. Zaanen , *Riesz spaces I,* North-Holland , Amsterdam-London 1971.
5. H.H. Schaeffer, *Banach lattices and positives operators*, Springer-Verlag 1974.
6. E. Scheffold , FF-Banach verbandsalgebras, *Math.Z.171* (1981) 193 – 295.
7. C.E. Rickart , *Banach algebras*, Von Nostrand 1960.
8. A.C. Zaanen , *Riesz spaces II* , North-Holland Amsterdam 1983.

A Strong Superdensity Property for Some Subspaces of $C(X)$

ALAIN BERNARD, Université de Grenoble I, Institut Fourier, URA 188 du CNRS, BP 74, 38402 St MARTIN D'HÈRES Cedex (France)

ABSTRACT . — Let X be a compact set. If a continuous non constant function φ operates from $C(X)$ into a subspace E of $C(X)$, then this subspace E is in a certain sense "superdense" in $C(X)$. We give an example of such a situation with $E \neq C(X)$.

I. DEFINITION AND REMARKS.

Let X be a compact Hausdorff space and $C(X)$ be the linear space of all real valued continuous functions on X. In §I, §III and §IV, φ will denote a continuous non-constant real valued function defined on the interval $[-1, +1]$, and E will denote a linear subspace of $C(X)$, equipped with the uniform norm.

DEFINITION. — We will say that φ *operates from* $C(X)$ *into* E if for each $u \in C(X)$, with values in $[-1, +1]$, we have $\varphi \circ u \in E$.

Remark 1. — By the fact that φ is not a constant, one can prove that if φ operates from $C(X)$ into E, then E has to be *uniformly dense in* $C(X)$ (exercise, or see lemma 1 below). The case φ is a constant is without interest.

Remark 2. — If φ is supposed to be *somewhere injective* (i.e. if there exists an open non empty subinterval I of $[-1, +1]$ with φ/I injective) then if φ operates from $C(X)$ into E, then E has to be all of $C(X)$ (very elementary exercise).

Remark 3. — On the view of remark 2 one can wonder if it is possible to have a situation where φ operates from $C(X)$ into E with $E \neq C(X)$! (remember φ is not a constant). Yes it is, as shown by the following example.

Example. — Let $E_0(X)$ be the set of all $u \in C(X)$ such that there exists a dense open subset U of X with u locally constant on U (*Cantor type functions on* X). Then $E_0(X)$ is a linear subspace of $C(X)$, and if $X = [0,1]$ (or, more generally, if X is locally connected as S.J. Sidney pointed out to me) then it gives us the desired example: precisely if φ is Cantor type on $[-1,+1]$, and if $E = E_0(X)$, then φ operates from $C(X)$ into E (simple exercise).

II. SOME FUNCTIONAL ANALYSIS PRELIMINARIES.

DEFINITION. — Let E be a normed space, and ν a semi-norm on E. We say that ν is approximatively bounded on E if there exists a dense subset D of the unit ball of E such that ν is bounded on D.

We shall need in §IV the following characterization :

PROPOSITION 1. — *A semi-norm ν on the normed space E is approximatively bounded if and only if the following property is satisfied :*

$$(S) \qquad \ell^\infty_\nu(\mathbf{N}, E) \text{ is dense in } \ell^\infty(\mathbf{N}, E)$$

where $\ell^\infty(\mathbf{N}, E)$ is the linear space of all bounded sequences of elements of E, and $\ell^\infty_\nu(\mathbf{N}, E)$ its subspace formed of those sequences which are in plus bounded for the semi-norm ν. And where dense means dense for the natural sup-norm on $\ell^\infty(\mathbf{N}, E)$.

Proof. — First suppose ν approximatively bounded : ν being bounded on D, $\ell^\infty_\nu(\mathbf{N}, E)$ contains all the sequences of elements of D, so contains the space they generate. D being dense in the unit ball of E, this generated space is dense in $\ell^\infty(\mathbf{N}, E)$, and (S) follows.

Conversely, suppose ν is not approximatively bounded. Then for each integer $n \geq 1$, the closure of the convex set $\Gamma_n = \{x \in E; \nu(x) \leq n\}$ does not contain the unit ball of E, so, using Hahn-Banach separation theorem, one can choose a continuous linear form ℓ_n on E, of norm 1, such that $\ell_n \leq 1$ on Γ_n, that is such that $|\ell_n| \leq \nu/n$. Choose now for each $n \geq 1$, $u_n \in E$ such that $\|u_n\| \leq 1$ and $\ell_n(u_n) > 2/3$. The sequence $u = (u_n, n \in \mathbf{N})$ is in $\ell^\infty(\mathbf{N}, E)$ and cannot be approximated at less than $1/3$ by any $v = (v_n, n \in \mathbf{N})$ of $\ell^\infty_\nu(\mathbf{N}, E)$: If $\|u_n - v_n\| < 1/3$ then $|\ell_n(u_n) - \ell_n(v_n)| < 1/3$, and so $\ell_n(v_n) > 1/3$, which gives $\nu(v_n) > n/3$. So (S) is not satisfied.

In §III we shall meet normed spaces E such that *each* semi-norm ν on E is approximatively bounded, and use the fact that such spaces have Banach-Steinhauss property, as shows the following:

PROPOSITION 2. — *Let E be a normed space. The following conditions are equivalent:*

(i) Each semi-norm ν on E is approximatively bounded ;

(ii) If $(\lambda_n, n \in \mathbf{N})$ is a sequence of continuous linear forms on E such that for each $x \in E$, $\sup\{|\lambda_n(x)|; n \in \mathbf{N}\} < +\infty$, then $\sup\{\|\lambda_n\|; n \in \mathbf{N}\} < +\infty$.

Proof. — *(i)* $\Longrightarrow$ *(ii)* is obvious: take $\nu = \sup\{|\lambda_n|,\ n \in \mathbf{N}\}$.

To prove *(ii)* $\Longrightarrow$ *(i)* act as in the second part of the proof of proposition 1. If there exists on E a semi-norm ν which is not approximatively bounded, then there exists for each integer $n \geq 1$ a continuous linear form ℓ_n on E, or norm 1, with $|\ell_n| \leq \nu/n$. And $\lambda_n = n\ell_n$ satisfies the hypothesis, but not the conclusion, of *(ii)*.

Remark 1. — Of course $\ell^\infty(\mathbf{N}, E)$ is always dense in $\ell^\infty(\mathbf{N}, \widehat{E})$, where $\widehat{E}$ is the completion of the normed space E. So (S) is equivalent to

$$(S') \qquad \ell^\infty_\nu(\mathbf{N}, E) \text{ is dense in } \ell^\infty(\mathbf{N}, \widehat{E})$$

and the property for a normed space to be such that each semi-norm ν on E is approximatively bounded can be interpretated as a strong *superdensity property* of E into $\widehat{E}$. In particular for such spaces there cannot exist any Banach space F between E and $\widehat{E}$, with inclusions $E \subset F$ and $F \subset \widehat{E}$ continuous, other than $\widehat{E}$ (apply proposition 1 of [1] to the Banach space F).

Remark 2. — These normed spaces on which *each* semi-norm is approximatively bounded share with complete spaces a lot of properties (we have proven here Banach-Steinhauss, for others see S.J. Sidney's paper in these proceedings). So I wanted to call them "approximatively complete". But I have to thank the participants to the conference who have pointed out to me that such normed spaces are already known in literature as "barrelled spaces".

III. STATEMENT OF THE RESULT.

In [2] I have proved (lemma 12, page 467) that if φ (non-constant) operates from $C(X)$ into E, and if $E \neq C(X)$, then there cannot exist a *Banach function space norm on* E (that is a norm, finer than the uniform norm, such that E be a Banach space). I want to prove here the more precise result:

THEOREM 1. — *If a continuous non-constant real valued function φ defined on $[-1,+1]$ operates from $C(X)$ into a linear subspace E of $C(X)$, then each semi-norm ν on E is approximatively bounded.*

A proof of theorem 1 will be given in §IV. Let us give here a corollary of this theorem:

COROLLARY. — *The space $E_0([0,1])$ of all Cantor type continuous functions on $[0,1]$ has the Banach-Steinhauss property. Precisely if a sequence $(\mu_n, n \in \mathbf{N})$ of Radon measures on $[0,1]$ is such that:*

(i) $\forall u \in E_0([0,1]),\ \int u d\mu_n \to 0$

then it converges to zero in the sense that:

(ii) $\forall u \in C([0,1]),\ \int u d\mu_n \to 0\ .$

Proof of the corollary, assuming theorem 1. — If φ is a Cantor type continuous function on $[-1,+1]$, then φ operates from $C([0,1])$ into $E_0([0,1])$, and so by theorem 1 above and proposition 2 of §II, $E_0([0,1])$ equipped with the uniform norm has the Banach-Steinhauss property. So if a sequence $(\mu_n,\ n \in \mathbf{N})$ of measures on $[0,1]$ satisfies *(i)*, it has to be norm-bounded. But $E_0([0,1])$ is uniformly dense in $C([0,1])$, so the norm-bounded sequence $(\mu_n,\ n \in \mathbf{N})$ satisfying *(i)* must satisfy *(ii)*.

Remark. — S.J. Sidney has found an ingenious direct proof of this corollary: see his paper in these proceedings.

IV. A PROOF OF THEOREM 1.

Let us remark first that, replacing if necessary φ by ψ defined by $\psi(t) = \varphi(\alpha t + \beta)$ for α and β well chosen, one can suppose that $\varphi(0) = 0$ and φ is non-constant on each neighborhood of 0 in $[-1,+1]$.

We shall begin with a lemma (the basic fact is that if φ operate from $C(X)$ into E, then E has to be dense in $C(X)$, but we shall have to deal with a situation where φ operates only from a *ball* of a *subspace of* $C(X)$, and for others X and E than the given ones).

LEMMA 1. — *Let Y be a compact Hausdorff space. Let φ be a continuous real function on $[-1,+1]$, such that $\varphi(0) = 0$ and φ is non-constant on every neighborhood of 0. Let F be a uniformly closed subspace of $C(Y)$. Let Y_0 be a compact subset of Y and denote $C_{Y_0}(Y) = \{u \in C(Y); u/Y_0 = 0\}$. Suppose that there exist $u_0 \in C_{Y_0}(Y)$ with $||u_0|| < 1/2$ and $r > 0$ with $r < 1/2$ such that φ operates from* *the ball of center u_0 and radius r in $C_{Y_0}(Y)$ into F (i.e. for each $u \in C_{Y_0}(Y)$ with $\|u - u_0\| < r$, then $\varphi \circ u \in F$). Then if we define $V(Y_0) = \{y \in Y; |u_0(y)| < r/2\}$, we have:*

$$F \supset \{v \in C_{Y_0}(Y);\ \ \text{Support}(v) \subset V(Y_0)\}$$

(where Support(v) *denotes the closed support of the function v).*

Proof of the lemma. — *First we regularize* φ: let $\varepsilon > 0$ be given and take k a C^∞ function on $]0, +\infty[$, with support in $[1-\varepsilon, 1]$, and define φ_k on $[-1, +1]$ by $\varphi_k(t) = \int_{1-\varepsilon}^{1} \varphi(st)k(s)ds$. Then φ_k is C^∞ on $]-1, +1[-\{0\}$, and, if ε is small enough (independently of k), φ_k operates from the ball of center u_0 and radius $r/2$ in $C_{Y_0}(Y)$ into F (remember F is uniformly closed). Now by a good choice of k one can suppose φ_k closed enough to φ to be non-constant on the interval $[-r/2, +r/2]$. So there exists a with $|a| < r/2$ such that the derivative $\varphi_k'(a)$ of φ_k in a is non zero.

To conclude take v in $C_{Y_0}(Y)$ such that Support$(v) \subset V(Y_0) - Y_0$ and take u in the ball of center u_0 and radius $r/2$ of $C_{Y_0}(Y)$, with $u = a$ on a neighborhood of Support(v). Then for θ small enough $\big(\varphi_k(u + \theta v) - \varphi_k(u)\big)/\theta$ is defined and belongs to F. But when $\theta \to 0$, we have $\big(\varphi_k(u+\theta v) - \varphi_k(u)\big)/\theta$ tends uniformly to $\psi_k'(a) \cdot v$ (look in and out of the support of v). But $\varphi_k'(a) \neq 0$ and F is uniformly closed, so $v \in F$. We end the proof of the lemma by a simple passage to the limit.

Proof of the theorem. — We assume φ to be such that $\varphi(0) = 0$ and non-constant on each neighborhood of 0 in $[-1, +1]$. Let ν be a semi-norm on E. We shall prove the *superdensity property* (S). Take $x_0 \in X$.

a) Let B be the ball of center 0 and radius 1/2 in $C_{x_0}(X) = \{u \in C(X);\ u(x_0) = 0\}$. For each $M > 0$, let $B_M = \{u \in B;\ \nu(\varphi \circ u) \leq M\}$. We have $\bigcup_{M>0} B_M = B$ and so, by a *Baire argument*, one of the B_M must have a closure of non empty interior. From that we deduce the following property: there exist $u_0 \in C_{x_0}(X)$, with $\|u_0\| < 1/2$, and $r > 0$, with $r < 1/2$, and *a uniformly dense subset Q of the ball of center u_0 and radius r in $C_{x_0}(X)$ such that ν/Q is bounded.*

b) Now define $\underline{u}_0 \in \ell^\infty(\mathbf{N}, C_{x_0}(X))$ as the constant sequence $(u_0, u_0, \ldots, u_0, \ldots)$ and $\widetilde{Q} = \{(u_n,\ n \in \mathbf{N});\ \forall n, u_n \in Q\}$. $\widetilde{Q}$ is a dense subset (for the sup-sup-norm) of the ball of center $\underline{u}_0$ and radius r in $\ell^\infty(\mathbf{N}, C_{x_0}(X))$. By a) *$\varphi$ operates from $\widetilde{Q}$ into $\ell_\nu^\infty(\mathbf{N}, E)$.*

c) Passing to the limit we obtain from b) *that φ operates from the ball of center $\underline{u}_0$ and radius r of $\ell^\infty(\mathbf{N}, C_{x_0}(X))$ into the closure F of $\ell_\nu^\infty(\mathbf{N}, E)$ in $\ell^\infty(\mathbf{N}, C(X))$,* for the sup-sup-norm.

d) Denote by Y *the Stone-Cečh compactification* $\beta(\mathbf{N} \times X)$ of the topological product $\mathbf{N} \times X$ ($\mathbf{N}$ discrete, X compact). $\ell^\infty(\mathbf{N}, C(X))$ is equal to the bounded continuous functions on $\mathbf{N} \times X$ and so can be identified to $C(Y)$. Denote by Y_0 the closure in Y of $\mathbf{N} \times \{x_0\}$. $\ell^\infty(\mathbf{N}, C_{x_0}(X))$ is equal, via the above identification, to $C_{Y_0}(Y) = \{u \in C(Y);\ u/Y_0 = 0\}$. By c) the hypothesis of lemma 1 are fullfilled and so, if $V(Y_0)$ is the neighborhood of Y_0 in Y defined by $V(Y_0) = \{y \in Y;\ |\underline{u}_0(y)| < r/2\}$, then each function $v \in C_{Y_0}(Y)$ whose closed support is included in $V(Y_0)$ belongs to F.

e) We shall prove now that *each function* $v \in C(Y)$ *whose closed support is included in* $V(Y_0)$ *belongs to* F. For that it is enough to prove *the fact* that each continuous function h on Y_0 can be extended by a $v_h \in F$ with closed support included in $V(Y_0)$ (because then if you take $v \in C(Y)$ with Support$(v) \subset V(Y_0)$, if you note $h = v/Y_0$, you can apply the conclusion of d) to $v - v_h$). And this *fact* can be seen as follow: remember first that Y_0 is the closure of $\mathbf{N} \times \{x_0\}$ in the Stone-Cěch compactification of $\mathbf{N} \times X$, so the continuous function h on Y_0 is determined by its restriction to $\mathbf{N} \times X$, which is nothing but a bounded sequence $(\lambda_n,\ n \in \mathbf{N})$ of real numbers. Now choose $w \in E$ such that $w(x_0) \neq 0$ and $w(x) = 0$ if $|u_0(x)| \geq r/4$ (it is possible by taking $w = \varphi \circ u$, for a well chosen $u \in C(X)$. Then the element v_h of $\ell^\infty(\mathbf{N}, E)$ defined by $v_h = \frac{1}{w(x_0)}(\lambda_1 w, \lambda_2 w, \ldots, \lambda_n w, \ldots)$ is an extension of h whose support is included in $V(Y_0)$, which show the desired *fact*.

f) From e) you deduce, by a *partition of unity argument*, that F is equal to $C(Y)$, or, in other terms, $\ell^\infty_\nu(\mathbf{N}, E)$ is dense in $\ell^\infty(\mathbf{N}, C(X))$. That ends the proof of theorem 1.

Some of the ideas used in the above proof are of the same type than some used in the papers [1], [2], [3], [4], [5].

REFERENCES.

[1] A. Bernard. — *Espace des parties réelles des éléments d'une algèbre de Banach de fonctions*, J. Funct. Anal. **10** (1972), 387–409.

[2] A. Bernard. — *Une fonction non lipschitzienne peut-elle opérer sur un espace de Banach de fonctions non trivial,*, J. Funct. Anal. **122** n°2 (June 1994), 451–477.

[3] O. Hatori. — *Separation properties and operating functions on a space of continuous functions*, International Journal of Mathematics **4** n°4) (1993), 551–600.

[4] E. Briem and K. Jarosz. — *Operating functions for Banach function spaces*, preprint, 1993.

[5] S.J. SIDNEY. — *Functions which operate on the real part of a uniform algebra*, Pacific J. Math. **80** I (1979), 265–272.

Ultraseparating Function Spaces, Operating Functions, and Spaces of Continuous Functions Whose *n*th Powers Span a Dense Subspace of *C*(*X*)

Eggert Briem, Science Institute, University of Iceland, Dunhaga 3, IS-107 Reykjavik, Iceland

Abstract. In this note we look at the ultraseparating property for spaces of continuous functions. We prove that spaces of finite codimension have this property. We show that if h is not a polynomial, then $h(B) = \text{lin}\{h \circ b : b \in B\}$ is dense in $C(X)$. We also show that if B is ultraseparating then there is a polynomial h such that $h(B)$ is dense in $C(X)$.

0 Introduction

The spaces we consider in this note will be normed function spaces. Let X be a compact Hausdorff space and let $C(X)$ denote the space of all continuous real-valued functions on X. A *normed function space* on X is a subspace of $C(X)$ separating the points of X, containing the constant functions and equipped with a norm $\|\cdot\|$ which dominates the sup-norm $\|\cdot\|_\infty$. Further, if the space is a Banach space in the given norm, we call it a *Banach function space*.

In [8] K. de Leeuw and Y. Katznelson extended a version of the Stone-Weierstrass Theorem in the following way: Let B be a sup-norm closed function space on X and suppose there is a continuous non-affine function h defined on some subinterval I of the real line which operates on B (i.e. $h \circ b \in B$ if $b \in B$ and $b(X) \subset I$). Then $B = C(X)$. (In the Stone-Weierstrass

Theorem h is the function $h(t) = t^2$).

An example of a Banach function space which is not sup-norm closed is the space of continuously differentiable functions on the interval [0,1] where the norm is given by $\|f\| = \|f\|_\infty + \|f'\|_\infty$. This example shows immediately that the result of K. de Leeuw and Y. Katznelson does not extend to arbitrary Banach function spaces.

A number of papers deals with the extension of the result of K. de Leeuw and Y. Katznelson to more general Banach function spaces (see for example [1] , [2] , [3], [5] , [6] and [9]). In [1] A. Bernard introduced the so called ultraseparating Banach function spaces. A normed function space with norm $\|\cdot\|$ on a compact Hausdorff space X is said to be *ultraseparating* on X if the space $\ell^\infty(B)$ of all $\|\cdot\|$-bounded sequences of functions from B separates the points of the Stone-Čech compactification $\beta(\mathbf{N} \times X)$ of $\mathbf{N} \times X$. The space $\ell^\infty(B)$ is a subspace of $\ell^\infty(C(X))$, the space of all bounded sequences of functions continuous on X. The space $\ell^\infty(C(X))$ can be identified with $C(\beta(\mathbf{N} \times X))$ in a natural manner: If $(f_n) \in \ell^\infty(C(X))$ and $(k,x) \in \mathbf{N} \times X$ then $(f_n)(k,x) = f_k(x)$. Thus $\ell^\infty(B)$ is a subspace of $C(\beta(\mathbf{N} \times X))$. There are also intrinsic characterizations of the ultraseparation property, one of which we shall look at further on.

In this note we look at the following question: Let B be a space of continuous functions on a compact Hausdorff space X, separating the points of X and containing the constant functions. For a function h defined and continuous on some interval I of the real line we ask whether the space $h(B)$ defined by

$$h(B) = \mathrm{lin}\{h \circ b : b \in B \quad \text{and} \quad b(X) \subseteq I\}$$

is dense in $C(X)$.

It is fairly easy to see (thm 1) that if h is not a polynomial on I then $h(B)$ is dense in $C(X)$. It is also easy to give examples of spaces B for which $h(B)$ is not dense in $C(X)$ for any polynomial h. However, if B is ultraseparating on X it turns out that there is a natural number n such that B^n is dense in $C(X)$, where

$$B^n = \mathrm{lin}\{b^n : b \in B\},$$

and thus, as we shall see, $h(B)$ is dense in $C(X)$ for all polynomials h of degree at least n.

1 The main results

We begin with the case when h is not a polynomial. The case $h(t) = |t|$ has already been considered by O. Hatori ([6], Lemma 7).

Theorem 1 *Let B be a subspace of $C(X)$, separating the points of X and containing the constant functions, and let h be a continuous function on some open interval I. If h is not a polynomial on I then $h(B)$ is dense in $C(X)$.*

Proof: Composing h with an affine function we may suppose that $0 \in I$ and that h is not a polynomial in any neighbourhood of 0. Let J be an open interval containing 0 whose closure is contained in I. If φ is a C_0^∞-function on $\mathbf{R}$ with sufficiently small support then $h_\varphi \circ b \in \overline{h(B)}$, the sup-norm closure of $h(B)$, if $b(X) \subset J$. Here h_φ denotes the convolution $h_\varphi = h * \varphi$. If $b, c \in B$ then, taking the limit of $1/t(h_\varphi \circ (b + tc) - h_\varphi \circ b)$ as t tends to 0, we find that $c \cdot h'_\varphi \circ b \in \overline{h(B)}$ and differentiating again, we find that if n is any natural number then $c^n h_\varphi^{(n)} \circ b \in \overline{h(B)}$. Since h is not a polynomial we can choose for any $\delta > 0$ a C_0^∞-function φ and a number $\lambda \in (-\delta, \delta)$ such that $h_\varphi^{(n)}(\lambda) \neq 0$. Letting $b = \lambda 1$ we find that $c^n \in \overline{h(B)}$ for any $c \in B$ and any natural number n. It follows that $h(B)$ is dense in $C(X)$.

Looking for spaces B for which $h(B)$ is dense in $C(X)$ for some polynomial h, the first spaces that come to mind are spaces of finite codimension.

Theorem 2 *Let B be a subspace of $C(X)$ of finite codimension, separating the points of X and containing the constant functions. Then there is a natural number n such that B^n is dense in $C(X)$.*

Proof: Since B contains the constant functions it follows that if p is a polynomial of degree n then $p(B) = B^n$ and that $B^n = \text{lin}\{b_1 b_2 \ldots b_k : b_i \in B$ and $k \leq n\}$. Thus $\{\overline{B^n}\}$ is an increasing sequence of closed spaces of finite codimension. Here as elsewhere "bar" denotes uniform closure. Since, by the Stone-Weierstrass theorem, the space $\cup B^n$ is dense in $C(X)$ it follows that there is a number n such that $\overline{B^n} = C(X)$.

As we mentioned in the proof above, if B contains the constant functions and if h is a polynomial of degree n then $h(B) = B^n$. In searching for spaces

B for which B^n is dense in $C(X)$ for some n, we look next at sequence spaces. But, before we proceed we need the following characterization of the ultraseparating property.

Theorem 3 *Let A be a subspace of $C(X)$ equipped with a norm $\|\cdot\|$ which dominates the sup-norm. Then A is ultraseparating on X if and only if there exists a polynomial p in k variables and a positive number M such that the following condition is satisfied:*

For each $f \in C(X)$ with $\|f\|_\infty \leq 1$ there are functions $a_1, \ldots, a_k \in A$, with $\|a_i\| \leq M$, such that

$$\|f - p(a_1, \ldots, a_k)\|_\infty \leq 1/2\,.$$

Proof: Suppose we can not find a polynomial p and a positive number M with the required properties. Then for each natural number n and for each polynomial p in n variables, there is a function $f_n \in C(X)$ with $\|f\|_\infty \leq 1$, such that if $a_1, \ldots, a_n \in A$ with $\|a_i\| \leq n$, then

$$\|f_n - p(a_1, \ldots, a_n)\|_\infty \geq 1/2\,.$$

Since A is ultraseparating on X, it follows from the Stone-Weierstrass theorem that the algebra generated by $\ell^\infty(A)$ is dense in $\ell^\infty(C(X)) = C(\beta(\mathbf{N} \times X))$, and thus we can find a polynomial p in k variables and $(a_n'), \ldots, (a_n^k) \in \ell^\infty(A)$, such that

$$\|(f_n) - p\left((a_n'), \ldots, (a_n^k)\right)\|_\infty \leq 1/2\,,$$

which means that

$$\|f_n - p(a_n', \ldots, a_n^k)\|_\infty \leq 1/2$$

for each n. Since

$$\sup_{i,n} \|a_n^i\| < \infty$$

we have reached a contradiction.

Conversely, if the condition of the theorem is satisfied then elements in $\ell^\infty(C(X))$ of the type $p\left((a_n^1), \ldots, (a_n^k)\right)$, where $(a_n^i) \in \ell^\infty(A)$, separate the points of $\beta(\mathbf{N} \times X)$ and hence $\ell^\infty(A)$ separates the points of $\beta(\mathbf{N} \times X)$.

One obvious necessary condition for B^n to be dense in $C(X)$ is that B separates the points of X. For a sequence space $B = \ell^\infty(A)$ this is also a sufficient condition.

Theorem 4 *Let A be a normed function space on X. Then there is a natural number n such that $\ell^\infty(A)^n$ is dense in $C(\beta(N \times X))$ if and only if A is ultraseparating on X.*

Proof: Suppose that A is ultraseparating on X. Then by the lemma there is a natural number k such that given $(f_n) \in \ell^\infty(C(X))$ with $\|(f_n)\|_\infty \leq 1$ there is an element $(g_n) \in \ell^\infty(A)^k$ such that $\|(f_n) - (g_n)\|_\infty < 1/2$. Hence $\ell^\infty(A)^k$ is dense in $\ell^\infty(C(X))$. Conversely if $\ell^\infty(A)^k$ is dense in $\ell^\infty(C(X)) = C(\beta(N \times X))$ then the elements of $\ell^\infty(A)$ separate the points of $\beta(N \times X)$.

Remark Theorem 4 says that for the normed function space $B = \ell^\infty(A)$ on $Y = \beta(N \times X)$ the property that B^n is dense in $C(Y)$ for some n is equivalent to the property that B separates the points of Y. It is of course not in general the case that B^n is dense in $C(X)$ for some natural number n if B separates the points of X. A simple counterexample is $X = [0,1]$ and $B = \text{lin}\{1, id\}$.

The next result is also a direct consequence of theorem 3.

Theorem 5 *Let A be a normed function space which is ultraseparating on X. Then A^n is dense in $C(X)$ for some natural number n.*

Remark This last result can in some instances be used to show that a given space is not ultraseparating. The Poulsen simplex is not ultraseparating on $\overline{\partial K} = K$ because, the restriction of $A(K)^n$ to some line segment I in K is not dense in $C(I)$ for any natural number n. A finite dimensional space B on an infinite set X is not ultraseparating on X because B^n is finite dimensional for any natural number n and hence not dense in $C(X)$. These results have already been obtained in [4].

Let us now look at subspaces B of $C(X)$ and consider them as normed spaces in the sup-norm only. This gives the largest sequence spaces. One might ask whether ultraseparation (w.r.t. the sup-norm) of B on X is equivalent to the property that B^n is dense in $C(X)$ for some natural number n. This is not so, we give a counterexample at the end of this note.

Looking for counterexamples, subspaces of $C(X)$ of finite codimension come to mind because of Theorem 2. But, as the next theorem shows, these are already ultraseparating.

Theorem 6 *Let B be a sup-norm closed subspace of $C(X)$ of finite codimension, separating the points of X and containing the constant functions. Then B is ultraseparating on X.*

Proof: As we observed in the proof of theorem 2 the spaces B^k are increasing and thus B^n is dense in $C(X)$ for some natural number n. Since B^n contains B, a closed space of finite codimension in $C(X)$, we conclude that $B^n = C(X)$, and hence there are functions $f_1, \ldots, f_p \in B^n$ such that $C(X) = B \bigoplus L$, where $L = \text{lin}\{f_1, \ldots, f_p\}$. Then each $f \in C(X)$ can be written as $f = b + t_i f_i + \ldots + t_k f_k$, where $b \in B$, and since the projections on B and L are continuous, it follows from theorem 3 that B is ultraseparating on X.

In [4], A.J. Ellis proved that if B is a sup-norm closed subspace of $C(X)$ of codimension 1, containing the constant functions and separating the points of X, then B is ultraseparating on X. In the same paper he also proved that if K is a simplex for which $\overline{\partial K} | \partial K$ is finite then $B = A(K)|\overline{\partial K}$ is ultraseparating on ∂K. Since B is of finite codimension in $C(\overline{\partial K})$ if $\overline{\partial K}|\partial K$ is finite the theorem above is an extension of these results of Ellis.

Example Let X be the subset of the plane given by

$$X = \{(\frac{1}{n}, 0), (\frac{1}{n}, \frac{1}{n^2}), (\frac{1}{n}, \frac{1}{n}) \, : n \in \mathbf{N}\} \cup \{(0,0)\}$$

with the induced topology and let B be the space of continuous functions on X given by

$$B = \{f \in C(X) \, : \, f(\frac{1}{n}, \frac{1}{n^2}) = \frac{1}{n} f(\frac{1}{n}, \frac{1}{n}) + \frac{n-1}{n} f(\frac{1}{n}, 0) \text{ for all } n \in \mathbf{N}\}.$$

The space B, equipped with the sup-norm, is a Banach function space on X and since the only restriction on the functions in B is that they should be affine on each of the sets $\{(\frac{1}{n}, 0), (\frac{1}{n}, \frac{1}{n^2}), (\frac{1}{n}, \frac{1}{n})\}$, it is clear that B^2 is dense in $C(X)$. However, B is not ultraseparating on X. Because, if $b \in B$ then

$$|b(\frac{1}{n}, 0) - b(\frac{1}{n}, \frac{1}{n^2})| \leq \frac{2\|b\|_\infty}{n}$$

and hence the condition of Theorem 3 is not satisfied.

References

[1] A. Bernard, *Espaces des parties réelles des éléments d'une algébre de Banach de fonctions*, J. Funct. Anal. **10** (1972), 387-409.

[2] E. Briem, *Ultraseparating function spaces and operating functions for the real part of a function algebra*, Proc. Amer. Math. Soc. **111** (1991), 55-59.

[3] E. Briem and Krzysztof Jarosz, *Operating functions for Banach function spaces.*

[4] A.J. Ellis, *Separation and ultraseparation properties for continuous function spaces*, J. London Math. Soc. **29** (1984), 552-532.

[5] O. Hatori, *Functions which operate on the real part of a uniform algebra*, Proc. Amer. Math. Soc. **83** (1981), 565-568.

[6] O. Hatori, *Range transformations on a Banach function algebra II*, Pacific J. Math. **138** (1989), 89-118.

[7] O. Hatori, *Separation properties and operating functions on a space of continuous function*, Internat. J. Math. **4** (1993), 551-600.

[8] K. de Leeuw and Y. Katznelson, *Functions that operate on non-selfadjoint algebras*, J. Analyse Math. **11** (1963), 207-219.

[9] S.J. Sidney, *Functions which operate on the real part of a uniform algebra*, Pacific J. Math. **80** (1979), 265-272.

Small-Bound Isomorphisms of Function Spaces

C.-H. CHU Goldsmiths College, University of London, London, England

H.B. COHEN Department of Mathematics and Statistics, University of Pittsburgh, Pittsburgh, PA 15260 USA

To the memory of Alan Ellis (1940-1992)

This note is motivated by Amir [2] and Cambern's [6, 7] generalization of the Banach-Stone Theorem and by Ellis's recent work [12, 13, 14] on *real* isometries between complex function spaces. Let $C(X)$ be the Banach space of real continuous functions on a compact Hausdorff space X and let $C_{\mathbb{C}}(X)$ be the space of complex continuous functions on X. Amir and Cambern's result states that if the Banach-Mazur distance between $C(X)$ and $C(Y)$, or between $C_{\mathbb{C}}(X)$ and $C_{\mathbb{C}}(Y)$, is less than 2, then X and Y are homeomorphic. Extensions of this result to complex function algebras can be found in [8, 15]. See also [4, 16] for related references.

Given complex function spaces A and B with dual balls $\sum_A$ and $\sum_B$ respectively, every affine homeomorphism between $\sum_A$ and $\sum_B$ induces a *real* linear isometry between A and B, and *vice versa*. This led Ellis [12, 13, 14] to a detailed investigation of the real linear isometries between complex function spaces. A natural sequel of this investigation is the study of *real linear homeomorphisms* between complex function spaces which is the subject of this note. Our objective is to show, as a generalization of the above result of Amir and Cambern, that if there is a small-bound *real* linear homeomorphism between two complex functions spaces A and B, then their Choquet boundaries $Ch\ A$ and $Ch\ B$ are homeomorphic under certain conditions. To achieve this result, we adapt the method in [8] for *real* functions spaces.

We first recall some basic definitions and results in [8]. We will always denote by $A(K)$ the Banach space of real continuous affine functions on a compact convex set K in some locally convex space. The *extreme boundary* ∂K of K is the set of all extreme points of K, with the relative topology. By a *real function space* on X we mean a real closed subspace A of $C(X)$ such that A contains the constant functions and separates points of X. The dual of A will be denoted by A'. Let $S_A = \{\mu \in A' : \mu(1) = 1 = \|\mu\|\}$ be its *state space.* Then X embeds homeomorphically into S_A via the evaluation map $x \to \tau_x$ which also identifies the Choquet boundary $Ch\ A$ of A with ∂S_A (cf. [17; p. 38]). Further A is isometrically isomorphic to $A(S_A)$ [3; Theorem I. 4.9]. By a *complex function space* on X we mean a complex closed subspace A of $C_{\mathbb{C}}(X)$ containing constants and separating points of X. We denote its (complex) dual by A^* and its *state space* by $S_A = \{\mu \in A^* : \mu(1) = 1 = \|\mu\|\}$. Note that the real restriction $(A^*)_r$ of A^* can be identified with the dual $(A_r)'$ of the real restriction A_r via $\mu \in (A^*)_r \to re\mu \in (A_r)'$ where $\|\mu\| = \|re\mu\|$.

Let reA be the real part of A. The uniform closure $\overline{re\ A}$ in $C(X)$ is a real function space and plainly $S_{reA} = S_{\overline{re\ A}}$. The affine homeomorpism $\mu \in S_A \to \tilde{\mu} \in S_{re\ A}$, where $\tilde{\mu}(re\ f) = re(\mu(f))$ for $f \in A$, identifies S_A with $S_{\overline{re\ A}}$. Therefore $\overline{re\ A} = A(S_A)$ and also the Choquet boundary $Ch\ A$ identifies with ∂S_A. Let $Z_A = co(S_A \cup -iS_A) \subset A^*$ be the so-called *complex state space* of A. The *real* linear homeomorphism

$$\theta_A : A \to A(Z_A)$$

defined by

$$(\theta_A f)(z) = re\ z(f) \quad (f \in A, \quad z \in Z_A) \tag{1}$$

satisfies $\|\theta_A f\| \leq \|f\| \leq \sqrt{2}\|\theta_A f\|$ [3; p.146]. This linear homeomorphism has been exploited frequently in [11, 12, 13, 14] to study to what extent Z_A determines A which, as remarked earlier, motivated Ellis's investigation of *real* linear isometries between complex function spaces. In what follows, a *function space* is either real or complex.

Let A be a function space on X. A point $x \in X$ is called a *weak peak point* for A if given $1 > \varepsilon > 0$ and an open set $U \subset X$ containing x, there exists $f \in A$ such that $\|f\| \leq 1$, $f(x) > 1-\varepsilon$ and $|f| < \varepsilon$ on $X \backslash U$. If A is a complex function *algebra*, then Bishop's peak point theorem (cf. [5; Theorem 2.3.4]) implies that this definition is equivalent to that given $1 > \varepsilon > 0$ and open set U containing x, there is an $f \in A$ with $\|f\| = 1 = f(x)$ and $|f| < \varepsilon$ on $X \backslash U$, and moreover, the Choquet boundary $Ch\ A$ consists of exactly the weak peak points for A.

A remark on terminology may be in order here. By a *peak point* for a function space A on X, it is usually meant a point $x \in X$ for which there is some $f \in A$ satisfying $|f(x)| > |f(y)|$ for $y \in X \backslash \{x\}$. For a complex function *algebra* A, peak points are indeed weak peak points, but unfortunately, for function *spaces*, a peak point need not be a weak peak point as defined above! In fact, peak points are always dense in the Choquet boundary [17; Corollary 8.5] whereas weak peak points may not even exist as Example 1 below shows. Since peak points will not figure in our discussion, this pitfall will not affect us.

Example 1. Let K be a square in the plane. The Choquet boundary of $A(K)$ is the set ∂K of extreme points and none of them is a weak peak point.

Example 2. Let A be the disc algebra on the unit circle Γ and let $A_1 = \{\bar{z}f(z) : f \in A\}$. Then A_1 is not an algebra. Since $A \subset A_1$, every point in Γ is a weak peak point for A_1. Note that A and A_1 are (complex) isometrically isomorphic [12; p. 362].

The following result has been obtained in [8].

Theorem 1. *Let A and B be separable real function spaces such that every point in $Ch\ A$ (and $Ch\ B$) is a weak peak point. If there is a linear isomorphnism $\varphi : A \to B$ with $\|\varphi\|\|\varphi^{-1}\| < 2$, then $Ch\ A$ is homeomorphic to $Ch\ B$.*

We have given an example in [8] to show that the above result is false without the assumption of weak peak points. It is also false if $\|\varphi\|\|\varphi^{-1}\| = 2$ [9].

We now consider real linear isomorphisms between complex function spaces.

Example 3. Let $\triangle$ be the unit disc in the complex plane and let $A = \{f \in C_{\mathbb{C}}(\triangle) : f(z) = \alpha z + \beta; \alpha, \beta \in \mathbb{C}\}$. Define $\varphi : A \to A$ by $(\varphi f)(z) = \bar{\alpha} z + \beta$. Then φ is a real linear isometry but not a complex isometry (cf. [14; p. 73]).

Example 4. Let X and Y be nonhomeomorphic compact Hausdorff spaces such that $X \times \{1,2\}$ is homeomorphic to $Y \times \{1,2\}$. (cf. [4; p.143 and 11.15]). Then there is a *real* linear isomorphism $\varphi : C_{\mathbb{C}}(X) \to C_{\mathbb{C}}(Y)$ such that $\|\varphi\|\|\varphi^{-1}\| = 2$. Indeed, let $\Phi : C(X \times \{1,2\}) \to C(Y \times \{1,2\})$ be an isometry and define $\psi : C_{\mathbb{C}}(X) \to C(X \times \{1,2\})$ by

$$\psi(re\ f + i\ im\ f)(x,k) = \begin{cases} re\ f(x) & \text{if} \quad k = 1 \\ im\ f(x) & \text{if} \quad k = 2. \end{cases}$$

Then ψ is a real linear isomorphism with $\|\psi\|\|\psi^{-1}\| = \sqrt{2}$ and therefore $\Phi_0\psi : C_{\mathbb{C}}(X) \to C(Y \times \{1,2\})$ is a real linear isomorphism with $\|\Phi_0\psi\|\|(\Phi_0\psi)^{-1}\| = \sqrt{2}$.

Let A be a complex function space on X. By restricting A to the Shilov boundary $\overline{Ch\ A}$, we will always assume, without loss of generality, that $X = \overline{Ch\ A}$. Let $\sum_A$ be the closed unit ball of A^* and let $x \mapsto \tau_x \in S_A \subset \sum_A$ be the evaluation map. Then $\partial \sum_A = \{\alpha\tau_x : x \in X, \alpha \in \Gamma\}$ where Γ is the unit circle in $\mathbb{C}$. It follows that $\sum_A$ is the w^*-closed convex hull of $\Gamma.\partial S_A$ under the assumption that $X = \overline{Ch\ A}$. If $Ch\ A$ is closed, then we have $\partial \sum_A = \Gamma.\partial S_A$. As usual, we embed A into A^{**} and we will denote by $< \cdot, \cdot >$ the bilinear functional on a pair of Banach spaces in duality.

Lemma 2. *Let A be a complex function space with closed Choquet boundary. Let $a^{**} \in A^{**}$ be such that $a^{**} = \sigma(A^{**}, A^*) - \lim_{n\to\infty} a_n$ for some sequence (a_n) in A. Then*

$$\|a^{**}\| = \sup\{| < \mu, a^{**} > | : \quad \mu \in \partial S_A\}.$$

Proof. Write $t = \sup\{| < \mu, a^{**} > | : \mu \in \partial S_A\}$. We need to show $\|a^{**}\| \leq t$. Note that $\sum_A$ is a w^*-compact convex set and $|a_n|$ is a real continuous convex function on $\sum_A$.

Since $Ch\ A$ is closed, we have $\partial \sum_A = \Gamma.\partial S_A$ as remarked above. For $\alpha \in \Gamma$ and $\mu \in \partial S_A$, we have

$$< \alpha\mu, \lim_{n\to\infty} |a_n| >=< \alpha\mu, |a^{**}| >= | < \alpha\mu, a^{**} > |$$
$$= |\alpha < \mu, a^{**} > | = | < \mu, a^{**} > | \leq t.$$

Therefore, by Meyer's maximum principle [1; Proposition I. 4.10], we have

$$t \geq \sup_{\nu \in \sum_A} < \nu, \lim_{n\to\infty} |a_n| >= \sup_{\nu\in\sum_A} < \nu, |a^{**}| >= \|a^{**}\|.$$

Let A be a complex function space on X with complex state space $Z_A = co(S_A \cup -iS_A)$ and let $\theta : A \to A(Z_A)$ be the real isomorphism defined in (1). If A is an algebra, then every extreme point of Z_A is a split face of Z_A [10; p.166]. Recall that a face $F \subset Z_A$ is called *split* if there is a complementary face F' such that every point in $Z_A \backslash (F \cup F')$ can be uniquely represented as a convex combination of a point in F and a point in F'. In this case, the envelope function

$$\hat{\chi}_F(z) = \inf\{a(z) : a \in A(Z_A), a > \chi_F\} \tag{2}$$

where χ_F is the characteristic function of F, is upper semi-continuous and affine on Z_A with $F = \hat{\chi}_F^{-1}(1)$ and $F' = \hat{\chi}_F^{-1}(0)$. Further, there is a decreasing net (a_β) in $A(Z_A)$ converging pointwise to $\hat{\chi}_F$ with $a_\beta > \chi_F$.

If A is a complex function space such that every point in $Ch\ A$ is a weak peak point, then every extreme point of Z_A is also a split face of Z_A. Indeed, given $z \in \partial Z_A = \partial S_A \cup -i\partial S_A$ with an open set $V \subset Z_A$ containing z and given $1 > \varepsilon > 0$, then either $z = \tau_x$ or $z = -i\tau_x$ for some $x \in Ch\ A$ and there exists $f \in A$ with $\|f\| \leq 1$, $f(x) > 1 - \varepsilon$ and $|f| < \varepsilon$ on $X \backslash \tau^{-1}(V)$. Now $\|\theta f\| \leq 1$ and $\|\theta(if)\| \leq 1$.

If $z = \tau_x$, then $(\theta f)(z) = re\ z(f) = f(x) > 1 - \varepsilon$ and $|\theta f| < \varepsilon$ on $\partial Z_A \backslash V$. If $z = -i\tau_x$, then $\theta(if)(z) = re\ z(if) = f(x) > 1 - \varepsilon$and $|\theta(if)| < \varepsilon$ on $\partial Z_A \backslash V$. So $z \in \partial Z_A$ is a weak peak point for $A(Z_A)$ in the sense of [8] and by [8; Proposition 1], $\{z\}$ is a split face of Z_A.

We now adapt Cambern's idea [6, 7] and use the real isomorphism $\theta : A \to A(Z_A)$ to prove an extension of Theorem 1.

Theorem 3. *Let A and B be separable complex function spaces with closed Choquet boundaries such that every point in $Ch\ A$ (and $Ch\ B$) is a weak peak point. If there is a real linear isomorphism $\varphi : A \to B$ with $\|\varphi\|\|\varphi^{-1}\| < \sqrt{2}$, then $Ch\ A$ is homeomorphic to $Ch\ B$.*

Proof. We will regard the complex duals A^*, A^{**} as real Banach spaces and we may assume that $\|\varphi\| < \sqrt{2}$ and $\|\varphi(f)\| > c\|f\|$ for all nonzero $f \in A$ where $1 < c < \sqrt{2}$. We need to construct a homeomorphism $\rho : \partial S_B \to \partial S_A$. Here is the gist. Consider the dual map

$$(\theta_A^{-1})' \circ \varphi' : B^* \to A^* \to A(Z_A)'$$

where we regard Z_A as a subset of $A(Z_A)'$ via evaluation. We will show that, for any $y \in \partial S_B$, there is a *unique* $x \in \partial S_A$ such that

$$(\theta_A^{-1})' \circ \varphi'(y) = \lambda x + \mu$$

with $|\lambda| > c$ and $\mu \in A(Z_A)'$. So we can define $\rho(y) = x$ which gives the homeomorphism.

Now the details. We will use the second dual map $\varphi'' : A^{**} \to B^{**}$. Fix $y \in \partial S_B$. Take any $x \in \partial S_A$. By the above remarks, $\{x\}$ is a split face of Z_A. Let $\chi_x = \hat{\chi}_{\{x\}}$ be the envelope function defined in (2). Then $\chi_x \in A(Z_A)''$ since the latter identifies with the space $A^b(Z_A)$ of bounded affine functions on Z_A (cf. [8; p. 75]). As in [8; p. 77], we have a unique expression

$$(\theta_A^{-1})' \circ \varphi'(y) = \lambda x + \mu \tag{3}$$

where $\lambda = < y, \varphi'' \circ (\theta_A^{-1})''(\chi_x) >$ and $2 > \|(\theta_A^{-1})' \circ \varphi'(y)\| = |\lambda| + \|\mu\|$ which implies that there is *at most one* $x \in \partial S_A$ with $|\lambda| > c$. We now show there is indeed one $x \in \partial S_A$ with $| < y, \varphi'' \circ (\theta_A^{-1})''(\chi_x) > | > c$.

By separability of $A(Z_B)$, there is a sequence (b_n) in $A(Z_B)$ such that $\chi_y = \lim_{n\to\infty} b_n$ pointwise on Z_B which gives $(\varphi^{-1})'' \circ (\theta_B^{-1})''(\chi_y) = \lim_{n\to\infty} \varphi^{-1} \circ \theta_B^{-1}(b_n)$ and by Lemma 2, we have $\|(\varphi^{-1})'' \circ (\theta_B^{-1})''(\chi_y)\| = \sup\{| < x, (\varphi^{-1})'' \circ (\theta_B^{-1})''(\chi_y) > | : x \in \partial S_A\}$. Since $c > 1$, we can choose $x \in \partial S_A$ such that

$$| < x, (\varphi^{-1})'' \circ (\theta_B^{-1})''(\chi_y) > | > \frac{1}{c}\|(\varphi^{-1})'' \circ (\theta_B^{-1})''(\chi_y)\|.$$

Let $\alpha \in \Gamma$ be such that

$$< x, \alpha(\varphi^{-1})''(\theta_B^{-1})''(\chi_y) > = \alpha < x, (\varphi^{-1})''(\theta_B^{-1})''(\chi_y) > = | < x, (\varphi^{-1})''(\theta_B^{-1})''(\chi_y) > |.$$

For such $x \in \partial S_A$, there exists $z \in \partial S_B$ with

$$| < z, \varphi''(\theta_A^{-1})''\chi_x > | > c. \tag{4}$$

Otherwise, as before, we have

$$\begin{aligned} c &\geq \sup\{| < z, \varphi''(\theta_A^{-1})''\chi_x > | : z \in \partial S_B\} = \|\varphi''(\theta_A^{-1})''\chi_x\| \\ &> c\|(\theta_A^{-1})''\chi_x\| \geq c\|\chi_x\| = c \end{aligned}$$

which is impossible.

It remains to show $z = y$. If not, then both z and $-iz$ belong to the complementary face $\{y\}' \subset Z_B$ and so $\chi_y(z) = 0 = \chi_y(-iz)$. Therefore $re < z, (\theta_B^{-1})''(\chi_y) > = \chi_y(z) = 0$ and $re - i < z, (\theta_B^{-1})''(\chi_y) > = re < -iz, (\theta_B^{-1})''(\chi_y) > = \chi_y(-iz) = 0$ which yields $< z, (\theta_B^{-1})''(\chi_y) > = 0$. So $0 = < z, (\theta_B^{-1})''(\chi_y) > = < \varphi'(z), (\varphi^{-1})''(\theta_B^{-1})''\chi_y >$ implies

$$\begin{aligned} 0 &= \alpha < \varphi'(z), (\varphi^{-1})''(\theta_B^{-1})''\chi_y > = < \varphi'(z), \alpha(\varphi^{-1})''(\theta_B^{-1})''\chi_y > \\ &= < (\theta_A^{-1})'\varphi'(z), \theta_A''(\alpha(\varphi^{-1})''(\theta_B^{-1})''\chi_y) > \\ &= < \eta x + \nu, \theta_A''(\alpha(\varphi^{-1})''(\theta_B^{-1})''\chi_y) > \end{aligned}$$

where

$$|\eta| = |< z, \varphi''(\theta_A^{-1})''\chi_x >| > c$$

by (3) and (4). Now

$$\begin{aligned}
&|< \eta x, \theta_A''(\alpha(\varphi^{-1})''(\theta_B^{-1})''\chi_y) >| \\
&= |\eta|| < x, \theta_A''(\alpha(\varphi^{-1})''(\theta_B^{-1})''\chi_y) > | \\
&= |\eta||re < x, \alpha(\varphi^{-1})''(\theta_B^{-1})''\chi_y > | \\
&= |\eta|| < x, (\varphi^{-1})''(\theta_B^{-1})''\chi_y > | > c \cdot \frac{1}{c}\|(\varphi^{-1})''(\theta_B^{-1})''(\chi_y)\|.
\end{aligned}$$

But also $2 > |\eta| + \|\nu\|$ and $| < \nu, \theta_A''(\alpha(\varphi^{-1})''(\theta_B^{-1})''\chi_y) > | \leq \|\nu\| \, \|\theta_A''(\alpha(\varphi^{-1})''(\theta_B^{-1})''\chi_y)\| < (2-c)\|\alpha(\varphi^{-1})''(\theta_B^{-1})''\chi_y\| < \|(\varphi^{-1})''(\theta_B^{-1})''\chi_y\|$ which gives a contradiction. So $z = y$ and we have shown there is a unique $x \in \partial S_A$ with $| < y, \varphi''(\theta_A^{-1})''\chi_x > | > c$. Therefore $\rho : \partial S_B \to \partial S_A$ can be defined.

Reversing the above arguments, one can show that given $x \in \partial S_A$, there is a unique $y \in \partial S_B$ such that $| < x, (\varphi^{-1})''(\theta_B^{-1})''\chi_y > | > \frac{1}{\sqrt{2}}$ and $| < y, \varphi''(\theta_A^{-1})''\chi_x > | > c$. Hence ρ is a bijection.

Finally, using the weak peak point property, one can show ρ is a homeomorphism as in [8; Theorem 7].

Remark. In the above proof, one can not apply Theorem 1 and the method in [8] directly to the real isomorphism $\theta_B\varphi\theta_A^{-1} : A(Z_A) \to A(Z_B)$ because its bound is too large and also, even if one can show that ∂Z_A and ∂Z_B are homeomorphic, it does not follow that ∂S_A and ∂S_B are homeomorphic (cf. Example 4).

Corollary 4. *Let A and B be separable complex function algebras with closed Choquet boundaries. If there is a real linear isomorphism $\varphi : A \to B$ with $\|\varphi\| \, \|\varphi^{-1}\| < \sqrt{2}$, then $Ch\ A$ is homeomorphic to $Ch\ B$.*

REFERENCES

1. E.M. Alfsen, *Compact convex sets and boundary integrals*, Springer-Verlag, 1971.
2. D. Amir, *On isomorphisms of continuous function spaces*, Israel J. Math. **3** (1965), 205-210.
3. L. Asimow and A.J. Ellis, *Convexity theory and its applications in functional analysis*, A.P. (1980).
4. E. Behrends, *M−structure and the Banach-Stone Theorem*, Lecture Notes in Math. 736, Springer-Verlag, 1979.
5. A. Browder, *Introduction to function algebras*, W.A Benjamin, 1969.
6. M. Cambern, *A generalized Banach-Stone theorem*, Proc. Amer. Math. Soc. **17** (1966), 396-400.
7. ______, *On isomorphism with small bounds*, Proc. Amer. Math. Soc. **18** (1967), 1062-1066.
8. C.-H. Chu and H.B. Cohen, *Isomorphisms of spaces of continuous affine functions*, Pacific J. Math **155** (1992), 71-85.
9. H.B. Cohen, *A bound-two isomorphism for $C(X)$ Banach spaces*, Proc. Amer. Math. Soc. **50** (1975), 215-217.
10. A.J. Ellis, *On split faces and function algebras*, Math. Ann. **195** (1972), 159-166.
11. A.J. Ellis and W.S. So, *Isometries and the complex state spaces of uniform algebras*, Math. Z **195** (1987), 119-125.

12. A.J. Ellis, *Equivalence for complex state spaces of function spaces*, Bull. London Math. Soc. **19** (1987), 359-362.
13. ———, *Real characterizations of function algebras amongst function spaces*, Bull. London Math. Soc. **22** (1990), 381-385.
14. ———, *Real linear isometries of complex functions*, Proceedings of "Function Spaces" Conference, Marcel- Dekker, 1992, pp. 71-78.
15. K. Jarosz, *Perturbations of Banach algebras*, Lecture Notes in Math., vol. 1120, Springer-Verlag, 1985.
16. K. Jarosz and V.D. Pathak, *Isometries and small bound isomorphisms of function spaces*, Proceedings of "Function Spaces" Conference, Marcel-Dekker, 1992, pp. 241-271.
17. R.R. Phelps, *Lectures on Choquet's Theorem*, Van Nostrand, Princeton, 1966.

Amenability, Weak Amenability, and the Close Homomorphism Property for Commutative Banach Algebras

P. C. CURTIS, JR. Department of Mathematics, University of California, Los Angeles, California 90024-1555

ABSTRACT

In this paper we would like to survey recent and not so recent work relating the notions of amenability, weak amenability, and the close homomorphism property for commutative Banach algebras. Our objective is to discuss results and pose open problems but not to provide complete details. At the end we introduce an example of a commutative weakly amenable radical Banach algebra and discuss its implications.

1 AMENABILITY

A locally compact group G is said to be amenable if there is a left invariant mean on the group G, that is, a continuous complex linear functional m, defined on the bounded continuous functions on G which is left translation invariant, $m(f) \geq 0$ if $f \geq 0$, and $m(1) = 1$. In 1972 B. E. Johnson [15] related this condition to a condition on the space of derivations on $L^1(G)$ in the following way. Let X be a Banach space which is a two sided Banach $L^1(G)$ module, that is, there exist bounded maps $(a,x) \to a \cdot x$, $(a,x) \to (x \cdot a)$, $a \in L^1(G)$, $x \in X$, which satisfy $a \cdot (x \cdot b) = (a \cdot x) \cdot b$. The conjugate space X^* is then a dual Banach bi-module under the module actions defined by $(a \cdot x, \lambda) = (x, \lambda \cdot a)$ and $(x \cdot a, \lambda) = (x, a \cdot \lambda)$, $x \in X, \lambda \in X^*$, $a \in L^1(G)$. Johnson showed that the locally compact group G was amenable if and only if every bounded derivation D from $L^1(G)$ to a dual $L^1(G)$ bi-module X^* is inner. That is, if $D : L^1(G) \to X^*$ satisfies $D(ab) = a \cdot Db + (Da) \cdot b$, then there exists $x^* \in X^*$ satisfying $Da = a \cdot x^* - x^* \cdot a$. The product ab is of course convolution product. A Banach algebra $\mathfrak{A}$ satisfying this condition on its space of derivations was defined by Johnson to be amenable.

Johnson showed in [16] that this property can be characterized by a condition on the second dual of the projective tensor product $\mathfrak{A}\hat{\otimes}\mathfrak{A}$. Namely, $\mathfrak{A}$ is amenable if and only if $\mathfrak{A}$ has a bounded approximate identity and there exists $M \in (\mathfrak{A}\hat{\otimes}\mathfrak{A})^{**}$ satisfying $a \cdot M = M \cdot a$ and if π^{**} is the second dual of the natural map $\pi : \mathfrak{A}\hat{\otimes}\mathfrak{A} \to \mathfrak{A}$ defined by $\pi(a \otimes b) = ab$, then $\pi^{**}(M \cdot a) = a$, $a \in \mathfrak{A}$. The module actions in $(\mathfrak{A}\hat{\otimes}\mathfrak{A})^{**}$ are the natural ones inherited from $\mathfrak{A}\hat{\otimes}\mathfrak{A}$ where

$$a \cdot (b \otimes c) \equiv ab \otimes c \quad \text{and}$$
$$(b \otimes c) \cdot a \equiv b \otimes ca, \quad a, b, c, \in \mathfrak{A}.$$

The element M is called a virtual diagonal.

With no topological assumptions, an algebra $\mathfrak{A}$ over the complex field with identity which has a diagonal, that is, an element $M \in \mathfrak{A} \otimes \mathfrak{A}$ satisfying $a \cdot M = M \cdot a$ and $\pi(M) = 1$, is known to be finite dimensional and is a direct sum of full matrix algebras [17]. For Banach algebras if the diagonal $M \in \mathfrak{A}\hat{\otimes}\mathfrak{A}$ rather than $\mathfrak{A} \otimes \mathfrak{A}$, then $\mathfrak{A}$ is finite dimensional and is a direct sum of full matrix algebras if $\mathfrak{A}$ is commutative, or is a C^* algebra, or generally if $\mathfrak{A}$ is semi simple and satisfies the compact approximation property. The last result is due to Taylor [21]. Whether some form of the approximation property for Banach spaces is necessary for the result to hold in general is an interesting open question.

In the case that the diagonal $M \in \mathfrak{A}\hat{\otimes}\mathfrak{A}$, the relationship between M and the derivation D is relatively easy to see. If $D : \mathfrak{A} \to X$, then the bilinear map $\Lambda(a \otimes b) \equiv aDb$ extends to a bounded linear operator Λ from $\mathfrak{A}\hat{\otimes}\mathfrak{A}$ to X. If $x = \Lambda(M)$, then a short calculation yields that

$$D(a) = a \cdot (x + y) - (x + y) \cdot a \quad \text{where} \quad y = -1 \cdot D(1).$$

If $\mathfrak{A}$ is amenable, we need a weak version of this calculation. Consult [16] for details.

In [15] Johnson developed several important consequences of amenability.

(1) Amenable algebras $\mathfrak{A}$ always posess bounded approximate identities, and if $\mathfrak{A}$ has a unit, then closed two sided ideals of finite co-dimension are amenable.

(2) If ν is a continuous homomorphism of an amenable algebra $\mathfrak{A}$, and $\mathfrak{B} = \overline{\nu(\mathfrak{A})}$, then $\mathfrak{B}$ is amenable.

(3) If I is a closed two sided ideal in an amenable algebra $\mathfrak{A}$, then I is amenable if I has a bounded approximate identity.

Condition (2) above yields that $C(\Omega), \Omega$ compact, is amenable, for if the discrete group $G = \{e^{ih} : h \in C_R(\Omega)\}$, then $C(\Omega) = \overline{\ell_1(G)}$.

Parallel to Johnson's work in the west, A. Helemskii in Moscow was developing similar results but from a homological algebra point of view. One striking result, which went unnoticed in the west for quite some time was the result of M. V. Sheinberg [18], a student of Helemskii, to the effect that $C(\Omega), \Omega$ compact Hausdorff, was the only uniform algebra which was amenable. Thus the absence of derivations into dual Banach $\mathfrak{A}$ bi-modules characterizes $C(\Omega)$ among uniform algebras.

Assuming $\mathfrak{A} \neq C(\Omega)$, the proof proceeds by defining $Y = L^2(d|\mu|)$ where $\mu \in M(\Omega)$, and $\mu \perp \mathfrak{A}$, $\mu \neq 0$. If $X = L^2$ closure of $\mathfrak{A}$ in Y, then the homological algebra consequences of amenability yield that there exists a projection P of Y onto X satisfying $P(fy) = fP(y)$, $f \in \mathfrak{A}$, $y \in Y$. Since multiplication by f is a normal operator, an application of Fuglede's theorem shows that $P(f^*y) = f^*P(y)$. Applying the Stone Weierstrass theorem and the fact that $\mathfrak{A}$ has an identity we infer that $P(f) = f$, if $f \in C(\Omega)$; therefore $C(\Omega) \subset X$. Since for $f \in C(\Omega)$, $\int f d\mu = \lim_{n\to\infty} \int (f - a_n) d\mu$, $a_n \in \mathfrak{A}$, an application of Schwarz's inequality shows that $\mu = 0$. The complete details may be found in [13].

For C^* algebras U. Haagerup has shown that amenability is equivalent to nuclearity, and much research has been conducted in this area. In what follows here, however, we shall be concerned only with commutative algebras.

The first question to be considered is the following. If T is a bounded operator in a Hilbert space $\mathcal{H}$, and the closed algebra of operator $\mathfrak{A}_T$ generated by T is amenable, what can be said? In particular, is $\mathfrak{A}_T$ similar to a C^* algebra? The first result in this direction is a result of Sz-Nagy [20] who proved that if an invertible operator $T \in B(\mathcal{H})$ satisfied

$$\|T^n\| = O(1), \ n \in \mathbf{Z},$$

then T is similar to a unitary operator. Greenleaf [11] generalized this result by showing that if a closed algebra of operators in Hilbert space is generated by a bounded amenable group of invertible operators, then the algebra is similar to a C^* algebra. In both cases, the algebra $\mathfrak{A} = \overline{\ell_1(G)}$ where G is a bounded discrete amenable group, hence $\mathfrak{A}$ is amenable. A corollary to this theorem is the result of Mackey and Wermer to the effect that if $\mathfrak{A}$ is generated by a uniformly bounded set of commuting idempotent operators P_α, then $\{P_\alpha\}$ are simultaneously similar to self adjoint projections and consequently $\mathfrak{A}$ is similar to a commutative C^* algebra. The bounded abelian group of invertible operators is generated by the operators $Q_\alpha = 2P_\alpha - I$.

Each of these results invokes an application of (2) which implies that the algebra $\mathfrak{A}$ is amenable. Whether amenability alone will guarantee that a commutative amenable operator algebra is similar to a C^* algebra seems to be quite hard. However, George Willis [24] has shown that if T is a compact operator in $\mathcal{H}$ and $\mathfrak{A}_T$ is amenable, then T is similar to a normal operator. This seems to be the best result so far.[1]

If $T \in B(\mathcal{H})$ is quasi nilpotent, can $\mathfrak{A}_T$ be amenable? The answer is not known[2], but condition (2) on the other hand cannot be satisfied in the same way as above. Gelfand [10] in 1941 showed that if N is a quasi nilpotent operator in a Banach space X and

$$\|(I+N)^n\| = O(1), \ n \in \mathbf{Z}$$

then $N = 0$. This was later improved by Hille [14, 4.10] to if $\|(I+N)^n\| = o(|n|), \ n \in \mathbf{Z}$, then $N = 0$.

Although the existence of a commutative radical amenable Banach algebra is not known, non-semisimple amenable algebras exist as quotient algebras of $L^1(G)$ for any abelian locally compact group with nondiscrete dual group Γ. These are the quotient algebras arising from sets $E \subset \Gamma$ of non-spectral synthesis.

2 WEAK AMENABILITY

Assume now that $\mathfrak{A}$ is commutative and that X is a commutative or one sided $\mathfrak{A}$ module, i.e. $a \cdot x = x \cdot a$, then if D is a derivation from $\mathfrak{A}$ to X, $D(a^n) = na^{n-1}D(a)$. Consequently if a is invertible and

$$\|a^n\| = O(1), \ n \in \mathbf{Z}, \ \text{then}$$
$$D(a) = \frac{1}{n} a^{1-n} D(a^n)$$

and

$$\|Da\| \le \frac{1}{n} \|a\| \ \|a^{-n}\| \ \|a^n\| \ \|D\|.$$

[1] See Section 4.

[2] See Section 4.

Therefore $Da = 0$. Thus if $\mathfrak{A}$ is the closed linear span of regular elements satisfying

$$(*) \qquad \frac{\|a^n\| \, \|a^{-n}\|}{n} \to 0 \quad \text{as} \quad n \to \infty,$$

then $D : \mathfrak{A} \to X$ implies $D = 0$. We call an algebra with this latter property weakly amenable. Every bounded derivation from a commutative Banach algebra to an arbitrary commutative $\mathfrak{A}$-module is trivial if and only if that is true for the special case $X = \mathfrak{A}^*$ [4]. Another condition guaranteeing weak amenability is that $\mathfrak{A}$ is the closed linear span of elements a satisfying

$$(**) \qquad \frac{\|e^{na}\| \, \|e^{-na}\|}{n} \to 0 \quad \text{as} \quad n \to \infty,$$

This is clear since $D(e^{na}) = ne^{na}D(a)$.

Examples of weakly amenable algebras include certain classes of Beurling algebras and Lipschitz algebras. Specifically if

$$\ell_1(\alpha) = \left\{ a = a(n) : n \in \mathbf{Z} : \sum_{-\infty}^{\infty} |a(n)|(1 + |n|)^{\alpha} < \infty \right\}$$

then $\ell_1(\alpha)$ is not amenable for $\alpha > 0$ since maximal ideals do not contain bounded approximate identities. If $0 \leq \alpha < \frac{1}{2}$, $\ell_1(\alpha)$ is weakly amenable since $\ell_1(\alpha)$ satisfies condition (*). If $\alpha \geq \frac{1}{2}$, $(Da)(n) = na(n)$ defines a bounded derivation from $\ell_1(\alpha)$ to

$$X = \left\{ x = x(n) : n \in \mathbf{Z} : \sum_{-\infty}^{\infty} \frac{|x(n)|(1 + |n|)^{\alpha}}{1 + |n|} < \infty \right\}.$$

For details see [4].

Another example considered in [4] is the algebra

$$\begin{aligned} \text{lip}_{\alpha}(\mathbf{T}) = \Bigg\{ & f : \sup_{s \neq t} \frac{|f(s) - f(t)|}{|s - t|^{\alpha}} < \infty, \; s, t \in \mathbf{T} \\ \text{and} \; & \frac{|f(s) - f(t)|}{|s - t|^{\alpha}} \to 0 \;\; \text{as} \;\; |s - t| \to 0 \Bigg\}. \end{aligned}$$

The algebra $\text{lip}_{\alpha}(\mathbf{T})$ is not amenable for $\alpha > 0$ since again maximal ideals do not contain bounded approximate identities. For $0 \leq \alpha < \frac{1}{2}$, $\text{lip}_{\alpha}(\mathbf{T})$ is weakly amenable, since the algebra is generated by elements satisfying (**). For $\alpha > \frac{1}{2}$, $(Df)(k) = k\hat{f}(-k)$, defines a bounded derivation from $\text{lip}_\alpha(\mathbf{T})$ to $\text{lip}_\alpha(\mathbf{T})^*$. In the case $\alpha = \frac{1}{2}$, $\text{lip}_{1/2}(\mathbf{T})$ is weakly amenable even though it is not generated by elements satisfying (*) or (**). In fact any such elements are identically constant [6]. However, if

$$D : \text{lip}_{1/2}(\mathbf{T}) \to \text{lip}_{1/2}(\mathbf{T})^*, \text{ then}$$
$$D(e_n) = ine_n \cdot \lambda$$

for some $\lambda \in \text{lip}_{1/2}(\mathbf{T})^*$ where $e_n(\theta) = e^{ni\theta}$. An argument of B. E. Johnson contained in [4] shows that this derivation is necessarily unbounded if $\lambda \neq 0$.

It is clear that if $\mathfrak{A}$ is weakly amenable, $\overline{\mathfrak{A}^2} = \mathfrak{A}$ for otherwise there exists a nonzero continuous linear functional δ on $\mathfrak{A}$ such that $\delta(\overline{\mathfrak{A}^2}) = 0$, and it is easily seen that δ defines a derivation from $\mathfrak{A}$ to a one dimensional module X for which $a \cdot x = 0$. If ν is a continuous homomorphism and $\mathfrak{B} = \overline{\nu(\mathfrak{A})}$, then it is easily seen that $\mathfrak{B}$ is weakly amenable if $\mathfrak{A}$ is. If I is a closed ideal in a weakly amenable algebra $\mathfrak{A}$, then N. Grønbæk [12] has shown that I is weakly amenable if $\overline{I^2} = I$. Thus analogues of conditions (1)-(3) for amenable algebras hold for commutative weakly amenable algebras as well.

An interesting question is whether the property of being weakly amenable characterizes $C(\Omega)$ among uniform algebras. At present the answer is unknown. A natural test case is the algebra $R(K)$, where K is a compact nowhere dense set in $\mathbf{C}$, $R(K)$ is the uniform closure of the rational functions with poles off K, and it is assumed that $R(K) \neq C(K)$. The best result so far is a theorem of Joel Feinstein [7] that if $R(K) \neq C(K)$, then there exists a measure $\mu \in M(K)$, $\mu \perp R(K), \mu \neq 0$, such that if h is the Cauchy transform of μ restricted to K, then for a rational function f with poles off K, the derivation $D(f) = f'h$ extends via the Beurling transform of μ to a continuous derivation from $R(K)$ to $L^1_*(K, dm)$, the space of functions of weak type L^1 with respect to plane Lebesgue measure. Unfortunately $L^1_*(K, dm)$ is not a Banach space although it is a locally bounded $R(K)$ module with respect to its (F) space topology. It is a classical theorem of Wermer [22] that $R(K)$ may be different from $C(K)$ and have no bounded point derivations. For a commutative Banach algebra $\mathfrak{A}$ an equivalent condition to having no bounded point derivations is that every bounded derivation to a finite dimensional bi-module is necessarily inner [4].

3 THE RADICAL OF $\mathfrak{A}$ AND THE CLOSE HOMOMORPHISM PROPERTY

We have already observed that non-semisimple amenable algebras arise naturally in connection with sets of non-spectral synthesis in locally compact groups. The context is the following. Let E be a closed set in a locally compact, non-discrete abelian group Γ with dual group G and regard $L^1(G) \cong A(\Gamma)$ via the Gelfand representation of $L^1(G)$. Then associated with E are two ideals of $A(\Gamma)$

$$\begin{aligned} I(E) &= \{f \in A(\Gamma) : f(E) = 0\} \\ \text{and } J(E) &= \{f \in A(\Gamma) : f = 0 \text{ on a neighorhood of } E\}. \end{aligned}$$

Necessarily $J(E) \subset I(E)$ and if $\overline{J(E)} \neq I(E)$, then E is called a set of non-synthesis, and the quotient algebra $A(E) = A(\Gamma)/\overline{J(E)}$ has radial $I(E)/\overline{J(E)}$. A fundamental question for non-semisimple algebras $\mathfrak{A}$ is when it is possible to write $\mathfrak{A} \simeq \mathfrak{B} \oplus \text{ rad } \mathfrak{A}$ where $\mathfrak{B} \simeq \mathfrak{A}/\text{rad}\mathfrak{A}$ and $\mathfrak{B}$ is closed. Such a decomposition is called a strong Wedderburn decomposition of $\mathfrak{A}$. Bachelis and Saeki [1] in 1987 observed first that such a decomposition is never possible in an algebra generated by doubly power bounded elements, and if E is compact in Γ, then $A(E)$ always has this property, hence the strong Wedderburn theorem always fails for these algebras. The essential tool was the theorem of Gelfand mentioned earlier.

If we substitute Hille's generalization of Gelfand's result and generalize slightly, we obtain the following.

THEOREM. *Let $\mathfrak{A}$ be a commutative Banach algebra with unit and let σ, τ be continuous homomorphisms of $\mathfrak{A}$ into a commutative Banach algebra $\mathfrak{B}$. Let x be an invertible element of $\mathfrak{A}$ satisfying $\|x^n\|\ \|x^{-n}\| = o(n)$ as $n \to \infty$. Then if $\sigma(x) - \tau(x) \in \text{rad}\mathfrak{B}$, it follows that $\sigma(x) = \tau(x)$.*

Proof: Without loss of generality we may assume that if e is the identity in $\mathfrak{A}$ and $e_1 = \sigma(e) = \tau(e)$, then $\mathfrak{B} = \{b \in \mathfrak{B} : e_1 b = b\}$. Set $\sigma(x) = \tau(x) + r$, $r \in \text{rad}\mathfrak{B}$ then $\tau(x^{-1})\sigma(x) = (e_1 + \tau(x^{-1})r)$ and $\tau(x^{-n})\sigma(x^n) = (e_1 + \tau(x^{-1})r)^n$. Hence if $\|x^{-n}\|\ \|x^n\| = o(n)$ as $n \to \infty$, $\tau(x^{-1}r) = 0$ by Hille's theorem. Therefore $r = 0$.

A similar conclusion holds if $\|e^{nx}\|\ \|e^{-nx}\| = o(n)$. See [6] for details. Therefore if $\mathfrak{A}$ is the closed linear span of elements satisfying either (*) or (**), $\mathfrak{A}$ has the following property: Whenever σ and τ are two continuous homomorphisms from $\mathfrak{A}$ into a commutative Banach algebra $\mathfrak{B}$ satisfying $\sigma \equiv \tau$ mod rad $\mathfrak{B}$, it follows that $\sigma = \tau$. Following M. C. White [23] we say then that $\mathfrak{A}$ has the close homomorphism property (c.h.p.). Clearly if $\mathfrak{A}$ has (c.h.p), $\mathfrak{A}$ is weakly amenable since if $D : \mathfrak{A} \to X$, X a commutative $\mathfrak{A}$-module and $\mathfrak{B} \equiv \mathfrak{A} \oplus X$ and we define $x_1 x_2 = 0$, $ax = a \cdot x$, then $\sigma(a) = a$ and $\tau(a) = a + Da$ are homomorphisms congruent mod $X \subset$ rad $\mathfrak{B}$, therefore $\sigma(a) = \tau(a)$ and $D = 0$. A similar construction shows that if $\mathfrak{A}$ has (c.h.p.) the strong Wedderburn theorem necessarily fails for $\mathfrak{A}$. More generally, if I is a closed ideal of $\mathfrak{A}$ contained in the radical of $\mathfrak{A}$, then if $\mathfrak{A}$ has (c.h.p.), there cannot exist a closed subalgebra $\mathfrak{B}$ of $\mathfrak{A}$ satisfying $\mathfrak{A} = \mathfrak{B} \oplus I$.

If $\mathfrak{A}$ has (c.h.p), then the radical may have finite dimension, since there are algebras generated by their idempotents with one dimensional radical [8]. However, the radical of $\mathfrak{A}$ must have infinite co-dimension since if $co(\text{rad}\mathfrak{A}) < \infty$, $\mathfrak{A}/\text{rad}\mathfrak{A}$ is isomorphic to a subalgebra of $\mathfrak{A}$ which is generated by finitely many orthogonal idempotents, and a strong Wedderburn decomposition must hold for $\mathfrak{A}$. Similarly it is easily seen that if $\mathfrak{A}$ has a unit and has (c.h.p.), then $\mathfrak{A}$ has no closed primary non-maximal ideals, that is, ideals contained in only one maximal ideal. This follows since if I is an ideal contained in one maximal ideal M, then $\mathfrak{A}/I$ has radical M/I and $co(M/I) = 1$, which is impossible.

It is easily seen that if $\mathfrak{A}$ has (c.h.p.), $\overline{\mathfrak{A}^2} = \mathfrak{A}$ and if ν is a continuous homomorphism of $\mathfrak{A}$ to $\mathfrak{B}$ and $\mathfrak{B} = \overline{\nu(\mathfrak{A})}$, then $\mathfrak{B}$ has (c.h.p.). However if I is a closed ideal in $\mathfrak{A}$, the condition that $\overline{I^2} = I$ is not sufficient to guarantee that I has (c.h.p.). We will provide an example which illustrates this at the end of this section. An appropriate necessary and sufficient condition on a closed ideal I for I to have (c.h.p.) is not known.

If G is a locally compact abelian group does $L^1(G)$ satisfy (c.h.p.)? The Beurling-Helson theorem shows that (*) cannot hold for $L^1(\mathbf{R})$ with unit adjoined. However (**) does hold for $L^1(\mathbf{R})$ since if f is piecewise linear, continuous, and has compact support, then $f \in A(\mathbf{R})$ and $\|e^{inf}\| = 0(\log|n|)$ where the norm is computed in the algebra with unit adjoined [5]. For general groups G the result is open. However, Bade and Dales [3] have shown that if E is a closed set of non-synthesis in the dual group Γ then $A(\Gamma)/\overline{J(E)}$ never has the strong Wedderburn property. M. C. White [23] has replaced (*) and (**) by the assumption that $\mathfrak{A}$ is generated by an analytic semigroup satisfying appropriate growth conditions. Specifically he proves the following

THEOREM. *Let $\mathfrak{A}$ be generated by a semigroup a^λ which is analytic for* $\text{Re}\lambda > 0$. *Assume there exists a weight function $\Omega : \mathbf{R} \to \mathbf{R}_+$ satisfying*

$$\Omega(t) = 0(|t|^k) \quad \textit{for some} \quad k \in N$$

and

$$\lim_{n\to\infty} \frac{\Omega(nt)}{n} = 0, \quad t \in R.$$

Then if

$$\|a^{1+it}\| \, \|a^{1-it}\| \le C\Omega(t),$$

$\mathfrak{A}$ has (c.h.p.).

For locally compact groups G this yields that if there exists a weight function $\omega : G \to R^+$ satisfying

$$\omega(ng)\omega(-ng) = o(|n|^k) \text{ for some } k \in N \text{ and all } g \in G$$

and

$$\lim_{n\to\infty} \frac{\omega(ng)\omega(-ng)}{n} = 0,$$

then $L^1(G,\omega)$ has (c.h.p.).

Earlier Galé [9] had shown that if $\mathfrak{A}$ was generated by an analytic semigroup a^λ and

$$\|a^{1+it}\| = 0(|t|^\alpha), t \to \infty \quad \text{and} \quad 0 \leq \alpha < \frac{1}{2}$$

then $\mathfrak{A}$ was weakly amenable. This condition was improved by White in [23] to

$$\lim_{t\to\infty} \frac{\|a^{1+it}\| \; \|a^{1-it}\|}{t} = 0.$$

We close with an example of a weakly amenable radical algebra R which shows that weak amenability does not imply the close homomorphism property since the identity homomorphism and the zero homomorphism on R are congruent mod R. If we adjoin an identity to R forming $R^\sharp$, then $R^\sharp$ is weakly amenable and has a strong Wedderburn decomposition. Therefore weak amenability does not imply the failure of the strong Wedderburn theorem.

To construct the example, let E be a Helson set of non-synthesis on the circle $\mathbf{T}$ or the line $\mathbf{R}$. For definiteness, let $E \subset \mathbf{T}$, then E is totally disconnected and $A(E) = A(\mathbf{T})/\overline{J(E)}$ has the close homomorphism property and has radical $R = I(E)/\overline{J(E)}$. Furthermore $A(E)/R \simeq C(E)$. We claim $\overline{R^2} = R$. If not, then

$$\tilde{A}(E) = A(E)/\overline{R^2}$$

has radical $R/\overline{R^2}$ with square zero. Since E is totally disconnected, it follows from [2, Theorem 4.2] that

$$\tilde{A}(E) \simeq C(E) \oplus R/\overline{R^2}$$

since the idempotents in $\tilde{A}(E)$ are uniformly bounded. This contradicts the fact that the strong Wedderburn decomposition must fail for algebras with the close homomorphism property. Then $\overline{R^2} = R$, and therefore R is weakly amenable by [12, Corollary 1.5].

An interesting question is whether $I(E)/\overline{J(E)}$ always is weakly amenable if E is a Helson set of non-synthesis in any locally compact group Γ. The same argument as the above would apply if it were known that if $\mathfrak{A}/R \simeq C(E)$ and $R^2 = 0$ implies $\mathfrak{A}$ has a strong Wedderburn decomposition. This is not known in general for non- totally disconnected sets E. Partial results can be found in [19].

Finally, does the above example provide an example of an amenable radical algebra?

4 ADDED IN PROOF

Since this paper was written, J. A. Gifford of the Australian National University has shown in an unpublished pre-print "Amenable subalgebras of $B(\mathcal{H})$ and $K(\mathcal{H})$", that if $\mathfrak{A}$ is a closed amenable operator algebra on a Hilbert space $\mathcal{H}$, then $\mathfrak{A}$ is similar to an operator algebra whose bicommutant is self-adjoint. If $\mathfrak{A}$ is either commutative or consists of compact operators, then $\mathfrak{A}$ is similar to a C^* algebra. This answers two questions posed in Section 1. It also follows immediately from this result that commutative amenable radical operator algebras on Hilbert space do not exist. The above example of a weakly amenable radical algebra is not constructed as an operator-algebra on some Hilbert space $\mathcal{H}$. Can it be so embedded, or do commutative weakly amenable operator algebras also fail to exist?

5 REFERENCES

[1] G. F. Bachelis and S. Saeki, 'Banach algebras with uncomplemented radical', *Proc. Amer. Math. Soc.* 100 (1987) 271-274.

[2] W. G. Bade and P. C. Curtis, Jr., 'The Wedderburn decomposition of commutative Banach algebras', *Amer. J. Math.* 82 (1960) 851-866.

[3] W. G. Bade and H. G. Dales, 'The Wedderburn decomposability of some commutative Banach algebras', *J. Funct. Anal.* 107 (1992) 105-121.

[4] W. G. Bade, P. C. Curtis, Jr. and H. G. Dales, 'Amenability and weak amenability in Beurling and Lipschitz algebras', *Proc. London Math. Soc.* (3) 55 (1987) 359-377.

[5] P. C. Curtis, Jr., 'Complementation problems concerning the radical of a commutative amenable Banach algebra', *Proc. Centre Math. Anal. A.N.U.* 21 (1989) 56-59.

[6] P. C. Curtis, Jr. and R. J. Loy, 'The structure of amenable Banach algebras', *J. Lond. Math. Soc.* (2) 40 (1989) 89-104.

[7] J. Feinstein, 'Weak (F) amenability of $R(X)$', *Proc. Centre Math. Anal. A.N.U.* 21 (1989) 97-125.

[8] C. Feldman, 'The Wedderburn principal theorem in Banach algebras', *Proc. Amer. Math. Soc.* 2 (1951) 771-777.

[9] J. Galé, 'Weak amenability of Banach algebras generated by some analytic semigroups', *Proc. Amer. Math. Soc.* (2) 104 (1988) 546.

[10] I. Gelfand, 'Zur Theorie der Charaktere der Abelschen topologischen Gruppen', *Rec. Math. N.S.* (9) 51 (1941) 49-50.

[11] F. P. Greenleaf, 'Invariant means on topological groups', van Nostrand, New York,

1969.
[12] N. Grønbæk, 'Commutative Banach algebras, module derivations and semigroups', *J. London Math. Soc.* (2) 40 (1989) 137-157.
[13] A. Ya. Helemskii, 'Flat Banach modules and amenable algebras', *Trans Moscow Math. Soc.* 47 (1984) 179-218; *Amer. Math. Soc. Translations* (1985) 199-224.
[14] E. Hille and R. S. Phillips, 'Functional analysis and semigroups', Colloquium Publication 31 (American Mathematical Society, Providence, 1957).
[15] B. E. Johnson, 'Cohomology in Banach algebras', Memoir 127 (American Mathematical Society, Providence, 1972).
[16] B. E. Johnson, 'Approximate diagonals and cohomology of certain annihilator Banach algebras', *Amer. J. Math.* 94 (1972) 685-698.
[17] A. Rosenberg and D. Zelinsky, 'Cohomology of infinite algebras', *Trans. Amer. Math. Soc.* 82 (1956) 85-98.
[18] M. V. Sheinberg, 'A characterisation of the algebra $C(\Omega)$ in terms of cohomology groups', *Uspeckhi Matem. Nauk* 32 (1977) 203-204.
[19] M. Solovej, 'Wedderburn decomposition of commutative Banach algebras', *Proc. Amer. Math. Soc.*, to appear.
[20] B. Sz.-Nagy, 'Uniformly bounded linear transformations in Hilbert space', *Acta Math.* (Szeged), 11 (1947) 152-157.
[21] J. L. Taylor, 'Homology and cohomology for topological algebras', *Advances in Mathematics* 9 (1972) 137-252.
[22] J. Wermer, 'Bounded point derivations on certain Banach algebras', *J. Functional Anal.* 1 (1967) 28-36.
[23] M. C. White, 'Strong Wedderburn decomposition of Banach algebras containing analytic semigroups', *J. London Math. Soc.* 49 (2) (1994) 331-342.
[24] G. Willis, 'When the algebra generated by an operator is amenable', Preprint.

Kadec-Klee Properties for $L(l_p, l_q)$

S. J. DILWORTH Department of Mathematics, University of South Carolina, Columbia, SC 29208, U.S.A.

and

DENKA KUTZAROVA Institute of Mathematics, Bulgarian Academy of Sciences, 1090 Sofia, Bulgaria.

ABSTRACT

It is proved that $\mathcal{L}(\ell_p, \ell_q)$ has the KK property if and only if it has the UKK property if and only if $1 < q < 2 < p < \infty$. It is also proved that $\mathcal{L}(c_0, \ell_1)$ has the UKK property and that $\mathcal{L}(c_0, \ell_q)$ can be renormed to have the weak-star UKK property if (and only if) $1 \leq q < 2$. In all other cases $\mathcal{L}(\ell_p, \ell_q)$ has no equivalent UKK norm. Finally, it is proved that none of these spaces is either strictly convex or uniformly convexifiable.

Denka Kutzarova was supported in part by the Bulgarian Ministry of Education and Science under contract MM-213/92

1 INTRODUCTION

Recall that a Banach space $\mathfrak{X}$ has the *Kadec-Klee* property, denoted KK, if $\|x_n - x\| \to 0$ whenever $\|x_n\| \to \|x\|$ and $x_n \to x$ weakly. A sequence $\langle x_n \rangle$ in the closed unit ball of $\mathfrak{X}$ (denoted $Ba(\mathfrak{X})$) is said to be ε-separated, if $\inf_{n \neq m} \|x_n - x_m\| \geq \varepsilon$. We say that $\mathfrak{X}$ has the *uniform Kadec-Klee property*, denoted UKK, if, given $\varepsilon > 0$, there exists $\delta > 0$ such that whenever $\langle x_n \rangle$ is a weakly-convergent ε-separated sequence in $Ba(\mathfrak{X})$ with weak-limit x, then $\|x\| < 1 - \delta$. The UKK property was introduced by Huff [Hu]. For a dual Banach space the weak-star KK and UKK properties are defined similarly, with weak-star sequential convergence replacing weak convergence.

Many classical non-reflexive dual spaces have the weak-star UKK property, e.g. the Hardy spaces H_1 of analytic functions on the ball or on the polydisk [BDDL], the Lorentz spaces $L_{p,1}(\mu)$ [CDLT, DH], and the trace class $\mathcal{C}_1$ [B,L,Hs].

In this paper we examine Kadec-Klee properties of the operator spaces $\mathcal{L}(\ell_p, \ell_q)$ and $\mathcal{L}(c_0, \ell_q)$, and we determine which of these spaces have any or all of the properties defined above. We also discuss the possibility of *renorming* them to have the UKK and the weak-star UKK properties. The most surprising conclusion concerns the space $\mathcal{L}(c_0, \ell_1)$, which fails the weak-star KK property while it has the UKK property and can be renormed to have the weak-star UKK property.

2 KK FOR $\mathcal{L}(\ell_p, \ell_q)$

Given two Banach spaces $\mathfrak{X}$ and $\mathfrak{Y}$, let $\mathcal{L}(\mathfrak{X}, \mathfrak{Y})$ denote the space of bounded linear operators T from $\mathfrak{X}$ into $\mathfrak{Y}$ equipped with the usual operator norm:

$$\|T\| = \sup\{\|Tx\| : \|x\| \leq 1\}.$$

Given $x^* \in \mathfrak{X}^*$ and $y \in \mathfrak{Y}$, let $x^* \otimes y$ denote the rank one operator $x \mapsto x^*(x)y$.

Throughout the paper, $\|.\|_p$ will denote the norm of the sequence space ℓ_p $(1 \leq p \leq \infty)$. We shall denote the canonical basis of ℓ_p and c_0 by $\langle e_n \rangle$. Often it will be convenient to identify an element $x \in \mathcal{L}(\ell_p, \ell_q)$ with the infinite matrix X which represents it with respect to the basis $\langle e_n \rangle$. In this case, the adjoint of x (denoted x') belongs to $\mathcal{L}(\ell_{q'}, \ell_{p'})$, where $1/p + 1/p' = 1$ and $1/q + 1/q' = 1$, and is represented by the adjoint matrix X'.

It is well-known (see e.g. [DU]) that the dual of the projective tensor product $\mathfrak{X} \hat{\otimes} \mathfrak{Y}$ is isometrically isomorphic to $\mathcal{L}(\mathfrak{X}, \mathfrak{Y}^*)$. This gives rise to the following identifications: $\mathcal{L}(\ell_p, \ell_q) = (\ell_p \hat{\otimes} \ell_{q'})^*$ and $\mathcal{L}(c_0, \ell_q) = (c_0 \hat{\otimes} \ell_{q'})^*$. The weak-star topology on $\mathcal{L}(\ell_p, \ell_q)$

associated to this duality corresponds to the topology of pointwise convergence of matrix elements. We shall often use the fact that the adjoint mapping defines an isometry from $\mathcal{L}(\ell_p, \ell_q)$ onto $\mathcal{L}(\ell_{q'}, \ell_{p'})$.

Several numerical constants arise in the course of the paper. They are labelled within each calculation simply as c, c_1, etc., because their precise values are not relevant to the task in hand. For standard Banach space terminology employed throughout the paper the reader is referred to [DU] and [LT].

Proposition 1. *Suppose that either (i) $1 \leq p \leq q < \infty$ or (ii) $1 < q \leq p \leq 2$ or (iii) $2 \leq q \leq p < \infty$. Then $\mathcal{L}(\ell_p, \ell_q)$ fails the KK property.*

Proof. (i) $1 \leq p \leq q < \infty$. In this case the sequence $\langle e_n \otimes e_n \rangle$ is isometrically equivalent to the unit vector basis of c_0, and c_0 is known to fail the KK property.

(ii) $1 \leq q \leq p \leq 2$. Consider the sequence

$$x_n = \frac{1}{2^{1/q}} \left(e_1 \otimes (e_1 + e_2) + e_n \otimes (e_1 - e_2) \right).$$

Clearly, $\langle x_n \rangle$ converges weakly to $x = (1/2^{1/q}) e_1 \otimes (e_1 + e_2)$ and $\|x\| = 1$. Moreover, $\langle x_n \rangle_{n \geq 3}$ is $2^{1-1/p}$-separated. To show that $\mathcal{L}(\ell_p, \ell_q)$ fails the KK property, it suffices to show that $\|x_n\| = 1$ for all n, which is a consequence of the following straightforward calculation included for the reader's convenience:

$$\begin{aligned} \|x_n(ae_1 + be_n)\|_q &= \left(\frac{|a+b|^q}{2} + \frac{|a-b|^q}{2} \right)^{1/q} \\ &\leq \left(\frac{(a+b)^2}{2} + \frac{(a-b)^2}{2} \right)^{1/2} \end{aligned}$$

(by Hölder's inequality since $q \leq 2$)

$$\begin{aligned} &= (a^2 + b^2)^{1/2} \\ &\leq (|a|^p + |b|^p)^{1/p} \end{aligned}$$

(since $p \leq 2$)

$$= \|ae_1 + be_n\|_p.$$

(iii) $2 \leq q \leq p < \infty$. This follows by from (ii) by taking adjoints, that is, by considering the sequence $\langle x_n' \rangle$. □

We now deal with the end-point spaces c_0 and ℓ_1.

Proposition 2. *Let $1 < p < \infty$. Then $\mathcal{L}(\ell_p, \ell_1)$ and $\mathcal{L}(c_0, \ell_p)$ fail the KK property. Moreover, $\mathcal{L}(c_0, \ell_1)$ fails the weak-star KK property.*

Proof. First we consider $\mathcal{L}(\ell_p, \ell_1)$. For $n \geq 1$, let $x_n = (1/2)(e_1 \otimes (e_1 + e_2) + e_n \otimes (e_1 - e_2))$. One checks readily that $\|x_n\| = 1$ and that $x_n \to x$ weakly, where $x = (1/2)e_1 \otimes (e_1 + e_2)$ and $\|x\| = 1$. Thus $\mathcal{L}(\ell_p, \ell_1)$ fails KK, and hence $\mathcal{L}(c_0, \ell_p)$ also fails KK (because of the adjoint mapping isometry). In $\mathcal{L}(c_0, \ell_1)$, however, the convergence of $\langle x_n \rangle$ to x is weak-star (and *not* weak), and so we may only conclude that $\mathcal{L}(c_0, \ell_1)$ fails weak-star KK. □

3 UKK FOR $\mathcal{L}(\ell_p, \ell_q)$

In the cases not covered by Propositions 1 and 2, namely $\mathcal{L}(c_0, \ell_1)$ and $\mathcal{L}(\ell_p, \ell_q)$ for $1 < q < 2 < p < \infty$, our next goal is to prove that the UKK property holds.

Lemma 3. *Let $1 < q < 2 < p < \infty$ and let*

$$A = \begin{bmatrix} B & E \\ C & D \end{bmatrix}$$

be a partitioned (infinite) matrix. Then

$$\|A\|^{r_\rho} \geq \|B + E\|^{r_\rho} + c_1 \|C + D\|^{r_\rho}$$

and

$$\|A\|^{r_\kappa} \geq \|B + C\|^{r_\kappa} + c_2 \|D + E\|^{r_\kappa},$$

where

$$\frac{1}{r_\rho} = \frac{1}{2} - \frac{1}{p} \quad \text{and} \quad \frac{1}{r_\kappa} = \frac{1}{2} - \frac{1}{q'}.$$

Proof. From the asymptotic behavior of the modulus of convexity in ℓ_q (see e.g. [LT, p.63]), we have

$$\max(\|x + y\|_q^2, \|x - y\|_q^2) \geq \|x\|_q^2 + c_q \|y\|_q^2 \tag{1}$$

for all $x, y \in \ell_q$. To prove the first inequality we may assume by homogeneity that $\|B +$

$E\| = 1$. First suppose that $\|C + D\| \geq 3$. Then, by the triangle inequality, we have

$$\begin{aligned}\|A\| &\geq \|C + D\| - \|B + E\| \\ &\geq \left(1 + \frac{2}{3}\|C + D\|\right) - \frac{1}{3}\|C + D\| \\ &= 1 + \frac{1}{3}\|C + D\| \\ &\geq \left(1 + \left(\frac{1}{3}\right)^{r_p} \|C + D\|^{r_p}\right)^{1/r_p} \\ &= \left(\|B + E\|^{r_p} + \left(\frac{1}{3}\right)^{r_p} \|C + D\|^{r_p}\right)^{1/r_p},\end{aligned}$$

as required. Now consider the case $\|C + D\| = \alpha \leq 3$. Given $\varepsilon > 0$, there exist disjointly suported unit vectors $u, v \in \ell_p$ such that $\|(B + E)(u)\|_q \geq 1 - \varepsilon$ and $\|(C + D)(v)\|_q \geq (1 - \varepsilon)\alpha$. Suppose that $a^p + b^p = 1$. Then $\|au \pm bv\| = 1$, and so

$$\|A\| \geq \|(B + E)(au) \pm (C + D)(bv)\|.$$

Applying (1) with $x = (B+E)(au)$ and $y = (C+D)(bv)$ we get $\|A\|^2 \geq (1-\varepsilon)F(b)$, where

$$F(b) = (1 - b^p)^{2/p} + c_q\alpha^2 b^2 \geq 1 - \frac{2}{p}b^p + c_q\alpha^2 b^2.$$

Since $\alpha \leq 3$ we may set $b = c_3\alpha^{2/(p-2)}$, which yields $F(b) \geq 1 + c_4\alpha^{r_p}$. Since $\varepsilon > 0$ is arbitrary, this implies that

$$\begin{aligned}\|A\| &\geq (1 + c_4\alpha^{r_p})^{1/2} \\ &\geq (1 + c_5\alpha^{r_p})^{1/r_p}\end{aligned}$$

(since $r_p \geq 2$)

$$= (\|B + E\|^{r_p} + c_5\|C + D\|^{r_p})^{1/r_p}$$

as required. The second inequality follows by taking adjoints. $\square$

Theorem 4. *Let $1 < q < 2 < p < \infty$. Then $\mathcal{L}(\ell_p, \ell_q)$ has the weak-star UKK property.*

Proof. Let $\langle x_n \rangle$ be weak-star null and let $a \in \mathcal{L}(\ell_p, \ell_q)$. Let A and X_n be the matrices representing a and x_n. First suppose that A has only finitely many non-zero entries. Since each matrix entry of X_n converges to zero as $n \to \infty$, we can apply the previous lemma to deduce that

$$\|A + X_n\|^r \geq \|A\|^r + c(\limsup \|X_n\|^r) \tag{2}$$

for $r = \max(r_\rho, r_\kappa)$. Since each member of $\mathcal{L}(\ell_p, \ell_q)$ is a compact operator, it follows that the matrices with only finitely many non-zero entries are norm-dense in $\mathcal{L}(\ell_p, \ell_q)$, and so (2) continues to hold even when $\|A\|$ has infinitely many non-zero entries. Clearly, (2) implies that $\mathcal{L}(\ell_p, \ell_q)$ has the weak-star UKK property. □

There is one space under consideration that is not covered by Theorem 4 for which the UKK property holds. This is a consequence of the next (probably folklore) result for which only a sketch of the proof will be indicated. Recall that $\mathfrak{X}$ has the *Schur property* if weak sequential convergence implies norm convergence. Clearly, the Schur property implies, vacuously, the UKK property.

Theorem 5. *$\mathcal{L}(c_0, \ell_1)$ has the Schur property (and hence its predual $c_0 \hat{\otimes} c_0$ has the Dunford-Pettis property).*

Sketch of Proof. It is readily verified that any sequence of "block diagonal" matrices in $\mathcal{L}(c_0, \ell_1)$ is isometrically equivalent to the unit vector basis of ℓ_1. Similarly, for any fixed N, the row and column sequences $\langle e_N \otimes e_i \rangle_{i=1}^\infty$ and $\langle e_i \otimes e_N \rangle_{i=1}^\infty$ are also isometrically equivalent to the ℓ_1-basis. Since weak-star convergence implies pointwise convergence of matrix elements, it follows readily from these observations that every weak-star null sequence is either norm-null or has a ℓ_1 subsequence. In particular, every weakly null sequence must be norm-null. This proves that $\mathcal{L}(c_0, \ell_1)$ has the Schur property. The parenthetical remark about the Dunford-Pettis property is a purely formal consequence (see e.g. [Di, p.212]). □

Recall that $\mathfrak{X}$ is *nearly uniformly convex* (NUC), if, given $\varepsilon > 0$, there exists $\delta > 0$ such that for every ε-separated sequence in $Ba(\mathfrak{X})$ there exists $x \in \text{conv}(x_n)$ such that $\|x\| > 1 - \delta$. It is proved in [Hu] that $\mathfrak{X}$ is NUC if and only if $\mathfrak{X}$ is reflexive and has the UKK property. Recall also that $\mathfrak{X}$ is *nearly uniformly smooth* (NUS), if, given $\varepsilon > 0$, there exists $\delta > 0$ such that for each $t \in [0, \delta)$ and each basic sequence $\langle x_n \rangle$ in $Ba(\mathfrak{X})$, there exists $k > 1$ such that $\|x_1 + tx_k\| < 1 + \varepsilon t$. Then, by a theorem of Prus [P], $\mathfrak{X}$ is NUC if and only if $\mathfrak{X}^*$ is NUS. Moreover, it is well-known (see e.g. [DU, p.248]) that $\mathcal{L}(\ell_p, \ell_q)$ is reflexive if (and only if) $1 < q < p < \infty$ Thus, combining Proposition 1 and Theorem 4, we obtain the following corollary.

Corollary 6. *(i) $\mathcal{L}(\ell_p, \ell_q)$ is nearly uniformly convex if and only if $1 < q < 2 < p < \infty$.*
(ii) $\ell_p \hat{\otimes} \ell_q$ is nearly uniformly smooth if and only if $1 < \min(p, q) \leq \max(p, q) < 2$ or $2 < \min(p, q) \leq \max(p, q) < \infty$.

Remark. It is proved in [B] that $\ell_p \hat{\otimes} \ell_q$ is NUC provided $1/p + 1/q = 1$.

Let K be a closed bounded convex subset of $\mathfrak{X}$. A mapping $T : K \to K$ is said to be *non-expansive* if $\|Tx - Ty\| \leq \|x - y\|$ for all x, y in K, and K is said to have the *fixed point property* if every non-expansive mapping on K has a fixed point. According to a theorem of van Dulst and Sims [DS], who utilized Kirk's important concept of ***normal structure*** [K], the UKK property implies the fixed point property for weakly compact convex sets. Thus Theorem 4 and reflexivity of $\mathcal{L}(\ell_p,\ell_q)$ yield the next corollary.

Corollary 7. *Let $1 < q < 2 < p < \infty$. Then $\mathcal{L}(\ell_p,\ell_q)$ has the fixed point property for closed bounded convex sets.*

Remark. It is most likely that $\mathcal{L}(\ell_p,\ell_q)$ has the fixed point property whenever $1 \le q < p < \infty$. But the results of this paper do not establish this.

Our next goal is to determine whether those spaces which fail the UKK or the weak-star UKK may be renormed to have either property. To investigate this question we shall rely on a result from [DGK] which we now describe. First, let us fix some notation. Let $\langle e_n\rangle$ be a basis for $\mathfrak{X}$ with coefficient functional sequence $\langle e_n^*\rangle$ in $\mathfrak{X}^*$. An element $x \in \mathrm{lin}\{e_n\}$ is said to be a ***block*** if $\mathrm{supp}(x) = \{n : e_n^*(x) \neq 0\}$ is finite. A family $\langle \mathfrak{X}_n\rangle$ of finite-dimensional subspaces of $\mathrm{lin}\{e_n\}$ is a ***blocking*** of $\langle e_n\rangle$ provided there exists an increasing sequence of integers $\langle n_k\rangle$, $n_1 = 1$, such that $\mathfrak{X}_k = \mathrm{lin}\{e_i : n_k \le i < n_{k+1}\}$ for each k. We say that the blocks $y_1, y_2, \ldots, y_n$ are disjoint (with respect to the blocking $\langle \mathfrak{X}_k\rangle$) and write $y_1 < y_2 < \cdots < y_n$ if

$$\min\left\{m : y_i \in \sum_{j=1}^{m} \mathfrak{X}_j\right\} < \max\left\{m : y_{i+1} \in \sum_{j=m}^{\infty} \mathfrak{X}_j\right\}$$

for $i = 1, 2, \ldots, n-1$. Finally, we say that, for $1 \le p, q \le \infty$, the blocking $\langle \mathfrak{X}_n\rangle$ satisfies a (p,q)–estimate provided there exist positive constants c and C such that

$$c\left(\sum_{i=1}^{n} \|y_i\|^p\right)^{1/p} \le \left\|\sum_{i=1}^{n} y_i\right\| \le C\left(\sum_{i=1}^{n} \|y_i\|^q\right)^{1/q}$$

for all disjoint blocks $y_1, y_2, \ldots, y_n$.

The following theorem was proved in [DGK].

Theorem A. *Let $\mathfrak{X}$ be a Banach space such that $\mathfrak{X}^*$ has a basis. Then the following are equivalent:*

(1) *$\mathfrak{X}^*$ admits an equivalent dual norm with the weak-star UKK property;*
(2) *Some (equivalently, every) basis $\langle e_n\rangle$ of $\mathfrak{X}^*$ whose coefficient functionals are weak-star continuous admits a blocking which satisfies a $(p,1)$-estimate for some $p < \infty$.*

For $n \ge 1$ and $(n-1)^2 + 1 \le i \le n^2$, let

$$b_i = \begin{cases} e_n \otimes f_{i-(n-1)^2}, & \text{for } (n-1)^2 + 1 \le i \le (n-1)^2 + n \\ e_{i-(n-1)^2-n} \otimes f_n, & \text{for } (n-1)^2 + n + 1 \le i \le n^2. \end{cases}$$

We shall prove the following theorem.

Theorem 8. *Let $1 \le q < 2 < p < \infty$. Then the basis $\langle b_i \rangle$ of $\mathcal{L}(\ell_p, \ell_q)$ satisfies an $(r,1)$-estimate, where $1/r = \min(1/2 - 1/q', 1/2 - 1/p)$. In $\mathcal{L}(c_0, \ell_q)$, $\langle b_i \rangle$ satisfies an $(r,1)$-estimate, where $1/r = 1/2 - 1/q'$.*

Combining Theorem A and Theorem 8, we obtain the following renorming result.

Theorem 9. *Let $1 \le q < 2$. Then $\mathcal{L}(c_0, \ell_q)$ admits an equivalent dual norm with the weak-star UKK property.*

Before embarking on the proof of Theorem 8, we introduce some more notation. For $n \in \mathbb{N}$, let $\mathcal{P} = \{S_k : 1 \le k \le n\}$ be any partition of $\mathbb{N}$ into n sets. Then, for each infinite matrix A, $\mathcal{P}$ defines two decompositions $A = \sum_{k=1}^n A_k$ (resp., $A = \sum_{k=1}^n A^k$) where the i-th row (resp., column) of A_k (resp. A^k) equals the i-th row (resp., column) of A if $i \in S_k$ and equals zero if $i \notin S_k$.

Lemma 10. *Let $1 \le q < 2 < p < \infty$ For every partition $\mathcal{P}$ and for every $A \in \mathcal{L}(\ell_p, \ell_q)$, we have*

$$\|A\| \ge c_1 \left(\sum_{k=1}^n \|A_k\|^{r_\rho} \right)^{1/r_\rho} \qquad \text{and} \qquad \|A\| \ge c_2 \left(\sum_{k=1}^n \|A^k\|^{r_\kappa} \right)^{1/r_\kappa},$$

where

$$\frac{1}{r_\rho} = \frac{1}{2} - \frac{1}{p} \qquad \text{and} \qquad \frac{1}{r_\kappa} = \frac{1}{2} - \frac{1}{q'}$$

Proof. Let $\varepsilon > 0$. For each $1 \le k \le n$, there exists $x_k \in \text{lin}\{e_i : i \in S_k\}$ such that $\|x_k\|_p = 1$ and $\|A_k(x_k)\|_q \ge (1-\varepsilon)\|A_k\|$. Choose non-negative scalars $a_1 \dots a_n$ with $\sum a_k^p = 1$. Then $\|\sum \pm a_k x_k\| = 1$ for all choices of signs, and so

$$\begin{aligned}
\|A\| &\ge \max_{\pm} \left\| A\left(\sum_{k=1}^n \pm a_k x_k \right) \right\|_q \\
&\ge 2^{-n} \sum_{\pm} \left\| \sum_{k=1}^n \pm a_k (A_k x_k) \right\|_q \\
&\ge c_3 \left(\sum_{k=1}^n a_k^2 \, \|A_k(x_k)\|_q^2 \right)^{1/2} \\
&\ge c_3 (1-\varepsilon) \left(\sum_{k=1}^n a_k^2 \, \|A_k\|^2 \right)^{1/2},
\end{aligned}$$

where the penultimate inequality follows from the fact that ℓ_q has cotype two (see e.g. [LT]). Choosing $a_k = \lambda \|A_k\|^{2/(p-2)}$, for $\lambda = (\sum \|A_k\|^{r_\rho})^{-1/p}$, yields

$$\left(\sum_{k=1}^n a_k^2 \, \|A_k\|^2 \right)^{1/2} = \left(\sum \|A_k\|^{r_\rho} \right)^{1/r_\rho},$$

and since ε is arbitrary the first result follows. For the second result we use duality. Note that $A' \in \mathcal{L}(\ell_{q'}, \ell_{p'})$, that $\|A'\| = \|A\|$, and that $(A')_k = (A^k)'$. So, by the first part,

$$\|A\| = \|A'\| \geq c_2 \left(\sum_{k=1}^{n} \|(A')_k\|^{r_\kappa} \right)^{1/r_\kappa} = c_2 \left(\sum_{k=1}^{n} \|A^k\|^{r_\kappa} \right)^{1/r_\kappa}.$$

□

Some more notation is required. Let $\|A\|_\rho = \sup\left(\sum_{k=1}^{n} \|A_k\|^{r_\rho}\right)^{1/r_\rho}$ and let $\|A\|_\kappa = \sup\left(\sum_{k=1}^{n} \|A^k\|^{r_\kappa}\right)^{1/r_\kappa}$, where the suprema are taken over all partitions $\mathcal{P}$. By Lemma 10, $\|A\| \leq \min(\|A\|_\rho, \|A\|_\kappa) \leq \max(\|A\|_\rho, \|A\|_\kappa) \leq c\|A\|$. Note that

$$\|A\|_\rho \geq \left(\sum_{k=1}^{n} \|A_k\|_\rho^{r_\rho} \right)^{1/r_\rho} \quad \text{and} \quad \|A\|_\kappa \geq \left(\sum_{k=1}^{n} \|A^k\|_\kappa^{r_\kappa} \right)^{1/r_\kappa}. \tag{3}$$

Finally, put $r = \max(r_\rho, r_\kappa)$ and let $|A| = (\|A\|_\rho^r + \|A\|_\kappa^r)^{1/r}$ (so that $\|A\| \leq |A| \leq c\|A\|$).

Lemma 11. *Suppose that*

$$A = \begin{bmatrix} B & E \\ C & D \end{bmatrix}$$

is a partitioned matrix. Then

$$|A|^r \geq |B|^r + c|C + D + E|^r.$$

Proof. By the triangle inequality and the obvious inequality $\|D + E\|_\kappa \geq \|E\|_\kappa$, we have

$$\begin{aligned} \|C + D + E\|^r &\leq \|C + D + E\|_\rho^r \\ &\leq 2^{1-1/r}(\|C + D\|_\rho^r + \|E\|_\rho^r) \\ &\leq 2^{1-1/r}(\|C + D\|_\rho^r + c_1\|E\|_\kappa^r) \\ &\leq 2^{1-1/r}(\|C + D\|_\rho^r + c_1\|D + E\|_\kappa^r) \\ &\leq c_2(\|C + D\|_\rho^r + \|D + E\|_\kappa^r) \end{aligned} \tag{4}$$

Now by (3)

$$\|A\|_\rho^r \geq \|B + E\|_\rho^r + \|C + D\|_\rho^r \geq \|B\|_\rho^r + \|C + D\|_\rho^r. \tag{5}$$

Similarly,

$$\|A\|_\kappa^r \geq \|B\|_\kappa^r + \|D + E\|_\kappa^r. \tag{6}$$

Adding (5) and (6) and then applying (4) yields

$$\begin{aligned} |A|^r &= \|A\|_\rho^r + \|A\|_\kappa^r \\ &\geq (\|B\|_\rho^r + \|B\|_\kappa^r) + (\|C + D\|_\rho^r + \|D + E\|_\kappa^r) \\ &\geq |B|^r + (1/c_2)\|C + D + E\|^r \\ &\geq |B|^r + c|C + D + E|^r. \end{aligned}$$

□

Lemma 12. *Let $F_n = lin\{b_i : (n-1)^2 < i \leq n^2\}$. Then the finite-dimensional decomposition satisfies an $(r,1)$-estimate.*

Proof. Since the norm $|\cdot|$ is equivalent to $\|\cdot\|$, it suffices to show that $\langle F_n \rangle$ satisfies an $(r,1)$-estimate for $|\cdot|$. Suppose that $x_1 < x_2 < \cdots < x_n$, and let $X_1, \ldots, X_n$ be the matrices corresponding to $x_1, \ldots, x_n$. Applying Lemma 11 with $A = X_1 + \cdots + X_n$, $B = X_1 + \cdots + X_{n-1}$, and $C + D + E = X_n$, we get

$$|X_1 + \ldots X_n|^r \geq |X_1 + \ldots X_{n-1}|^r + c|X_n|^r,$$

and a further $n-2$ applications of Lemma 11 give

$$|X_1 + \ldots X_n|^r \geq |X_1|^r + c\sum_{i=2}^{n} |X_i|^r,$$

which gives the result. □

Proof of Theorem 8. We shall only prove the result for $\mathcal{L}(\ell_p, \ell_q)$ as the proof for $\mathcal{L}(c_0, \ell_q)$ is very similar. Consider the bases $\langle b_i \rangle_{(n-1)^2+1}^{n^2}$ of the F_n's. In view of Lemma 12 it obviously suffices to show that each of these bases satisfies an $(r,1)$-estimate with a uniform lower constant. To that end, observe that for any scalars a_i, we have

$$\left\| \sum_{(n-1)^2+1}^{n^2} a_i b_i \right\| \geq \frac{1}{2}\left(\left\| \sum_{(n-1)^2+1}^{(n-1)^2+n} a_i b_i \right\| + \left\| \sum_{(n-1)^2+n+1}^{n^2} a_i b_i \right\| \right) \geq \frac{1}{2}\left\| \sum_{(n-1)^2+1}^{n^2} a_i b_i \right\|,$$

and

$$\left\| \sum_{(n-1)^2+1}^{(n-1)^2+n} a_i b_i \right\| + \left\| \sum_{(n-1)^2+n+1}^{n^2} a_i b_i \right\| = \left(\sum_{(n-1)^2+1}^{(n-1)^2+n} |a_i|^q \right)^{1/q} + \left(\sum_{(n-1)^2+n+1}^{n^2} |a_i|^{p'} \right)^{1/p'}.$$

Hence the bases $\langle b_i \rangle_{(n-1)^2+1}^{n^2}$ of the F_n's are uniformly equivalent to the standard bases of $\ell_q^n \oplus \ell_{p'}^{n-1}$, and since $r > \max(q, p')$, they satisfy the required $(r,1)$-estimate with a uniform lower constant. □

In the cases that are not covered by Theorems 4, 5 and 9 we shall now show there is no *equivalent* norm with the UKK property. To do this we require the following necessary condition for $\mathfrak{X}$ to be 'UKK-able', which is due to Huff [Hu].

Fact. For $\varepsilon > 0$, let $B_\varepsilon^{(0)}(\mathfrak{X}) = Ba(\mathfrak{X})$, and for $n \geq 1$ define $B_\varepsilon^{(n)}(\mathfrak{X})$ inductively thus:

$$B_\varepsilon^{(n)}(\mathfrak{X}) = \{x \in \mathfrak{X} : x = w - \lim\langle x_k \rangle, x_k \in B_\varepsilon^{(n-1)}(\mathfrak{X}), \|x_j - x_k\| \geq \varepsilon \quad (j \neq k)\}.$$

Suppose that $\mathfrak{X}$ admits an equivalent UKK norm. Then, for each $\varepsilon > 0$, $B_\varepsilon^{(n)}(\mathfrak{X}) = \emptyset$ for all sufficiently large n.

Theorem 13. *Suppose that (i) $1 \leq p \leq q < \infty$ or (ii) $1 \leq q \leq p \leq 2$ or (iii) $2 \leq q \leq p < \infty$. Then $\mathcal{L}(\ell_p, \ell_q)$ does not admit an equivalent UKK norm.*

Proof. (i) $1 \leq p \leq q < \infty$. In this case $\mathcal{L}(\ell_p, \ell_q)$ contains a subspace isomorphic to c_0. It is well-known and it follows easily from the Fact that c_0 cannot be renormed to be UKK.
(ii) $1 \leq q \leq p \leq 2$. Using the Fact with $\varepsilon = 2^{-1/p}$ it suffices to show that $B_\varepsilon^{(n)}(\mathcal{L}(\ell_p, \ell_q))$ is non-empty for every $n \geq 1$. By Dvoretzky's Theorem [Dv] there exist vectors $y_1, \ldots y_n \in \ell_q$ such that

$$\frac{1}{2}\left(\sum_{i=1}^{n} a_i^2\right)^{1/2} \leq \left\|\sum_{i=1}^{n} a_i y_i\right\| \leq \left(\sum_{i=1}^{n} a_i^2\right)^{1/2}$$

for all scalars $a_1, \ldots, a_n$. Let $m_1, \ldots, m_{n-1}$ be arbitrary distinct positive integers. For $k \geq 1$, let

$$x_k = \sum_{j=1}^{n-1} e_{m_j} \otimes y_j + e_k \otimes y_n.$$

Since $p \leq 2$, $\|x_k\| \leq 1$ as soon as $k \geq \max(m_1, \ldots, m_{n-1})$. For $k \neq l$, we have

$$\|x_k - x_l\| = \|(e_k - e_l) \otimes y_n\| = 2^{1-1/p}\|y_n\| \geq 2^{-1/p} = \varepsilon.$$

Moreover, $\langle x_k \rangle$ converges weakly to $x = \sum_{j=1}^{n-1} e_{m_j} \otimes y_j$. This shows that $B_\varepsilon^{(1)}(\mathcal{L}(\ell_p, \ell_q))$ contains every operator of the form $\sum_{j=1}^{n-1} e_{m_j} \otimes y_j$. Hence we may repeat the argument above to deduce that $B_\varepsilon^{(2)}(\mathcal{L}(\ell_p, \ell_q))$ contains every operator of the form $\sum_{j=1}^{n-2} e_{m_j} \otimes y_j$. Continuing in this way a total of n times, we see that $B_\varepsilon^{(n)}(\mathcal{L}(\ell_p, \ell_q))$ contains the zero operator. The Fact now implies that $\mathcal{L}(\ell_p, \ell_q)$ does not admit a norm with the UKK property.
(iii) $2 \leq q \leq p < \infty$. This follows from (ii) by taking adjoints. □

4 ROTUNDITY

Finally, we put our main results in context by showing that none of the spaces under consideration is either strictly convex or uniformly convexifiable (cf. Corollary 6, above).

Proposition 14. *None of the spaces $\mathcal{L}(\ell_p, \ell_q)$ or $\mathcal{L}(c_0, \ell_q)$ is strictly convex.*

Proof. The only case that is not totally obvious is the case $1 < q < p < \infty$. For $r > 1$, let $f_1^r = 2^{-1/r}(e_1 + e_2)$ and $f_2^r = 2^{-1/r}(e_1 - e_2)$. Observe that f_1^r and f_2^r are unit vectors in ℓ_r^2 and that $\|f_1^q + af_2^q\|_q \leq \|f_1^p + af_2^p\|_p$ for $1 < q < p < \infty$ and all $a \in \mathbb{R}$ (by Hölder's inequality). For $a > 0$

$$\|f_1^q \pm af_2^q\|_q^q = \|f_1^q + af_2^q\|_q^q = \frac{1}{2}((1+a)^q + |1-a|^q),$$

which is an increasing function of a. Hence, for all $\alpha \in [-1, 1]$, we have

$$\|f_1^q + a\alpha f_2^q\|_q \leq \|f_1^q + af_2^q\|_q \leq \|f_1^p + af_2^p\|_p. \tag{7}$$

We can reinterpret (7) as saying that the operators from ℓ_p^2 to ℓ_q^2 which are represented by the matrices

$$A(\alpha) = \begin{bmatrix} 1 & 0 \\ 0 & \alpha \end{bmatrix}$$

with respect to the bases (f_1^p, f_2^p) and (f_1^q, f_2^q) have norm 1 for all $\alpha \in [-1, 1]$. Hence the unit sphere of $\mathcal{L}(\ell_p^2, \ell_q^2)$ contains a line segment (of length 2). □

Remark. The proof of Proposition 14 actually shows that $\mathcal{L}(\ell_p^2, \ell_q^2)$ contains an isometric copy of ℓ_∞^2 for all p, q.

Proposition 15. *All of the spaces $\mathcal{L}(\ell_p, \ell_q)$ and $\mathcal{L}(c_0, \ell_q)$ contain ℓ_∞^n uniformly. In particular, none of them is uniformly convexifiable.*

Proof. We prove the result for $\mathcal{L}(c_0, \ell_q)$. By Dvoretzky's Theorem, ℓ_p contains subspaces that are uniformly isomorphic to ℓ_2^n. Hence $\mathcal{L}(c_0, \ell_q)$ contains subspaces that are uniformly isomorphic to $\mathcal{L}(\ell_\infty^n, \ell_2^n)$. By duality, $\mathcal{L}(\ell_\infty^n, \ell_2^n)$ is isometric to $\mathcal{L}(\ell_2^n, \ell_1^n)$, and by Dvoretzky's Theorem (now applied to ℓ_1^n) the latter spaces contain subspaces that are uniformly isomorphic to $\mathcal{L}(\ell_2^n, \ell_2^n)$, which in turn contain ℓ_∞^n uniformly. The other cases are similar (the argument is shorter when $1 < p < \infty$ because then ℓ_p contains complemented ℓ_2^n's uniformly). □

ACKNOWLEDGEMENT

We thank the referee for suggesting several ways to improve the presentation of this paper.

REFERENCES

[B]. M. Besbes, *Points fixes dans les espaces des opérateurs nucléaires*, Bull. Austral. Math. Soc. **46** (1992), 287–294.

[BDDL]. M. Besbes, S. J. Dilworth, P. N. Dowling and C. J. Lennard, *New convexity and fixed point properties in Hardy and Lebesgue-Bochner spaces*, J. Funct. Anal. **119** (1994), 340–357.

[CDLT]. N. L. Carothers, S. J. Dilworth, C. J. Lennard and D. A. Trautman, *A fixed point property for the Lorentz space $L_{p,1}(\mu)$*, Indiana Univ. Math. J. **40** (1991), 345–352.

[DGK]. S. J. Dilworth, Maria Girardi and Denka Kutzarova, *Banach spaces which admit a norm with the uniform Kadec-Klee property*, Studia Math. (to appear).

[Di]. J. Diestel, *Sequences and series in Banach spaces*, Springer-Verlag, New York, 1983.

[DU]. J. Diestel and J. J. Uhl, *Vector Measures*, American Mathematical Society, 1977.

[DH]. S. J. Dilworth and Yu-Ping Hsu, *The uniform Kadec-Klee property for the Lorentz spaces $L_{w,1}$*, J. Austral. Math. Soc. (to appear).

[DS]. D. van Dulst and B. Sims, *Fixed points of non-expansive mappings and Chebyshev centers in Banach spaces with norms of type (KK)*, Banach Space Theory and its Applications, Lecture Notes in Mathematics, no. 991, Springer-Verlag, 1983, pp. 35–43.

[Dv]. A. Dvoretzky, *Some results on convex bodies and Banach spaces*, Proceedings of the International Symposium on Linear spaces, Jerusalem, 1961, pp. 123–160.

[Hs]. Yu-Ping Hsu, *The lifting of the UKK property from E to C_E*, Proc. Amer. Math. Soc. (to appear).

[Hu]. R. Huff, *Banach spaces which are nearly uniformly convex*, Rocky Mountain J. Math. **10** (1980), 743–749.

[K]. W. A. Kirk, *A fixed point theorem for mappings which do not increase distances*, Amer. Math. Monthly **72** (1965), 1004–1006.

[L]. C. J. Lennard, *C_1 is uniformly Kadec-Klee*, Proc. Amer. Math. Soc. **109** (1990), 71–77.

[LT]. J. Lindenstrauss and L. Tzafriri, *Classical Banach Spaces II: Function Spaces*, Springer-Verlag, Berlin-Heidelberg, 1979.

[P]. Stanislaw Prus, *Nearly uniformly smooth Banach spaces*, Boll. Un. Mat. Ital. **7** (1989), 507–521.

Regular Retractions onto Finite Dimensional Convex Sets

TADEUSZ DOBROWOLSKI Department of Mathematics, Pittsburg State University, Pittsburg, Kansas 66762

JERZY MOGILSKI Departamento de Matematicas, Universidad Autonoma Metropolitana, Unidad Iztapalapa, Av. Michoacan y La Purisima, Apartado Postal 55-534, México D.F. 09340, MÉXICO

1. Introduction

Let us recall that a continuous map $r : X \to Y$ of a metric space X onto a closed subset Y is a retraction if $r(x) = x$ for $x \in Y$. We say that Y is an absolute retract (or that Y has the AR-property) if for every embedding $v : Y \to X$ of Y onto a closed subset of a metric space X there is a retraction of X onto $v(Y)$.

A retraction $r : X \to Y$ of a metric space (X, ρ) onto a closed subset Y is regular (with respect to the metric ρ) if for every $\epsilon > 0$ there is a $\delta > 0$ such that $\rho(r(x), x) < \epsilon$ whenever $\operatorname{dist}_\rho(x, Y) < \delta$. This (metric) concept is due to Toruńczyk (see [15] and [16]). It is easy to see that every retraction $r : X \to Y$ of a metric space (X, ρ) onto a compact subset Y is regular. On the other hand, there exists a subset A in the plane (with the Euclidean metric) which is a retract of the plane but there is no regular retraction of the plane onto A (see [16]).

In [16] Toruńczyk proved that every absolute retract admits a closed embedding into a normed linear space onto a regular retract. The starting point for this result was the observation that for a given closed convex subset K of a normed linear space $(E, \|\cdot\|)$ the Dugundji extension formula (see [2, p. 58]) defines a regular retraction of E onto K. In fact, there exists a constant C such that, for $x \in X$, the following Lipschitz type condition is satified

$$\|r(x) - x\| \leq C \operatorname{dist}_{\|\cdot\|}(x, Y).$$

Calling the infimum of such constants C the constant of regularity of the retraction r, one can additionally say that for an arbitrary locally convex metric linear space E equipped with a translation invariant metric whose balls are convex, the Dugundji formula yields a regular retraction of E onto a closed convex subset K with a constant of regularity $< 1 + \epsilon$.

The situation apparently dramatically changes when we drop the above assumptions on the metric, in particular, if E is not locally convex. Here we show that for every closed finite-dimensional convex subset K of a metric linear space $(E, \|\cdot\|)$ the Dugundji formula yields a retraction with a constant of regularity $\leq \dim(K) + 2$. Next we apply this fact to investigate the AR-property of convex subsets K in metric linear spaces E.

We notice that if K admits a base of neighborhoods (relative to K) consisting of convex sets, then the Dugundji formula establishes the AR-property of K. Convex sets with the above property will be called locally convex (this terminology is not in wide use, but it is self-explanatory). It is known that a convex, locally convex, compact set K can be affinely embedded into a locally convex space (in fact, into a Hilbert space), and therefore not only it is an absolute retract but also by the Krein-Milman Theorem it has an extreme point.

Roberts was the first who constructed a convex compact set which was not locally convex. In fact, he provided a general method of constructing compact convex sets without extreme points (see [11] and [12] cf. [6],[7], [13] and section 4 of this paper). Such sets can be located in the standard nonlocally convex Lebesgue spaces L^p, $0 \leq p < 1$. We show that some class of Roberts convex sets have the AR-property.

The results of this note were known to us for several years. There are two reasons we decided to write them up right now. The first reason is Cauty's recent example of a σ-compact metric linear space without the AR-property (see [3]). Although Cauty's construction does not bring much light on the compact case, it emotionally prepares us for a possibility of the existence of a compact convex set without the AR-property. If this is going to be the case, then it is important to find the border line between convex compacta K which are and those which are not absolute retracts (cf. the discussion included in [7, p. 218]). We show that the local convexity of K cannot be such a border line.

The second reason for writing this note is to rectify some statements of [7] on pages 216-217. Theorem 9.8 therein lists five conditions (1)-(5) on approximation of compact convex sets K by finite-dimensional ones. The condition (4) easily implies the AR-property of K. Without providing proofs, it is claimed in [7] that: (i) all five conditions are equivalent, and (ii) all compact convex sets obtained via Roberts' method satisfy the condition (1). If (i) and (ii) were true this would imply that every compact convex set obtained by Roberts' method has the AR-property. After several conversations with the authors of [7], we are now in a position to state that no proof of the equivalence of the conditions (3) and (4) exists. Consequently, the statement of Theorem 9.8 in [7] must be treated merely as a conjecture for now.

In [7, p. 217] it also is claimed that: (iii) if K satisfies the condition (1) of Theorem 9.8 then K has the fixed point property. (An easy verification of (iii) is provided in the Appendix to the present paper.) The proof of the fact (ii) is not very difficult and is due to Roberts (see also the Appendix of the present paper). Of course, the facts (ii) and (iii) yield that all Roberts sets have the fixed point property.

Let us mention that (apparently not being aware of the above statements in [7]) Nhu and Tri [10] have modified the Roberts construction and obtained a compact convex set with the AR-property which is not locally convex.

We would like also to point out that in [9] Nhu and Tri provided their own rather lengthy proof of the main statement on the page 217 of [7] that all Roberts sets have the fixed point property (they used a loose statement of a referee of one of their papers that concerned only the AR-property as a basis to disqualify facts concerning the fixed point property discussed in [7].) In Appendix we decided to give a short proof of this fact which, according to our best knowledge, has to be attributed to Roberts and has been provided in the early eigthties.

2. Regular retractions onto finite-dimensional convex sets

We shall use the terminology of Rolewicz [13]. Let E be a vector space. A non-decreasing F-norm on E is a function $x \to \|x\|$ from E into $[0,\infty)$ so that $\|x\| > 0$ for $x \in E \setminus \{0\}$; $\|x+y\| \leq \|x\| + \|y\|$ for $x,y \in E$; $\|t_n x\| \to 0$ provided $x \in E$, $t_n \in \mathbf{R}$ and $t_n \to 0$; and $\|tx\| \leq \|x\|$ for $x \in E$, $t \in \mathbf{R}$ and $|t| \leq 1$. If $\|\cdot\|$ is a non-decreasing F-norm, then it induces a metrizable vector topology on E, where an invariant metric $\rho(x,y)$ on E is given by the formula $\rho(x,y) = \|x-y\|$.

In this paper by an F-norm we always understand a non-decreasing F-norm. If $(E, \|\cdot\|)$ is a metric linear space then $B_{\|\cdot\|}(x,\epsilon) = \{y \in E : \|x-y\| \leq \epsilon\}$ for each $x \in E$ and $\epsilon > 0$.

2.1. Lemma. *Let K be a finite dimensional convex subset of a metric linear space $(E, \|\cdot\|)$. Then*

$$\operatorname{conv}(B_{\|\cdot\|}(x,\epsilon) \cap K) \subset B_{\|\cdot\|}(x, (\dim K + 1)\epsilon)$$

for each $x \in E$ and $\epsilon > 0$.

Proof. Let $\dim K = n$ and let $y \in \operatorname{conv}(B_{\|\cdot\|}(x,\epsilon) \cap K)$. Then there are $n+1$ vectors $y_1, y_2, \ldots, y_{n+1}$ in the set $B_{\|\cdot\|}(x,\epsilon) \cap K$ and $n+1$ real numbers $t_1, t_2, \ldots, t_{n+1}$, with $\sum_{i=1}^{n+1} t_i = 1$, and $0 \leq t_i \leq 1$ for $i = 1, 2, \ldots, n+1$, and such that $y = \sum_{i=1}^{n+1} t_i y_i$ (cf. [14, p. 73]). Hence $\|x - y\| = \|x - \sum_{i=1}^{n+1} t_i y_i\| = \|\sum_{i=1}^{n+1} t_i(x - y_i)\| \leq \sum_{i=1}^{n+1} \|t_i(x-y_i)\| \leq \sum_{i=1}^{n+1} \|x - y_i\| \leq (n+1)\epsilon$.

2.2. Proposition. *Let K be a closed, finite-dimensional convex subset of a metric linear space $(E, \|\cdot\|)$. Then there exists a retraction $r : E \to K$ such that*

$$\|r(x) - x\| \leq (\dim K + 2)\operatorname{dist}_{\|\cdot\|}(x, K)$$

for $x \in E$.

Proof. The retraction $r : E \to K$ will be defined by the Dugundji extension formula (see [2, p. 58]). Let $\epsilon > 0$ and let $\{U_s, a_s\}_{s\in S}$, where $U_s \subset E \setminus K$ and $a_s \in K$ for $s \in S$, be a Dugundji system for $E \setminus K$ and K. It means that $\{U_s\}_{s\in S}$ is a locally finite open cover of $E \setminus K$ and if $x \in U_s$ then $\|x - a_s\| \leq (1+\epsilon)\operatorname{dist}_{\|\cdot\|}(x, K)$ for $s \in S$. Define $r : E \to K$ by

$$r(x) = \begin{cases} x, & \text{for } x \in K \\ \sum_{s\in S} b_s(x) a_s, & \text{for } x \in E \setminus K \end{cases},$$

where $\{b_s\}_{s\in S}$ is a locally finite partition of unity inscribed into $\{U_s\}_{s\in S}$.

Then

$$r(x) \in \operatorname{conv}(K \cap B_{\|\cdot\|}(x, (1+\epsilon)\operatorname{dist}_{\|\cdot\|}(x, K)))$$

for every x in E. By Lemma 2.1, we have that

$$r(x) \in B_{\|\cdot\|}(x, (1+\epsilon)(\dim K + 1)\operatorname{dist}_{\|\cdot\|}(x, K)).$$

Hence, if $\epsilon(\dim K + 1) < 1$ then $\|r(x) - x\| \leq (\dim K + 2)\operatorname{dist}_{\|\cdot\|}(x, K)$.

Let us mention that Toruńczyk showed that if F is a one-dimensional linear subspace in a metric linear space $(E, \|\cdot\|)$ and $p : E \to E/F$ is a quotient map then there exists a selection map $q : E/F \to E$ such that $pq = \text{id}$ and $\|qp(x)\| \leq 4\operatorname{dist}_{\|\cdot\|}(x, F)$ (see the lemma in the proof of Theorem 1.7 in [17]). Let us point out that $r(x) = x - qp(x)$ defines a retraction of E onto F satisfying $\|r(x) - x\| \leq 4\operatorname{dist}_{\|\cdot\|}(x, F)$. Moreover, it is possible to generalize Toruńczyk's proof for higher dimensional linear spaces F. However, the proof of the existence of regular retractions onto finite-dimensional convex sets with the use of the Dugundji extension formula seems to be more direct and works for convex sets as well.

Finally, let us mention an interesting and useful observation of the Dugundji extension formula.

2.3. Propositon. *Let K be a closed, convex and locally convex subset of metric linear space $(E, \|\cdot\|)$. Then K is an absolute retract.*

Proof. Let $f : A \to K$ be a continuous map of a closed subset A of a metric space (X, d) into K. An extension $F : X \to K$ will again be defined by the Dugundji extension formula (see [2, p. 58]). Let us choose $\epsilon > 0$ (e.g. $\epsilon = 1$) and let $\{U_s, a_s\}_{s\in S}$ be a Dugundij system for $X \setminus A$ and A. This time it means that $\{U_s\}_{s\in S}$ is a locally finite open cover of $X \setminus A$ and if $x \in U_s$ then $d(x, a_s) \leq (1+\epsilon)\operatorname{dist}_d(x, A)$ for $s \in S$. Define $F : X \to K$ by

$$F(x) = \begin{cases} f(x), & \text{for } x \in A \\ \sum_{s\in S} b_s(x) f(a_s), & \text{for } x \in X \setminus A \end{cases},$$

where $\{b_s\}_{s\in S}$ is a locally finite partition of unity inscribed into $\{U_s\}_{s\in S}$.

It is enough to check the continuity of the map F at each point $x \in A$. Fix $x \in A$ and $\delta > 0$. By the local convexity of K there exists a convex subset W of K such that for some $\eta > 0$ we have

$$K \cap B_{\|\cdot\|}(f(x), \eta) \subseteq W \subseteq K \cap B_{\|\cdot\|}(f(x), \delta).$$

Since the map f is continuous at the point x there exists $\sigma > 0$ such that if $d(z, x) < \sigma$ then $\|f(z) - f(x)\| < \eta$. Let us observe that if $y \in U_s$ for some $s \in S$ then $d(a_s, x) \leq d(a_s, y) + d(y, x) \leq (1 + \epsilon) \operatorname{dist}_d(y, A) + d(y, x) \leq (2 + \epsilon)d(y, x)$. Hence, if $(2 + \epsilon)d(y, x) < \sigma$ then

$$F(y) \in K \cap B_{\|\cdot\|}(f(x), \eta) \subseteq W \subseteq K \cap B_{\|\cdot\|}(f(x), \delta).$$

The last means that $\|F(y) - F(x)\| < \delta$. Thus the map F is continuous at x.

3. AR-property of convex sets

In this section we shall use regular retractions to provide several criteria for convex sets to be absolute retracts.

3.1. Theorem. *Let K be a convex subset of a metric linear space $(E, \|\cdot\|)$ such that there exists a sequence $\{K_n\}_{n=1}^{\infty}$ of finite dimensional closed convex subsets of K satisfying $\operatorname{cl} \bigcup_{n=1}^{\infty} K_n = K$ and*

$$\lim_{n\to\infty} \dim K_n \cdot \sup_{x\in L} \operatorname{dist}_{\|\cdot\|}(x, K_n) = 0$$

for every compact subset L of K. Then K is an absolute retract.

Proof. According to Proposition 2.2 there are retractions $r_n : K \to K_n$ of K onto K_n such that

$$\|r_n(x) - x\| \leq (\dim K_n + 2) \operatorname{dist}_{\|\cdot\|}(x, K_n)$$

for $x \in K$. Let L be a compact subset of K. Since

$$\lim_{n\to\infty} (\dim K_n + 2) \cdot \sup_{x\in L} \operatorname{dist}_{\|\cdot\|}(x, K_n) = 0$$

the sequence $\{r_n|L\}$ converges uniformly to the identity map on L. Thus the identity map on K is the limit in the compact-open topology of the sequence $\{r_n\}$. By [4, Theorem 1] the set K is an absolute retract.

3.2. Theorem. *Let K be a convex subset of a separable metric linear space $(E, \|\cdot\|)$ such that there exist a positive number α and an increasing sequence $\{K_n\}_{n=1}^{\infty}$ of finite dimensional closed convex subsets of K, with $\operatorname{cl} \bigcup_{n=1}^{\infty} K_n = K$, and satisfying*

$$\lim_{n\to\infty} (\dim K_n)^{1+\alpha} \cdot \operatorname{dist}_{\|\cdot\|}(x, K_n) = 0$$

for every $x \in K$. Then K is an absolute retract.

Proof. According to Proposition 2.2 there are retractions $r_n : K \to K_n$ of K onto K_n such that

$$\|r_n(x) - x\| \leq (\dim K_n + 2) \operatorname{dist}_{\|\cdot\|}(x, K_n)$$

for $x \in K$. We can assume that K is infinite-dimensional and therefore that the sequence $\dim K_n$ increases so fast that $\Sigma_{n=1}^{\infty} \frac{1}{(\dim K_n)^\alpha} = a < \infty$. Let ϵ be a positive number. We shall construct a map $f : K \to K$ which is locally finite-dimensional (i.e. every point of K has a neighborhood V with $\dim(\operatorname{span} f(V)) < \infty$), and such that $\|f(x) - x\| < \epsilon$. For each point $x \in K$ there are an open neighborhood U_x of x and a positive integer n_x such that

$$(\dim K_{n_x})^{1+\alpha} \operatorname{dist}_{\|\cdot\|}(y, K_{n_x}) < \frac{\epsilon}{2a}$$

for $y \in U_x$. Since K is separable there exists a countable star-finite open cover $\mathcal{V}$ inscribed in $\{U_x\}_{x \in K}$. For each $V \in \mathcal{V}$ we pick $x \in K$ such that $V \subseteq U_x$ and define $n_V = n_x$. We can write $\{n_V\}_{V \in \mathcal{V}} = \{n_1, n_2, \dots\}$, where the sequence $\{n_i\}_{i=1}^{\infty}$ is a strictly increasing. Using the star-finiteness of $\mathcal{V}$ we see that writing $W_i = \bigcup\{V \in \mathcal{V} : n_V = n_i\}$, $\{W_i\}_{i=1}^{\infty}$ is an open locally finite cover of K. Now, we define

$$f(x) = \Sigma_{i=1}^{\infty} \lambda_i(x) r_{n_i}(x),$$

where $\{\lambda_i\}_{i=1}^{\infty}$ is a partition of unity inscribed into $\{W_i\}_{i=1}^{\infty}$. We have

$$\begin{aligned}
&\|f(x) - x\| = \|\Sigma_{i=1}^{\infty} \lambda_i(x) r_{n_i}(x) - x\| = \|\Sigma_{i \in \{j : x \in W_j\}} \lambda_i(x) r_{n_i}(x) - x\| \\
&\le \Sigma_{i \in \{j : x \in W_j\}} \|r_{n_i}(x) - x\| \le \Sigma_{i \in \{j : x \in W_j\}} (\dim K_{n_i} + 2) \operatorname{dist}_{\|\cdot\|}(x, K_{n_i}) \\
&\le \Sigma_{i \in \{j : x \in W_j\}} 2 \dim K_{n_i} \operatorname{dist}_{\|\cdot\|}(x, K_{n_i}) \\
&\le \Sigma_{i \in \{j : x \in W_j\}} 2 (\dim K_{n_i})^{1+\alpha} \operatorname{dist}_{\|\cdot\|}(x, K_{n_i}) \frac{1}{(\dim K_{n_i})^\alpha} \\
&\le \Sigma_{i \in \{j : x \in W_j\}} \frac{\epsilon}{a} \frac{1}{(\dim K_{n_i})^\alpha} \le \epsilon.
\end{aligned}$$

Hence the identity map on K is the limit in the compact-open topology of a sequence of locally finite-dimensional maps of K into itself. By Theorem 5 in [4] K is an absolute retract.

3.3. Theorem. *Let $(E, \|\cdot\|)$ be a metric linear space and let $\{K_n\}_{n=1}^{\infty}$ be a sequence of finite dimensional closed convex subsets in E, with $\operatorname{cl} \bigcup_{n=1}^{\infty} K_n = E$. Then every compact convex subset K of E satisfying*

$$\lim_{n \to \infty} (\dim K_n)^2 \cdot \sup_{x \in K} \operatorname{dist}_{\|\cdot\|}(x, K_n) = 0$$

is an absolute retract.

Proof. Let $\{\epsilon_n\}$ be a sequence of positive numbers such that $\lim_{n \to \infty} \epsilon_n = 0$. Then, for $n = 1, 2, \dots$, there exists a countable open cover $\mathcal{V}_n$ of K_n such that $\operatorname{ord} \mathcal{V}_n \le \dim K_n + 1$ and

$$\operatorname{mesh} \mathcal{V}_n = \sup\{\operatorname{diam} V : V \in \mathcal{V}_n\} < \epsilon_n.$$

According to Proposition 2.2 there are retractions $r_n : K \to K_n$ of K onto K_n such that

$$\|r_n(x) - x\| \le (\dim K_n + 2) \operatorname{dist}_{\|\cdot\|}(x, K_n)$$

for $x \in K$. Let $\{\lambda_V^n\}_{V \in \mathcal{V}_n}$ be a partition of unity inscribed into the cover $r_n^{-1}(\mathcal{V}_n)$ and let $\{x_V^n\}_{V \in \mathcal{V}_n}$ be a set of points such that $x_V^n \in K$ and $r_n(x_V^n) \in V$ for $V \in \mathcal{V}_n$. We define maps $f_n : K \to \operatorname{conv}\{x_V^n : V \in \mathcal{V}_n\}$ by

$$f_n(x) = \Sigma_{V \in \mathcal{V}_n} \lambda_V^n(x) x_V^n.$$

The map f_n is locally finite dimensional. Moreover, we have

$$\begin{aligned}
&\|f_n(x) - x\| = \|\Sigma_{V \in \mathcal{V}_n} \lambda_V^n(x) x_V^n - x\| \\
&= \|\Sigma_{V \in \mathcal{V}_n} \lambda_V^n(x)(x_V^n - x)\| \leq \Sigma_{V \in \mathcal{V}_n} \lambda_V^n(x) \|x_V^n - x\| \\
&\leq \Sigma_{V \in \mathcal{V}_n} \lambda_V^n(x)(\|x_V^n - r_n(x_V^n)\| + \|r_n(x_V^n) - r_n(x)\| + \|r_n(x) - x\|) \\
&\leq \Sigma_{V \in \mathcal{V}_n} \lambda_V^n(x)[(\dim K_n + 2)(\operatorname{dist}_{\|\cdot\|}(x, K_n) + \operatorname{dist}_{\|\cdot\|}(x_V^n, K_n)) + \epsilon_n] \\
&\leq (\dim K_n + 1)[2(\dim K_n + 2) \sup_{y \in K} \operatorname{dist}_{\|\cdot\|}(y, K_n) + \epsilon_n].
\end{aligned}$$

Hence the identity map on K is the uniform limit of the sequence $\{f_n\}_{n=1}^\infty$. By Theorem 5 in [4] K is an absolute retract.

3.4. Theorem. *Let K be a convex subset of a separable metric linear space $(E, \|\cdot\|)$ such that there exists an increasing sequence $\{K_n\}_{n=1}^\infty$ of finite dimensional closed convex subsets of E, with $\operatorname{cl} \bigcup_{n=1}^\infty K_n = E$. Then if K is a convex subset of E such that there exists a positive number α satisfying*

$$\lim_{n \to \infty} (\dim K_n)^{2+\alpha} \cdot \operatorname{dist}_{\|\cdot\|}(x, K_n) = 0$$

for all $x \in K$ then K is an absolute retract.

Proof. According to Proposition 2.2 there are retractions $r_n : K \to K_n$ of K onto K_n such that

$$\|r_n(x) - x\| \leq (\dim K_n + 2) \operatorname{dist}_{\|\cdot\|}(x, K_n)$$

for $x \in K$. We can assume that K is infinite-dimensional and that the sequence $\dim K_n$ increases so fast that $\Sigma_{n=1}^\infty \frac{1}{(\dim K_n)^\alpha} = a < \infty$. Let ϵ be a positive number. We shall construct a locally finite-dimensional map $f : K \to K$ with $\|f(x) - x\| < \epsilon$. For each point $x \in K$ there are an open neighborhood U_x of x and a positive integer n_x such that

$$(\dim K_{n_x})^{2+\alpha} \operatorname{dist}_{\|\cdot\|}(y, K_{n_x}) < \frac{\epsilon}{12a}$$

for $y \in U_x$. Since K is separable there exists a countable star-finite open cover $\mathcal{V}$ inscribed in $\{U_x\}_{x \in K}$. For each $V \in \mathcal{V}$ we pick $x \in K$ such that $V \subseteq U_x$ and define $n_V = n_x$. We can write $\{n_V\}_{V \in \mathcal{V}} = \{n_1, n_2, \ldots\}$, where the sequence $\{n_i\}_{i=1}^\infty$ is a strictly increasing. Writing

$$U_i = \bigcup \{V \in \mathcal{V} : n_V = n_i\},$$

and using the fact that $\mathcal{V}$ is star-finite, we can infer that $\{U_i\}_{i=1}^\infty$ is an open locally finite cover of K.

Let $\{\epsilon_n\}_{n=1}^{\infty}$ be a sequence of positive numbers such that

$$\epsilon_n < \frac{\epsilon}{6a(\dim K_n)^{1+\alpha}}.$$

Then, for $n = 1, 2, \ldots$, there exists a countable open cover $\mathcal{W}_n$ of K_n such that $\operatorname{ord} \mathcal{W}_n \leq \dim K_n + 1$ and

$$\operatorname{mesh} W_n = \sup\{\operatorname{diam} W : W \in \mathcal{W}_n\} < \epsilon_n.$$

Let $\{\{\lambda_W^{n_i}\}_{W \in \mathcal{W}_{n_i}}\}_{i=1}^{\infty}$ be a partition of unity associated with the cover

$$\bigcup_{i=1}^{\infty} r_{n_i}^{-1}(\mathcal{W}_{n_i}) \cap U_i$$

and let $\{\{x_W^{n_i}\}_{W \in \mathcal{W}_{n_i}}\}_{i=1}^{\infty}$ be a set of points such that $x_W^{n_i} \in K \cap U_i$ and $r_{n_i}(x_W^{n_i}) \in W$ for $W \in \mathcal{W}_{n_i}$.

Now, we define

$$f(x) = \Sigma_{i=1}^{\infty}(\Sigma_{W \in \mathcal{W}_{n_i}} \lambda_W^{n_i}(x) x_W^{n_i}).$$

We have

$$\begin{aligned}
&\|f(x) - x\| = \|\Sigma_{i=1}^{\infty}(\Sigma_{W \in \mathcal{W}_{n_i}} \lambda_W^{n_i}(x) x_W^{n_i}) - x\| \\
&= \|\Sigma_{i \in \{j : x \in U_j\}}(\Sigma_{W \in \{W' \in \mathcal{W}_{n_i} : x \in W'\}} \lambda_W^{n_i}(x) x_W^{n_i}) - x\| \\
&\leq \Sigma_{i \in \{j : x \in U_j\}}(\Sigma_{W \in \{W' \in \mathcal{W}_{n_i} : x \in W'\}} \|x_W^{n_i} - x\|) \\
&\leq \Sigma_{i \in \{j : x \in U_j\}}(\Sigma_{W \in \{W' \in \mathcal{W}_{n_i} : x \in W'\}}(\|x_W^{n_i} - r_{n_i}(x_W^{n_i})\| \\
&+ \|r_{n_i}(x_W^{n_i}) - r_{n_i}(x)\| + \|r_{n_i}(x) - x\|)) \\
&\leq \Sigma_{i \in \{j : x \in U_j\}}(\Sigma_{W \in \{W' \in \mathcal{W}_{n_i} : x \in W'\}}((\dim K_{n_i} + 2) \operatorname{dist}_{\|\cdot\|}(x_W^{n_i}, K_{n_i}) \\
&+ \epsilon_{n_i} + (\dim K_{n_i} + 2) \operatorname{dist}_{\|\cdot\|}(x, K_{n_i})) \\
&\leq \Sigma_{i \in \{j : x \in U_j\}}(\Sigma_{W \in \{W' \in \mathcal{W}_{n_i} : x \in W'\}}(2 \dim K_{n_i} \operatorname{dist}_{\|\cdot\|}(x_W^{n_i}, K_{n_i}) \\
&+ \epsilon_{n_i} + 2 \dim K_{n_i} \operatorname{dist}_{\|\cdot\|}(x, K_{n_i})) \\
&\leq \Sigma_{i \in \{j : x \in U_j\}}(\Sigma_{W \in \{W' \in \mathcal{W}_{n_i} : x \in W'\}}(2 \frac{(\dim K_{n_i})^{2+\alpha}}{(\dim K_{n_i})^{1+\alpha}} \operatorname{dist}_{\|\cdot\|}(x_W^{n_i}, K_{n_i}) \\
&+ \epsilon_{n_i} + 2 \frac{(\dim K_{n_i})^{2+\alpha}}{(\dim K_{n_i})^{1+\alpha}} \operatorname{dist}_{\|\cdot\|}(x, K_{n_i})) \\
&\leq \Sigma_{i \in \{j : x \in U_j\}}(\Sigma_{W \in \{W' \in \mathcal{W}_{n_i} : x \in W'\}}(\frac{\epsilon}{6a(\dim K_{n_i})^{1+\alpha}} \\
&+ \epsilon_{n_i} + (\frac{\epsilon}{6a(\dim K_{n_i})^{1+\alpha}})) \\
&\leq \Sigma_{i \in \{j : x \in U_j\}}(\Sigma_{W \in \{W' \in \mathcal{W}_{n_i} : x \in W'\}} \frac{\epsilon}{2a(\dim K_{n_i})^{1+\alpha}}) \\
&\leq \Sigma_{i \in \{j : x \in U_j\}}(\dim K_{n_i} + 1) \frac{\epsilon}{2a(\dim K_{n_i})^{1+\alpha}} \\
&\leq \Sigma_{i \in \{j : x \in U_j\}} 2 \dim K_{n_i} \frac{\epsilon}{2a(\dim K_{n_i})^{1+\alpha}} \\
&\leq \Sigma_{i \in \{j : x \in U_j\}} \frac{\epsilon}{a(\dim K_{n_i})^{\alpha}} \\
&< \epsilon.
\end{aligned}$$

Hence the identity map on K is the limit in the compact-open topology of a sequence of locally finite-dimensional maps of K into itself. By Theorem 5 in [4] K is an absolute retract.

3.5. Corollary. *Let E be a metric linear space such that there are a compact convex subset K, a sequence $\{U_n\}_{n=1}^{\infty}$ of neighborhoods of zero in E, and a sequence $\{K_n\}_{n=1}^{\infty}$ of finite-dimensional closed convex subsets of E satisfying the following properties:*

(1) $E = \operatorname{span} K$;
(2) $tU_n \subseteq U_n$ *for* $t \in \mathbf{R}$, $|t| \leq 1$ *and* $n = 1, 2, \ldots$;
(3) $U_{n+1} + U_{n+1} \subseteq U_n$ *for* $n = 1, 2, \ldots$;
(4) $\operatorname{cl} \bigcup_{n=1}^{\infty} K_n = E$;
(5) $\lim_{n \to \infty} 2^{-n} (\dim K_n)^2 = 0$;
(6) $\bigcap_{n=1}^{\infty} U_n = \{0\}$;
(7) $K \subset K_n + U_n$, *for* $n = 1, 2, \ldots$.

Then every convex subset L of E is an absolute retract.

Proof. We shall first prove that every compact convex subset L of the convex set $K - K$ is an absolute retract. We have $K - K \subset (K_n - K_n) + (U_n - U_n) \subset (K_n - K_n) + U_{n-1}$ for $n = 1, 2, \ldots$. By Kakutani's Theorem (see [13, p. 2]) there exists a continuous non-decreasing F-norm $\|\cdot\|$ on the space E such that $U_n \subseteq \{x \in E : \|x\| \leq 2^{-n}\}$. The F-norm $\|\cdot\|$ defines the same topology on the compact set L. We have $\sup_{x \in L} \operatorname{dist}_{\|\cdot\|}(x, K_n - K_n) \leq 2^{-n+1}$ and $\dim(K_n - K_n) \leq \dim K_n + 1$. By the condition (5)

$$\lim_{n \to \infty} (\dim K_n + 1)^2 \sup_{x \in L} \operatorname{dist}_{\|\cdot\|}(x, K_n - K_n) = 0.$$

By Theorem 3.3, the convex set L is an absolute retract. More generally, let L be a compact convex subset of the set $l(K - K)$, where l is a positive integer. Then L being homeomorphic to a compact convex subset of the set $K - K$ is an absolute retract as well. Now, let L be any convex subset of E. If $\bar{L}$ denotes the closure of L in E then $\bar{L} = \bigcup_{l=1}^{\infty} (\bar{L} \cap (l(K - K))$, where $\bar{L} \cap (l(K - K))$ is a compact convex absolute retract for $l = 1, 2, \ldots$. By [4, Corollary 1] the set $\bar{L}$ is an absolute retract, and by [5, Lemma 3.1] the set L itself is an absolute retract.

4. Roberts compact convex sets that are absolute retracts

In this section we show that many of Roberts' examples of nonlocally convex compact convex sets satisfy the conditions elaborated in section 3 and consequently have the AR-property, (see Lemma 4.1 and Proposition 4.2 below). We shall start with recalling Roberts construction as described in [7] (we refer the reader to the other sources like [11], [12], [6] and [13]).

Let x be a nonzero vector in a metric linear space $(E, \|\cdot\|)$, and let ϵ be a positive number. Then F is an ϵ-needle set for the vector x if

(1) F is a finite set;
(2) $x \in \operatorname{conv} F$;
(3) $F \subset B_{\|\cdot\|}(0, \epsilon)$;
(4) $\operatorname{conv}(0 \cup F) \subseteq \operatorname{conv}\{0, x\} + B_{\|\cdot\|}(0, \epsilon)$.

If a point $x \in E$ possesses an ϵ-needle set for every $\epsilon > 0$ then x is called a needle point. If every point in E is a needle point then we say that E is a needle point space. A large class of spaces (in particular, Lebesgue spaces L^p, $0 \leq p < 1$) are needle point spaces (see [7] and [6]).

We say that $(\{\epsilon_i\}_{i=1}^{\infty}, \{F_i\}_{i=0}^{\infty})$ is a Roberts system in a metric linear space $(E, \|\cdot\|)$ if $\{\epsilon_i\}_{i=1}^{\infty}$ is a sequence of positive numbers, $\{F_i\}_{i=0}^{\infty}$ is a sequence of finite subsets of E and the following conditions are satisfied

(5) $F_0 = \{x_0\}$, where $x_0 \neq 0$;

(6) $F_{i+1} = \bigcup_{x \in F_i} F_x^i$, where F_x^i is an ϵ_{i+1}-needle set for the vector x;

(7) $\Sigma_{i=0}^{\infty} \epsilon_{i+1} \cdot \operatorname{card} F_i < \infty$, where $\operatorname{card} F_i$ denotes the cardinality of the set F_i.

Roberts proved that if $(E, \|\cdot\|)$ is a needle point space and $\{\eta_i\}_{i=1}^{\infty}$ is a sequence of positive numbers with $\Sigma_{i=1}^{\infty} \eta_i < \infty$, then there exists a Roberts system $(\{\epsilon_i\}_{i=1}^{\infty}, \{F_i\}_{i=0}^{\infty})$ in E with $\epsilon_{i+1} \leq \frac{\eta_{i+1}}{\operatorname{card} F_i}$.

For a Roberts system $(\{\epsilon_i\}_{i=1}^{\infty}, \{F_i\}_{i=0}^{\infty})$ we define the Roberts set K by $K = L - L$, where $L = \operatorname{cl} \bigcup_{i=1}^{\infty} L_i$, and $L_i = \operatorname{conv}(\{0\} \cup F_i)$. According to Roberts such set K is a nonlocally convex compact convex set without extreme points.

4.1. Proposition. *If a Roberts system $(\{\epsilon_i\}_{i=1}^{\infty}, \{F_i\}_{i=0}^{\infty})$ satisfies the condition*

$$\lim_{n \to \infty} \operatorname{card} F_n \cdot (\Sigma_{i=n}^{\infty} \epsilon_{i+1} \cdot \operatorname{card} F_i) = 0,$$

then the corresponding Roberts compact convex set is an absolute retract.

Proof. Let $K_i = L_i - L_i$. Then $\dim K_i \leq \dim L_i + 1 \leq \operatorname{card} F_i + 2$ and

$$K_{i+1} \subseteq K_i + B_{\|\cdot\|}(0, 2\epsilon_{i+1} \cdot \operatorname{card} F_i).$$

Hence

$$K \subseteq K_n + B_{\|\cdot\|}(0, \Sigma_{i=n}^{\infty} 2\epsilon_{i+1} \cdot \operatorname{card} F_i).$$

Thus

$$\sup_{x \in K} \operatorname{dist}_{\|\cdot\|}(x, K_n) \leq \Sigma_{i=n}^{\infty} 2\epsilon_{i+1} \cdot \operatorname{card} F_i$$

and

$$\lim_{n \to \infty} \sup_{x \in K} \operatorname{dist}_{\|\cdot\|}(x, K_n) = 0.$$

By our assumption,

$$\lim_{n \to \infty} \dim K_n \sup_{x \in K} \operatorname{dist}_{\|\cdot\|}(x, K_n) = 0.$$

By Theorem 3.1 the convex set K is an absolute retract.

In particular, we obtain

4.2. Corollary. *If a Roberts system $(\{\epsilon_i\}_{i=1}^{\infty}, \{F_i\}_{i=0}^{\infty})$ satisfies the condition*

$$\Sigma_{i=0}^{\infty} \epsilon_{i+1} \cdot (\operatorname{card} F_i)^2 < \infty,$$

then the corresponding Roberts compact convex set is an absolute retract.

One can easily derive from the construction of Roberts that in each needle point space $(E, \|\cdot\|)$ there are Roberts systems $(\{\epsilon_i\}_{i=1}^{\infty}, \{F_i\}_{i=0}^{\infty})$ satisfying $\Sigma_{i=0}^{\infty} \epsilon_{i+1} \cdot (\operatorname{card} F_i)^2 < \infty$. Thus the class of Roberts sets with the AR-property is large.

5. APPENDIX

Let us recall after [7] that an infinite dimensional compact convex subset K of a metric linear space $(E, \|\cdot\|)$ is said to have the simplicial approximation property if

(*) for every $\epsilon > 0$ there exists a finite dimensional compact convex set K_ϵ in K such that if S is any finite dimensional simplex in K (i.e. S is a convex hull of finitely many vectors in K) then there exists a continuous map $\gamma : S \to K_\epsilon$ with $\|\gamma(x) - x\| < \epsilon$.

We have (c.f. Remark (2) on the page 217 in [7]):

5.1. Lemma. *If a compact convex set K has the simplicial approximation property then it also has the fixed point property.*

Proof. Let $f : K \to K$ be a continuous map. Since K is compact, in order to prove that K has the fixed point property, it is enough to show that for every $\epsilon > 0$ there exists a point $x_\epsilon \in K$ such that $\|x_\epsilon - f(x_\epsilon)\| < 2\epsilon$. Let $\epsilon > 0$. By the simplicial approximation property of K there exists a finite dimensional compact convex set K_ϵ in K such that if S is any finite dimensional simplex in K then there exists a continuous map $\gamma : S \to K_\epsilon$ with $\|\gamma(x) - x\| < \epsilon$. Let us consider the restriction $f|K_\epsilon$. There exists a map $g : K_\epsilon \to K$ such that $g(K_\epsilon)$ is contained in a finite dimensional simplex S and $\|g(x) - f(x)\| < \epsilon$ for $x \in K_\epsilon$. By the simplicial approximation property there exists a map $\psi : g(K_\epsilon) \to K_\epsilon$ satisfying $\|\psi(x) - x\| < \epsilon$. The map $\psi g : K_\epsilon \to K_\epsilon$ has a fixed point x_ϵ. Moreover, we have $\|\psi(g(x)) - f(x)\| \le \|\psi(g(x)) - g(x)\| + \|g(x) - f(x)\| \le 2\epsilon$. Thus $\|x_\epsilon - f(x_\epsilon)\| < 2\epsilon$.

Let us recall that if $(\{\epsilon_i\}_{i=1}^{\infty}, \{F_i\}_{i=0}^{\infty})$ is a Roberts system in a metric linear space $(E, \|\cdot\|)$ then $F_{i+1} = \bigcup_{x \in F_i} F_x^i$, where F_x^i is an ϵ_{i+1}-needle set for the vector x and $\Sigma_{i=0}^{\infty} \epsilon_{i+1} \cdot \operatorname{card} F_i < \infty$. We let $F_x^i = \{y(x,1), y(x,2), \ldots, y(x, m_x)\}$, $L_i = \operatorname{conv}(\{0\} \cup F_i)$ and $K_i = L_i - L_i$.

5.2. Proposition. *If $(\{\epsilon_i\}_{i=1}^{\infty}, \{F_i\}_{i=0}^{\infty})$ is a Roberts system in a metric linear space $(E, \|\cdot\|)$ then the corresponding Roberts compact convex set K has the fixed point property.*

Proof. This is a consequence of Lemma 5.1 and the three lemmas below.

5.3. Lemma. *Let X and X' be compact finite dimensional convex sets in metric linear spaces $(E, \|\cdot\|)$ and $(E', \|\cdot\|')$, respectively, and let f be an affine map of X' onto X. Then there exists a map $g : X \to X'$ such that $fg(x) = x$ for $x \in X$.*

Proof. Without loss of generality we can assume that the origins of $\operatorname{span} X' \subseteq E'$ and of $\operatorname{span} X \subseteq E$ are interior points in X' and X, respectively. This allows us to identify f with the restriction $\bar{f}|X'$ of a linear map $\bar{f}$ of $\operatorname{span} X'$ onto $\operatorname{span} X$. Now, it easily follows that f is an open map. Consequently, the convex carrier $\Phi(x) = f^{-1}(x)$, $x \in X$, is lower semi-continuous. Applying Michael's Selection Theorem (see [8], cf. [2, p. 85]) we conclude that there exists a continuous selection g for the carrier Φ. The map g is as required.

5.4. Lemma. *There are continuous maps* $\gamma_i : K_{i+1} \to K_i$ *satisfying*

$$\|\gamma_i(y) - y\| \le 14\epsilon_{i+1} \operatorname{card} F_i$$

for $y \in K_{i+1}$.

Proof. We shall start with constructing maps $\phi_i : L_{i+1} \to L_i$ satifying $\|\phi_i(x) - x\| \le 7\epsilon_{i+1} \operatorname{card} F_i$. If $x \in E$ then $r_x : E \to \operatorname{conv}\{0, x\}$ denotes a regular retraction satisfying $\|r_x(y) - y\| \le 3 \operatorname{dist}_{\|\cdot\|}(y, \operatorname{conv}\{0, x\})$ for $y \in E$ (see Proposition 2.2).

Let us choose vectors $z(x, k)$, where $x \in F_{i+1}$ and $k = 1, 2, \ldots, m_x$, such that $\|z(x, k) - y(x, k)\| < \frac{\epsilon_{i+1}}{\operatorname{card} F_{i+1}}$ and the set

$$G_{i+1} = \bigcup_{x \in F_i} \{z(x, 1), z(x, 2), \ldots, z(x, m_x)\}$$

consists of linearly independent vectors.

Let us denote $L'_{i+1} = \operatorname{conv}(\{0\} \cup G_{i+1})$. Then if $z \in L'_{i+1}$ then $z = \Sigma_{x \in F_i} \Sigma_{k=1}^{m_x} t(x, k) z(y,$ where $t(x, k)$ are nonnegative numbers such that $\Sigma_{x \in F_i} \Sigma_{k=1}^{m_x} t(x, k) \le 1$. We let

$$\alpha_i(\Sigma_{x \in F_i} \Sigma_{k=1}^{m_x} t(x, k) z(y, k)) = \Sigma_{x \in F_i} \Sigma_{k=1}^{m_x} t(x, k) y(x, k)$$

and

$$\sigma_i(\Sigma_{x \in F_i} \Sigma_{k=1}^{m_x} t(x, k) z(y, k)) = \Sigma_{x \in F_i} r_x(\Sigma_{k=1}^{m_x} t(x, k) z(x, k)).$$

Then α_i is an affine map of L'_{i+1} onto L_{i+1} and σ_i is a continuous map of L'_{i+1} into L_i. We have

$$\begin{aligned}
&\|\alpha_i(\Sigma_{x \in F_i} \Sigma_{k=1}^{m_x} t(x, k) z(y, k)) - \sigma_i(\Sigma_{x \in F_i} \Sigma_{k=1}^{m_x} t(x, k) z(y, k))\| \\
&= \|\Sigma_{x \in F_i} \Sigma_{k=1}^{m_x} t(x, k) y(x, k) - \Sigma_{x \in F_i} r_x(\Sigma_{k=1}^{m_x} t(x, k) z(x, k))\| \\
&\le \|\Sigma_{x \in F_i} \Sigma_{k=1}^{m_x} t(x, k) y(x, k) - \Sigma_{x \in F_i} \Sigma_{k=1}^{m_x} t(x, k) z(x, k)\| \\
&+ \|\Sigma_{x \in F_i} \Sigma_{k=1}^{m_x} t(x, k) z(x, k) - \Sigma_{x \in F_i} r_x(\Sigma_{k=1}^{m_x} t(x, k) z(x, k))\| \\
&\le \Sigma_{x \in F_i} \Sigma_{k=1}^{m_x} \|y(x, k) - z(x, k)\| \\
&+ \Sigma_{x \in F_i} \|\Sigma_{k=1}^{m_x} t(x, k) z(x, k) - r_x(\Sigma_{k=1}^{m_x} t(x, k) z(x, k))\| \\
&\le \operatorname{card} F_{i+1} \frac{\epsilon_{i+1}}{\operatorname{card} F_{i+1}} + \Sigma_{x \in F_i} 3 \operatorname{dist}_{\|\cdot\|}(\Sigma_{k=1}^{m_x} t(x, k) z(x, k), \operatorname{conv}\{0, x\}) \\
&\le \epsilon_{i+1} + 3\Sigma_{x \in F_i} \|\Sigma_{k=1}^{m_x} t(x, k) y(x, k) - \Sigma_{k=1}^{m_x} t(x, k) z(x, k)\| \\
&+ 3\Sigma_{x \in F_i} \operatorname{dist}_{\|\cdot\|}(\Sigma_{k=1}^{m_x} t(x, k) y(x, k), \operatorname{conv}\{0, x\}) \\
&\le \epsilon_{i+1} + 3\Sigma_{x \in F_i} \Sigma_{k=1}^{m_x} \|y(x, k) - z(x, k)\| \\
&+ 3\Sigma_{x \in F_i} \operatorname{dist}_{\|\cdot\|}(\Sigma_{k=1}^{m_x} t(x, k) y(x, k), \operatorname{conv}\{0, x\}) \\
&\le \epsilon_{i+1} + 3 \frac{\epsilon_{i+1}}{\operatorname{card} F_{i+1}} \operatorname{card} F_{i+1} + 3\epsilon_{i+1} \operatorname{card} F_i \\
&\le 4\epsilon_{i+1} + 3\epsilon_{i+1} \operatorname{card} F_i \le 7\epsilon_{i+1} \operatorname{card} F_i.
\end{aligned}$$

By Lemma 5.3 there exists a map $\beta_i : L_{i+1} \to L'_{i+1}$ such that $\alpha_i \beta_i(y) = y$ for $y \in L_{i+1}$. We define a map $\phi_i : L_{i+1} \to L_i$ by $\phi_i = \sigma_i \circ \beta_i$. Let $x \in F_{i+1}$. We have $\|\phi_i(x) - x\| = \|\sigma_i \beta_i(x) - \alpha_i \beta_i(x)\| \le 7\epsilon_{i+1} \operatorname{card} F_i$.

By Lemma 5.3 there exists a map $q_{i+1} : K_{i+1} \to L_{i+1} \times L_{i+1}$ such that $q_{i+1}(x) = (q^1_{i+1}(x), q^2_{i+1}(x))$ and $x = q^1_{i+1}(x) - q^2_{i+1}(x)$ for every $x \in K_{i+1}$. We let $\gamma_i(x) = \phi_i(q^1_{i+1}(x)) - \phi_i(q^2_{i+1}(x))$. Then

$$\begin{aligned}\|\gamma_i(x) - x\| &= \|\phi_i(q^1_{i+1}(x)) - \phi_i(q^2_{i+1}(x)) - (q^1_{i+1}(x) - q^2_{i+1}(x))\| \\ &\le \|\phi_i q^1_{i+1}(x) - q^1_{i+1}(x)\| + \|\phi_i q^2_{i+1}(x) - q^2_{i+1}(x)\| \\ &\le 14\epsilon_{i+1} \operatorname{card} F_i.\end{aligned}$$

5.5. Lemma. *If $(\{\epsilon_i\}_{i=1}^{\infty}, \{F_i\}_{i=0}^{\infty})$ is a Roberts system in a metric linear space $(E, \|\cdot\|)$ then the corresponding Roberts compact convex set K has the simplicial approximation property.*

Proof. Let $\epsilon > 0$ be given. We pick a positive integer n such that $\Sigma_{i=n}^{\infty} 14\epsilon_{i+1} \operatorname{card} F_i < \frac{\epsilon}{2}$ and we let $K_\epsilon = K_n$. Let S be a finite dimensional simplex in K. Since $\bigcup_{i=1}^{\infty} K_i$ is a dense convex subset of K there is a positive integer m and a map $\sigma : S \to K$ such that $\sigma(S) \subseteq K_m$ and $\|\sigma(x) - x\| < \frac{\epsilon}{2}$. We define

$$\psi = \gamma_n \circ \cdots \circ \gamma_{m-2} \circ \gamma_{m-1} \circ \sigma.$$

Then

$$\begin{aligned}\|\psi(x) - x\| &\le \|\psi(x) - \sigma(x)\| + \|\sigma(x) - x\| \\ &\le \|\gamma_n \circ \cdots \circ \gamma_{m-2} \circ \gamma_{m-1} \circ \sigma(x) - \sigma(x)\| + \|\sigma(x) - x\| \\ &\le \Sigma_{i=n}^{\infty} 14\epsilon_{i+1} \operatorname{card} F_i + \frac{\epsilon}{2} < \frac{\epsilon}{2} + \frac{\epsilon}{2} = \epsilon.\end{aligned}$$

It is claimed in [7] that the simplicial approximation property of a compact convex set K of a metric linear space $(E, \|\cdot\|)$ is equivalent to the following condition (which corresponds to (4) of Theorem 9.8 in [7]):

(**) for every $\epsilon > 0$ there exists a simplex S in K and a continuous map $\gamma : K \to S$ such that for every $x \in K$ $\|\gamma(x) - x\| < \epsilon$.

The condition (**) easily implies the AR-property of the convex set K (see [4, Corollary 1]). Unfortunately, no proof of the equivalence of the conditions (*) and (**) exists and it is not clear whether (*) implies (**). This is not a case in the general topological setting. We shall recall an example of a compactum from the theory of shape (so called the "Warsaw circle") which shows that nonconvex counterparts of the conditions (*) and (**) are not equivalent.

5.6. Example. *Let A, B, C and C_δ be subsets of the plane $(\mathbf{R}^2, \|\cdot\|)$ such that $A = \{(0,t) : 0 \le t \le 1\}$, $B = \{(\cos t, \sin t) : \frac{\pi}{2} \le t \le 2\pi\}$, $C = \{(t, \sin\frac{\pi}{t}) : 0 < t < 1\}$, and $C_\delta = \{(t, \sin\frac{\pi}{t}) : \delta < t < 1\}$, where $0 < \delta < 1$. We denote $X = A \cup B \cup C$ and $X_\delta = A \cup B \cup C_\delta$. Then for every $\epsilon > 0$ there exists $\delta > 0$ such that if S is any arc in X (i.e. S is a continuous image of the interval $[0,1]$) then there exists a continuous map $\gamma : S \to X_\delta$ with $\|\gamma(x) - x\| < \epsilon$. However, if $\epsilon < \frac{1}{2}$ then there is no continuous map γ of X into X_δ with $\|\gamma(x) - x\| < \epsilon$ (this means that X fails to* *satisfy the condition: for every* $\epsilon > 0$ *there exists an arc* S *in* X *and a continuous map* $\gamma : X \to S$ *such that for every* $x \in X$ $\|\gamma(x) - x\| < \epsilon$.*)*

The Example 5.6 suggests that the simplicial approximation property, being in its nature finite-dimensional, may be satisfied for every compact convex set K (and consequently every K would have the fixed point property). On the other hand, property (**) involves maps defined on the whole K; these maps may fail to exist (and consequently yield an example of a K without the AR-property).

The Example 5.6 also suggests that topological generalizations of the simplicial approximation property are worth considering. Let us recall that, according to [7], the simplicial approximation property of a compact convex set K is equivalent to the following condition (which corresponds to (3) of Theorem 9.8 in [7]):

(* * *) there exists a sequence $\{S_n\}_{n=1}^{\infty}$ of simplices in K and continuous maps $\gamma_n : S_{n+1} \to S_n$ so that:

(a) $S_1 \subset S_2 \subset \ldots;$

(b) $K = \mathrm{cl} \bigcup_{n=1}^{\infty} S_n;$

(c) $\|\gamma_n(x) - x\| < 2^{-n}$ for $x \in S_{n+1}$.

Let (X, ρ) be a metric compact space; then the condition (* * *) may be generalized as follows:

(* * *)′ there exists a sequence $\{X_n\}_{n=1}^{\infty}$ of compact AR-sets in X and a continuous maps $\gamma_n : X_{n+1} \to X_n$ so that:

(a) $X_1 \subset X_2 \subset \ldots;$

(b) $X = \mathrm{cl} \bigcup_{n=1}^{\infty} X_n;$

(c) $\rho(\gamma_n(x), x) < \epsilon_n$, where $x \in X_{n+1}$ and $\Sigma_{n=1}^{\infty} \epsilon_n < \infty$;

(d) for every continuous map $f : X_n \to X$ and for every $\epsilon > 0$ there exists m and a continuous map $f' : X_n \to X_m$, with $\rho(f(x), f'(x)) < \epsilon$ for $x \in X_n$.

An argument similar to the one used in the proof of Lemma 5.1 shows that if a compactum X satisfies (* * *)′ then X has the fixed point property. We presume that the class of compacta satisfying condition (* * *)′ is wide. It is interesting that the "Warsaw circle" whose fixed point property, though elementary to determine by an ad hoc method, is however hard to get by the above aproach i.e., by isolating a class of compacta with the fixed point property and then verifying that this class contains the object.

References

1. C. Bessaga and T. Dobrowolski, *Some problems in the border of functional anlysis and topology*, Proc. Internat. Conf. Geometric Topology (Warsaw, 1978), PWN, Warsaw, 1980, pp. 39–42.
2. C. Bessaga and A. Pełczyński, *Selected topics in infinite-dimensional topology*, PWN, Warszawa, 1975.
3. R. Cauty, *Un espace métrique linéare qui n'est pas un rétracte absolu*, Fund. Math. (to appear).
4. T. Dobrowolski, *On extending mappings into nonlocally convex linear metric spaces*, Proc. Amer. Math. Soc. **93** (1985), 555–560.
5. T. Dobrowolski and J. Mogilski, *Certain sequence and function spaces homeomorphic to the countable product of* l_f^2, J. London Math. Soc. **(2) 45** (1992), 566–576.

6. N. J. Kalton and N. T. Peck, *A re-examination of the Roberts example of a compact convex sets without extreme points*, Math. Ann. **253** (1981), 89–101.
7. N. J. Kalton, N. T. Peck and J. W. Roberts, *An F-space sampler*, Lond. Math. Soc. Lecture Note Ser. 89, Cambridge University Press, Cambridge, 1984.
8. E. Micheal, *Continuous selection I*, Canad. J. Math. **63** (1956), 361–382.
9. N.T. Nhu and L. H. Tri, *No Roberts space is a conterexample to Schauder's conjecture*, Topology **33** (1994), 371–378.
10. N.T. Nhu and L. H. Tri, *Every needle point space contains a compact convex AR-set with no extreme points*, Proc. Amer. Soc. Math. Soc. (to appear).
11. J. W. Roberts, *Pathological compact convex sets in the spaces L_p, $0 < p < 1$*, Altgeld Book, University of Illinois (1976).
12. J. W. Roberts, *A compact convex set with no extreme points*, Studia Math. **60** (1977), 255–266.
13. S. Rolewicz, *Metric Linear Spaces*, PWN and D. Reidel Publishing Company, Warszawa and Dordrecht/Boston/Lancaster, 1984.
14. W. Rudin, *Functional Analysis*, McGraw-Hill Book Company, New York, 1973.
15. H. Toruńczyk, *Compact absolute retracts as factors of the Hilbert space*, Fund. Math. **83** (1973), 75–84.
16. H. Toruńczyk, *Absolute retracts as factors of normed linear spaces*, Fund. Math. **86** (1974), 53–67.
17. H. Toruńczyk, *On Cartesian factors and the topological classification of linear metric spaces*, Fund. Math. **88** (1975), 71–86.

Contractive Projections on Lebesgue–Bochner Spaces

IAN DOUST[1] School of Mathematics, University of New South Wales, Sydney, NSW 2052, Australia

ABSTRACT

We survey what is known about the class of contractive projections on Lebesgue-Bochner function spaces. Even in the simplest nonscalar case, there are many open questions. Various connections are made to potential applications in the geometry of Banach spaces and in spectral theory.

1. INTRODUCTION

On a Hilbert space the orthogonal projections occupy a special place amongst the set of all idempotent operators. When one moves to working on Banach spaces, the concept of orthogonality is no longer available, so the question arises as to whether there is some other class of projections which can replace the orthogonal ones. One approach has been to look at the projections of norm 1. Study of these *contractive* projections has proved fruitful in many areas of mathematics. As well as leading to results in the geometry of Banach spaces [DO] and in spectral theory [D1], contractive projections have arisen naturally in approximation theory [CP].

Identifying the contractive projections on a particular Banach space has frequently proven to be a difficult task. One class of spaces where we now know a

[1] This research was supported by the Australian Research Council.

good deal about the contractive projections is the class of L^p spaces. A natural next step would be to look at Lebesgue-Bochner spaces. The characterization of the contractive projections on these spaces is still an open problem. In this paper our aim is to make the little that is known in this area available to a wider audience. The study of contractive projections has by now generated a rather large literature, which we shall not attempt to cover. Readers interested in sampling this wider topic should perhaps start at some of the more recent papers, such as [AF2][BG][CF][P][Ra][Sp][W], and iterate through their bibliographies.

Throughout X and Y denote Banach spaces, and $\mathcal{H}$ denotes a separable Hilbert space. The triple $(\Omega, \mathcal{A}, \mu)$ denotes a finite measure space. For $1 \leq p \leq \infty$ we shall denote the usual Lebesgue space of (equivalence classes of) p-integrable functions on a measure space $(\Omega, \mathcal{A}, \mu)$ by $L^p(\Omega, \mathcal{A}, \mu)$ or $L^p(\Omega)$ or L^p. The corresponding Lebesgue-Bochner space of X-valued functions will be denoted by $L^p(\Omega, \mathcal{A}, \mu; X)$ or $L^p(\Omega; X)$. We refer the reader to [DU,Chapter II] for more background on these spaces. The set of all bounded operators on X will be denoted by $B(X)$; the set of idempotent operators by $\mathrm{Proj}(X)$. A *contractive projection* on X is an idempotent element of $B(X)$ of norm at most 1. The set of all contractive projections on X will be denoted by $\mathrm{Proj}_1(X)$.

An *increasing sequence of projections* is a finite or infinite sequence of projections such that $P_i P_j = P_j P_i = P_i$ for all $i \leq j$. For a sequence $\{P_j\}_{j=1}^{\infty}$ we shall employ the convention that $P_0 = 0$. Given a suitably measurable scalar or operator-valued map $\{H(t)\}_{t \in \Omega}$, we shall denote the corresponding multiplication operator on $L^p(\Omega; X)$ by M_H. That is,

$$(M_H)f(t) = H(t)f(t), \qquad t \in \Omega.$$

If $\Omega_0 \in \mathcal{A}$, then we shall write M_{Ω_0} rather than $M_{\chi_{\Omega_0}}$. Also, we shall write $(\Omega_0, \mathcal{A}, \mu)$ rather than $(\Omega_0, \mathcal{A}|\Omega_0, \mu|\Omega_0)$, and we shall identify $L^p(\Omega_0, \mathcal{A}, \mu)$ with the elements of $L^p(\Omega, \mathcal{A}, \mu)$ which vanish almost everywhere on $\Omega \setminus \Omega_0$.

2. THE SCALAR CASE

Work identifying the contractive projections on L^p goes back to Grothendieck [Gr] and Douglas [Dg]. In 1966 Ando showed that for $1 < p < \infty$, $p \neq 2$, all such operators were similar to conditional expectation operators. Recall that if $\mathcal{A}_0$ is a sub-sigma-algebra of $\mathcal{A}$ and $f \in L^1(\Omega, \mathcal{A}, \mu)$, then there exists a unique $g \in L^1(\Omega, \mathcal{A}_0, \mu)$, called the *conditional expectation* of f with respect to $\mathcal{A}_0$ and μ, such that

$$\int_A f \, d\mu = \int_A g \, d\mu, \qquad \text{for all } A \in \mathcal{A}_0.$$

We shall denote the function g by $\mathbb{E}(f|\mathcal{A}_0)$. It is well-known that the linear transformation $\mathbb{E}(\cdot|\mathcal{A}_0)$ is a contractive projection on L^p for $1 \leq p \leq \infty$. Let $\mathcal{A}_0$ denote a sub-sigma-ring of $\mathcal{A}$ with greatest element Ω_0. It is a matter of notational convenience as to whether you wish to consider $\mathbb{E}(\cdot|\mathcal{A}_0)$ as being an operator on $L^p(\Omega_0, \mathcal{A}, \mu)$ or as an operator on $L^p(\Omega, \mathcal{A}, \mu)$. In most of what follows we shall omit the projection $M_{\Omega_0} : L^p(\Omega, \mathcal{A}, \mu) \to L^p(\Omega_0, \mathcal{A}, \mu)$. Further background on conditional expectation operators may be found in [DU, Chapter V], [St, Chapter 4] or [EG, Chapter 5].

Theorem 2.1. *Suppose that $(\Omega, \mathcal{A}, \mu)$ is a finite measure space, that $1 < p < \infty$, $p \neq 2$, and that $P \in Proj_1(L^p(\Omega, \mathcal{A}, \mu))$. Then there exists a sub-sigma-ring $\mathcal{A}_0 \subset \mathcal{A}$, finite measure ν on $(\Omega, \mathcal{A})$ and an invertible isometry $S : L^p(\Omega, \mathcal{A}, \mu) \to L^p(\Omega, \mathcal{A}, \nu)$ such that*

$$P = S^{-1}\mathbb{E}(\cdot|\mathcal{A}_0)S.$$

The proof progresses by showing that the range of P must contain a positive function k of maximal support (that is, a function whose support Ω_0 contains the supports of all other functions in the range of P). The operator $Qf = P(kf)/k$ is then clearly a contractive projection on $L^p(\Omega_0, \mathcal{A}, \nu)$ where ν_0 is the measure given by $\nu_0(A) = \int_A k^p \, d\mu$. Furthermore Q leaves the constant functions invariant. This is enough to ensure that Q is a conditional expectation operator. The map S is given by

$$(Sf)(t) = \begin{cases} f(t)/k(t), & t \in \Omega_0, \\ f(t), & t \in \Omega \setminus \Omega_0. \end{cases}$$

The measure is given by $\nu = \nu_0|\Omega_0 + \mu|(\Omega \setminus \Omega_0)$.

Indeed, even more is true. The following theorem of Dor and Odell [DO] (see also [PR]) says that we can handle increasing sequences of contractive projections in a similar manner.

Theorem 2.2. *Suppose that $(\Omega, \mathcal{A}, \mu)$ is a finite measure space, that $1 < p < \infty$, $p \neq 2$, and that $\{P_j\}_{j=1}^{\infty}$ is an increasing sequence of contractive projections on $L^p(\Omega, \mathcal{A}, \mu)$. Then there exists a sequence of sub-sigma-rings of $\mathcal{A}$, $\mathcal{A}_1 \subset \mathcal{A}_2 \subset \cdots \subset \mathcal{A}$, a finite measure ν on $(\Omega, \mathcal{A})$ and an invertible isometry S of $L^p(\Omega, \mathcal{A}, \mu)$ onto $L^p(\Omega, \mathcal{A}, \nu)$ such that*

$$P_j = S^{-1}\mathbb{E}_j S, \qquad j = 1, 2, \ldots,$$

where $\mathbb{E}_j$ is the conditional expectation operator with respect to $\mathcal{A}_j$ and ν.

Before turning to the vector-valued case, we shall briefly describe some of the applications of these results. The first is that increasing sequences of contractive projections on L^p possess an unconditionality property similar to that possessed by increasing sequences of orthogonal projections on a Hilbert space.

Definition 2.3. An increasing sequence of projections $\{P_j\}_{j=1}^{\infty}$ is said to have the *unconditionality property* if there exists a constant K such that for any sequence of scalars $a = \{a_j\}_{j=1}^{\infty}$

$$\left\| \sum_{j=1}^{n} a_j(P_j - P_{j-1}) \right\| \leq K \, \|a\|_{\infty}$$

for all $n \in \mathbb{N}$.

Definition 2.4. A Banach space Y has the *unconditionality property for contractive projections (UPCP)* if every increasing sequence of contractive projections has the unconditionality property. If the constant K is independent of the sequence $\{P_j\}$ we say that Y has *uniform UPCP*.

Theorem 2.5. *If $(\Omega, \mathcal{A}, \mu)$ is an arbitrary measure space and $1 < p < \infty$ then $L^p(\Omega, \mathcal{A}, \mu)$ has uniform UPCP.*

The proof of this (in the case $p \neq 2$) relies on writing

$$\sum a_j(P_j - P_{j-1}) = S^{-1} \sum a_j(\mathbb{E}_j - \mathbb{E}_{j-1})S.$$

If $f \in L^p$, then the sequence $\{(\mathbb{E}_j - \mathbb{E}_{j-1})f\}$ forms a martingale difference sequence, and the sum $\sum a_j(\mathbb{E}_j - \mathbb{E}_{j-1})f$ is a martingale transform of f. It is a deep result of Burkholder [B1] that martingale transforms are bounded on L^p with

$$\left\|\sum_{j=1}^{n} a_j(\mathbb{E}_j - \mathbb{E}_{j-1})\right\| \leq (\min\{p, (p-1)/p\} - 1) \|a\|_\infty, \qquad (n \in \mathbb{N}).$$

For details of proving the sharp bound in this inequality see [B2] (for the real case) and [B3] (for the complex case).

An immediate corollary of the above result is the following theorem of Dor and Odell [DO].

Theorem 2.6. *If $1 < p < \infty$ then all monotone Schauder decompositions (and in particular all monotone Schauder bases) of L^p are unconditional.*

These results also have applications in spectral theory. It is a standard result in operator theory that $T \in B(\mathcal{H})$ is selfadjoint if and only if there exists a compact interval $[a, b] \subset \mathbb{R}$ such that

$$\|g(T)\| \leq \|g\|_{C[a,b]}$$

for all polynomials g. More recently Fong and Lam [FL] proved that a weaker functional calculus condition suffices. Let $AC[a, b]$ denote the Banach algebra of all absolutely continuous functions on $[a, b]$. Then T is selfadjoint if and only if

$$\|g(T)\| \leq |g(a)| + \int_a^b |g'(t)| \, dt = \|g\|_{AC[a,b]}$$

for all polynomials g. The author showed that the analogous result holds on the reflexive L^p spaces [D1]. The natural generalization of self-adjoint operators to Banach spaces is the class of scalar-type spectral operators. Loosely speaking, an operator is scalar-type spectral if it can be represented as an integral with respect to a countably additive spectral measure; the precise definitions may be found in [DS3] or [Dow].

Theorem 2.7. *Suppose that $1 < p < \infty$ and that $T \in L^p(\Omega, \mathcal{A}, \mu)$ satisfies $\|g(T)\| \leq \|g\|_{AC[a,b]}$ for all polynomials g. Then T is a scalar-type spectral operator.*

Indeed, for a reflexive Banach space Y, this contractive AC functional calculus condition is equivalent to a slight variant of UPCP. The details may be found in [D2].

Finding natural nonreflexive Banach spaces without UPCP, or on which Theorem 2.7 fails, is easy. A simple renorming argument allows one to construct reflexive examples as well. What has proven much more difficult has been finding any other natural infinite-dimensional spaces which do have these properties.

3. THE VECTOR-VALUED CASE

Some of the natural candidates for spaces which might have similar properties for their contractive projections are the Lebesgue-Bochner spaces. Of course, there are many other directions in which you could head. Arazy and Friedman [AF1,AF2] have completely characterized the contractive projections on the von Neumann-Schatten p-classes of compact operators C_p. It is still unknown however whether every monotone Schauder decomposition in C_p $(1 < p < \infty, p \neq 2)$ is unconditional (see [AF2, p. 101]).

The main questions that we would like to answer about contractive projections on $L^p(\mu; X)$ are:

(1) What do they look like;
(2) What must their ranges look like;
(3) Do they have the same sort of unconditionality properties as those on scalar L^p spaces.

Even for the simplest cases, such as when X is the two dimensional Hilbert space ℓ_2^2, these questions remain open.

Much of the scalar theory goes over directly to the vector-valued case. One can define conditional expectation operators on Lebesgue-Bochner spaces just as one does for the scalar-valued L^p spaces.

Theorem 3.1. [S1, Proposition 4] *Suppose $1 \leq p \leq \infty$ and that X is a Banach space. Suppose also that $(\Omega, \mathcal{A}, \mu)$ is a finite measure space, and that $\mathcal{A}_0$ is a sub-sigma-ring of $\mathcal{A}$ with greatest element Ω_0. Then there exists a unique $\mathbb{E} \in$ Proj$_1(L^p(\Omega, \mathcal{A}, \mu; X))$ such that for all $f \in L^p(\Omega, \mathcal{A}, \mu; X)$*

(1) *$\operatorname{supp} \mathbb{E}f \subset \Omega_0$, and;*
(2) *$\int_A \mathbb{E}f \, d\mu = \int_A f \, d\mu$, for all $A \in \mathcal{A}_0$.*

This gives a large class of contractive projections on $L^p(\Omega, \mathcal{A}, \mu; X)$. Clearly however there are many elements of $\mathrm{Proj}_1(L^p(\Omega; X))$ which are not even similar to conditional expectation operators.

Example 3.2. Let $X = \ell_2^2$ and define $Q \in \mathrm{Proj}_1(X)$ by $Q(x_1, x_2) = (x_1, 0)$. Then the operator $P \in B(L^p(\Omega; X))$ defined by

$$(Pf)(t) = Q(f(t)), \qquad t \in \Omega$$

is a contractive projection on $L^p(\Omega; X)$ which, if $p \neq 2$, is not similar to a conditional expectation operator.

Following Douglas [Dg], Sundaresan [S2] characterized the contractive projections which are conditional expectation operators. For $x \in X$ let

$$L^p(\Omega)x = \left\{ f \in L^p(\Omega; X) \,:\, f = \tilde{f}x \text{ for some } \tilde{f} \in L^p(\Omega) \right\}.$$

An operator $T \in B(L^p(\Omega; X))$ is said to be *direction preserving* if for all $x \in X$, and all $T(L^p(\Omega)x) \subset L^p(\Omega)x$.

Theorem 3.3. *Suppose that $1 \le p < \infty$, $p \ne 2$, and that $P \in Proj_1(L^p(\Omega; X))$. Then P is a conditional expectation operator if and only if P is a direction preserving operator leaving the constant functions invariant.*

You can of course construct contractive projections on $L^p(\Omega; X)$ by combining elements of $\mathrm{Proj}_1(L^p)$ and elements of $\mathrm{Proj}_1(X)$. There are restrictions however. A *projection multiplier function* is an element $P \in \mathrm{Proj}(L^p(\Omega; X))$ such that for all $t \in \Omega$ there exists $Q(t) \in \mathrm{Proj}(X)$ such that for all $f \in L^p(\Omega; X)$

$$(Pf)(t) = Q(t)(f(t)), \qquad (a.e.).$$

In general, the projections $Q(t)$ need not be uniquely determined. However, if X is separable then by considering the constant functions it is not hard to prove that the operators $Q(t)$ are uniquely determined up to a set of measure zero. We shall say that a projection multiplier function is *contractive* if $Q(t) \in \mathrm{Proj}_1(X)$ for almost all t. If $\mathcal{A}_0$ is a sub-sigma-ring of $\mathcal{A}$, we shall say that P is $\mathcal{A}_0$*-measurable* if Pf is $\mathcal{A}_0$-measurable for all $f \in L^p(\Omega_0, \mathcal{A}_0, \mu; X)$. As usual, if the set $\{Q(t)\}_{t\in\Omega}$ defines a projection multiplier function on $L^p(\Omega; X)$ we shall denote this operator by M_Q. That is,

$$(M_Q f)(t) = Q(t)(f(t)), \qquad f \in L^p(\Omega; X).$$

Theorem 3.4. *Let $(\Omega, \mathcal{A}, \mu)$ be a finite measure space, and let $\mathcal{A}_0$ be a sub-sigma-ring of $\mathcal{A}$. If M_Q is a contractive $\mathcal{A}_0$-measurable projection multiplier function, then the operator*

$$Pf = M_Q \mathbb{E}(f|\mathcal{A}_0)$$

is a contractive projection on $L^p(\Omega; X)$ for $1 \le p \le \infty$.

Proof. The only thing to note is that if f is $\mathcal{A}_0$-measurable, then clearly $\mathbb{E}(f|\mathcal{A}_0) = f$. Thus

$$\mathbb{E}(M_Q \mathbb{E}(f|\mathcal{A}_0)|\mathcal{A}_0) = M_Q \mathbb{E}(f|\mathcal{A}_0).$$

We can generate further examples of contractive projections by introducing isometries from $L^p(\Omega; X)$ to itself. More generally, if J is an isometric isomorphism from $L^p(\Omega; X)$ to a Banach space Z, and $P \in \mathrm{Proj}_1(Z)$, then

$$\widetilde{P} = J^{-1} P J \tag{3-1}$$

is a contractive projection on $L^p(\Omega; X)$. Identifying all possible such maps is of course a hopeless task. More realistically, one might aim to find a suitably large class of such maps so that all the contractive projections on $L^p(\mu; X)$ could be written in the form (3-1) where P is of the type appearing in Theorem 3.4. If we restrict the class of spaces to which X and Z map belong, there is much that can be said about these isometries. In particular, if $\mathcal{H}$ denotes a separable Hilbert space, then a surjective isometry from $L^p(\Omega; \mathcal{H})$ to itself can be written in the form

$$Jf(t) = h(t)U(t)(\Phi f)(t), \qquad t \in \Omega.$$

Here Φ denotes a regular set isomorphism of $\mathcal{A}$ onto itself. Such a map induces a linear transformation $f \mapsto \Phi f$ which is characterized by the property that $\Phi\chi_A = \chi_{\Phi(A)}$ for all $A \in \mathcal{A}$. The scalar-valued function h is such that $|h(x)|^p = d\nu/d\mu$ where $\nu(A) = \mu(\Phi^{-1}(A))$. Finally, U is a weakly measurable operator-valued function such that $U(t)$ is an isometry of $\mathcal{H}$ (a.e.). Full details of this can be found in [Ca]. A relatively recent survey of what is known about isometries of more general Lebesgue-Bochner spaces can be found in [K].

One way of interpreting the above result is that isometries of $L^p(\Omega;\mathcal{H})$ decompose into a combination of isometries of $L^p(\Omega)$ ($J_1 : f \mapsto h\Phi f$) and isometries of $\mathcal{H}$ ($J_2 : f \mapsto M_U f$). For the scalar L^p spaces it turns out to be simpler to consider isometries of the form $J : L^p(\Omega,\mathcal{A},\mu) \to L^p(\Omega,\mathcal{A},\nu)$ where $\nu(A) = \int_A |h|^{-p}\, d\mu$ and $Jf = hf$. Ando's result (Theorem 2.1) characterising contractive projections says that if $P \in \mathrm{Proj}_1(L^p(\Omega,\mathcal{A},\mu))$ then it can be factored as follows:

$$\begin{array}{ccc} L^p(\Omega,\mathcal{A},\mu) & \xrightarrow{M_h} & L^p(\Omega,\mathcal{A},\nu) \\ {\scriptstyle P}\downarrow & & \downarrow{\scriptstyle \mathbb{E}} \\ L^p(\Omega,\mathcal{A},\mu) & \xleftarrow[M_{h^{-1}}]{} & L^p(\Omega,\mathcal{A},\nu) \end{array}$$

The analogue of this in the vector-valued case would be a factorization of the form

$$\begin{array}{ccccc} L^p(\Omega,\mathcal{A},\mu;X) & \xrightarrow{M_h} & L^p(\Omega,\mathcal{A},\nu;X) & \xrightarrow{M_U} & L^p(\Omega,\mathcal{A},\nu;X) \\ {\scriptstyle P}\downarrow & & & & \downarrow{\scriptstyle M_Q\mathbb{E}} \\ L^p(\Omega,\mathcal{A},\mu;X) & \xleftarrow{M_{h^{-1}}} & L^p(\Omega,\mathcal{A},\nu;X) & \xleftarrow{M_{U^{-1}}} & L^p(\Omega,\mathcal{A},\nu;X) \end{array} \tag{3-2}$$

In other words, contractive projections on $L^p(\mu;X)$ would split into combinations of isometries and contractive projections on X, and isometries and contractive projections associated with $L^p(\mu)$. It is easy to show that all operators constructed in this way are in fact contractive projections on $L^p(\Omega,\mathcal{A},\mu;X)$. Furthermore no part of this representation is redundant.

Open Question. *Does every $P \in Proj_1(L^p(\Omega,\mathcal{A},\mu;X))$ admit a factorization as in Figure (3-2)?*

Of course, as stated, this is a little too general. If p equals to 1, 2, or ∞ then other contractive projections exist. Even for the remaining values of p, if X is chosen appropriately, it is possible to find contractive projections which are not of the above form.

Example 3.5. Let $X = \ell_2^p$, the two-dimensional ℓ^p space. We shall denote an element of $L^p([0,1];X)$ by (f_1, f_2). Let $\mathbb{E}$ be the conditional expectation operator on $L^p[0,1]$ corresponding to the trivial sub-sigma-algebra, and define $P(f_1, f_2) = (f_1, \mathbb{E}f_2)$. It is easy to show that $P \in \mathrm{Proj}_1(L^p([0,1];X))$. On the other hand, P does not factor as in Figure (3-2). The problem here of course is that $L^p([0,1];X)$ has a rather more complicated set of isometries (see for example [G] and [So]). Note that as $L^p([0,1];X)$ is isometrically isomorphic to $L^p[0,2]$ we know that it has uniform UPCP and the other properties discussed in section 2.

Apart from these exceptional cases, however, the answer to the above question is unknown, even in the case that X is two-dimensional.

Even more ambitiously one might hope to show that given an increasing family of contractive projections $\{P_j\}$ on $L^p(\Omega, \mathcal{A}, \mu; X)$ you can mimic the scalar situation by writing

$$P_j = M_{h^{-1}} M_{U^{-1}} M_{Q_j} \mathbb{E}_j M_U M_h.$$

In order to make sure that the projections commute, it is necessary here to assume that each projection multiplier function is $\mathcal{A}_1$-measurable. If such a representation is possible, then, at least for finite-dimensional X, this would allow you to establish the UPCP.

REFERENCES

[AF1] J. Arazy and Y. Friedman, *Contractive projections in $\mathcal{C}_1$ and $\mathcal{C}_\infty$*, Mem. Amer. Math. Soc **13** (1978), No. 200.

[AF2] J. Arazy and Y. Friedman, *Contractive projections in C_p*, Mem. Amer. Math. Soc **95** (1992), No. 459.

[An] T. Ando, *Contractive projections in L_p spaces*, Pacific J. Math. **17** (1966), 391-405.

[B1] D.L. Burkholder, *Martingale transforms*, Ann. Math. Stat. **37** (1966), 1494-1504.

[B2] D.L. Burkholder, *Boundary value problems and sharp inequalities for martingale transforms*, Ann. Prob. **12** (1984), 647-702.

[B3] D.L. Burkholder, *A proof of Pełczyński's conjecture for the Haar system*, Studia Math. **91** (1988), 79-83.

[BG] A.P. Bosznay and B.M. Garay, *On norms of projections*, Acta Sci. Math. **50** (1986), 87-90.

[Ca] M. Cambern, *The isometries of $L^p(X, K)$*, Pacific J. Math. **55** (1974), 9-17.

[CF] B. Calvert and S. Fitzpatrick, *Nonexpansive projections onto two-dimensional subspaces of Banach spaces*, Bull. Austral. Math. Soc. **37** (1988), 149-160.

[CP] E.W. Cheney and K.H. Price, *Minimal projections*, Approximation theory, Proceedings of a symposium held at Lancaster, July 1969 (A. Talbot, ed.), Academic Press, London, 1970, pp. 261-290.

[D1] I. Doust, *Well-bounded and scalar-type spectral operators on L^p spaces*, J. London Math. Soc. (2) **39** (1989), 525-534.

[D2] I. Doust, *Contractive projections on Banach spaces*, Miniconference on Functional Analysis/Optimization, Canberra, 1988, Proc. Centre. Math. Anal. Austral. Nat. Univ. **20** (1988), 50-58.

[Dg] R.G. Douglas, *Contractive projections on an $\mathcal{L}_1$-space*, Pacific J. Math. **15** (1965), 443-462.

[DO] L.E. Dor and E. Odell, *Monotone bases in L_p*, Pacific J. Math **60** (1975), 51-61.

[Dow] H.R. Dowson, *Spectral theory of linear operators*, London Mathematical Society Monographs 12, Academic Press, London, 1978.

[DS3] N. Dunford and J.T. Schwartz, *Linear operators, Part III: Spectral operators*, Wiley Interscience, New York, 1971.

[DU] J. Diestel and J.J. Uhl, *Vector measures*, Mathematical surveys 15, American Mathematical Society, Providence, 1977.

[EG] R.E. Edwards and G.I. Gaudry, *Littlewood-Paley theory and multiplier theory*, Ergebnisse der Mathematik und ihrer Grenzgebiete 90, Springer-Verlag, Berlin, 1977.

[FL] C.K. Fong and L. Lam, *On spectral theory and convexity*, Trans. Amer. Math. Soc. **264** (1981), 59-75.

[G] P. Greim, *Isometries and L_p-structure of separably valued Bochner L_p-spaces*, Lect. Notes in Math. **1033** (1983), 209-218.

[Gr] A. Grothendieck, *Une caractérisation vectorielle-metrique des espaces L^1*, Canad. J. Math. **7** (1955), 552-561.

[K] A.L. Koldobskiĭ, *Isometries of $L_p(X; L_q)$ and equimeasurability*, Indiana Univ. Math. J. **40** (1991), 677-705.

[P] G. Pisier, *Projections from a von Neumann algebra onto a subalgebra*, preprint.

[PR] A. Pełczynski and H.P. Rosenthal, *Localization techniques in L^p spaces*, Studia Math. **52** (1975), 263-289.

[Ra] B. Randrianantoanina, *Contractive projections and isometries in sequence spaces*, preprint.
[S1] K. Sundaresan, *Banach lattices of Lebesgue-Bochner function spaces and conditional expectation operators I*, Bull. Inst. Math. Acad. Sinica **2** (1974), 165-184.
[S2] K. Sundaresan, *Lebesgue-Bochner function spaces and conditional expectation operators II*, Bull. Inst. Math. Acad. Sinica **3** (1975), 249-257.
[So] A.R. Sourour, *The isometries of $L_p(\Omega, X)$*, J. Funct. Anal. **30** (1978), 276-285.
[Sp] M. Spivack, *Contractive projections on Banach space*, Bull. Austral. Math. Soc. **34** (1986), 271-274.
[St] E.M. Stein, *Topics in harmonic analysis related to the Littlewood-Paley theory*, Annals of Mathematical Studies No. 63, Princeton University Press, Princeton, 1970.
[W] U. Westphal, *Contractive projections on ℓ^1*, Constr. Approx. **8** (1992), 223-231.

Some Measures of Convexity in Banach Spaces

PATRICK N. DOWLING Department of Mathematics and Statistics, Miami University, Oxford, Ohio 45056

ZHIBAO HU Department of Mathematics and Statistics, Miami University, Oxford, Ohio 45056

DOUGLAS MUPASIRI Department of Mathematics, University of Northern Iowa, Cedar Falls, Iowa 50614

1 INTRODUCTION

In the theory of complex quasi-normed spaces, there are two well-known measures of complex uniform convexity, namely, uniform PL-convexity [1] and uniform H-convexity [17, 18]. It is still unknown whether these two notions are equivalent, even in Banach spaces. The aim of this note is to consider the real analogues of these properties and to show that the real analogues are equivalent, and they are each equivalent to the usual notion of uniform convexity. If we consider the localized versions of these properties, we get two distinct properties, one equivalent to that of a strongly extreme point of the unit ball of a Banach space and the other equivalent to a denting point of the unit ball of a Banach space. We will also consider localized non-uniform versions of these properties and we will show that the resulting properties are both equivalent to that of an extreme point of the unit ball of a Banach space. The advantage of using these new measures of convexity is that they describe the notions of uniform convexity of a Banach space, and

The first author was supported in part by a Miami University Research Appointment.
The first author would like to thank Mark Smith for some helpful suggestions.

denting point, strongly extreme point, and extreme point of the the unit ball of a Banach space in a standard manner. Moreover, the description of these properties can be made to depend on a specific value of p with $1 < p < \infty$, and so can be used to give unified proofs of some of the lifting properties of X to $L^p(\mu, X)$ that can be found in [5, 6, 8, 9, 13, 14, 15]. Complex analogues of these lifting results have been obtained in [1, 2, 3, 7, 12, 18].

2 PRELIMINARIES AND DEFINITIONS

A complex quasi-normed space X is said to be uniformly PL-convex if $h_X^p(\varepsilon) > 0$ for each $\varepsilon > 0$, for some (equivalently, for all) $0 < p < \infty$, where

$$h_X^p(\varepsilon) = inf\{\left(\int_0^{2\pi} \|x + e^{i\theta}y\|^p \frac{d\theta}{2\pi}\right)^{\frac{1}{p}} - 1 : \|x\| = 1 \text{ and } \|y\| \geq \varepsilon\}.$$

A complex quasi-normed space X is called uniformly H-convex if $H_X^p(\varepsilon) > 0$ for each $\varepsilon > 0$, for some (equivalently, for all) $0 < p < \infty$, where

$$H_X^p(\varepsilon) = inf\{\|f\|_p - 1 : f \in H^p(\mathbb{T}, X), \|\hat{f}(0)\| = 1, \text{ and } \|f - \hat{f}(0)\|_p \geq \varepsilon\}.$$

One can consider the h_X^p-modulus of X as a modulus that is built from the X-valued analytic polynomials of degree 1 on $\mathbb{T}$, and since the space $H^p(\mathbb{T}, X) = \{f \in L^p(\mathbb{T}, X) : \hat{f}(n) = 0 \text{ for all } n < 0\}$ is the $L^p(\mathbb{T}, X)$-closure of the X-valued analytic polynomials on $\mathbb{T}$, one can consider the H_X^p-modulus of X as a modulus that is built from all the X-valued analytic polynomials on $\mathbb{T}$. If X is a real Banach space and we replace analytic polynomials by the real part of analytic polynomials, i.e., harmonic polynomials, then we obtain the following moduli of convexity in X,

$$\delta_X^p(\varepsilon) = inf\{\left(\int_0^{2\pi} \|x + (\cos\theta)y\|^p \frac{d\theta}{2\pi}\right)^{\frac{1}{p}} - 1 : \|x\| = 1 \text{ and } \|y\| \geq \varepsilon\}$$

and

$$\Delta_X^p(\varepsilon) = inf\{\|f\|_p - 1 : f \in L^p(\mathbb{T}, X), \|\hat{f}(0)\| = 1, \text{ and } \|f - \hat{f}(0)\|_p \geq \varepsilon\}.$$

We will say that a real Banach space X is δ-uniformly convex (respectively, Δ-uniformly convex) if $\delta_X^p(\varepsilon) > 0$ for all $\varepsilon > 0$ (respectively, $\Delta_X^p(\varepsilon) > 0$ for all $\varepsilon > 0$), for some (equivalently, for all) $1 < p < \infty$. Note that Δ-uniformly convexity is the real analogue of uniform H-convexity because the X-valued harmonic polynomials on $\mathbb{T}$ are dense in $L^p(\mathbb{T}, X)$.

We can localize these properties in the following manner:

Definition 1 Let X be a real Banach space and let x be a norm 1 element of X.

(1) We say that x is a strong δ-point if $\delta_X^p(x, \varepsilon) > 0$ for all $\varepsilon > 0$, for some (equivalently, for all) $1 < p < \infty$, where

$$\delta_X^p(x, \varepsilon) = inf\{\left(\int_0^{2\pi} \|x + (\cos\theta)y\|^p \frac{d\theta}{2\pi}\right)^{\frac{1}{p}} - 1 : \|y\| \geq \varepsilon\}.$$

(2) We say that x is a strong Δ-point if $\Delta_X^p(x,\varepsilon) > 0$ for all $\varepsilon > 0$, for some (equivalently, for all) $1 < p < \infty$, where

$$\Delta_X^p(x,\varepsilon) = inf\{\|f\|_p - 1 : f \in L^p(\mathbb{T}, X), \hat{f}(0) = x, \text{ and } \|f - \hat{f}(0)\|_p \geq \varepsilon\}.$$

We also have the following non-uniform versions of the above properties:

Definition 2 Let X be a real Banach space and let x be a norm 1 element of X.

(1) We say that x is a δ-point if $\int_0^{2\pi} \|x + (\cos\theta)y\|^p \frac{d\theta}{2\pi} > 1$ for all non-zero $y \in X$, for some (equivalently, for all) $1 < p < \infty$.

(2) We say that x is a Δ-point if $\|f\|_p > 1$ for each $f \in L^p(\mathbb{T}, X)$ with $\hat{f}(0) = x$ and $\|f - \hat{f}(0)\|_p > 0$, for some (equivalently, for all) $1 < p < \infty$.

3 THE RESULTS

Theorem 1 Let X be a real Banach space. Then the following conditions are equivalent:

(1) X is uniformly convex,
(2) X is δ-uniformly convex,
(3) X is Δ-uniformly convex.

Proof. If X is uniformly convex, then for each $\varepsilon > 0$ there is a $\delta(\varepsilon) > 0$ so that whenever x is a norm 1 element of X and $y \in X$ has $\|y\| \geq \varepsilon$, then either $\|x + y\| > 1 + \delta(\varepsilon)$ or $\|x - y\| > 1 + \delta(\varepsilon)$. Therefore, since the function $\phi_p(t) = \frac{(1-t)^p + (1+t)^p}{2}$ is a strictly increasing function on $[0, \infty)$ for each $1 < p < \infty$, we have

$$\frac{\|x - y\|^p + \|x + y\|^p}{2} \geq \phi_p(\delta(\epsilon)).$$

Thus, if $x, y \in X$ with $\|x\| = 1$ and $\|y\| \geq \varepsilon$ then

$$\begin{aligned} \int_0^{2\pi} \|x + (\cos\theta)y\|^p \frac{d\theta}{2\pi} &= 4 \int_0^{\frac{\pi}{2}} \frac{\|x + (\cos\theta)y\|^p + \|x - (\cos\theta)y\|^p}{2} \frac{d\theta}{2\pi} \\ &= 4 \left[\int_0^{\frac{\pi}{3}} + \int_{\frac{\pi}{3}}^{\frac{\pi}{2}} \left(\frac{\|x + (\cos\theta)y\|^p + \|x - (\cos\theta)y\|^p}{2} \right) \frac{d\theta}{2\pi} \right] \\ &\geq \frac{2}{3}\phi_p(\delta(\varepsilon/2)) + \frac{1}{3}. \end{aligned}$$

Consequently, we obtain that $\delta_X^p(\varepsilon) \geq \left(\frac{2}{3}\phi_p(\delta(\varepsilon/2)) + \frac{1}{3}\right)^{\frac{1}{p}} - 1$, and so X is δ-uniformly convex if X is uniformly convex.

On the other hand, if X is δ-uniformly convex and $x, y \in X$ satisfy $\|x\| = 1$ and $\|y\| \geq \varepsilon$, then clearly either $\|x + y\| \geq 1 + \delta_X^p(\varepsilon)$ or $\|x - y\| \geq 1 + \delta_X^p(\varepsilon)$, and so X is uniformly convex.

It is obvious that Δ-uniformly convex spaces are δ-uniformly convex, so it remains to show that uniformly convex spaces are Δ-uniformly convex. Note that X is Δ-uniformly convex if $\alpha_X^p(\varepsilon) > 0$ for all $\varepsilon > 0$, for some (equivalently, for all) $1 < p < \infty$, where

$$\alpha_X^p(\varepsilon) = inf\{1 - \|\hat{f}(0)\| : f \in L^p(\mathbb{T}, X), \text{ with } \|f\|_p \leq 1, \text{ and } \|f - \hat{f}(0)\|_p \geq \varepsilon\}.$$

Using this modulus and the facts that $\|\hat{f}(0)\| \leq \|\frac{f+\hat{f}(0)}{2}\|_p$ and $L^p(\mathbb{T}, X)$ is uniformly convex whenever X is uniformly convex and $1 < p < \infty$, we easily see that X is Δ-uniformly convex whenever X is uniformly convex.

Theorem 2 Let X be a real Banach space and let x be a norm 1 element of X.

(1) x is a strong δ-point of X if and only if x is a strongly extreme point of the unit ball of X.

(2) x is a strong Δ-point of X if and only if x is a denting point of the unit ball of X.

Proof. A close analysis of the proof of the equivalence of (1) and (2) in Theorem 1 proves part (1).

Strong Δ-points of X are denting points of the unit ball of X was proved in [4, Proposition 3.6]. The fact that denting points of the unit ball of X are strong Δ-points of X follows from the results of [16] for the case $p = 2$, and from [10] for the general case $1 < p < \infty$.

Theorem 3 Let X be a real Banach space and let x be a norm 1 element of X. Then the following conditions are equivalent:

(1) x is an extreme point of the unit ball of X,

(2) x is a δ-point of X,

(3) x is a Δ-point of X.

Proof. (3) implies (2) is obvious. (2) implies (1) follows from the proof of (2) implies (1) in Theorem 1.

For (1) implies (3), suppose that x is an extreme point of the unit ball of X and $f \in L^p(\mathbb{T}, X)$ satisfies $\hat{f}(0) = x$ and $\|f\|_p = 1$. Then $1 = \|\hat{f}(0)\| \leq \|f\|_1 \leq \|f\|_p = 1$, by Hölder's inequality. We therefore have $\|f\|_1 = \|f\|_p$, and by Hölder's inequality again we see that $\|f(z)\| = 1$ for almost all $z \in \mathbb{T}$. For each $t \in (0, 2\pi)$

$$x = \frac{t}{2\pi}\left(\frac{\int_0^t f(e^{i\theta})\frac{d\theta}{2\pi}}{t/2\pi}\right) + \frac{2\pi - t}{2\pi}\left(\frac{\int_t^{2\pi} f(e^{i\theta})\frac{d\theta}{2\pi}}{(2\pi - t)/2\pi}\right),$$

and x is an extreme point of the unit ball of X, so $\int_0^t f(e^{i\theta})\frac{d\theta}{2\pi} = \frac{t}{2\pi}x = \int_0^t x\frac{d\theta}{2\pi}$ for all $t \in [0, 2\pi]$. Thus $\int_0^t (f(e^{i\theta}) - x)\frac{d\theta}{2\pi} = 0$ for all $t \in [0, 2\pi]$. Since f is measurable it has an essentially separable range. We therefore can assume that there is a separable closed subspace Y of X, so that $f(z) \in Y$ for all $z \in \mathbb{T}$. Since Y is separable Y^* contains a countable total set $(x_n^*)_n$. Therefore, $\int_0^t x_n^*(f(e^{i\theta}) - x)\frac{d\theta}{2\pi} = 0$ for all $t \in [0, 2\pi]$ and for all $n \in \mathbb{N}$. Hence there exists a sequence $(A_n)_n$ of subsets of $[0, 2\pi]$, each of measure 0, such that $x_n^*(f(e^{i\theta}) - x) = 0$ for all $[0, 2\pi] - A_n$. Therefore, since $(x_n^*)_n$ is a total subset of Y^*, $f(e^{i\theta}) = x$ for all $\theta \in [0, 2\pi] - A$, where $A = \bigcup_{n=1}^{\infty} A_n$. Finally, since A has measure 0, f is almost everywhere constant and so x is a Δ-point of X.

REFERENCES

1. W.J. Davis, D.J.H. Garling and N. Tomczak-Jaegermann, The complex convexity of quasi-normed spaces, *J. Funct. Anal.* **55** (1984), 110–150.
2. S.J. Dilworth, Complex convexity and the geometry of Banach spaces, *Math. Proc. Camb. Phil. Soc.* **99** (1986), 495–506.
3. P.N. Dowling, Z. Hu and D. Mupasiri, Complex convexity in Lebesgue-Bochner function spaces, *preprint* (1994).
4. P.N. Dowling, Z. Hu and M.A. Smith, Geometry of spaces of vector-valued harmonic functions, *Can. J. Math.* **46(2)** (1994), 274–283.
5. P. Greim, An extremal vector-valued L^p-function taking no extremal vectors as values, *Proc. Amer. Math. Soc.* **84** (1982), 65–68.
6. P. Greim, A note on strong extreme and strongly exposed points in Bochner L^p-spaces *Proc. Amer. Math. Soc.* **93** (1985), 65–66.
7. Z. Hu and D.Mupasiri, Complex strongly extreme points in quasi-normed spaces, *preprint* (1994).
8. J.A. Johnson, Extreme measurable selections, *Proc. Amer. Math. Soc.* **44** (1974), 107–112.
9. Bor-Luh Lin and Pei-Kee Lin, Denting points in Bochner L^p-spaces, *Proc. Amer. Math. Soc.* **97** (1986), 629-633.
10. Bor-Luh Lin, Pei-Kee Lin and S.L. Troyanski, Some geometric and topological properties of the unit sphere in a Banach space, *Math. Ann.* **274** (1986), 613–616.
11. Bor-Luh Lin, Pei-Kee Lin and S.L. Troyanski, Characterizations of denting points, *Proc. Amer. Math. Soc.* **102** (1988), 526–528.
12. D. Mupasiri, Some results on complex convexity and the geometry of complex vector spaces, *Dissertation, Northern Illinois University* (1992).
13. M.A. Smith, Strongly extreme points in $L^p(\mu, X)$, *Rocky Mountain J. Math.* **16** (1986), 1–5.
14. M.A. Smith, Rotundity and extremity in $\ell^p(X_i)$ and $L^p(\mu, X)$, *Contemporary Math.* **52** (1986), 143–162.
15. M.A. Smith and B. Turett, Rotundity in Lebesgue-Bochner function spaces, *Trans. Amer. Math. Soc.* **257** (1980), 105–118.
16. S.L. Troyanski, On a property of the norm which is close to local uniform rotundity, *Math. Ann.* **271** (1985), 305–313.
17. Q. Xu, Inégalités pour les martingales de Hardy et renormage des espaces quasi-normés, *C. R. Acad. Sci. Paris, Sér. I* **306** (1988), 601–604.
18. Q. Xu, Convexités uniformes et inégalités de martingales, *Math. Ann.* **287** (1990), 193–211.

Regularity Conditions for Banach Function Algebras

J. F. FEINSTEIN Department of Mathematics, University of Nottingham, Nottingham NG7 2RD, England.

ABSTRACT

There are many forms of regularity condition which a Banach function algebra (i.e. a commutative, semisimple Banach algebra) may satisfy. Many of these conditions have their origins in Harmonic Analysis, and now have applications in the theory of Automatic Continuity, and the theory of Wedderburn (and strong Wedderburn) decomposability of commutative Banach algebras. We shall discuss the relationships between these conditions in the general setting and also for uniform algebras. We shall include in our discussion strong regularity, Ditkin's condition, bounded relative units, and maximal ideals which have a bounded approximate identity.

1 INTRODUCTION

We shall consider only unital Banach algebras, although most of the definitions and results also make sense in the non-unital setting.

Terminology and notation. In our terminology a *compact space* is a compact, Hausdorff topological space. For a compact space X we denote the algebra of all continuous, complex-valued functions on X by $C(X)$.

Let A be a commutative, unital Banach algebra. We denote the character space of A by Φ_A .

Definition 1.1. Let X be a compact space. A *Banach function algebra* on X is a complete normed subalgebra of $C(X)$ which contains the constant functions and separates the points of X. A *uniform algebra* on X is a Banach function algebra on X whose norm is the uniform norm on X.

Every commutative, semisimple Banach algebra can be regarded, via the Gelfand transform, as a Banach function algebra on its character space.

Notation. Let A be a Banach function algebra on Φ_A, and let $\phi \in X$. We define two ideals, J_ϕ and M_ϕ, as follows:

$$M_\phi = \ker(\phi) = \{f \in A : f(\phi) = 0\};$$

$$J_\phi = \{f \in A : f^{-1}(\{0\}) \text{ is a neighbourhood of } \phi\}.$$

Definition 1.2. Let A be a uniform algebra on Φ_A, and let $\phi \in \Phi_A$. The *Gleason part of* ϕ, P_ϕ, is defined by

$$P_\phi = \{\psi \in \Phi_A : \|\psi - \phi\| < 2\},$$

using the usual norm on linear functionals.

It is standard that the Gleason parts partition Φ_A, and that if M_ϕ has a bounded approximate identity, then $P_\phi = \{\phi\}$ (see Chapter VI of [7], for example).

Definition 1.3. Let A be a Banach function algebra on Φ_A. Then A is *normal* if, for all closed sets $E \subseteq \Phi_A$ and for all compact sets $F \subseteq \Phi_A \backslash E$, there exists $f \in A$ with $f(E) \subseteq \{0\}$ and $f(F) \subseteq \{1\}$.

Let $\phi \in \Phi_A$. Then A is *strongly regular at* ϕ if $\overline{J_\phi} = M_\phi$: A satisfies *Ditkin's condition at* ϕ if for all $f \in M_\phi$ and all $\epsilon > 0$ there exists $g \in J_\phi$ with $\|gf - f\| < \epsilon$. Let $C \geq 1$. We say that A has *bounded relative units at* ϕ *with bound* C if, for all compact sets $E \subseteq \Phi_A \backslash \{\phi\}$, there exists $f \in J_\phi$ with $\|f\| \leq C$ such that $f(E) \subseteq \{1\}$.

The algebra A is *strongly regular* if it is strongly regular at every point; A is a *Ditkin algebra* if it satisfies Ditkin's condition at every point; A is a *strong Ditkin algebra* if it is strongly regular, and every maximal ideal has a bounded approximate identity.

There are two definitions in print for a Banach function algebra A to have bounded relative units. One is that, for every $\phi \in \Phi_A$, A has bounded relative units at ϕ. In this case we shall say that A has *bounded relative units in the sense of Bade* (which we shall abbreviate to "A has b.r.u. (B)"). The other definition is stronger, insisting that the constant C involved does not depend on ϕ. In this case we shall say that A has *bounded relative units in the sense of Dales* ("A has b.r.u. (D)").

The following elementary result is well known. The proof is essentially the same as that of the corresponding result for uniform algebras, proved by Chalice in [2].

Proposition 1.4. *Let A be a Banach function algebra on Φ_A, and let $\phi \in \Phi_A$. Suppose that A is strongly regular at ϕ and that M_ϕ has a bounded approximate identity. Then M_ϕ has a bounded approximate identity consisting of elements of J_ϕ.* ∎

It follows immediately that every strong Ditkin algebra is indeed a Ditkin algebra. In [1], W. G. Bade asked the following question: does every strong Ditkin algebra have b.r.u. (B)? This question is resolved positively in [6]. In fact, the following stronger result is proved there. The proof is short and elementary, and we include it for convenience.

Theorem 1.5. *Let A be a normal Banach function algebra on Φ_A, and let $\phi \in \Phi_A$. Suppose that A is strongly regular at ϕ and that M_ϕ has a bounded approximate identity. Then A has bounded relative units at ϕ.*

Proof. By Proposition 1.4, M_ϕ has a bounded approximate identity (e_α) consisting of elements of J_ϕ. Set $C = \sup_\alpha \|e_\alpha\|$. Let $\epsilon > 0$. We show that A has bounded relative units at ϕ with bound $C + \epsilon$. To see this, let E be a compact subset of $\Phi_A \backslash \{\phi\}$. Choose $g \in J_\phi$ such that $g(E) \subseteq \{1\}$. Then the functions $1 - (1 - e_\alpha)(1 - g)$ are also in J_ϕ and are identically 1 on E. But

$$\|1 - (1 - e_\alpha)(1 - g)\| = \|e_\alpha + g - e_\alpha g\|$$

$$\leq C + \|g - e_\alpha g\|.$$

Choosing α such that $\|g - e_\alpha g\| < \epsilon$, set $f = e_\alpha + g - e_\alpha g$. Then $f \in J_\phi$, $\|f\| \leq C + \epsilon$ and $f(E) \subseteq \{1\}$. ∎

It follows that every strong Ditkin algebra has b.r.u. (B). However, there is an example in [6] of a strong Ditkin algebra which does not have b.r.u. (D).

2 KNOWN RELATIONSHIPS BETWEEN REGULARITY CONDITIONS

Let us list some of the conditions which a Banach function algebra A may satisfy.

(a) $A = C(X)$.
(b) A has b.r.u (D).
(c) A has b.r.u (B).
(d) A is a strong Ditkin algebra.
(e) A is a Ditkin algebra.
(f) A is strongly regular.
(g) A is normal.

For general Banach function algebras A the following relationships between these conditions are well known.

$$\text{(a)} \Rightarrow (\text{(b) and (f)}) \Rightarrow (\text{(c) and (f)}) \Rightarrow \text{(d)} \Rightarrow \text{(e)} \Rightarrow \text{(f)} \Rightarrow \text{(g)}.$$

All but one of the reverse implications are false, and (b) does not imply (f). See [4] for a comprehensive list of results and counterexamples. The one implication which may be reversed comes from the result we gave earlier: (d) $\Rightarrow$ (c), and so a Banach function algebra A is a strong Ditkin algebra if and only if it is strongly regular and has b.r.u. (B).

The story for uniform algebras is a little different. If A is a uniform algebra, then

$$(a) \Rightarrow (b) \Leftrightarrow (c) \Leftrightarrow (d) \Rightarrow (e) \Rightarrow (f) \Rightarrow (g)$$

That (f) and (g) are not equivalent follows from an example of O'Farrell ([8]). Examples in [5] show that (f) does not imply (d), and (b) does not imply (a). These examples are obtained by modifying a construction due to Cole (see [3]). It is not known whether (f) implies (e) or (e) implies (d).

We may also ask about conditions at a single point. If A is a Banach function algebra on Φ_A and $\phi \in \Phi_A$, consider the following possible statements.

(i) For all $C > 1$, A has bounded relative units at ϕ with bound C.

(ii) A has bounded relative units at ϕ.

(iii) A is strongly regular at ϕ and M_ϕ has a bounded approximate identity.

(iv) A satisfies Ditkin's condition at ϕ.

(v) A is strongly regular at ϕ.

In general, we have ((ii) and (v)) $\Rightarrow$ (iii) $\Rightarrow$ (iv) $\Rightarrow$ (v), and, of course, (i) $\Rightarrow$ (ii). If A is normal, then Theorem 1.5 tells us that the first implication may be reversed. If A is a normal *uniform* algebra, however, then considerably more is true:

$$\text{(i)} \Leftrightarrow \text{(ii)} \Leftrightarrow \text{(iii)} \Rightarrow \text{(iv)} \Rightarrow \text{(v)} .$$

The first of these equivalences is stated in a footnote in [5]. The proof is elementary. The normality assumption on A is only used in order to obtain (iii) $\Rightarrow$ (ii).

Wang gives an example in [9] to show that (v) does not imply (iii). But whether or not either of the implications (v) $\Rightarrow$ (iv) or (iv) $\Rightarrow$ (iii) is true for uniform algebras appears to be open.

We have already mentioned that there are strongly regular uniform algebras which are not strong Ditkin algebras. The example constructed in [5], however, has the property that each of its Gleason parts contain only a single point. We conclude this section by showing that this need not happen.

Theorem 2.1. There exists a strongly regular uniform algebra B on Φ_B such that all Gleason parts of B have exactly one point, except for one part which has exactly two points.

Proof. Choose any normal uniform algebra A_0 on Φ_{A_0} such that A_0 has a Gleason part P containing more than one point (for example, the example of O'Farrell mentioned above has this property). Choose $\phi_0, \psi_0 \in P$ with $\phi_0 \neq \psi_0$. A simple modification of the construction used in Section 5 of [5] allows us to construct a normal uniform algebra A_1 on Φ_{A_1} and a continuous, linear surjection $T : A_1 \longrightarrow A_0$, and to find ϕ_1, ψ_1 in Φ_{A_1} with $\phi_1 \neq \psi_1$ satisfying the following conditions:

(1) $\|T\| = 1$;

(2) A_1 has bounded relative units at every point of $\Phi_{A_1} \backslash \{\phi_1, \psi_1\}$;

(3) for all $f \in A_1$, $(Tf)(\phi_0) = f(\phi_1)$ and $(Tf)(\psi_0) = f(\psi_1)$.

Now set $B = \left(\overline{J_{\phi_1}} \oplus \mathbb{C} \cdot 1\right) \cap \left(\overline{J_{\psi_1}} \oplus \mathbb{C} \cdot 1\right)$. By the results in Section 4 of **[5]**, B is a strongly regular uniform algebra, $\Phi_B = \Phi_{A_1}$ and, for all $\phi \in \Phi_B \backslash \{\phi_1, \psi_1\}$, M_ϕ has a bounded approximate identity, and so $P_\phi = \{\phi\}$. Finally, for $f \in B$ with $\|f\| \leq 1$,

$$\begin{aligned} |f(\psi_1) - f(\phi_1)| &= |(Tf)(\psi_0) - (Tf)(\phi_0)| \\ &\leq \|\psi_0 - \phi_0\|. \end{aligned}$$

Taking the supremum over all such f gives us $\|\psi_1 - \phi_1\| \leq \|\psi_0 - \phi_0\|$, and so $\{\phi_1, \psi_1\}$ must be a two-point part. ∎

3 OPEN QUESTIONS

As we have seen, the relationships between the various regularity conditions for Banach function algebras are now fairly well understood. In this Section we bring together the questions mentioned earlier, and mention some others.

(A) Is every strongly regular uniform algebra a Ditkin algebra?

(B) Is every uniform algebra which is a Ditkin algebra also a strong Ditkin algebra?

The examples mentioned in Section 2 show that (A) and (B) can not both have positive answers. We can also ask versions of (A) and (B) at one point at a time.

(C) Does every uniform algebra satisfy Ditkin's condition at every point at which it is strongly regular?

(D) Let A be a uniform algebra, and let $\phi \in \Phi_A$. Suppose that A satisfies Ditkin's condition at ϕ. Must M_ϕ have a bounded approximate identity?

The example of Wang mentioned in Section 2 shows that (C) and (D) can not both have the answer "yes".

Finally we mention two questions which are of interest both in the general case and for uniform algebras.

(E) Let A be a normal Banach function algebra on Φ_A, and let $\phi \in \Phi_A$. Suppose that M_ϕ has a bounded approximate identity. Does it follow that A must be strongly regular at ϕ?

(F) Let A be a normal Banach function algebra on Φ_A, and suppose that, for all $\phi \in \Phi_A$, M_ϕ has a bounded approximate identity. Does it follow that A must be strongly regular?

In [5] it is shown that (E) and (F) must have the same answer for uniform algebras.

REFERENCES

[1] W. G. Bade, *Proceedings of the Centre for Mathematical Analysis, Australian National University*, Volume **21**, 1989, Conference on Automatic Continuity and Banach Algebras, Open Problems.

[2] Donald R. Chalice, S-algebras on sets in $\mathbf{C}^n$, *Proc. Amer. Math. Soc.*, **39** (1973), 300–304.

[3] B. J. Cole, One point parts and the peak point conjecture, Ph.D. Dissertation, Yale University 1968.

[4] H. G. Dales, *Banach algebras and automatic continuity*, Oxford University Press, to appear.

[5] Joel F. Feinstein, A non-trivial, strongly regular uniform algebra, *J. London Math. Soc.* **45** (1992), 288–300.

[6] J. F. Feinstein, A note on strong Ditkin algebras, *Nottingham Mathematics Preprint* #94.12.

[7] T. W. Gamelin, *Uniform Algebras*, Prentice-Hall, Englewood Cliffs, New Jersey, 1969.

[8] A. G. O'Farrell, A regular uniform algebra with a continuous point derivation of infinite order, *Bull. London Math. Soc.* **11** (1979) 41-44.

[9] James Li-Ming Wang, Strong regularity at nonpeak points, *Proc. Amer. Math. Soc.* **51** (1975) 141-142.

Some Bargmann Spaces of Analytic Functions

D. J. H. Garling
Dept. of Mathematics & Stat.
University of Cambridge
Cambridge CB2 1SB, England

P. Wojtaszczyk
Institute of Mathematics
Polish Academy of Sciences
00-950 Warsaw, Poland

1. *Introduction*

Let the d-fold Cartesian product $\mathcal{C}^d$ of the complex numbers $\mathcal{C}$ be given its usual inner product

$$\langle z, w\rangle = \sum_{i=1}^{d} z_i \bar{w}_i$$

and associated norm

$$|z| = (\sum_{i=1}^{d} |z_i|^2)^{1/2}.$$

Let $dv(z) = dx_1 dy_1 \dots dx_n dy_n$ be Lebesgue measure on $\mathcal{C}^d$. Then the measure

$$d\varpi(z) = \pi^{-d} e^{-|z|^2} dv(z)$$

is a probability measure on $\mathcal{C}^d$ which is invariant under the group $\mathcal{U}(\mathcal{C}^d)$ of unitary linear transformations of $\mathcal{C}^d$.

Suppose that $0 < p < \infty$. Let $B_p = B_p(\mathcal{C}^d)$ be the set of entire functions on $\mathcal{C}^d$ which are in $L_p(\mathcal{C}^d, \varpi)$. B_p is a linear subspace of L_p: the purpose of this paper is to investigate the properties of B_p as a Banach (or quasi-Banach) space, and the properties of the scale of spaces $\{B_p : 0 < p < \infty\}$.

The Hilbert space B_2 was studied by Bargmann [**1**] in connection with the commutation relations of quantum mechanics and representations of the Heisenberg group. Spaces corresponding to B_p were considered in some detail by Janson, Peetre and Rochberg [**5**] in their study of Hankel forms, and some of their results are reproduced below. Their setting is somewhat different from ours: they introduce weights which depend upon p. These are very natural in many respects, but this means that the scale of spaces that they consider is different from ours.

This work was carried out while the second-named author was a Central and East European Fellow under the Human Capital and Mobility Programme of the Commission of the European Communities, and while he was a Visiting Scholar of St. John's Colle ge, Cambridge. He would particularly like to thank St. John's College, Cambridge for the hospitality that was extended to him and to his family.

2. *Terminology and notation*

If $\alpha = (\alpha_1, \ldots, \alpha_d) \in (\mathcal{Z}^+)^d$ and $z \in \mathcal{C}^d$ we set

$$z^\alpha = z_1^{\alpha_1} \ldots z_d^{\alpha_d}, \quad \alpha! = \alpha_1! \ldots \alpha_d!, \quad |\alpha| = \alpha_1 + \ldots + \alpha_d \quad \text{and} \quad |\alpha|_\infty = \max \alpha_i.$$

We set $\bar{z} = (\bar{z}_1, \ldots, \bar{z}_d)$. If z and w are in $\mathcal{C}^d$, we write $z.w$ for $(z_1 w_1, \ldots, z_d w_d)$.

We denote the vector space of linear mappings of $\mathcal{C}^d$ to itself by $L(\mathcal{C}^d)$.

If $1 < p < \infty$, we denote the conjugate index $p/(p-1)$ by p'.

We denote normalised Haar measure on the torus $\mathcal{T}^d$ by σ_d and normalised Haar measure on the unit sphere S^{2d-1} by s_{2d-1}. Let the area of S^{2d-1} be κ_{2d-1}.

If $r \in (\mathcal{R}^+)^d$ and $t \in \mathcal{T}^d$ then $r.t = (r_1 t_1, \ldots, r_d t_d) \in \mathcal{C}^d$, and

$$d\varpi = 2^d (r_1 e^{-r_1^2} dr_1) \ldots (r_d e^{-r_d^2} dr_d) d\sigma_d(t).$$

Similarly, if $r \in \mathcal{R}^+$ and $u \in S^{2d-1}$ then $ru \in \mathcal{C}^d$, and

$$d\varpi = \kappa_{2d-1} r^{2d-1} e^{-dr^2} dr ds_{2d-1}.$$

Let $\mathcal{P} = \mathcal{P}[z_1, \ldots, z_d]$ be the vector space of all polynomial functions on $\mathcal{C}^d$ with complex coefficients, and let $\mathcal{P}_n$ be the subspace of polynomials of degree at most n. $\mathcal{P}$ is a linear subspace of B_p, for $0 < p < \infty$.

Let $\mathcal{E} = \mathcal{E}(\mathcal{C}^d)$ be the Fréchet space of entire functions on $\mathcal{C}^d$, with the topology of local uniform convergence. If $f \in \mathcal{E}$, we write its Taylor series expansion as

$$f(z) = \sum_\alpha \frac{f_\alpha}{\alpha!} z^\alpha = \sum_\alpha \psi_\alpha(f) z^\alpha.$$

If $w \in \mathcal{C}^d$, we define the *dilation* $D_w(f)$ of f by

$$D_w(f)(z) = f(w.z) = \sum_\alpha \frac{f_\alpha}{\alpha!} w^\alpha z^\alpha.$$

In particular, if $r > 0$, we write $d_r(f)$ for $D_{(r,\ldots,r)}(f)$, so that

$$d_r(f)(z) = \sum_\alpha f_\alpha r^{|\alpha|} z^\alpha / \alpha!.$$

If $f \in \mathcal{E}$ and $r \in (\mathcal{R}^+)^d$, we define $f_{(r)}$ as a function on $\mathcal{T}^d$ by $f_{(r)}(t) = f(r.t)$.

We shall use the following notation for certain functions:

$$\begin{aligned}
\gamma_\alpha(z) &= z^\alpha, \\
l_h(z) &= \langle z, h \rangle, \\
c_i(z) &= \langle z, e_i \rangle = z_i, \\
exp(z) &= \sum_\alpha \frac{z^\alpha}{\alpha!} \\
exp_n(z) &= \sum_{|\alpha|_\infty \le n} \frac{z^\alpha}{\alpha!}
\end{aligned}$$

$$\begin{aligned} e_w(z) &= e^{\langle z,w\rangle} = \sum_\alpha \frac{(z.\bar{w})^\alpha}{\alpha!} = \sum_\alpha \frac{z^\alpha \bar{w}^\alpha}{\alpha!} = exp(\bar{w}.z) \\ \lambda_{p,w}(z) = \lambda_p(w,z) &= e^{-|w|^2/p} e^{2\langle z,w\rangle/p}. \end{aligned}$$

3. *Some endomorphisms of B_p*

Suppose that ϕ is an analytic mapping of $\mathcal{C}^d$ to itself. Then if $f \in \mathcal{E}$ we can define

$$\tilde{\phi}(f)(z) = f(\phi(z)).$$

$\tilde{\phi}$ is a continuous linear mapping of $\mathcal{E}$ into itself. When does $\tilde{\phi}$ map B_p into itself? We begin by considering linear mappings.

PROPOSITION 1. *Suppose that $0 < p < \infty$. If $T \in L(\mathcal{C}^d)$, then $\tilde{T}$ maps B_p into itself if and only if $\|T\| \leq 1$. If so, $\tilde{T}$ is continuous and $\left\|\tilde{T}\right\| = 1$.*

Proof. If T is unitary then clearly $\tilde{T}$ is an isometry of B_p onto itself with inverse $\tilde{T}^*$. Otherwise, we can write $T = |T|U$, where U is unitary and $|T| \geq 0$, so that it is enough to consider the case where $T \geq 0$. We can choose an orthonormal basis of $\mathcal{C}^d$ so that T is given by a diagonal matrix $\text{diag}(t_1, \ldots, t_d)$.

Suppose first that $\|T\| \leq 1$, so that $0 \leq t_i \leq 1$, for $1 \leq i \leq d$. Then if $f \in B_p$,

$$\begin{aligned} \int |\tilde{T}(f)|^p \, d\varpi &= \int |f(t_1 z_1, \ldots, t_d z_d|^p \, d\varpi(z) \\ &= \int\int \cdots \left(\int\int |f(t_1 r_1 e^{i\theta_1}, \ldots, t_d r_d e^{i\theta_d}|^p \, d\theta_1 \, 2r_1 e^{-r_1^2} dr_1\right) \ldots \, d\theta_d \, 2r_d e^{-r_d^2} dr_d \\ &\leq \int\int \cdots \left(\int\int |f(r_1 e^{i\theta_1}, \ldots, r_d e^{i\theta_d}|^p \, d\theta_1 \, 2r_1 e^{-r_1^2} dr_1\right) \ldots \, d\theta_d \, 2r_d e^{-r_d^2} dr_d \\ &= \int |f|^p \, d\varpi, \end{aligned}$$

since $\int |f(r_1 e^{i\theta_1}, \ldots, r_d e^{i\theta_d}|^p \, d\theta_j$ is an increasing function of r_j, by the plurisubharmonicity of $|f|^p$. Thus $\tilde{T}(f) \in B_p$ and $\left\|\tilde{T}(f)\right\|_p \leq \|f\|_p$. Since the constants are in B_p and $\tilde{T}(1) = 1$, it follows that $\left\|\tilde{T}\right\| = 1$.

Conversely suppose that $\|T\| > 1$, so that $t_j > 1$ for some j. Let $f(z) = e^{z_j^2/pt_j}$. Then $f \in B_p$, while $\tilde{T}(f) \notin B_p$.

Next we consider affine transformations of $\mathcal{C}^d$. Since every such mapping is the composition of a linear mapping and a translation, it is enough to consider translations.

PROPOSITION 2. *Suppose that $0 < p < \infty$ and that $c \in \mathcal{C}^d$. If $f \in B_p$, let*

$$V_c^p(f)(z) = \lambda_p(z,c)f(z-c).$$

Then V_c^p is an isometry of B_p onto itself, with inverse V_{-c}^p.

Proof. Certainly $V_c^p(f) \in \mathcal{E}$. Changing variables,

$$\begin{aligned}\int |V_c^p(f)(z)|^p \, d\varpi(z) &= \pi^{-d} \int |e^{\langle 2z-c,c\rangle}||f(z-c)|^p e^{-|z|^2} \, dv(z) \\ &= \pi^{-d} \int |f(z-c)|^p e^{-|z-c|^2} \, dv(z) \\ &= \int |f(z)|^p \, d\varpi(z).\end{aligned}$$

Thus V_c^p is an isometry of B_p into itself. Elementary calculations show that it is invertible, with inverse V_{-c}^p.

We shall return to this topic later, in Theorem 2.

Let us now see how the complex Euclidean group $Euc(\mathcal{C}^d)$ (generated by translations and unitary linear mappings) acts on B_p. If $g \in Euc(\mathcal{C}^d)$ and $g(z) = U(z) + c$, then $g = T_c U$, where $T_c(z) = z + c$. We therefore define $V_g^p = V_c^p \tilde{U}^*$. Then if $f \in B_p$,

$$V_g^p(f)(z) = V_c^p(f(U^*(z))) = \lambda_p(z,c)f(g^{-1}(z)).$$

Now suppose that $g_1 = T_{c_1}U_1$ and that $g_2 = T_{c_2}U_2$. Let $g_1g_2 = g_3 = T_{c_3}U_3$, so that $U_3 = U_1U_2$ and $c_3 = U_1c_2 + c_1$. Then

$$\begin{aligned}V_{g_1}^p V_{g_2}^p(f)(z) &= \lambda_p(z,c_1)V_{g_2}^p f(g_1^{-1}(z)) \\ &= \lambda_p(z,c_1)\lambda_p(g^{-1}(z)z,c_2)f(g_2^{-1}g_1^{-1}(z)) \\ &= e^{-(|c_1|^2+|c_2|^2)/p}e^{2c_1/p}(z)e^{2U_1(c_2)/p}(z-c_1)f(g_3^{-1}(z)) \\ &= e^{2i\alpha/p}\lambda_p(z,c_3)f(g_3^{-1}(z)) \\ &= e^{2i\alpha/p}V_{g_3}^p(f)(z),\end{aligned}$$

where $\alpha = \text{Im } \langle c_1, U_1(c_2)\rangle$. Let $\tilde{B}_p$ denote the projective space of one-dimensional linear subspaces of B_p, and let $\tilde{V}_g^p$ be the automorphism of $\tilde{B}_p$ induced by V_g^p. Then $\tilde{V}_{g_1}^p\tilde{V}_{g_2}^p = \tilde{V}_{g_1g_2}^p$, so that the mapping $g \to \tilde{V}_g^p$ is a projective representation of $Euc(\mathcal{C}^d)$.

Alternatively, let H be the cross product $Euc(\mathcal{C}^d) \times \mathcal{T}$, with composition defined for $(g_1, e^{i\theta_1}) = (T_{c_1}U_1, e^{i\theta_1})$ and $(g_2, e^{i\theta_2}) = (T_{c_2}U_2, e^{i\theta_2})$ by

$$(g_1, e^{i\theta_1})(g_2, e^{i\theta_2}) = (g_1g_2, e^{2i \text{ Im } \langle c_1, U_1(c_2)\rangle}e^{i(\theta_1+\theta_2)}).$$

Then the mapping $(g, e^{i\theta}) \to e^{i\theta}V_g^p$ is a faithful representation of H as a group of isometries of B_p. We shall see later (Theorem 2) that when $p \neq 2$ these are *all* the isometries of B_p.

If $f \in B_p$, where $0 < p < \infty$, then $|f|^p$ is plurisubharmonic, and so $|f(0)| \leq \|f\|_p$. Thus the functional $\delta_0(f) = f(0)$ is continuous on B_p, and $\|\delta_0\| \leq 1$. Since the constants are in B_p, $\|\delta_0\| = 1$. Since

$$\delta_0(V_{-a}^p(f)) = e^{-|a|^2/p} f(a) = e^{-|a|^2/p} \delta_a(f),$$

the evaluation functional δ_a is also continuous on B_p, and $\|\delta_a\| = e^{|a|^2/p}$. Note also that

$$\delta_0(f) = \lim_{R\to\infty} \int_{|z|\leq R} f(z)\, d\varpi(z) \quad \left(= \int f(z)\, d\varpi(z), \text{ when } p \geq 1\right),$$

and

$$\begin{aligned} \delta_a(f) &= \lim_{R\to\infty} \int_{|z|\leq R} e^{-2\langle z,a\rangle/p} f(z+a)\, d\varpi(z) \\ \Big(&= \int e^{-2\langle z,a\rangle/p} f(z+a)\, d\varpi(z), \text{ when } p \geq 1\Big). \end{aligned}$$

As a consequence we have

PROPOSITION 3. *The inclusion mapping $B_p \to \mathcal{E}$ is continuous.*

By Fatou's Lemma, the closed unit ball of B_p is closed in $\mathcal{E}$. We therefore have the following corollary:

COROLLARY 1. *B_p is a closed linear subspace of L_p.*

COROLLARY 2. *For each $\alpha \in (\mathcal{Z}^+)^d$, the mapping: $f \to f_\alpha$ is a continuous linear functional on B_p.*

Let

$$G_p = \{f \in \mathcal{E} : \|f\| = \sup_z |f(z)| e^{-|z|^2/p} < \infty\},$$

and let $B_{p,\infty} = \mathcal{E} \cap L_{p,\infty}$. Then $B_p \subseteq G_p \subseteq B_{p,\infty}$.

A trivial calculation establishes the following

PROPOSITION 4. *If $w \in \mathcal{C}^d$, D_w is a norm-decreasing linear mapping of G_p into $G_{p/|w|^2}$.*

COROLLARY 3. *If $r > 0$ then d_r is an isometry of G_p onto G_{p/r^2}.*

COROLLARY 4. *D_w maps B_p continuously into B_q for $0 < |w|^2 q < p < \infty$.*

4. *Polynomial approximation*

We now consider the density of $\mathcal{P}$ in B_p.

We begin with some elementary calculations.

LEMMA 1. *Suppose that $f \in B_p$.*
(i) $|\psi_\alpha(f)| \leq \prod_{i=1}^d (\frac{2e}{\alpha_i p})^{\alpha_i/2} \|f\|_p$.
(ii) $\|\gamma_\alpha\|_p = (\prod_{i=1}^d \Gamma(\alpha_i p/2+1))^{1/p} \sim \prod_{i=1}^d (\frac{\alpha_i p}{2e})^{\alpha_i/2} (\frac{\alpha_i p}{2})^{1/2p}$.
(iii) There exists K_p such that

$$\|\psi_\alpha(f)\gamma_\alpha\|_p \leq K_p(\prod_{i=1}^d (\frac{\alpha_i p}{2}))^{1/2p} \|f\|_p .$$

Proof. By Cauchy's integral formula,

$$\psi_\alpha(f) = \int_{T^d} \frac{f(R.z)}{R^\alpha z^\alpha}\, d\sigma_d(z),$$

so that, since $|f(R.z)| \leq e^{|R.z|^2/p} \|f\|_p$, if we set $R_i^2 = \alpha_i p/2$ we obtain

$$|\psi_\alpha(f)| \leq \prod_{i=1}^d \frac{e^{R_i^2/p}}{R_i^{\alpha_i}} \|f\|_p = \prod_{i=1}^d (\frac{2e}{\alpha_i p})^{\alpha_i/2} \|f\|_p .$$

This gives (i). (ii) is obtained by direct calculation and the use of Stirling's formula, and (iii) follows by combining (i) and (ii).

PROPOSITION 5. *$\mathcal{P}$ is dense in B_p, for $0 < p < \infty$.*

Proof. Suppose that $f \in B_p$ and that $\varepsilon > 0$. There exists $R > 0$ such that

$$\int_{|z|>R} |f(z)|^p \, d\varpi(z) < \varepsilon/2^{d+1}.$$

If $1/2 < r < 1$,

$$\begin{aligned}\int_{|z|>2R} |f(rz)|^p \, d\varpi(z) &= \frac{1}{\pi^d} \int_{|z|>2R} |f(rz)|^p e^{-|z|^2} \, dv(z) \\ &= \frac{1}{(r\pi)^d} \int_{|z|>2R/r} |f(z)|^p e^{-|z|^2/r^2} \, dv(z) < \varepsilon/2.\end{aligned}$$

Since $f(rz) \to f(z)$ uniformly on $|z| \leq 2R$ as $r \nearrow 1$, it follows that $d_r(f) \to f$ as $r \nearrow 1$. But $\sum_\alpha r^{|\alpha|}\psi_\alpha(f)\gamma_\alpha$ converges absolutely, by Lemma 1, and $d_r(f)(z) = \sum_\alpha r^{|\alpha|}\psi_\alpha(f) z^\alpha$. Thus, since the evaluation functionals are continuous, $d_r(f) = \sum_\alpha r^{|\alpha|}\psi_\alpha(f)\gamma_\alpha$, and $d_r(f)$ can be approximated in norm by elements of $\mathcal{P}$.

COROLLARY 5. $B_p \subseteq g_p = \{f \in G_p : f(z)e^{-z^2/p} \to 0 \text{ as } z \to \infty\}$.

Similarly, we have the following.

PROPOSITION 6. *d_r is a nuclear mapping of B_p into B_q for $0 < qr < p$.*

Proof. $\sum_\alpha r^{|\alpha|} \|\psi_\alpha\|_{B_p^*} \|\gamma_\alpha\|_q < \infty$, by Lemma 1, and $d_r(f) = \sum_\alpha r^{|\alpha|}\psi_\alpha(f)\gamma_\alpha$.

PROPOSITION 7. *If $1 < p < \infty$ and $d = 1$ then (γ_n) is a basis for B_p, so that* $f = \sum_{n=0}^{\infty} \psi_n(f)\gamma_n$.

Proof. By the continuity of the Riesz projection, there exists a constant C_p such that

$$\left\| \sum_{n=0}^{N} f_n r^n e^{in\theta} \right\|_{L_p(\mathcal{T})}^p \leq C_p^p \left\| f_{(r)} \right\|_{L_p(\mathcal{T})}^p,$$

for $f(z) = \sum_{n=0}^{\infty} \psi_n(f) z^n \in B_p$, $n \in \mathcal{Z}^+$ and $r > 0$. Multiplying by $2re^{-r^2}$ and integrating, we see that

$$\left\| \sum_{n=0}^{N} \psi_n(f)\gamma_n \right\|_p \leq C_p \|f\|_p .$$

Thus (γ_n) is a basic sequence. The result now follows from Proposition 5.

A similar argument shows, when $d > 1$, that

$$f = \lim_{n\to\infty} \sum_{|\alpha|_\infty \leq n} \psi_\alpha(f)\gamma_\alpha,$$

and that $\{\gamma_\alpha : \alpha \in (\mathcal{Z}^+)^d\}$, suitably ordered, is a basis for $B_p(\mathcal{C}^d)$.

Since $\int \gamma_\alpha \bar{\gamma}_\beta \, d\varpi = \alpha!\delta_{\alpha\beta}$, we have the following corollary (of which, *(i)* is of course well-known):

COROLLARY 6. *(i) The functions $\{\gamma_\alpha/(\alpha!)^{1/2} : \alpha \in (\mathcal{Z}^+)^d\}$ form an orthonormal basis for B_2.*

(ii) If $1 < p < \infty$ and $f(z) = \sum_\alpha \psi_\alpha(f) z^\alpha \in B_p$, $g = \sum_\alpha \psi_\alpha(g) z^\alpha \in B_{p'}$ then

$$\int f\bar{g}\, d\varpi = \lim_{n\to\infty} \sum_{|\alpha|_\infty \leq n} \frac{\psi_\alpha(f)\bar{\psi_\alpha}(g)}{\alpha!}.$$

PROBLEM 1. Is γ_n a basis for B_1?

Gröchenig and Walnut [4] assert, surely over-optimistically, that the Taylor series expansion of a function in B_1 converges in B_1.

5. *Projections onto B_p*

We now show that when $p \geq 1$ then B_p is complemented in L_p, by constructing a suitable projection.

Proposition 8. *Suppose that $p \geq 1$. Suppose that $k \in \mathcal{E}$ has the property that there exists C such that*

$$\|D_{2w}(k)\|_1 \leq Ce^{|w|^2}$$

for each $w \in C^d$. Let $k_p(w,z) = (2/p)^d k(2w.\bar{z}/p)$. If $f \in L_p$, let

$$\begin{aligned} K_p(f)(w) &= \int k_p(w,z) e^{(1/p'-1/p)|z|^2} f(z)\, d\varpi(z) \\ &= \frac{1}{\pi^d} \int k_p(w,z) e^{-2|z|^2/p} f(z)\, dv(z). \end{aligned}$$

Then K_p maps L_p continuously into B_p, with norm at most $2^d C$.

Proof. Suppose first that $1 < p < \infty$. We apply Schur's Lemma [**12, Proposition III.A.9 (page 87)**]. Let $g(z) = e^{|z|^2/p'p}$. Then

$$\begin{aligned} &\int |k_p(w,z)| e^{(1/p'-1/p)|z|^2} (g(z))^{p'}\, d\varpi(z) \\ &= \left(\frac{2}{p}\right)^d \int |k(2w.\bar{z}/p)| e^{|z|^2/p'}\, d\varpi(z) \\ &= \left(\frac{2}{p\pi}\right)^d \int |k(2w.\bar{z}/p)| e^{-|z|^2/p}\, dv(z) \\ &= \left(\frac{2}{\pi}\right)^d \int |k(2w.\bar{z}/\sqrt{p})| e^{-|z|^2}\, dv(z) \\ &= 2^d \left\|D_{2w/\sqrt{p}}(k)\right\|_1 \leq 2^d C e^{|w|^2/p} = 2^d C (g(w))^{p'}. \end{aligned}$$

Similarly,

$$\begin{aligned} &\int |k_p(w,z)| e^{(1/p'-1/p)|z|^2} (g(w))^p\, d\varpi(w) \\ &= \left(\frac{2}{p\pi}\right)^d e^{(1/p'-1/p)|z|^2} \int |k(2w.\bar{z}/p)| e^{-|w|^2/p}\, dv(w) \\ &= 2^d e^{(1/p'-1/p)|z|^2} \left\|D_{2w/\sqrt{p}}(k)\right\|_1 \\ &\leq 2^d C e^{|z|^2/p'} = 2^d C (g(z))^p. \end{aligned}$$

Secondly, suppose that $p = 1$, so that

$$K_1(f)(w) = 2^d \int k(2w.z) e^{-|z|^2} f(z)\, d\varpi(z).$$

Thus

$$\begin{aligned} \|K_1(f)\|_1 &\leq \int \left(2^d \int |k(2w.\bar{z})| e^{-|z|^2} |f(z)|\, d\varpi(z)\right) d\varpi(w) \\ &= 2^d \int e^{-|z|^2} |f(z)| \left(\int |k(2w.\bar{z})|\, d\varpi(w)\right) d\varpi(z) \end{aligned}$$

$$\leq \; 2^d C \int |f(z)| \, d\varpi(z) = 2^d C \, \|f\|_1 .$$

Thus K_p maps L_p into L_p, with norm at most $2^d C$. For each z, $k_p(w,z)$ is an entire function of w. Thus if f is bounded and of compact support, $K_p(f) \in B_p$. Since such functions are dense in L_p, the result follows.

Let us now see in more detail how the operator K_p acts.

PROPOSITION 9. *Let k be as in Proposition 8, and let $k(z) = \sum_\beta \psi_\beta z^\beta$. Then $K_p(\gamma_\alpha) = \psi_\alpha \gamma_\alpha$ and $K_p(exp) = k$.*

Proof. By Corollary 4, $k \in B_p$ for $0 < p < \infty$. As $k_p(w,z) = 2^d k(2w.\bar{z}/p)/p^d$,

$$\begin{aligned} K_p(\gamma_\alpha) &= \left(\frac{2}{p\pi}\right)^d \int \left(\sum_\beta \psi_\beta \frac{2^{|\beta|} w^\beta \bar{z}^\beta}{p^{|\beta|} \beta!} \right) z^\alpha e^{-2|z|^2/p} \, dv(z) \\ &= \sum_\beta \psi_\beta \left(\left(\frac{2}{p\pi}\right)^d \int \frac{2^{|\beta|} \bar{z}^\beta}{p^{|\beta|}\beta!} z^\alpha e^{-2|z|^2/p} \, dv(z) \right) w^\beta \\ &= \psi_\alpha w^\alpha . \end{aligned}$$

Thus $K_p(\gamma_\alpha) = \psi_\alpha \gamma_\alpha$.

It remains to show that $K_p(exp) = k$. When $p > 1$ this follows from Proposition 7 and the remarks following it. When $p = 1$, we argue as follows. Let $f_n(z) = exp_n(z/2)$ and $f(z) = exp(z/2)$. Then $f_n \to f$ in B_2, and so $f_n^2 \to exp$ in B_1. But if $K_1(f_n^2) = \sum_\alpha c_{n,\alpha} z^\alpha/\alpha!$, then $c_{n,\alpha} = \psi_\alpha$ for $n \geq |\alpha|_\infty$, so that $K_1(exp) = k$, by Proposition 3.

We now have the following special case [**5, Theorem 7.1**].

THEOREM 1. *Suppose that $p \geq 1$. If $f \in B_p$, let*

$$\begin{aligned} P_p(f)(w) &= \left(\frac{2}{p}\right)^d \int \lambda_p(z,w) e^{|z|^2/p'} f(z) \, d\varpi(z) \\ &= \left(\frac{2}{p}\right)^d \int e^{2\langle w,z\rangle/p} e^{(1/p'-1/p)|z|^2} f(z) \, d\varpi(z) \\ &= \left(\frac{2}{p\pi}\right)^d \int e^{2\langle w,z\rangle/p} e^{-2|z|^2/p} f(z) \, dv(z). \end{aligned}$$

Then P_p is a projection of L_p onto B_p of norm at most 2^d.

Proof. Let $k(z) = e^{z_1+\dots+z_d}$. Then $D_{2w}(k)(z) = e^{2(w_1 z_1 + \dots + w_d z_d)}$, so that $\|D_{2w}(k)\|_1 = e^{|z|^2}$, and we can apply Proposition 8: P_p maps L_p into B_p with norm at most 2^d. By Proposition 9, $P_p(\gamma_\alpha) = \gamma_\alpha$ for each α, and so P_p is a projection of L_p onto B_p, since the polynomials are dense in B_p (Proposition 5).

It is easy to check that P_2 is the orthogonal projection of L_2 onto B_2, so that $||P_2||=1$, and (by considering the characteristic function of a small ball near the origin) that $||P_1|| = 2^d$. Straightforward calculations, considering the images of the characteristic functions of small balls in $\mathcal{C}^d$, show that P_p maps L_q continuously into B_q only if $p = q$. (This is [**5, Theorem 9.1**].)

Corollary 7. *If $0 < p < \infty$, the linear span of $\{e_w : w \in \mathcal{C}^d\}$ is dense in B_p.*

When $0 < p < \infty$, this follows from the fact that B_1 is dense in B_p.

Proposition 10. *If $0 < p < \infty$ then B_p is linearly isomorphic to l_p.*

Proof. When $p \geq 1$, Proposition 3 and Theorem 1 provide the two necessary ingredients for the argument of [**12, Theorem III.A.11 (page 89)**] to apply.

When $0 < p < 1$, we use the result of Theorem 3.1 of [**11**] (not stated in the theorem, but stated explicitly in the proof) that B_p is isomorphic to an (infinite-dimensional) complemented subspace of l_p; such spaces are isomorphic to l_p ([**9**]; see also [**6**]).

Note that this implies that d_r is not an isomorphism of B_p onto B_{p/r^2}.

Galbis [**3**] has shown that g_p is isomorphic to c_0. The inclusion: $B_p \to g_p$ is therefore strict.

Gröchenig and Walnut [**4**] have given an explicit construction of an unconditional basis for B_p, in the case where $d = 1$.

Theorem 1 does not extend to the case $0 < p < 1$.

Proposition 11. *There are no non-zero continuous linear mappings from L_p to B_p, for $0 < p < 1$.*

Proof. The evaluation functionals $\{\delta_c : c \in \mathcal{C}^d\}$ are continuous on B_p and separate the points of B_p, while there are no non-zero continuous linear functionals on L_p.

6. *The isometries of B_p for $p \neq 2$*

We shall now show that the isometries described in Section 3 are the only isometries of B_p when $0 < p < \infty$, $p \neq 2$.

Theorem 2. *Suppose that $0 < p < \infty$ and that $p \neq 2$. If T is an isometry of B_p into itself, there exist $e^{i\theta} \in \mathcal{T}$ and $g \in Euc(\mathcal{C}^d)$ such that $T = e^{i\theta} V_g^p$.*

Corollary 8. *Every isometry of B_p into itself is surjective.*

REMARK 1. The isometries of $H_p(\mathcal{D})$ were determined by Forelli [**2**]. Of course, not every isometry of $H_p(\mathcal{D})$ is surjective: a simple example is given by $T(f)(z) = f(z^2)$.

Proof. We proceed in several steps.

LEMMA 2. *Suppose that $0 < p < \infty$ and that $p \neq 2$. If T is an isometry of B_p into itself, there exists a holomorphic mapping $\phi : \mathcal{C}^d \to \mathcal{C}^d$ such that*

$$T(f)(z) = g(z)f(\phi(z)),$$

where $g(z) = T(1)$.

Proof. Let $d\nu = |g|^p d\varpi$. If $f \in B_p$, let $S(f)(z) = T(f)(z)/g(z)$. Since $g \in B_p$, $S(f)$ is defined ϖ-almost everywhere, and S is an isometry of B_p into $L_p(\mathcal{C}^d, d\nu)$.

We now apply a fundamental theorem of Rudin. By [**8, Theorem II**], S is multiplicative on the polynomials $\mathcal{P}$. Let

$$\phi(z) = (\phi_1(z), \ldots, \phi_d(z)) = (S(c_1)(z), \ldots S(c_d)(z)).$$

Then if $h \in \mathcal{P}$, $S(h)(z) = h(\phi(z))$, and so $T(h)(z) = g(z)h(\phi(z))$. Since the polynomials are dense in B_p, and since the inclusion: $B_p \to \mathcal{E}$ is continuous, it follows that

$$T(f)(z) = g(z)f(\phi(z)),$$

for $f \in B_p$.

It remains to show that ϕ is holomorphic. By Hartogs' Theorem [**7, Theorem 1.2.5**], it is enough to show that each ϕ_i is holomorphic in each variable separately. This follows from the following elementary lemma, whose proof we leave to the reader.

LEMMA 3. *Suppose that g is a non-zero entire function on $\mathcal{C}$ and that f is a complex-valued function on $\mathcal{C}$ for which $g(z)f(z)^n$ is entire for all n. Then f is entire.*

PROPOSITION 12. *Suppose that T is a non-zero bounded linear mapping of B_p into itself of the form*

$$T(f)(z) = g(z)f(\phi(z)),$$

where g and ϕ are as in Lemma 2. Then $\phi(z) = Az + c$, where $A \in L(\mathcal{C}^d)$ and $\|A\| \leq 1$.

Proof. First observe that if $b \in \mathcal{C}^d$ then

$$(T^*(\delta_b))(f) = |g(b)f(\phi(b))| \leq \|T\| \, \|\delta_b\| \, \|f\| \leq \|T\| \, e^{|b|^2/p} \, \|f\| \, .$$

Applying this to $\lambda_{p,\phi(b)}$, we see that

$$|g(b)|e^{-|\phi(b)|^2/p}\left|e^{2\langle\phi(b),\phi(b)\rangle/p}\right| \leq \|T\|\, e^{|b|^2/p},$$

so that

$$|g(b)|e^{|\phi(b)|^2/p} \leq \|T\|\, e^{|b|^2/p}. \tag{6.1}$$

Now let z_0 be a fixed unit vector in C^d. let

$$\psi(\lambda) = g(\lambda z_0), \text{ and } \theta_i(\lambda) = g(\lambda z_0)e^{(\phi_i(\lambda z_0))^2/p}.$$

These are entire functions of order 2, which share the same zeros. Let $P(\lambda)$ be the canonical product defined by these zeros (see **[10, page 250]**). Then using Hadamard's Factorization Theorem **[10, page 250–251]**, there exist polynomials Q and R_i of degree at most 2 such that

$$\psi(\lambda) = P(\lambda)e^{Q(\lambda)} \text{ and } \theta_i(\lambda) = P(\lambda)e^{R_i(\lambda)}.$$

Thus

$$e^{(\phi_i(\lambda z_0))^2/p} = e^{R_i(\lambda)-Q(\lambda)},$$

so that $(\phi_i(\lambda z_0))^2$ is a polynomial of degree at most 2, and so $\phi_i(\lambda z_0)$ is an affine function of λ. Since this holds for all such z_0, $\phi_i(z)$ is an affine function of z. Thus $\phi(z) = Az + c$ for suitable A and c.

It remains to show that $\|A\| \leq 1$. Let $S = TV_c^p$. Then S is bounded and $S(f)(z) = k(z)f(Az)$. Let $U = \{z : |Az| > |z|\}$. By (6.1), if $z \in U$ then

$$|k(\lambda z)| \leq \|S\|\, e^{|\lambda|^2(|z|^2-|Az|^2)/p} \to 0$$

as $\lambda \to \infty$. Thus $k(z) = 0$, by Liouville's theorem. Since U is open, this implies that U must be empty: $|Az| \leq |z|$ for all z.

PROPOSITION 13. *Suppose that A is a 1-1 linear map on C^d with $\|A\| \leq 1$ and that g is entire. The operator T defined by $T(f)(z) = g(z)f(Az)$ is continuous on B_p if and only if there exists C such that*

$$|g(z)| \leq Ce^{(|z|^2-|Az|^2)/p}$$

for all $z \in C^d$.

Proof. The condition is necessary, by the inequality (6.1) of Proposition 12. If it is satisfied,

$$\begin{aligned}\|T(f)\|_p^p &= \int |g(z)|^p|f(Az)|^p\, d\varpi(z)\\ &\leq C\int e^{|z|^2}e^{-|Az|^2}|f(Az)|^p\, d\varpi(z)\\ &= C\pi^{-d}\int e^{-|Az|^2}|f(Az)|^p\, dv(z)\end{aligned}$$

$$= \quad C\,\|f\|_p^p\,/|\det A|.$$

We need one more lemma.

LEMMA 4. *Suppose that $\phi(z)$ is a measurable function on $\mathcal{C}^d$ and that $|\phi(z)| \leq Ce^{\beta|z|^2}$ for some $0 \leq \beta < 1$. Then if*

$$\int \phi(z)|e^{\langle z,w\rangle}|\,d\varpi(z) = 0$$

for all $w \in \mathcal{C}^d$, $\phi = 0$.

Proof. Writing everything in real co-ordinates, the condition becomes

$$\int_{\mathcal{R}^{2d}} \phi(x)e^{\langle x,a\rangle}e^{-|x|^2}\,dx = 0.$$

Because of the growth condition on ϕ, we can differentiate repeatedly with respect to a, to obtain

$$\int_{\mathcal{R}^{2d}} p(x)\phi(x)e^{-|x|^2}\,dx = 0,$$

for all polynomials p. We can write this as

$$\int_{\mathcal{R}^{2d}} p(x)\phi(x)e^{-\beta|x|^2}e^{-(1-\beta)|x|^2}\,dx = 0,$$

for all polynomials p. Since $\phi(x)e^{-\beta|x|^2} \in L^2(\mathcal{R}^{2d}, e^{-(1-\beta)|x|^2}dx)$, and since the polynomials are dense in $L^2(\mathcal{R}^{2d}, e^{-(1-\beta)|x|^2}dx)$, $\phi(x)e^{-\beta|x|^2} = 0$, and so $\phi = 0$.

Proof of Theorem 2. By Proposition 12, and the remarks in it, it is enough to consider the case where $T(f)(z) = g(z)f(Az)$. We need to show that A is unitary and that $|g(z)| = 1$.

First we observe that A is 1-1. For if $a \neq 0$ and $A^*a = 0$ then

$$T(l_a)(z) = g(z)\,\langle Az, a\rangle = 0$$

for all z, so that T is not 1-1. Applying Proposition 13, we conclude that

$$|g(z)|^p \leq Ce^{|z|^2-|Az|^2} \leq e^{\beta|z|^2}$$

for some $0 \leq \beta < 1$.

Now $\|T(\lambda_{p,w})\|_p = 1$, and so

$$\int |g(z)|^p|e^{2\langle z,A^*w\rangle}|\,d\varpi(z) = e^{|w|^2}.$$

But

$$\frac{1}{|\det A|}\int e^{|z|^2-|Az|^2}|e^{2\langle z,A^*w\rangle}|\,d\varpi(z) = e^{|w|^2},$$

so that, applying Lemma 4,

$$|g(z)|^p = \frac{1}{|\det A|} e^{|z|^2-|Az|^2}. \tag{6.2}$$

Suppose that $|Az_0| < |z_0|$. Then $|g(\lambda z_0)| \to \infty$ as $\lambda \to \infty$. This implies that the entire function $h(\lambda) = g(\lambda z_0)$ is a polynomial, and this in turn contradicts (6.2), since the left-hand side has polynomial growth, while the right-hand side grows exponentially. Consequently, A is unitary and $|g(z)| = 1$.

7. *Concluding remarks*

We conclude by listing some open problems suggested by our study.

PROBLEM 2. In Problem 1 (at the end of Section 4) we asked if (γ_n) is a basis for $B_1(\mathcal{C})$. Gröchenig and Walnut [**4**] have constructed an unconditional basis in $B_p(\mathcal{C})$, $1 \le p < \infty$ equivalent to the unit vector basis in l_p. Are there similar explicit constructions for $d > 1$ and for $0 < p < \infty$?

PROBLEM 3. After describing the isometries of a Banach space, it is natural to ask what the projections of norm 1 are. When $p \ge 1$, there are some natural norm 1 projections given by taking conditional expectations. For $0 \le s \le d$, let $P_s(f)(z_1, \ldots, z_d) =$

$$= \frac{1}{\pi^{d-s}} \int f(z_1, \ldots, z_s, w_{s+1}, \ldots, w_d) e^{-|w_{s+1}|^2 - \cdots - |w_d|^2} \, du_{s+1} dv_{s+1} \ldots du_d dv_d.$$

Then P_s is a norm one projection on B_p, as are the mappings IP_sI^{-1}, where I is an isometry of B_p.

If E is a one-dimensional subspace of B_p, where $1 \le p < \infty$, there is a norm 1 projection of B_p onto E. This is unique if $1 < p < \infty$, since B_p is uniformly smooth. It is also unique when $p = 1$, since the zeros of a non-zero element of B_p have measure 0.

We have not been able to find any other norm 1 projections. Are these all the norm 1 projections on B_p when $1 \le p < \infty$, $p \ne 2$? In particular, when $d = 1$ does every norm 1 projection other than the identity have rank 1?

Are there any non-trivial norm 1 projections when $0 < p < 1$?

PROBLEM 4. As we remarked after Theorem 1, P_p maps L_p into itself only if $p = q$. Does there exist a projection Q which is continuous from L_p onto B_p for more that one value of p?

PROBLEM 5. Writing $\alpha = 2/p$, we can write the projection P_p as

$$P_p(f)(\omega) = Q_\alpha(f)(\omega) = \left(\frac{\alpha}{\pi}\right)^d \int e^{\alpha\langle w,z\rangle} e^{(1-\alpha)|z|^2} f(z) \, d\varpi(z).$$

Algebraically, this is a projection for all $\alpha > 0$. What mapping properties does Q_α possess, for $\alpha > 2$?

PROBLEM 6. In Propositions 12 and 13 we considered operators of the form

$$T(f)(z) = g(z)f(\phi(z))$$

which are bounded on B_p. This class of operators is a very natural one - it forms a semigroup of operators - and we believe that it merits further study. It follows easily from our arguments that if such an operator is invertible, then it must be a scalar multiple of an isometry. What if T is an isomorphism *into*? To be precise, suppose that there exist $0 < c \leq C < \infty$ such that $T(f)(z) = g(z)f(Az)$ satisfies $c\,\|f\| \leq \|T(f)\| \leq C\,\|f\|$ for all $f \in B_p$. Is T a scalar multiple of an isometry?

If this is the case, it provides another proof that every isometry of B_p into itself is surjective.

It is a straightforward matter to check that if $0 < \mu < 1$ and $\lambda = (1 - \mu^2)/p$ then the operator defined by

$$T(f)(z) = e^{\lambda z^2} f(\mu z)$$

is continuous on B_p, but is neither compact nor invertible.

References

1. V. BARGMANN, 'On a Hilbert space of analytic functions and an associated integral transform', *Comm. Pure Appl. Math.* 14 (1961) 187–214.
2. F. FORELLI, 'The isometries of H^p' *Canad. J. Math.* 16 (1964) 721–728.
3. A. GALBIS, 'Weighted Banach spaces of entire functions', *Arch. Math.* 62 (1994) 58–64.
4. K. GRÖCHENIG and D. WALNUT, 'A Riesz basis for Bargmann-Foch space related to sampling and interpolation', *Archiv for matematik* 30 (1992) 283–295.
5. S. JANSEN, J. PEETRE and R. ROCHBERG, 'Hankel forms and the Fock space', *Revista Mat. Iberoamericana* 3 (1987) 61–138.
6. N. J. KALTON, N. T. PECK and J. W. ROBERTS, *An F-space sampler* (London Math. Soc. Lecture Notes Series 89, Cambridge University Press, 1984).
7. S. G. KRANTZ, *Function theory of several complex variables* (Wadsworth and Brooks/-Cole, Second Edition, 1992).
8. W. RUDIN, 'L_p isometries and equimeasurability', *Indiana University Math. J.* 25 (1976) 215–228.
9. W. J. STILES, 'Some properties of $l_p, 0 < p < 1$', *Studia Math.* 42 (1972) 109–119.
10. E. C. TITCHMARSH, *The theory of functions* (Oxford University Press, Second Edition, 1939).
11. R. WALLSTÉN, 'The S^p-criterion for Hankel forms on the Fock space, $0 < p < 1$', *Math. Scand.* 64 (1989) 123–132.
12. P. WOJTASZCZYK, *Banach spaces for analysts* (Cambridge University Press, 1991).

Characters of Function Algebras on Banach Spaces

M.I. GARRIDO Departamento de Matemáticas, Universidad de Extremadura, 06071 Badajoz, Spain.

J. GÓMEZ GIL[1] Departamento de Análisis Matemático, Facultad de Matemáticas, Universidad Complutense, 28040 Madrid, Spain.

J.A. JARAMILLO[1] Departamento de Análisis Matemático, Facultad de Matemáticas, Universidad Complutense, 28040 Madrid, Spain.

Let A be an algebra of continuous real functions on a real Banach space E. We consider the problem of whether every character $\varphi : A \to \mathbb{R}$ is given by evaluation at some point of E. We are especially interested in the cases where A is the algebra of rational functions, or real-analytic functions, or smooth functions on E. In this note, we briefly survey some recent results on this topic, and we also provide a new, natural proof of one of these results, in a special case.

Let $C(E)$ denote the algebra of all continuous real functions on E. If A is a subalgebra of $C(E)$, we denote by Hom$A$ the set of all characters (nonzero multiplicative linear functionals) on A. We shall write Hom$A$=E when every $\varphi \in$ Hom$A$ is given by evaluation at some point $a \in E$, that is, $\varphi(f) = f(a)$ for every $f \in A$.

[1] Partially supported by DGICYT PB 90-0044

Now let $\mathcal{P}(E)$ denote the algebra of all continuous polynomials on E and, for $j = 0, 1, 2, \ldots$, let $\mathcal{P}(^jE)$ denote the space of all continuous j-homogeneous polynomials on E. Each $P_j \in \mathcal{P}(^jE)$ is a function of the form $P_j(x) = T_j(x, \ldots, x)$, where T_j is a continuous j-linear functional on $E \times \cdots \times E$ (for $j = 0$, P_0 is constant), and each $P \in \mathcal{P}(E)$ is a finite sum $P = P_0 + P_1 + \ldots + P_m$, where $P_j \in \mathcal{P}(^jE)$ for $j = 0, 1, 2, \ldots m$. We denote by $\mathcal{R}(E)$ the algebra of all rational functions on E, that is, the functions of the form P/Q, where $P, Q \in \mathcal{P}(E)$ and $Q(x) \neq 0$ for every $x \in E$. Recall that a real function f on E is said to be real-analytic if, for every $x \in E$ there exist a neighbourhood W of 0 and a sequence (P_j) with each $P_j \in \mathcal{P}(^jE)$, such that $f(x + h) = \sum_{j=0}^{\infty} P_j(h)$, for every $h \in W$. We denote by $\mathcal{A}(E)$ (respectively, $C^\infty(E)$), the algebra of all real-analytic functions (respectively, infinitely Fréchet differentiable functions) on E. Note that $\mathcal{R}(E) \subset \mathcal{A}(E) \subset C^\infty(E)$.

When E is finite-dimensional then $\mathrm{Hom}\mathcal{R}(E) = \mathrm{Hom}\mathcal{A}(E) = \mathrm{Hom}C^\infty(E) = E$ and, as we shall see, the same holds for a wide class of Banach spaces. Nevertheless this is not true in general, if we assume the existence of measurable cardinals. Here, we say that a set X has measurable cardinal if there exists a nontrivial, countably additive, two-valued measure defined on the power set of X. So, if we assume that the index set I has measurable cardinal, then $E = \ell_2(I)$ is not realcompact ([GJ], [B]), and it is not difficult to show that there exist characters on $\mathcal{R}(E)$, on $\mathcal{A}(E)$ and on $C^\infty(E)$ that are not given by evaluation at any point of E [GGJ].

We recall that the existence of measurable cardinals is not provable with the usual axioms of set theory (ZFC). On the other hand, it is consistent with ZFC that there are no measurable cardinals. Also note that these cardinals, whether they exist at all, must be extremely big: in fact, bigger than $\aleph_0, 2^{\aleph_0}, 2^{2^{\aleph_0}}, \ldots$ (see [Je], [GJ]).

Next we give the results concerning the algebra $C^\infty(E)$. The algebras $\mathcal{R}(E)$ and $\mathcal{A}(E)$ will be considered later.

1.Theorem. [KMS, J] Suppose that the Banach space E admits C^∞-partitions of unity and has non-measurable cardinal. Then $\mathrm{Hom}C^\infty(E)$=E.

The existence of smooth partitions of unity is a quite restrictive condition for a Banach space. Nevertheless, it is fulfilled by some remarkable spaces, such as $\ell_2(I)$ and $c_0(I)$ (see

[T]), and this fact will be useful in the sequel.

In order to extend Theorem 1 to more general spaces, we shall use the technique developed in [BL], [GGJ] and [J], where the existence of an injection from E into F allows us to transfer the information from subalgebras of $C(F)$ to subalgebras of $C(E)$. First we need to introduce the following concept.

We say that a subalgebra A of $C(E)$ is sequentially evaluating if, for every $\varphi \in \mathrm{Hom}A$ and every sequence (f_n) in A, there exists a point $a \in E$ such that $\varphi(f_n) = f_n(a)$ for all n.

2.Theorem. [BBL, KMS] For every Banach space E, the algebra $C^\infty(E)$ is sequentially evaluating.

3.Theorem. [B, BL] Let E and F be Banach spaces and let $A_E \subset C(E)$ and $A_F \subset C(F)$ be subalgebras. Suppose that:

(1) $\mathrm{Hom}A_F{=}F$.

(2) There exists a one-to-one map $T : E \to F$ such that $f \circ T \in A_E$ for every $f \in A_F$.

(3) For each $b \in F$ there exists a sequence (f_n) in A_F separating b from all other points in F.

(4) A_E is sequentially evaluating.

Then, $\mathrm{Hom}A_E{=}E$.

Finally, combining the preceding results, we arrive at the main theorem in this part.

4.Theorem. [B, BL] Let E be a Banach space such that there exists a one-to-one, continuous linear map $T : E \to c_0(I)$, for some index set I with non-measurable cardinal. Then $\mathrm{Hom}C^\infty(E){=}E$.

A wide class of spaces satisfy the conditions of the above theorem, which extends results of [AdR], [GL] and [J].

5.Examples. [B, BL] The hypotheses of Theorem 4 are fulfilled in the following cases:

(1) When E is separable or E is a subspace of the dual of a separable space.

(2) When E is a subspace of a WCD-space (in particular, when E is a reflexive space) with non-measurable cardinal.

(3) When E is a subspace of $C(K)$, where K is a Corson-compact with non-measurable cardinal.

We now turn our attention to the algebras $\mathcal{R}(E)$ and $\mathcal{A}(E)$, which will receive a common treatment. We start with a result for some specific spaces.

6.Theorem. [GGJ] Suppose that the index set I has non-measurable cardinal, and let $1 < p < \infty$. Then Hom $\mathcal{R}(\ell_p(I))$=Hom $\mathcal{A}(\ell_p(I))$=$\ell_p(I)$.

The relevance of the preceding theorem comes from the fact that it can be combined with Theorem 7 below in order to give a more general result.

7.Theorem. [J, GGJ] Let E and F be Banach spaces and let $A_E \subset C(E)$ and $A_F \subset C(F)$ be subalgebras. Suppose that

(1) HomA_F=F.

(2) There exists a one-to one map $T : E \to F$ such that $f \circ T \in A_E$ for every $f \in A_F$.

(3) For each $b \in F$ there exists $f_b \in A_F$ such that $f_b^{-1}(0) = \{b\}$.

Then, HomA_E=E.

Next we give the main result of this part.

8.Theorem. [GGJ] Let E be a real Banach space such that there exists a one-to-one, continuous linear map $T : E \to \ell_p(I)$, for some $1 < p < \infty$ and some index set I with non-measurable cardinal. Then Hom$\mathcal{R}(E)$= Hom$\mathcal{A}(E)$=E.

We know of no characterization of Banach spaces for which there exists an injection into $\ell_p(I)$ for some $1 < p < \infty$. Nevertheless, there are many spaces satisfying this condition.

9.Examples. The hypotheses of Theorem 8 are satisfied in the following cases:

(1) When E is separable or E is a subspace of the dual of a separable space.

(2) When E is super-reflexive with non-measurable cardinal [JTZ].

It would be interesting to know whether Theorem 8 holds if we merely suppose the existence of an injection $T : E \to c_0(I)$, thus obtaining a unified result for the algebras $\mathcal{R}(E)$, $\mathcal{A}(E)$ and $C^\infty(E)$.

We finish this note with an easy and natural proof of Theorem 6 in the special case that the index set I has non real-valued measurable cardinal. Here we say that a set X has real-valued measurable cardinal if there exists a nontrivial, countably additive, real-valued measure defined on the power set of X. Real-valued measurable cardinals were introduced by Banach in [Ba]. It is clear that every measurable cardinal is real-valued measurable, and it follows from the results of Ulam [U] that, under the Continuum Hypothesis, both classes coincide.

10. Proof of Theorem 6 in the case that I has non real-valued measurable cardinal: Let $E = \ell_p(I)$ and let $A = \mathcal{R}(\ell_p(I))$ or $A = \mathcal{A}(\ell_p(I))$. Let $\varphi : A \to \mathbb{R}$ be a nonzero algebra homomorphism. Consider the coordinate functionals on E, given by

$$\pi_i((x_j)_{j\in I}) = x_i,$$

for each $i \in I$, and let $a_i = \varphi(\pi_i)$. We shall see that $a = (a_i)_{i\in I} \in E$ and φ is given by evaluation at a.

Choose $m \in \mathbb{N}$ such that $2m \geq p$. Since A is inverse closed (that is, $1/f$ belongs to A if $f \in A$ and $f > 0$), it is not difficult to see that φ is monotone. Then for each finite subset $F \subset I$,

$$\sum_{i\in F} a_i^{2m} = \varphi\left(\sum_{i\in F} \pi_i^{2m}\right) \leq \varphi\left(\sum_{i\in I} \pi_i^{2m}\right) < +\infty,$$

and therefore $a \in \ell_{2m}(I)$. Now we can define a real-valued measure on the power set of I as follows. For each $J \subset I$, consider the polynomial $P_J = \sum_{i\in J}(\pi_i - a_i)^{2m}$ on E, and define

$$\mu(J) = \varphi(P_J).$$

Here, we understand that $\mu(\varnothing) = 0$. In order to prove that μ is countably additive, fix a sequence $(J_n)_{n\in\mathbb{N}}$ of pairwise disjoint subsets of I, and let $J = \bigcup_{n\in\mathbb{N}} J_n$.

Claim. There exists some $b \in E$ such that $\varphi(P_J) = P_J(b)$ and $\varphi(P_{J_n}) = P_{J_n}(b)$ for all $n \in \mathbb{N}$.

Indeed, consider the polynomials $R = (P_J - \varphi(P_J))^2 + \sum_{n=1}^{\infty} \frac{1}{n2^n} (P_{J_n} - \varphi(P_{J_n}))^2$ and $Q = \sum_{n=1}^{\infty} \frac{1}{2^n} (P_{J_n} - \varphi(P_{J_n}))^2$ on E.

For each $N \in \mathbb{N}$, we have that

$$\varphi(R) = \varphi\left(\sum_{n>N} \frac{1}{n2^n}\left(P_{J_n} - \varphi(P_{J_n})\right)^2\right) \leq$$
$$\leq \varphi\left(\sum_{n>N} \frac{1}{N2^n}\left(P_{J_n} - \varphi(P_{J_n})\right)^2\right) =$$
$$= \frac{1}{N}\varphi\left(\sum_{n>N} \frac{1}{2^n}\left(P_{J_n} - \varphi(P_{J_n})\right)^2\right) = \frac{1}{N}\varphi(Q),$$

and therefore $\varphi(R) = 0$. This establishes the claim, since if $R(x) \neq 0$ for every $x \in E$, then $\frac{1}{R} \in A$, and this leads to a contradiction.

As a consequence,

$$\mu\left(\bigcup_{n\in\mathbb{N}} J_n\right) = \varphi(P_J) = P_J(b) = \sum_{n=1}^{\infty}\sum_{i\in J_n}(b_i - a_i)^{2m} =$$
$$= \sum_{n=1}^{\infty} P_{J_n}(b) = \sum_{n=1}^{\infty}\varphi(P_{J_n}) = \sum_{n=1}^{\infty}\mu(J_n).$$

Since $\mu(\{i\}) = 0$ for each $i \in I$, and I has non real-valued measurable cardinal, we obtain that $\mu(I) = \varphi(P_I) = 0$.

Finally, given $f \in A$, we have as before that there exists some $c = (c_i)_{i\in I} \in E$ such that $\varphi(f) = f(c)$ and $\varphi(P_I) = P_I(c) = 0$. Then $c_i = a_i$ for every $i \in I$. Hence $a = c \in E$ and φ is given by evaluation at a.

REFERENCES

[AdR] Arias de Reyna, J. *A real valued homomorphism on algebras of differentiable functions.* Proc. Amer. Math. Soc. 104 (1988), 197-229.

[Ba] Banach, S. *Über additive Massfunctionen in abstrakten Mengen.* Fund. Math. 15 (1930), 97-101.

[B] Biström, P. *The Homomorphisms on Algebras of Real Valued Functions Defined on Locally Convex Spaces and Bounding Sets.* Thesis. Acta Academiae Aboensis, Ser. B 53, 1 (1993).

[BBL] Biström, P., Bjon, S. and Lidström, M. *Function algebras on which homomorphisms are point evaluations on sequences.* Manuscripta Math. 73 (1991), 179-185.

[BL] Biström, P. and Lindström. M. *Homomorphisms on $C^\infty(E)$ and C^∞-bounding sets.* Monatsh. Math. 115 (1993), 257-266.

[GGJ] Garrido, M. I., Gómez, J. and Jaramillo, J. A. *Homomorphisms on function algebras.* Can. J. Math. 46 (1994), 734-745 (see also Extracta Math. 7 (1992), 46-52).

[GJ] Gillman, L. and Jerison, M. *Rings of continuous functions.* Springer, 1960.

[GL] Gómez, J. and Llavona, J. G. *Multiplicative functionals on function algebras.* Rev. Mat. Univ. Complut. Madrid, 1 (1988), 19-22.

[J] Jaramillo, J. A. *Multiplicative functionals on algebras of differentiable functions.* Archiv Math. 58 (1992), 384-387.

[Je] Jech, T. *Set Theory.* Academic Press, 1978.

[JTZ] John, K., Toruńczyk, H. and Zizler, V. *Uniformly smooth partitions of unity on super-reflexive Banach spaces.* Studia Math. 70 (1981), 129-137.

[KMS] Kriegl, A., Michor, P. and Schachermayer, W. *Characters on algebras of smooth functions.* Ann. Global Anal. Geom. 7 (1989), 85-92.

[T] Toruńczyc, H. *Smooth partitions of unity on some non-separable Banach spaces.* Studia Math. 46 (1973), 43-51.

[U] Ulam, S. *Zur Masstheorie in der allgemeinen Mengelehre.* Fund. Math. 16 (1930), 140-150.

On the Corona Theorem

PAMELA GORKIN Department of Mathematics, Bucknell University, Lewisburg, PA 17837, PGORKIN@BUCKNELL.EDU

RAYMOND MORTINI Mathematisches Institut I, Universität Karlsruhe D-76128 Karlsruhe Germany, AB05@DKAUNI2.Bitnet

0. Introduction

Let $H^\infty = H^\infty(\mathbb{D})$ be the Banach algebra of bounded analytic functions in the open disk $\mathbb{D} = \{z \in \mathbb{C} : |z| < 1\}$. Its maximal ideal space is denoted by $M(H^\infty)$. We may think of $M(H^\infty)$ as the set of maximal ideals of H^∞, or we may identify this space with the set of nonzero multiplicative linear functionals on $M(H^\infty)$. When we do this identification, $M(H^\infty)$ is given the weak star topology; that is, a net $\varphi_\alpha \to \varphi$ if and only if $\varphi_\alpha(f) \to \varphi(f)$ for all $f \in H^\infty$. Identifying a point z_0 of the disc with the linear functional that is evaluation at the point z_0 means that the disc may be homeomorphically embedded in $M(H^\infty)$. The Corona Theorem answers the question of how the disc fits inside the maximal ideal space. It says that $\mathbb{D}$ is dense in $M(H^\infty)$. In the first section of this survey paper, we provide some history of the Corona Problem, some insight (we hope) into what is happening in a generalized form of the Corona Theorem, and an open question. In the second section, we look at Wolff's method for solving certain $\overline{\partial}$ equations. We also give a new proof of a lemma on Carleson measures induced by logarithmic derivatives of interpolating Blaschke products that is essential in solving the generalized corona theorem.

1. Algebraic Techniques

We now return to the Corona Theorem. An equivalent algebraic formulation of this theorem is the following:

Theorem 1.1 (Corona Theorem) [3]. *Let $f_1, \ldots, f_N \in H^\infty$ and suppose that there exists $\delta > 0$ such that*

$$|f_1(z)| + |f_2(z)| + \cdots + |f_N(z)| \geq \delta$$

on $\mathbb{D}$, then there are H^∞ functions $g_1, g_2, \ldots, g_N$ such that

$$1 = \sum_{j=1}^{N} f_j\, g_j$$

Carleson [3] proved this theorem in 1962. However, a simpler proof existed for a certain special case which we now describe. Recall that a Blaschke product with zeros $\{z_n\}$ is a function of the form

$$B(z) = z^m \prod_{z_n \neq 0} \frac{-\overline{z}_n}{|z_n|} \frac{z - z_n}{1 - \overline{z}_n z}, \qquad \sum_{n=1}^{\infty} 1 - |z_n| < \infty.$$

We say that a sequence $\{z_n\}$ is an interpolating sequence if whenever (w_n) is a bounded sequence of complex numbers, there exists a function f in H^∞ such that $f(z_k) = w_k$ for every k. If B is a Blaschke product and the zeros of B form an interpolating sequence, then B is an interpolating Blaschke product. One easy way to check that B is an interpolating Blaschke product comes from a description of interpolating sequences for H^∞. This description says that B is an interpolating Blaschke product with zero sequence $\{z_n\}$ if and only if there exists $\delta > 0$ such that

$$\inf_n (1 - |z_n|^2)|B'(z_n)| \geq \delta. \tag{1.1}$$

If some f_j in the statement of the Corona Theorem

$$|f_1| + |f_2| + \cdots + |f_N| \geq \delta$$

on $\mathbb{D}$ is an interpolating Blaschke product B, say $f_N = B$, then the conclusion trivially holds. In fact, one has just to solve the interpolation problem

$$g_j(z_n) = \frac{\overline{f_j(z_n)}}{\sum_{k=1}^{N} |f_k(z_n)|^2} \qquad (n \in \mathbb{N},\ j = 1, \ldots, N-1).$$

In 1980, Wolff gave a very accessible proof of Carleson's theorem. His methods are quite different from Carleson's original proof. They also allowed him to prove a generalized form of the Corona Theorem:

Theorem 1.2 (Generalized Corona Theorem) [6]. *Let $f_1, f_2, \ldots, f_N \in H^\infty$ and suppose $g \in H^\infty$ satisfies*

$$|g| \leq |f_1| + |f_2| + \cdots + |f_N|$$

on the unit disc, then there are functions $g_1, g_2, \ldots, g_N \in H^\infty$ such that

$$g^3 = \sum_{j=1}^{N} f_j\, g_j$$

An obvious question to ask is the following: can we replace g^3 above by g? If the answer is no, can we replace g^3 by g^2?

It turns out that the answer to the first question above is no, and a counterexample was provided by Rao in 1967 [20]. His example is so easy and had such a great influence upon what happened in the solution of ideal problems related to the Corona Theorem that we review it briefly here. To do so, we need very little background information. We only use a well- known factorization property of H^∞ functions.

Theorem 1.3 (Riesz's Factorization Theorem). *If B is a Blaschke product with zeros $\{z_n\}$ and $f \in H^\infty$ satisfies $f(z_n) = 0$ for all n, then there exists $h \in H^\infty$ such that $f = Bh$.*

Now we are ready to look at Rao's example.

Example 1.4 (Rao [20]). There exist H^∞ functions g, f_1, f_2 such that $|g| \leq |f_1| + |f_2|$ on $\mathbb{D}$ but g is not in the ideal generated by f_1 and f_2.

To find such functions, let B_1 and B_2 be Blaschke products with distinct zeros such that $\inf_{z\in D}(|B_1(z)| + |B_2(z)|) = 0$. This is quite easy to do. For example, take two Blaschke products with distinct zeros but such that the zeros tend pseudohyperbolically to zero. Schwarz's lemma implies that $\inf_{z\in D}(|B_1(z)| + |B_2(z)|) = 0$.

Now take $g = B_1B_2, f_1 = B_1^2, f_2 = B_2^2$. Then

$$|B_1B_2| \le \max(|B_1|^2, |B_2|^2) \le |B_1|^2 + |B_2|^2,$$

So the hypotheses above are clearly satisfied. Suppose

$$B_1B_2 = g_1B_1^2 + g_2B_2^2.$$

Then $B_2(z) = 0$ clearly implies that $g_1B_1^2 = 0$. Thus, since the zeros of B_1 and B_2 are distinct, $g_1(z) = 0$. Since g_1 vanishes on the zeros of B_2, we see that $g_1 = B_2h_2$ for some $h_2 \in H^\infty$. Similarly, $g_2 = B_1h_1$ for some $h_1 \in H^\infty$. Using this information, we obtain

$$B_1B_2 = B_1^2B_2h_2 + B_2^2B_1h_1.$$

Canceling, $1 = B_1h_2 + B_2h_1$. But we arranged things so that the right hand side of this equation tends to zero, while the left hand side cannot. This contradiction completes the example.

Rao's example turns out to be, in some sense, the only way things can go wrong. To discuss this, we introduce some notation and some more background on the problem. Let $I = I(f_1, \ldots, f_n) = \{\sum_{j=1}^n g_jf_j : g_j \in H^\infty, j = 1, \ldots, n\}$ and $J = J(f_1, \ldots, f_n) = \{f \in H^\infty :$ there exists M with $|f| \le M(|f_1| + |f_2| + \cdots + |f_n|)$ on $\mathbb{D}\}$.

Clearly, $I \subseteq J$ and, Rao's example shows that, in general, $I \ne J$. With this notation, the Corona Theorem tells us that if $1 \in J$, then $1 \in I$ (and $I = J = H^\infty$), and the Generalized Corona Theorem says that if $f \in J$ then $f^3 \in I$. The first question we will look at is the following:

Question 1.5. *When is $I = J$?*

In 1975, Von Renteln [21] showed that if I contains an interpolating Blaschke product, then $I = J$. In 1984, Tolokonnikov [22] showed that if J contains an interpolating Blaschke product, then $I = J$. Since $I \subseteq J$, Tolokonnikov's result is an improvement of von Renteln's. However, his techniques are deep and difficult. Why does $I = J$ when J contain an interpolating Blaschke product, but $I \subset J$ in Rao's example? The explanation we offer here relies on a generalized notion of the order of a zero of an analytic function.

Recall that for $x, y \in M(H^\infty)$ the pseudohyperbolic distance from x to y is

$$\rho(x,y) = \sup\{|f(y)| : f \in H^\infty,\ \|f\| \leq 1,\ \hat{f}(x) = 0\},$$

where $\hat{f}$ denotes the Gelfand transform of f.

Clearly, the pseudohyperbolic distance between any two points in $M(H^\infty)$ is less than or equal to one. We can define an equivalence relation by saying that $x \sim y$ if and only if $\rho(x,y) < 1$. In this case, the equivalence classes are called Gleason parts. Thus the Gleason part of x is given by $P(x) = \{y : \rho(x,y) < 1\}$.

Kenneth Hoffman [11] studied the Gleason parts of points in $M(H^\infty)$ extensively, and he showed that for $x \in M(H^\infty)$ there exists a map $L_x : \mathbb{D} \to P(x)$ such that $L_x(0) = x$ and $\hat{f} \circ L_x \in H^\infty$ for all $f \in H^\infty$ (where $(\hat{f} \circ L_x)(z) = \widehat{L_x(z)}(f))$. Thus if f is a function in H^∞ and $\hat{f}$ vanishes at x, then $\hat{f} \circ L_x$ is a bounded analytic function vanishing at the origin. In this context, it makes sense to talk about the order of the zero of $\hat{f}$ at x, which we denote by ord (f, x). As usual, the Gelfand transform $\hat{f}$ of f will be identified with f.

If α is a point in the disc, define L_α by

$$L_\alpha(z) = \frac{z + \alpha}{1 + \overline{\alpha} z}.$$

Hoffman showed that for any $f \in H^\infty$, if $\{\alpha_n\}$ is a net in $\mathbb{D}$ converging to a point x of $M(H^\infty)$, then

$$f \circ L_{\alpha_n} \to f \circ L_x. \tag{1.2}$$

Replacing f above by an interpolating Blaschke product B, and letting α_n denote a subnet of the zero sequence of B we see that

$$(B \circ L_{\alpha_n})'(0) \to (B \circ L_x)'(0).$$

But $(B \circ L_{\alpha_n})'(0) = (1 - |\alpha_n|^2)B'(\alpha_n)$ Combining this with the fact that B is interpolating (so that (1.1) holds) we see that there exists $\delta > 0$ such that $|(B \circ L_{\alpha_n})'(0)| > \delta$ for all n. In particular we have:

Proposition 1.6 [11]. *Let B be an interpolating Blaschke product and let $x \in M(H^\infty)$ be in the closure of the zeros of B. Then $B(x) = 0$ and there exists $\delta > 0$ such that $|(B \circ L_x)'(0)| > \delta$.*

Hence, Hoffman's work [11] implies that interpolating Blaschke products have zeros of order **at most** one. The generators in Rao's example have zeros of order **at least two**. In fact, we have the following theorem.

Theorem 1.7 [9]. *Suppose $f_1, f_2 \in H^\infty$ have no common factors. Let $I = I(f_1, f_2)$ and $J = J(f_1, f_2)$. Then $I = J$ if and only if J contains an interpolating Blaschke product.*

The proof of this theorem is joint work with Artur Nicolau and will appear in Math. Ann., and we will not repeat it in detail here. One direction is Tolokonnikov's result and simpler proofs now exist (see [14], [17]). We now indicate here a sketch of the proof in the remaining direction and as we go along we will point out the way in which Rao's example influences it. The proof has four main steps, which we describe below. Before proceeding to the proof, we define, for any $f \in H^\infty$ the set $Z(f) = \{x \in M(H^\infty) : f(x) = 0\}$. For any ideal I, the zero set or hull of I is given by

$$Z(I) = \bigcap_{f \in I} Z(f).$$

and the order of I at a zero x by

$$\text{ord}\,(I, x) = \min\{\text{ord}\,(f, x) : f \in I\}.$$

Now we proceed to the idea of the proof:

Step 1: We show that $I(h_1k_1, h_2k_2) = J(h_1k_1, h_2k_2)$ implies $I(h_1, h_2) = J(h_1, h_2)$.

Step 2: Next, we show that the ideal *cannot* have zeros of finite order ≥ 2. So suppose that $2 \leq \text{ord}\,(I, x) < \infty$. We apply some well known factorization theorems due to Hoffman [11] to factor our generators as $f_j = b_j c_j g_j$ for $j = 1, 2$ where b_j, c_j are interpolating Blaschke products vanishing at x. Then

$$I(b_1c_1g_1, b_2c_2g_2) = J(b_1c_1g_1, b_2c_2g_2)$$

and, by step 1, we have

$$I(b_1c_1, b_2c_2) = J(b_1c_1, b_2c_2).$$

Thus, we may assume that each of our generators are the products of two interpolating Blaschke products, both of which vanish at x. (Note that if $b_j = c_j$ for $j = 1, 2$ we would have precisely the same situation that occurred in Rao's example and we could easily obtain the desired contradiction.)

The object is to try to return to something close to Rao's example. We do this in step 3.

Step 3: Wiggle the zeros of the interpolating Blaschke products by refactoring them as a product of two interpolating Blaschke products vanishing at x and satisfying

$$BB^* = b_1c_1\,,\ CC^* = b_2c_2\,,\ \frac{B}{B^*} \text{ bounded on the zeros of } C^* \text{ in } \mathbb{D},$$

$$\text{and } \frac{C}{C^*} \text{ bounded on the zeros of } B^* \text{ in } \mathbb{D}.$$

(In truth, we need to be a bit more careful than this, and the factorization actually represents subfactors of the interpolating Blaschke products obtained in step 2. All this is described carefully in the proof of the theorem [9].)

Let $\{z_n\}$ denote the zero sequence of C^*. Now $\dfrac{B(z_n)}{B^*(z_n)}$ is a bounded sequence of complex numbers. Since $\{z_n\}$ is an interpolating sequence, there exists $f \in H^\infty$ such that $f(z_n) = \dfrac{B(z_n)}{B^*(z_n)}$ on the zeros of C^*. Therefore the factorization theorem (Theorem 1.3) implies that $B^*f - B = C^*h$, for some $h \in H^\infty$. So, $B = B^*f + C^*h$ for some $f, h \in H^\infty$. We may repeat the process above, replacing B by C to obtain

$$I(BB^*, CC^*) = J(BB^*, CC^*)$$

and

$$(1) \quad B = B^*f_1 + C^*h_1 \quad ,(2) \quad C = C^*f_2 + B^*h_2.$$

Now let $z \in D$. If $|C(z)| \leq |B(z)|$, using equations (1) and (2) above we find that there exists a constant M such that

$$|B(z)C(z)| \leq M(|B^*(z)C(z)| + |C(z)C^*(z)|) \leq M(|B^*(z)B(z)| + |C(z)C^*(z)|).$$

If $|B(z)| \leq |C(z)|$, everything is symmetric in B and C, so we may switch their roles above and we obtain, for all $z \in \mathbb{D}$

$$|BC| \leq M(|B^*B| + |CC^*|).$$

Thus $BC \in J(B^*B, CC^*)$. Since we are assuming that $I = J$, we see that $BC = B^*Bh + C^*Ck$. We may now proceed as in Rao's example to show that k must vanish on the zeros of B and h must vanish on the zeros of C. Hence we can cancel BC from both sides to obtain $1 = B^*h' + C^*k'$ for some $h', k' \in H^\infty$. But we arranged things so that the right hand side vanishes at

x while the left hand side does not. This completes step 3 and we know that I cannot have a zero of finite order unless it is a zero of order one.

Step 4: We now need to show that I cannot have zeros of infinite order. In this step, we prove a factorization theorem for Blaschke products having a zero of infinite order somewhere. This factorization is based on Izuchi's refinement [12] of Hoffman's work [11]. We then wiggle our zeros again to reduce to the case above.

Step 5: Thus we know that any zero of I is a zero of order one. We also know that interpolating Blaschke products have zeros of order one only. We now need a theorem that combines these two facts in some useful way and here it is:

Theorem 1.8 (Mortini [17]). *If an ideal I has only zeros of order one, then I contains an interpolating Blaschke product.*

Mortini's theorem is exactly what we need to conclude that I (or J, since they are equal in this case) contains an interpolating Blaschke product. ∎

Our theorem really depends on the fact that there are only two generators. Thus one should consider the following question: Is it true that $I(f_1, \ldots, f_n) = J(f_1, \ldots, f_n)$ if and only if J contains an interpolating Blaschke product? Unfortunately, a counterexample exists. Again, the motivation for this comes from Rao's example.

Example 1.9. *Let B and C be two interpolating Blaschke products with no common zeros such that*

$$\inf_{z \in D}(|B(z)|, |C(z)|) = 0.$$

Then $I(\underbrace{B^2}_{f_1}, \underbrace{C^2}_{f_2}, \underbrace{BC}_{f_3}) = J(\underbrace{B^2}_{f_1}, \underbrace{C^2}_{f_2}, \underbrace{BC}_{f_3})$.

So $I = J$ but since functions in either ideal must have zeros of order two, neither contains an interpolating Blaschke product. The justification for this result is easy and fun to work through! Complete justification for this is provided in [9].

Mortini has shown that more is true:

Theorem. 1.10 [18]. *Let b_1, b_2 be interpolating Blaschke products having no common zeros in D. Then*

$$J(b_1^n, b_2^n) = I(b_1^n, b_1^{n-1} b_2, \ldots, b_2^n)$$

It also known that, whenever f and g have no common factor, then $I(f,g) = J(f,g)$ if and only if $(1-|z|^2)(|f'|+|g'|)(z)+(|f|+|g|)(z) \geq \delta > 0$ (see [9]).

In view of all this, we ask for an answer to the problem below.

Problem 1.11. *Assume that the f_j have no common factor. Find necessary and sufficient conditions on $f_1, \ldots, f_N$ such that $J(f_1, \ldots, f_N)$ is finitely generated.*

We conjecture that $J = J(f_1, \ldots, f_N)$ is finitely generated if and only if ord $(J, m) < \infty$ *for every $m \in Z(J)$.*

We now return to the Generalized Corona Problem and ask when g^2 belongs to the ideal.

2. $\overline{\partial}$ techniques

Suppose that for $f_1, \ldots, f_N$, $g \in H^\infty$ and all $z \in \mathbb{D}$ we have

$$|g(z)| \leq |f_1(z)| + |f_2(z)| + \cdots + |f_N(z)|,$$

then Wolff [25] showed that

$$g^3 = \sum_{j=1}^{N} f_j\, g_j$$

(see also [6]). When is g^2 in the ideal generated by $f_1, \ldots, f_N$? Cegrell [4], [5], Lin [15] and Amar et al. [1] have proved special cases by introducing restrictions on the functions and generators. They use $\overline{\partial}$ techniques. In fact, the following is true.

Theorem 2.1 (Cegrell [4]). *Suppose g, f_1, f_2 satisfy*

$$|g| \leq (|f_1|^2 + |f_2|^2)^{1+\epsilon}$$

on D. Then $g \in I(f_1, f_2)$ whenever $\varepsilon > 0$.

Note that the case $\epsilon = 0$ is the question we are interested in here. Recently, Amar, Bruna and Nicolau [1] proved the following theorem.

Theorem 2.2 [1]. *Let $g, f_1, f_2 \in H^\infty$ such that*

$$|g| \leq \frac{|f_1|^2 + |f_2|^2}{|\log(|f_1| + |f_2|)|^{1+\epsilon}} \quad \textit{for some } \varepsilon > 0.$$

Then $g \in I(f_1, f_2)$.

Once again, the authors used $\bar{\partial}$ techniques. Using related techniques, we were able to show that the theorem below holds.

Theorem 2.3 [9]. *Let $I = I(f_1, \ldots, f_N)$. If* ord $(I, x) < \infty$ *for every $x \in Z(I)$, then*

$$|f| \leq |f_1| + \cdots + |f_N|$$

implies that

$$f^2 \in I(f_1, \ldots, f_N).$$

The first step of our proof (which is not repeated here) is to show that our ideal $I(f_1, \ldots, f_N)$ is generated by $N+1$ functions, each of which is a finite product of interpolating Blaschke products. This follows easily from well-known results in [11] and [10]. We call these generators $b_0, b_1, \ldots, b_N$.

The second step of our proof begins as all others do, but we need a key proposition, which appears here as Proposition 2.5. The proof of this proposition in our original paper uses an interplay between pseudohyperbolic discs and Carleson squares. Here we use Hoffman's theory to give a proof using pseudohyperbolic discs only.

The proof of Theorem 2.3 proceeds as follows: As in [6], § 8, without loss of generality we may assume that $\bigcap_{j=0}^{N} Z_{\mathbb{D}}(b_j) = \emptyset$. Let

$$\varphi_j = \frac{\bar{f}_j}{\sum_{k=0}^{N} |f_k|^2}, \qquad G_{jk} = \varphi_j \frac{\partial \varphi_k}{\partial \bar{z}} \quad (j, k = 0, \ldots, N). \tag{2.1}$$

We can use a normal families argument to obtain our final solutions. (For technical reasons, functions and their dilations are represented by the same symbol.) Suppose we can solve the $\bar{\partial}$-equations

$$\frac{\partial h_{jk}}{\partial \bar{z}} = g^2 G_{jk} \quad (j, k = 0, \ldots, N) \tag{$\star$}$$

with $||h_{jk}||_{L^\infty(\partial\mathbb{D})} \leq M$. Then $g_j = g^2\varphi_j + \sum_{k=0}^{N}(h_{jk} - h_{kj})b_k$ satisfy

$$\sum_{j=0}^{N} g_j b_j = g^2 \quad \text{and} \quad \frac{\partial g_j}{\partial \bar{z}} = 0 \quad (j = 0, \ldots, N)$$

(see [6], p. 329). Thus we have only to show that these $\overline{\partial}$-equations admit bounded solutions.

To do so, we recall some facts about Carleson measures and some of Hoffman's theory. First let us recall that a positive Borel measure μ on $\mathbb{D}$ is called a *Carleson measure* if there exists a constant C so that

$$\int_{\mathbb{D}} |f| \, d\mu \leq C||f||_1$$

for every f in the Hardy space

$$H^1 = \left\{ f : \ f \text{ analytic in } \mathbb{D}, \ ||f||_1 = \sup_{0<r<1} \frac{1}{2\pi} \int_0^{2\pi} |f(re^{i\theta})| \, d\theta < \infty \right\}.$$

We will use two results, due to Carleson and Wolff, on the existence of bounded solutions of $\overline{\partial}$-equations.

Theorem 2.4 ([6], p. 320–322). (a) *Let $G \in C(\mathbb{D})$ induce the Carleson measure μ on $\mathbb{D}$. If G is bounded then the $\overline{\partial}$-equation $\overline{\partial}h = G$ admits a solution $h \in C(\overline{\mathbb{D}}) \cap C^1(\mathbb{D})$ with $||h||_{L^\infty(\partial\mathbb{D})} \leq C_1$, where C_1 only depends on the Carleson norm of μ.*

(b) *Let G be a bounded C^1 function on $\mathbb{D}$ and assume that the two measures*

$$\mu_1 = |G(z)|^2(1 - |z|) \, dx \, dy, \qquad \mu_2 = \left| \frac{\partial G}{\partial z}(z) \right| (1 - |z|) \, dx \, dy$$

are Carleson measures. Then the $\overline{\partial}$-equation $\overline{\partial}h = G$ admits a solution $h \in C(\overline{\mathbb{D}}) \cap C^1(\mathbb{D})$ with $||h||_{L^\infty(\partial\mathbb{D})} \leq C_2$ for some constant C_2 depending only on the Carleson norms of μ_1 and μ_2.

Note that whenever G induces a Carleson measure μ, then the dilation G_r defined by $G_r(z) = G(rz)$ induces a Carleson measure μ_r for which $N(\mu_r) \leq N(\mu)$.

The following is the key result for the proof of our Theorem 2.3.

Proposition 2.5. *Let $\{z_n\}$ be a finite union of interpolating sequences and let*

$$U = \bigcup_{n=1}^{\infty} D(z_n, \eta)$$

for some $0 < \eta < 1$. Let Ψ_U denote the characteristic function of U. Then, for any interpolating Blaschke product B, the measure

$$\mu = \frac{|B'|}{|B|} \Psi_U \, dx \, dy$$

is a Carleson measure.

First we sketch a proof of Theorem 2.3 assuming Proposition 2.5 holds. Let G be any of the functions G_{jk} previously defined (see 2.1) and recall that $b_0, \ldots, b_N$ generate our ideal. An elementary calculation (see [6], p. 330) yields that

$$|g^2 G|^2 \leq c_0 \frac{\sum_{l=0}^{N} |b_l'|^2}{\sum_{l=0}^{N} |b_l|^2}. \tag{2.2}$$

Since b_0 is a finite product of interpolating Blaschke products with zeros $\{z_n\}$, for $\varepsilon > 0$ small enough, there exists a constant $c(\varepsilon)$ such that one has $|b_0(z)| \geq c(\varepsilon)$ if $\rho(z, \{z_n\}) \geq \varepsilon$ (see [11], or Lemma 2.10). Let $U = \bigcup_{n=1}^{\infty} D(z_n, \varepsilon)$. Then

$$\psi_{U^c}(z) |g^2 G(z)|^2 (1 - |z|) \leq \frac{c_0}{c(\varepsilon)^2} \sum_{l=0}^{N} |b_l'(z)|^2 (1 - |z|), \tag{2.3}$$

which is a Carleson measure (see [6], p. 327). Also ([6], p. 330)

$$\psi_{U^c}(z) |\partial (g^2 G)(z)| (1 - |z|) \tag{2.4}$$

is a Carleson measure. By (2.2) and the fact that $\sum_{j=1}^{n} |a_j|^2 \leq \left(\sum_{j=1}^{n} |a_j| \right)^2 \leq n \sum_{j=1}^{n} |a_j|^2$ for any complex numbers a_j, we have

$$|g^2 G| \leq c_1 \frac{\sum_{l=0}^{N} |b_l'|}{\sum_{l=0}^{N} |b_l|} \leq c_1 \sum_{l=0}^{N} \frac{|b_l'|}{|b_l|}.$$

Since the b_j are finite products of interpolating Blaschke products, there exist interpolating Blaschke products B_l $(l = 0, \ldots, L)$ so that

$$|g^2 G| \le c_1 \sum_{l=0}^{L} \left| \frac{B_l'}{B_l} \right| .$$

By Proposition 2.5 we have that

$$\psi_U |g^2 G(z)|\, dx\, dy \tag{2.5}$$

is a Carleson measure.

Hence, by (2.3), (2.4), (2.5) and Theorem 2.4, the $\overline{\partial}$-equation $\overline{\partial} h = \psi_U G g^2 + (1 - \psi_U) G g^2 = g^2 G$ admits bounded solutions. This finishes the proof. ∎

Before we present the proof of Proposition 2.5, we need some additional lemmas.

Lemma 2.6 ([6], p. 310). *Let $\{z_n\}$ be an interpolating sequence in $\mathbb{D}$ with $\inf_n \prod_{j \ne n} \rho(z_j, z_n) \ge \delta > 0$. Let $0 < \eta < (1 - \sqrt{1 - \delta^2})/\delta$, and let $w_n \in \mathbb{D}$ satisfy $\rho(z_n, w_n) < \eta$ for every $n \in \mathbb{N}$. Then $\{w_n\}$ is an interpolating sequence with*

$$\inf_n \prod_{j \ne n} \rho(w_j, w_n) \ge \frac{\delta - \frac{2\eta}{1+\eta^2}}{1 - \delta \frac{2\eta}{1+\eta^2}} .$$

Lemma 2.7 [16]. *A sequence $\{z_n\}$ is a finite union of interpolating sequences if and only if the measure $\mu = \sum_{n=1}^{\infty} (1 - |z_n|^2) \delta_{z_n}$ is a Carleson measure.*

In fact, more is true. In what follows for a Blaschke product B with zeros $\{z_n\}$, let $\delta(B) = \inf (1 - |z_n|^2) |B'(z_n)|$.

Lemma 2.8. *Let $\{z_n\}$ be a finite union of interpolating sequences. Fix an arbitrary $0 < \eta < 1$. For every $n \in \mathbb{N}$ let $\xi_n \in D(z_n, \eta)$. Then $\mu = \sum_{n=1}^{\infty} (1 - |\xi_n|^2) \delta_{\xi_n}$ is a Carleson measure.*

Proof. By our assumptions, $\{z_n\}$ is a finite union of interpolating sequences. Let B_j $(j = 1, \ldots, n)$ be the associated interpolating Blaschke products. By Hoffman's factorization theorem (see [6], p. 407), there exist factors $B_j^{(k)}$

($k = 1, \dots, 2^N$) of B_j so that $\delta(B_j^k) \geq \delta$ ($j = 1, \dots, n$; $k = 1, \dots, 2^N$), where $0 < \delta < 1$ is chosen so big that $\eta < (1 - \sqrt{1-\delta^2})/\delta$.

By Lemma 2.6 we can conclude that (ξ_n) is a union of $n \cdot 2^N$ interpolating sequences. The result now follows from the previous lemma. ■

In our case, all our generators are assumed to be finite products of interpolating Blaschke products. Using Hoffman's theory we can learn much more about the behavior of these functions on pseudohyperbolic discs in $\mathbb{D}$.

Lemma 2.9 ([11], p. 86, and [6], p. 404). *Let $0 < \delta < 1$, $0 < \eta < (1 - \sqrt{1-\delta^2})/\delta$, (that is, $0 < \eta < \rho(\delta, \eta)$), and let*

$$0 < \varepsilon \leq \varepsilon(\delta) := \frac{\delta - \eta}{1 - \delta\eta}\eta.$$

If b is any interpolating Blaschke product with zeros $\{z_n\}$ such that

$$\delta(b) = \inf_{n \in \mathbb{N}} (1 - |z_n|^2)|b'(z_n)| \geq \delta,$$

then

$$\begin{aligned} \{z \in \mathbb{D} : |b(z)| < \varepsilon\} &\subseteq \{z \in \mathbb{D} : \rho(z, Z(b)) < \eta\} \\ &\subseteq \{z \in \mathbb{D} : |b(z)| < \eta\}. \end{aligned} \tag{2.6}$$

Notice that $(1-\sqrt{1-\delta^2})/\delta$ is a monotone increasing function of $\delta \in (0,1)$, that $\varepsilon < \eta < \delta$ and that $0 < (1 - \sqrt{1-\delta^2})/\delta < \delta$. We shall also use the fact that $\eta < 2\eta/(1+\eta^2) < \delta$ is equivalent to $0 < \eta < (1 - \sqrt{1-\delta^2})/\delta$.

As a consequence, if b_0 is a finite product of interpolating Blaschke products $b_1, \dots, b_n$, then for any $\eta_j = \eta(b_j)$ above, we see that $|b_j| > \epsilon(b_j)$ on $\{z \in \mathbb{D} : \rho(z, Z(b_j)) > \eta_j\}$. In particular,we obtain the result below.

Lemma 2.10. *Let b_0 be a finite product of interpolating Blaschke products with zeros z_n, then for ϵ sufficiently small, there exists $c(\varepsilon)$ such that $|b(z)| > c(\varepsilon)$ on $\{z \in \mathbb{D} : \rho(z, z_n) > \varepsilon\}$.*

Lemma 2.11 [6]. *Let $f : \mathbb{D} \to \mathbb{D}$ be analytic with $f(0) = 0$, $|f'(0)| \geq \delta > 0$. Then f is schlicht in $|z| < \tau^* = (1 - \sqrt{1-\delta^2})/\delta$.*

We are now ready to prove Proposition 2.5.

Proof. Let $\delta = \delta(B)$ and $\tau^* = (1 - \sqrt{1-\delta^2})/\delta$. Choose $\eta^* \in (0,1)$ so that $D(z_n, \eta^*) \subseteq D(\xi, \tau^*)$ for all $\xi \in D(z_n, \eta^*)$ and $n \in \mathbb{N}$. (For example, take η^*

so that $2\eta^*/(1+\eta^{*2}) < \tau^*$.) Now choose $\eta \in (0,1)$ so that $2\eta/(1+\eta^2) = \eta^*$. We obviously have $0 < \eta < \eta^* < \tau^*$. Let $0 < \varepsilon \leq (\delta - \eta)/(1-\delta\eta)\eta$ be as in Lemma 2.9. Finally, let us note that the disks $D(z_n, \eta)$ are in general *not* disjoint.

Step 1: We first consider the set $I_1 = \{n : |B| \geq \varepsilon > 0 \text{ on } D(z_n, \eta)\}$. Let $D(z_n, \eta) = D_n$.

For every $h \in H^1$, we obtain:

$$\begin{aligned}
&\iint_{D_n} \left| h\,\frac{B'}{B} \right| dx\,dy \\
&\quad \leq \frac{1}{\varepsilon} \iint_{D_n} |hB'|\,dx\,dy \\
&\quad \leq \frac{1}{\varepsilon} \iint_{D_n} |hB'|(1-|z|^2)(1-|z|^2)^{-1}\,dx\,dy \\
&\quad \leq \frac{1}{\varepsilon} \iint_{D_n} |h|(1-|z|^2)^{-1}\,dx\,dy \\
&\quad = A_n.
\end{aligned}$$

We have used Schwarz's lemma (which tells us that $(1-|z|^2)|B'(z)| \leq 1 - |B^2(z)|$) to go from the third to fourth inequality. Now one can easily calculate ([6], p. 3) that for any z in $\mathbb{D}$,

$$\rho(z_0, z) < \eta \Longrightarrow \frac{1}{4}(1-\eta^2) < \frac{1-|z|^2}{1-|z_0|^2} < \frac{4}{1-\eta^2}. \tag{2.7}$$

and that area $(D_n) = \pi\eta^2 \left[\dfrac{1-|z_n|^2}{1-\eta^2|z_n|^2}\right]^2$. Letting $C(\eta)$ denote (a changing) constant depending on η. We have

$$\begin{aligned}
A_n &\leq \frac{1}{\varepsilon} \iint_{D_n} (\sup_{D_n} |h|)((1-|z_n|^2)^{-1})((1-|z_n|^2)^1(1-|z|^2)^{-1})\,dx\,dy \\
&\leq \frac{4/(1-\eta^2)}{\varepsilon} \iint_{D_n} (\sup_{D_n} |h|)(1-|z_n|^2)^{-1}\,dx\,dy \\
&\leq \frac{C(\eta)}{\varepsilon} (\sup_{D_n} |h|)(1-|z_n|^2)^{-1}(area D_n) \\
&\leq \frac{C(\eta)}{\varepsilon} (\sup_{D_n} |h|)(1-|z_n|^2)^{-1}\left(\pi\eta^2 \left[\frac{1-|z_n|^2}{1-\eta^2|z_n|^2}\right]^2\right) \\
&\leq \frac{C(\eta)}{\varepsilon} |h(\xi_n)|(1-|z_n|^2) \quad \text{for some } \xi_n \in \overline{D}_n
\end{aligned}$$

$$\leq \frac{C^*(\eta)}{\varepsilon}|h(\xi_n)|(1-|\xi_n|^2), \text{ by (2.7)}.$$

In view of Lemma 2.8, the measure $\sum_{n\in I_1}(1-|\xi_n|^2)\delta_{\xi_n}$ is a Carleson measure. Therefore we have

$$\iint_{\bigcup_{n\in I_1} D_n}\left|h\frac{B'}{B}\right| dx\, dy \leq C\|h\|_1 \text{ for some } C>0. \tag{2.8}$$

Step 2: Let $I_2 = \{n : |B(z)| < \varepsilon$ for some $z \in D(z_n,\eta)$ and some $n \in \mathbb{N}\}$.

Claim 1: B has a zero in $D(z_n,\eta^*)$ for each $n \in I_2$.

In fact, by Lemma 2.9, there exists $\zeta_m \in Z_{\mathbb{D}}(B)$ with $\rho(z,\zeta_m) < \eta$. Thus $\rho(z,z_n) < \eta$ and $\rho(z,\zeta_m) < \eta$ imply that

$$\rho(z_n,\zeta_m) \leq \frac{\rho(z_n,z)+\rho(z,\zeta_m)}{1+\rho(z_n,z)\rho(z,\zeta_m)} \leq \frac{2\eta}{1+\eta^2} = \eta^*.$$

Hence $\zeta_m \in D(z_n,\eta^*)$.

Claim 2: There exists $\lambda > 0$ (independent of n) such that

$$(1-|z|^2)|B'(z)| \geq \lambda > 0 \tag{2.9}$$

whenever $z \in D(z_n,\eta)$ and $Z(B)\cap D(z_n,\eta^*) \neq \emptyset$.

Proof of claim 2. Let $\xi_n \in Z(B)\cap D(z_n,\eta^*)$. By the choice of η^* we see that $D(z_n,\eta^*) \subseteq D(\xi_n,\tau^*)$. By Lemma 2.11 and the choice of our constants, B is schlicht on $D(\xi_n,\tau^*)$, hence on $D(z_n,\eta^*)$. Thus $(1-|z|^2)|B'(z)| \geq \lambda_n > 0$ on $D(z_n,\eta)$. Assume that the claim fails. Then there exist $x_n \in D(z_n,\eta)$ so that $(1-|x_n|^2)|B'(x_n)| \to 0$. Note that $\rho(x_n,\xi_n) < \tau^*$. Using the compactness of $M(H^\infty)$ and passing to subnets, we may assume that $(\xi_{n(\alpha)})$ converges to some $\xi \in M(H^\infty)\cap Z(B)$ and $(x_{n(\alpha)})$ converges to some $x \in M(H^\infty)$. Since the pseudohyperbolic distance ρ is lower semi-continuous on $M(H^\infty)\times M(H^\infty)$ (see [11], p. 103), we have $\rho(x,\xi) \leq \tau^*$. By Proposition 1.6 and (1.2) of this paper, $B\circ L_{\xi_{n(\alpha)}} \to B\circ L_\xi$ and $|(B\circ L_\xi)'(0)| > \delta$. Thus, by Lemma 2.11, $B\circ L_\xi$ is schlicht in $D(0,\tau^*)$. Moreover, $B\circ L_{x_{n(\alpha)}} \to B\circ L_x$ and

$$|(B\circ L_x)'(0)| = |\lim_{n\to\infty}(B\circ L_{x_n})'(0)| = \lim_{n\to\infty}(1-|x_n|^2)|B'(x_n)| = 0. \tag{2.10}$$

By Hoffman's theory, there exists a Möbius transform L mapping $\mathbb{D}$ onto itself so that $L_x\circ L = L_\xi$ (see [6], p. 414). Hence $B\circ L_x$ is schlicht on $L(D(0,\tau^*))$. Because x and ξ lie in the same Gleason part, there exists $z_0 \in \mathbb{D}$ with $L_\xi(z_0) = x$. Hence $L(z_0) = 0$. But $|z_0| = \rho(z_0,0) =$

$\rho(L_\xi(z_0), L_\xi(0)) = \rho(x,\xi) < \tau^*$; thus $0 \in L(D(0,\tau^*))$. So $(B \circ L_x)'(0) \neq 0$. This contradicts (2.10). Thus our Claim 2 is proven.

Taking any $h \in H^1$ we now obtain by (2.9) for any $n \in I_2 \subseteq \{n \in \mathbb{N} : Z(B) \cap D(z_n, \eta^*) \neq \emptyset\}$ the estimate

$$\begin{aligned}
&\iint_{D_n} \left| h\,\frac{B'}{B} \right| \,dx\,dy \\
&\le \iint_{D_n} |h| \frac{|B'(z)|^2(1-|z|^2)}{|B(z)|} \cdot \frac{1}{|B'(z)|(1-|z|^2)}\,dx\,dy \\
&\le \frac{1}{\lambda} \iint_{D_n} |h| \frac{|B'(z)|^2}{|B(z)|}(1-|z|^2)\,dx\,dy. \qquad (2.11)
\end{aligned}$$

But by [6], p. 327, the measure $\frac{|B'(z)|^2}{|B(z)|}(1-|z|^2)\,dx\,dy$ is a Carleson measure. Hence by (2.11)

$$\begin{aligned}
\iint_{\bigcup\limits_{n\in I_2} D_n} \left| h\,\frac{B'}{B} \right| \,dx\,dy &\le \frac{1}{\lambda} \iint_{\mathbb{D}} |h| \frac{|B'(z)|^2}{|B(z)|}(1-|z|^2)\,dx\,dy \\
&\le C\|h\|_1 \text{ for some constant } C>0.
\end{aligned} \qquad (2.12)$$

The estimates (2.8) and (2.12) now yield that

$$\iint_U \left| h\,\frac{B'}{B} \right| \,dx\,dy \le C\|h\|_1 \text{ for some constant } C>0.$$

Hence $\left|\frac{B'}{B}\right| \Psi_U \,dx\,dy$ is a Carleson measure. ■

The question of whether or not g^2 must *always* belong to the ideal remains open. The functions not covered by our theorem are usually the most difficult to handle. Hoffman studied such functions in his work, but the results are very different than those for interpolating Blaschke products. Whether or not these techniques will be useful in solving the complete problem remains to be seen.

References

[1] Amar, E., Bruna, J., and Nicolau, A.: *On H^p- solutions of the Bezout equation.* Centre de Recerca Matemàtica, Inst. d'Est. Catalans, Preprint No. 190 (1993), 1–11. (To appear in Pacific J. Math.)

[2] Bourgain, J.: *On finitely generated closed ideals in H^∞.* Ann. Inst. Fourier (Grenoble) **35** (1985), 163–174.

[3] Carleson, L.: *Interpolation by bounded analytic functions and the corona problem.* Ann. of Math. **76** (1962), 547– 559.

[4] Cegrell, U.: *A generalization of the corona theorem in the unit disc.* Math. Z. **203** (1990), 255–261.

[5] Cegrell, U.: *Generalizations of the corona theorem in the unit disc.* Proc. Royal Irish Acad. To appear.

[6] Garnett, J.B.: *Bounded Analytic Functions.* Academic Press, New York, 1981.

[7] Gorkin, P.: *Functions not vanishing on trivial Gleason parts of Douglas algebras.* Proc. Amer. Math. Soc. **104** (1988), 1086–1090.

[8] Gorkin, P., and Mortini, R.: *Interpolating Blaschke products and factorization in Douglas algebras.* Michigan Math. J. **38** (1991), 147–160.

[9] Gorkin, P., Mortini, R., and Nicolau, A.: *The generalized corona theorem.* To appear in Math. Ann.

[10] Guillory, C., Izuchi, K., and Sarason, D.: *Interpolating Blaschke products and division in Douglas algebras.* Proc. Royal Irish Acad. Sci. A **84** (1984), 1–7.

[11] Hoffman, K.: *Bounded analytic functions and Gleason parts.* Ann. of Math. **86** (1967), 74–111.

[12] Izuchi, K.: *Factorization of Blaschke products.* Michigan Math. J. **40** (1993), 53–75.

[13] Laroco, L.: *Stable rank and approximation theorems in H^∞.* Trans. Amer. Math. Soc. **327** (1991), 815–832.

[14] Lin, Kai-Ching: *The corona theorem and interpolating Blaschke products.* Indiana Univ. Math. J. **41** (1992), 851–859.

[15] Lin, Kai-Ching: *On the constants in the corona theorem and ideals of H^∞.* Houston J. Math. **19** (1993), 97–106.

[16] McKenna, P.J.: *Discrete Carleson measures and some interpolating problems.* Michigan Math. J. **24** (1977), 311–319.

[17] Mortini, R.: *Ideals generated by interpolating Blaschke products.* Analysis **14** (1994), 67–74.

[18] Mortini, R.: *On an ideal theoretic phenomenon of Jean Bourgain.* Preprint.

[19] Nakazi, T.: *Notes on interpolation by bounded analytic functions.* Canad. Math. Bull. **31** (1988), 404–408.

[20] Rao, K.V.R.: *On a generalized corona problem.* J. Analyse Math. **18** (1967), 277–278.

[21] von Renteln, M.: *Finitely generated ideals in the Banach algebra H^∞.* Collectanea Math. **26** (1975), 3–14.

[22] Tolokonnikov, V.: *Interpolating Blaschke products and ideals of the algebra H^∞.* Soviet J. Math. **27** (1984), 2549– 2553.

[23] Tolokonnikov, V.: *Blaschke products satisfying the Carleson-Newman condition and ideals of the algebra H^∞.* Soviet J. Math. **42** (1988), 1603–1610.

[24] Vinogradov, S.A., Gorin, E.A., and Hrushchev, S.V.: *Free interpolation in H^∞ à la P.W. Jones.* Soviet J. Math. **22** (1983), 1838–1839.

[25] Wolff, T.H.: *A refinement of the corona theorem.* In: Linear and Complex Analysis Problem Book, 199 Research Problems. Ed.: V.P. Havin, S.V. Hrushchev, N.K. Nikolskij. Lecture Notes Math. **1043** (1984), 399–400, Springer Verlag, Berlin – New York.

Spectral Synthesis of Ideals in Classical Algebras of Smooth Functions

LEONID G. HANIN Department of Mathematical Sciences, Michigan Technological University, 1400 Townsend Drive, Houghton, MI 49931, U.S.A.

INTRODUCTION

The purpose of the present work is to describe the state of art and to briefly review classical, new, and very recent results on spectral synthesis of ideals (SSI) in regular algebras of smooth functions.

A topological algebra is said to admit SSI, if every closed proper ideal in this algebra is an intersection of closed primary ideals. Assertions of this type go back to classical algebraic works of E. Noether [33] and E. Lasker [26] on primary decomposition of ideals in Noetherian rings. Later, theorems on SSI were established for a number of algebras of smooth functions (for more extensive discussion, see Section 2).

As discovered by G.E. Shilov [38], *every* proper ideal in a regular semisimple commutative Banach algebra is an intersection of primary ideals. This shows that the collection of *all* primary ideals (not necessarily closed) of a regular functional algebra is excessively ample. Therefore, it is reasonable to confine oneself to *closed* primary ideals. Accordingly, every ideal which is an intersection of closed primary ideals is a *closed* ideal.

Due to the influence of classical works of L. Schwartz [35], [36], P. Malliavin [28]-[30], G.E. Shilov [38] and others, problems of spectral synthesis in harmonic analysis attracted a lot of interest and gave rise to plenty of publications from the late 40's to the middle of 70's, see [2], [24], Chapter 10, [25], Chapter 5, and references therein.

By contrast, in non-harmonic setting, the problem of SSI is much less studied, probably because it was commonely believed that all classical algebras of smooth functions endowed with "local" norms (or collections of "local" seminorms) enjoy SSI though it is usually not easy to prove. Amazingly, as shown in [12], [13] this is not true even for some Sobolev algebras $W_p^l(R^n)$ (see Section 3), the algebra $W_2^2(R^2)$ providing apparently the simplest known "natural" counterexample to SSI.

Thus, very little is known to date on the structure of ideals in algebras of smooth functions, and the main problems in this field are still awaiting their solution.

Another direction of investigation in spectral synthesis which, along with harmonic analysis, is left beyond our present consideration, is the structure of ideals in algebras of analytic functions in a domain Ω in C^n. These algebras are non-regular and cospectra of their ideals form a very special class of closed subsets in Ω (for $n = 1$, the cospectrum of an ideal is a finite set or a countable set without limit points in Ω). By contrast, the cospectrum of an ideal in a semisimple regular algebra of functions is an arbitrary closed subset of the underlying topological space. These properties render algebras of analytic functions quite specific as far as both results on SSI and methods of their proof are concerned.

The circle of questions discussed in this paper is closely related to the *problem of spectral approximation* which has a lot of implications in analysis and mathematical physics. This problem consists of describing for a certain linear topological space $\mathcal{A}$ of functions defined on a topological space the closure of the set of functions in $\mathcal{A}$ vanishing in a neighborhood of a given closed set. Indeed, the statement of this problem does not require that $\mathcal{A}$ is an algebra with respect to pointwise multiplication. However, in the latter case the problem of spectral approximation is equivalent to characterizing minimal closed ideals with a given cospectrum, i.e., to a particular case of the problem of SSI. Stimulated by the classical Malliavin's theorem on lack of spectral synthesis in Wiener algebras for noncompact abelian groups [28] - [30], the problem of spectral approximation was originally studied mainly in the framework of harmonic analysis. However, profound results are also possible in non-harmonic setting. As an example, one can mention the solution of the spectral approximation problem for Sobolev spaces [21], [22].

In Section 1, we give main definitions related to SSI. Basic examples of algebras with the property of SSI are shown in Section 2. A theorem providing complete solution to the problem of SSI for Sobolev algebras is a matter of our concern in Section 3. This statement was announced in [12] and proved in [13]; a refined version of the proof is contained in [20].

The proofs of these results are very individual. This suggests the following obvious goal: to create a unified approach to SSI which enables to treat a number of essentially different cases from a single point of view. As an attempt at such unified approach, we discuss in Section 4 the concept of D-algebras and state a theorem claiming that all D-algebras admit SSI [17], [16]. The class of D-algebras is defined in terms of point derivations and embraces a number of algebras of functions with the "order of smoothness" ≤ 1, among them the following ones:

- algebras C^1 of continuously differentiable functions in several variables;
- algebras $Lip(X,\rho)$ of Lipschitz functions on a compact metric space (X,ρ);
- Zygmund algebras Λ_φ which are not contained in C^1.

Accordingly, the theorem on SSI for D-algebras not only simplifies and unifies the proofs of known results as in the first two cases [46], [39], [27], [44], [10] but also gives rise to new results as in the last case.

Zygmund algebras Λ_φ are treated in greater detail in Section 5. They are generalizations of the classical algebra Λ that was introduced by A. Zygmund [47] and appears in a very natural way as the limiting space in the Lipschitz scale $Lip\,\alpha$, $0<\alpha<1$, rather than $Lip\,1$.

All algebras studied below are supposed to be real. However all results can be carried over almost word-for-word to the complex case.

1. MAIN DEFINITIONS

Let X be a locally compact Hausdorff space, $C(X)$ be the space of all continuous functions on X, and let $\mathcal{A}$ be a linear subspace in $C(X)$ which is a topological algebra with respect to pointwise multiplication of functions. We assume that algebra $\mathcal{A}$ is Shilov regular, that is, for every closed subset F of X and for each point $x \in X\backslash F$ there exists a function $f \in \mathcal{A}$ such that f vanishes on F and $f(x)=1$. Also, it is supposed that the space of maximal ideals of algebra $\mathcal{A}$ coincides with X, i.e., every maximal ideal in $\mathcal{A}$ is of the form $M_x = \{f \in \mathcal{A} : f(x)=0\}$ for some point $x \in X$. If X is compact then algebra $\mathcal{A}$ is assumed to contain the unity function.

An *ideal* I in $\mathcal{A}$ is a linear subspace of $\mathcal{A}$ that stands multiplication on functions of $\mathcal{A}$, i.e. $fg \in I$ whenever $f \in I$ and $g \in \mathcal{A}$. For every ideal I in $\mathcal{A}$, we define a closed subset in X, $\sigma(I) := \cap\{f^{-1}(0) : f \in I\}$, called the *cospectrum* of I. An ideal I is *primary* at a point $x \in X$, if $\sigma(I)=\{x\}$. In other words, a primary ideal is that contained in exactly one maximal ideal.

For each closed subset F in X, we define the set M_F of all functions in $\mathcal{A}$ vanishing on F, and the closure J_F of the set of functions in $\mathcal{A}$ vanishing in a neighborhood (depending on a function) of the set F. It follows from the above assumptions on $\mathcal{A}$ (see [38]) that M_F and J_F are the maximal and the minimal closed ideals in $\mathcal{A}$ with the cospectrum F, respectively. In particular, M_x is the maximal (closed) primary ideal at x and J_x is the minimal closed primary ideal at x. Thus, for every closed ideal I in $\mathcal{A}$ with cospectrum F we have $J_F \subset I \subset M_F$.

The *primary component* I_x of an ideal I at a point $x \in \sigma(I)$ is defined to be the smallest closed primary ideal at x containing I. It is easily seen that

$$I_x = clos_{\mathcal{A}}(I + J_x) \ . \tag{1}$$

We say that algebra $\mathcal{A}$ admits SSI (notation: $\mathcal{A} \in Synt$), if for every closed proper ideal I in $\mathcal{A}$,

$$I = \cap\{I_x : x \in \sigma(I)\} \ . \tag{2}$$

Alternatively, $\mathcal{A} \in Synt$ if every closed proper ideal in $\mathcal{A}$ is an intersection of closed primary ideals.

In the simplest case when $J_x = M_x$ for all $x \in X$, (2) reduces to

$$I = M_F, \text{ where } F = \sigma(I) . \tag{3}$$

This means that every closed ideal in $\mathcal{A}$ is completely determined by its cospectrum. In this case we write $\mathcal{A} \in synt$.

2. BASIC EXAMPLES

The following algebras of smooth functions were known to admit SSI.

1. The algebra $C(X)$ of all continuous functions on a compact Hausdorff space X supplied with the norm $\|f\|_X := sup\{| f(x) |: x \in X\}$, $f \in C(X)$. The fact that $C(X) \in synt$ was first established in the pathbreaking work of M. Stone [42]. A simpler proof along the lines of the theory of Banach algebras was suggested by G.E. Shilov [38] while for a proof based on a duality argument the reader is referred e.g. to [11], Section 4.10.6. In like manner, the spectral synthesis property (3) holds for the algebra $C_0(X)$ of continuous functions on a

locally compact Hausdorff space X vanishing at "infinity", see [23], Appendix C, Theorem 30.

To infer that $C(X) \in synt$ we have to show the following: if a function $f \in C(X)$ vanishes on a closed subset F of X, then for every $\varepsilon > 0$ there is a function g vanishing in a neighborhood of F such that $\|g - f\|_X \leq \varepsilon$. For $r > 0$, define the truncation

$$T_r u = \begin{cases} u\,, & \text{if } | u | \leq r\,; \\ r\, sign(u)\,, & \text{if } | u | > r\,. \end{cases}$$

Then the function $g_\varepsilon(x) := f(x) - T_\varepsilon f(x)$, $x \in X$, is continuous, vanishes in a neighborhood of F, and $\|g - f\|_X = \|T_\varepsilon f\|_X \leq \varepsilon$ as required.

2. The "small" Lipschitz algebra $lip(X, \rho)$ constituted by functions f on a compact metric space (X, ρ) with the finite norm

$$\|f\|_{X,\rho} := max\{\|f\|_X\,, | f |_{X,\rho}\}\,,$$

where

$$| f |_{X,\rho} := sup\{\frac{| f(x) - f(y) |}{\rho(x,y)} : x, y \in X\,, x \neq y\}\,,$$

and satisfying the condition

$$\lim_{\rho(x,y) \to 0} \frac{f(x) - f(y)}{\rho(x,y)} = 0. \tag{4}$$

As shown in [37], $lip(X, \rho) \in synt$. To prove this, suppose a function $f \in lip(X, \rho)$ vanishes on a closed set $F \subset X$, and fix $\varepsilon > 0$. In view of (4) there exists $\delta > 0$ such that $| f(x) - f(y) | \leq \varepsilon\rho(x,y)$ for all $x, y \in X$ with $\rho(x,y) \leq \delta$. Observe that for each $r > 0$, $| T_r u - T_r v | \leq | u - v |$ for all $u, v \in R$. Therefore, $T_r f \in lip(K, \rho)$. The function $g_r := f - T_r f$ vanishes in a neighborhood of F , and assuming that $\rho(x,y) < \delta$ we have for every r

$$| (f - g_r)(x) - (f - g_r)(y) | = | T_r f(x) - T_r f(y) | \leq | f(x) - f(y) | \leq \varepsilon\rho(x,y)\,. \tag{5}$$

Now choose $r := \varepsilon\delta/2$. If $\rho(x,y) \geq \delta$, then

$$| (f - g_r)(x) - (f - g_r)(y) | \leq 2r = \varepsilon\delta \leq \varepsilon\rho(x,y)\,. \tag{6}$$

Combining (5) and (6) we conclude that for such r, $\|f - g_r\| \leq \varepsilon$, which completes the proof.

3. The "big" Lipschitz algebra $Lip(X, \rho)$ of *all* functions on X with the finite norm $\|\cdot\|_{X,\rho}$. Note that the algebra $lip(X, \rho)$ is a closed separable subalgebra of $Lip(X, \rho)$. The spectral synthesis theorem for algebras $Lip(X, \rho)$ is due to L. Waelbroeck [44]. In the particular case of a compact subset of R^n with the metric $\rho(x, y) = | x - y |^\alpha, 0 < \alpha \leq 1$, this was established by G. Glaeser [10]. The proofs of these results are appreciably harder as compared to the case of the "small" Lipschitz algebras. This is primarily conditioned by the fact that in "big" Lipschitz algebras for every cluster point $x \in X$ the quotient space M_x/J_x is infinite-dimensional (and even non-separable) while in "small" Lipschitz algebras this space is trivial.

4. Algebra $C^m(\Omega)$ of m times continuously differentiable functions in an open subset Ω of R^n with the topology of uniform convergence of functions and their derivatives up to order m on compact subsets of Ω. The classical theorem of H. Whitney [46] claims that $C^m(\Omega) \in Synt$. Indeed, the same is also true for Banach algebras $C^m([0,1]^n)$ (for $m = 1$ and arbitrary n this was established in [39], and for $n = 1$ with arbitrary m - in [38]) as well as for Banach algebras $C_0^m(R^n)$ of C^m-functions on R^n vanishing at infinity together with their derivatives of order $\leq m$. A simpler proof of the Whitney's theorem and its extension to the case $m = \infty$ can be found in [27]. A similar argument yields the theorem on SSI for algebras $C^m lip\varphi(\Omega)$ of functions in $C^m(\Omega)$ with the derivatives of order m in $lip\varphi(K)$ for every compact subset K of Ω [12]. Here φ is any majorant, i.e. a nondecreasing function on R_+ such that $\varphi(0) = \varphi(0+) = 0$ and $\varphi(t) > 0$ for $t > 0$, satisfying the condition $\lim_{t \to 0+} \varphi(t)/t = +\infty$.

Let A_n^m be the algebra of polynomials of degree at most m with multiplication defined as the truncation of the usual product to degree m. For $x \in \Omega$, the mapping $\pi_x : f \to T_x^m f$, where $T_x^m f$ is the Taylor expansion of order m of a function f at the point x, identifies the quotient algebra $C^m(\Omega)/J_x$ with A_n^m. Hence, for *every* ideal I in $C^m(\Omega)$, $\pi_x^{-1}(\pi_x(I)) = I + J_x$ is the smallest *closed* primary ideal in $C^m(\Omega)$ at the point x containing I, that is,

$$I_x = I + J_x \tag{7}$$

(compare with (1)). Also, the minimal closed primary ideal of the algebra $\mathcal{A} = C^m(\Omega)$ at a point $x \in \Omega$ has the following representation

$$J_x = clos_{\mathcal{A}} M_x^{m+1} = \{f \in \mathcal{A} : D^\alpha f(x) = 0 \,, \mid \alpha \mid \leq m\} \,. \tag{8}$$

Relation (7) suggests the following equivalent formulation of the property of SSI for algebras $\mathcal{A} = C^m(\Omega)$, see [46], [27]:

Let I be a closed ideal in $\mathcal{A}$ and $f \in \mathcal{A}$. Suppose that for every point $x \in \Omega$ there is a function $g_x \in I$ such that $T_x^m g_x = T_x^m f$. Then $f \in I$.

5. Algebras $C^m Lip\varphi([a,b])$ of C^m- functions on an interval $[a,b]$ with the derivatives of order m in $Lip\varphi([a,b])$. As shown in [20], $C^m Lip\varphi([a,b]) \in Synt$ for any majorant φ (for power majorants, this was established earlier in [15]). It can be conjectured that this remains true also for algebras $C^m Lip\varphi(Q)$ on any closed cube Q in R^n.

6. Non quasi-analytic Denjoy-Carleman classes $C(M_n)$ of functions of one variable with regular weights $\{M_n\}_{n=0}^{\infty}$, see [3] for details.

Examples 1-6 reveal a remarkable universality of the property of SSI in a wide range of smoothness and topological assumptions on function algebras.

3. CLOSED IDEALS IN SOBOLEV ALGEBRAS

In this section, we study the scale of classical Sobolev spaces $W_p^l(R^n), 1 \leq p < +\infty$ which consist of functions f on R^n having generalized derivatives $D^\alpha f \in L_p(R^n)$, $| \alpha | \leq l$, and are endowed with the norm $\|f\| := \sum_{|\alpha| \leq l} \|D^\alpha f\|_p$. Recall the following well-known continuous imbeddings [41], [8]:

$$W_p^l(R^n) \subset \begin{cases} CB(R^n), \text{ if } p > n/l \text{ or } p = n/l = 1 \; ; & (9.1) \\ L_r(R^n), \; p \leq r < +\infty, \text{ if } p = n/l > 1 \; ; & (9.2) \\ L_s(R^n), \; s^{-1} = p^{-1} - l/n, \text{ if } 1 \leq p < n/l \; , & (9.3) \end{cases}$$

where $CB(R^n)$ is the space of bounded continuous functions on R^n. It is known that in (and only in) the case (9.1) Sobolev spaces are algebras with respect to pointwise multiplication. Usually, this fact is derived from the theory of multipliers for Sobolev spaces [43], [31]. A direct proof bases solely on imbeddings (9.1)-(9.3) can be found in [13], [20] (for a proof of the sufficiency part in the case $p > n/l$, see also [1], Theorem 5.23).

Inclusion (9.1) and the density in Sobolev spaces of the class of $C^\infty-$ functions with compact support imply that in case (9.1) the space $W_p^l(R^n)$ is continuously imbedded in $C_0(R^n)$. Thus, the space of maximal ideals of a Sobolev algebra coincides with R^n.

We assume hereafter that $p > n/l$ or $p = n/l = 1$. Let m be the maximal integer for which $W_p^l(R^n) \subset C_0^m(R^n)$. It follows from (9.1) that $m = l - n$ for $p = 1$ and $m = [l - n/p] - 1$ for $p > 1$, where $[\cdot]$ stands for the integer part.

It turns out that ideals J_x in Sobolev algebras $\mathcal{A} = W_p^l(R^n)$ have exactly the same structure (8) as in the corresponding algebras $C_0^m(R^n)$ [12], [13], [20], and primary components of an ideal I in $\mathcal{A}$ are given by (7).

A complete solution of the problem of SSI for Sobolev algebras is contained in the forthcoming theorem. The general scheme of the proof of its sufficiency part follows in the whole the argument used by B. Malgrange [27] in his proof of the Whitney's spectral synthesis theorem. A crucial role in the proof of its necessity part is played by (p,1)-capacities, see e.g. [21].

Theorem 1 [12], [13]. *$W_p^l(R^n) \in Synt$ iff $m = l - 1$, i.e., iff $n = 1$ or $2 \leq n < p$.*

Remark 1. In the univariate case, the property of SSI in Sobolev algebras can be established by a simple argument using local contractions. A proof of the spectral synthesis theorem for Sobolev algebras of periodic functions on the line for $p = 2$ based on the Shilov techniques can be found in [34].

Remark 2. It follows from Theorem 1 that if $n = 1$ or $2 \leq n < p$ then for every closed set F in R^n

$$J_F = \{f \in W_p^l(R^n) : D^\alpha f(x) = 0\,, x \in F, \mid \alpha \mid \leq l - 1\}\,.$$

This solves the problem of spectral approximation in the case $m = l-1$, see also [4]. For arbitrary Sobolev spaces with $1 < p < +\infty$ this problem was settled in [21], [22].

Remark 3. Suppose that $m = 0$. If $n \geq 2$ and $l \geq 2$ then by Theorem 1 there exist closed subsets F in R^n such that for Sobolev algebra $W_p^l(R^n)$, $J_F \neq I_F$. It follows from the proof of Theorem 1 [13], [20] that this is the case for any compact subset F in R^n contained in an $(n-1)$-dimensional hyperplane and having there positive Lebesgue measure. This suggests the following problem.

Problem 1. To describe all sets of spectral synthesis for Sobolev algebras $W_p^l(R^n)$ with $m = 0$, i.e. closed subsets F in R^n such that $J_F = M_F$.

Remark 4. In the case $m < l - 1$ intersections of closed primary ideals fail to cover the totality of closed ideals in a Sobolev algebra. This leads us to

Problem 2. To describe closed ideals in Sobolev algebras for $m < l - 1$, i.e. beyond spectral synthesis.

Remark 5. In the case $p = +\infty$ the Sobolev space $W^l_\infty(R^n)$ coincides with the space $C^{l-1}Lip1(R^n)$ and is obviously a Banach algebra which space of maximal ideals is however larger than R^n. For a similar algebra defined on a closed cube Q in R^n this difficulty does not arise, and for $n = 1$, $W^l_\infty(Q) \in Synt$ [20] (see Example 5 in Section 2).

4. D - ALGEBRAS

Let $\mathcal{A}$ be a Shilov regular Banach algebra of continuous functions on a compact Hausdorff space X. We assume additionally that it possesses the following inversion property: if $f \in \mathcal{A}$ and $f(x) \neq 0$ for all $x \in X$, then $1/f \in \mathcal{A}$. Hence (see [38]) the space of maximal ideals of the algebra $\mathcal{A}$ coincides with X.

A bounded linear functional $D \in \mathcal{A}^*$ is called *a point derivation* of algebra $\mathcal{A}$ at a point $x \in X$, if

$$D(fg) = f(x)Dg + g(x)Df \qquad \text{for all } f, g \in \mathcal{A}.$$

Let $\mathcal{D}_x$ be the linear space of all point derivations of $\mathcal{A}$ at a point $x \in X$. It follows from the above definition that $\mathcal{D}_x = (1 \cup M_x^2)^\perp$. Also, due to the regularity of $\mathcal{A}$, $D(J_x) = \{0\}$ for all $D \in \mathcal{D}_x$.

For a closed subset $F \subset X$, define

$$K_F := \{D : D \in \mathcal{D}_x, \, x \in F, \, \|D\| \leq 1\},$$

where $\|\cdot\|$ stands for the norm on $\mathcal{A}^*$. Obviously, the set K_F is compact in the weak* topology on $\mathcal{A}^*$.

To each function $f \in \mathcal{A}$, we associate a function $\hat{f} \in C(K_F)$ by setting $\hat{f}(D) = Df, D \in K_F$. This formula determines a linear mapping $d_F : \mathcal{A}/J_F \to C(K_F)$ which stands to the Gel'fand transform in approximately the same position as the derivative of a function to the function itself. We see readily that

$$\|\hat{f}\|_{K_F} \leq \|\dot{f}\|_F \, , \quad f \in \mathcal{A},$$

where $\|\cdot\|_{K_F}$ is the supremum norm on $C(K_F)$, and $\|\dot{f}\|_F$ is the quotient norm of the element $\dot{f} \in \mathcal{A}/J_F$ corresponding to f. Now we are in a position to define the class of $D-$algebras.

An algebra $\mathcal{A}$ as above is called a *$D-$algebra*, if for every closed subset F in X there is a constant $A(F) \geq 0$ such that

$$\|\dot{f}\|_F \leq A(F)\|\hat{f}\| \text{ for all } f \in M_F/J_F\,. \tag{10}$$

Equivalently, the class of $D-$algebras can be characterized by claiming that the mapping d_F is an injection onto a closed subspace in $C(K_F)$. Note that (10) implies that $J_F = clos_{\mathcal{A}} M_F^2$ for every closed subset F of X. Thus, every $D-$algebra consists of functions which "order of smoothness" is less than 2. Observe also that condition (10) is a kind of an extension theorem which says in particular that the "trace" of a function from a $D-$algebra to a closed set $F \subset X$ is completely determined by the values at points of F of the function and its point derivations.

The following result is of prime importance in the context of this section.

Theorem 2 [17], [16]. *Every D−algebra admits SSI.*

Here are a few basic examples of $D-$algebras.

0. Every algebra with the property (3) provides a trivial example of a $D-$algebra without nonzero point derivations.

1. Lipschitz algebras $Lip(X,\rho)$ for which inequality (10) is valid with $A(F) = 1$ and thus turns into equality [37],[44]. The main reason standing behind this fact is the existence of isometric extensions of *real* Lipschitz functions from arbitrary subsets of X [32]. The space $\mathcal{D}_x$ of all point derivations of the algebra $Lip(X,\rho)$ at a cluster point $x \in X$ coincides with the weak* closure of the linear span of the set of all weak* limits of the functionals $f \mapsto \frac{f(a)-f(b)}{\rho(a,b)}$ as $a, b \to x$, $a \neq b$ [37].

2. Algebras $C^1([0,1]^n$ in which every point derivation at a point $x \in [0,1]^n$ has the form $f \mapsto (grad\, f(x), \eta)$ for some vector $\eta \in R^n$, i.e., is a directional derivative. For these algebras, estimate (10) with a constant depending only on n follows from the Whitney extension theorem [45] ,[27], [9].

3. However the most interesting new examples of $D-$algebras are found among Zygmund algebras discussed in the next section.

5. ZYGMUND ALGEBRAS

For a function f on the cube $Q_0 := [-1,1]^n$ we denote $\Delta_h^2 f(x) := f(x+h) - 2f(x) + f(x-h)$, where $x \pm h \in Q_0$, and define the second modulus of continuity of f by setting

$$\omega_2(f;t) := sup\{|\ \Delta_h^2 f(x)\ |:\ x \pm h \in Q_0, |\ h\ |\leq t\}\ , t \geq 0\ .$$

The Zygmund space Λ_φ consists of all bounded functions f on Q_0 with the finite norm

$$\|f\|_{\Lambda_\varphi} := max\{\|f\|_{Q_0},\ sup_{t>0}\frac{\omega_2(f;t)}{\varphi(t)}\}.$$

Here φ is an arbitrary majorant; however we may assume without loss of generality that the function $\varphi(t)/t^2$ is nonincreasing. The space Λ_φ is a Banach algebra with respect to pointwise multiplication of functions [19]. We have obviously $Lip(\varphi) \subset \Lambda_\varphi$ while these two spaces coincide iff for some constant $C > 0$

$$t\int_t^1 \frac{\varphi(s)}{s^2}ds \leq C\varphi(t)\,, \quad 0 < t \leq 1.$$

Note also that the space Λ_φ is not imbedded in C^1 iff

$$\int_0^1 \frac{\varphi(t)}{t^2}dt = +\infty\ , \tag{11}$$

see [19], [20]. In particular, for power majorants, $\varphi(t) = t^\alpha\,, 0 < \alpha \leq 2\,,$

$$\Lambda_\alpha = \begin{cases} Lip\alpha\ , & 0 < \alpha < 1; \\ \Lambda\ , & \alpha = 1; \\ C^1 Lip(\alpha - 1)\ , & 1 < \alpha \leq 2. \end{cases}$$

Functions $f \in \Lambda_\varphi$ satisfying the condition

$$\lim_{t\to 0+} \frac{\omega_2(f;t)}{\varphi(t)} = 0$$

constitute a closed separable subalgebra of Λ_φ which is denoted by λ_φ and is referred to as the small Zygmund algebra. In the limit case $\lim_{t\to 0+}\varphi(t)/t^2 < +\infty$, i.e., for majorants $\varphi(t)$ equivalent to t^2, the small Zygmund algebra is trivial, that is, consists only of constants and linear functions.

The structure of closed primary ideals and of the spaces $\mathcal{D}_x$ of point derivations at points $x \in Q_0$ in Zygmund algebras Λ_φ with majorants φ satisfying (11) is parallel to that in Lipschitz algebras (compare the work [37] which treats Lipschitz spaces with the papers [14], [20] dealing with the spaces Λ and Λ_φ , respectively). In particular, the space $\mathcal{D}_x$ for the algebra Λ_φ can be obtained as the weak* closure of the linear space spanned by the set of all limits in the weak* topology of the functionals $f \mapsto \Delta_h^2 f(y)/\varphi(|\,h\,|)$ as $y \to x$ and $|\,h\,| \to 0+$.

For Zygmund algebras Λ_φ with majorants φ subject to (11), condition (10) which defines the class of $D-$algebras is equivalent to the following extension property:

For every closed subset $F \in Q_0$, for each function f vanishing on F, and for every $\varepsilon > 0$, there exist $\delta > 0$ and a function $g \in \Lambda_\varphi$ such that $g_{|F_\delta} = f_{|F_\delta}$ and $\|g\|_{\Lambda_\varphi} \leq A[M_F(f)+\varepsilon]$, where $F_\delta :=$ $\{x \in Q_0 : dist(x, F) \leq \delta\}$, $M_F(f) := limsup_{dist(x,F)\to 0,\ |h|\to 0+} \frac{|\Delta_h^2 f(x)|}{\varphi(|h|)}$, and A is a positive constant independent of f, ε, and δ. **(Ext)**

Observe that (Ext) implies that $\lambda_\varphi \in synt$ [16],[19].

In the univariate case, the extension theorem (Ext) was proved in [19] (the case $\varphi(t) = t$ was treated earlier in [15],[16]). For $n = 2$, the property (Ext) was established in [20]. The proofs of (Ext) in these two cases are based on explicit descriptions of the trace spaces $\Lambda_\varphi(F)$ of Zygmund classes Λ_φ for $n = 1,\ 2$ to arbitrary closed subsets $F \subset Q_0$ (see [40] where these descriptions are given for the seminormed Zygmund classes; for the normed modifications, the reader is referred to [16], [18]). Intrinsic characterization of the trace spaces $\Lambda_\varphi(F)$ for $n > 2$ remains unknown. Nevertheless, it was shown in [7] using quite different approach involving quasiharmonic extensions of functions [5], [6] that property (Ext) is true in any number of variables for majorants φ satisfying along with (11) the following two regularity conditions:

$$t\int_0^t \frac{\varphi(s)}{s}ds \leq C\varphi(t)\ ,\quad 0 < t \leq 1\ , \tag{12.1}$$

$$t^2\int_t^1 \frac{\varphi(s)}{s^3}ds \leq C\varphi(t)\ ,\quad 0 < t \leq 1\ . \tag{12.2}$$

We summarize these results in

Theorem 3. *Suppose that $\int_0^1[\varphi(t)/t^2]dt = +\infty$. Then for $n = 1\,,\ 2$ with any φ and for $n \geq 3$ with φ satisfying (12.1) and (12.2), $\Lambda_\varphi \in Synt$ and $\lambda_\varphi \in synt$.*

This is indeed a natural conjecture that the conclusion of this theorem remains true for all n and for every majorant φ with the property (11).

References

[1] R.A. Adams, *Sobolev Spaces*, Academic Press 1975.

[2] J.J. Benedetto, *Spectral Synthesis*, New York 1975.

[3] J. Bruna, Spectral synthesis in non quasi-analytic classes of infinitely differentiable functions, *Bull. Sci. Math.*, 2-e serie, **104** (1980), 65-78.

[4] V.I. Burenkov, On the approximations of functions in Sobolev spaces by finite functions on an arbitrary open set, *Soviet Math. Doklady* **13, 2** (1972), 60-63.

[5] E.M. Dyn'kin, Asymptotic Cauchy problem for Laplace equation, Preprint, Department of Mathematics, Technion, Haifa, Israel, 1994.

[6] E.M. Dyn'kin, Asymptotic Cauchy problem for Laplace equation, II. Applications, Preprint, Department of Mathematics,Technion, Haifa, Israel, 1994.

[7] E.M. Dyn'kin and L.G. Hanin, Spectral synthesis of ideals in Zygmund algebras and quasiharmonic extensions, Preprint, Department of Mathematics, Technion, Haifa, Israel, 1993.

[8] E. Gagliardo, Proprietà di alcune classi di funzioni in più variabili, *Ricerche di Mat. Napoli* **7** (1958), 102-137.

[9] G. Glaeser, Étude de quelques algèbres tayloriennes, *Journal d'Analyse Math.* **6** (1958), 1-124.

[10] G. Glaeser, Synthèse spectrale des ideaux de fonctions lipschitziennes, *C. R. Acad. Sci. Paris* **260, 6** (1965), 1539-1542.

[11] R.E. Edwards, *Functional Analysis*, Holt, Rinehart and Winston, New York 1965.

[12] L.G. Hanin, Spectral synthesis of ideals in algebras of functions having generalized derivatives, *Russian Math. Surveys* **39, 2** (1984), 167-168.

[13] L.G. Hanin, Sobolev spaces as Banach algebras, *Studies in the Theory of Functions of Several Real Variables*, Yaroslavl', Yaroslavl' State University, 1984, 127-146 (in Russian).

[14] L.G. Hanin, Closed primary ideals and point derivations in Zygmund algebras, *Constructive Theory of Functions*, Proceedings, Publishing House of the Bulgarian Academy of Sciences, Sofia 1984, 397-402.

[15] L.G. Hanin, *Closed Ideals in Algebras of Smooth Functions*, Ph.D. thesis, Yaroslavl', Yaroslavl' State University, 1984, 133 p. (in Russsian).

[16] L.G. Hanin, The structure of closed ideals in some algebras of smooth functions, *Studies in the Theory of Functions of Several Real Variables*, Yaroslavl', Yaroslavl' State University, 1986, 97-114 (in Russian); English transl. in: *AMS Translations* **149** (1991), 97-113.

[17] L.G. Hanin, A theorem on spectral synthesis of ideals for a class of Banach algebras, *Soviet Math. Doklady* **35, 1** (1987), 108-112.

[18] L.G. Hanin, Description of traces of functions with higher order derivatives satisfying generalized Zygmund condition to an arbitrary closed set, *Studies in the Theory of Functions of Several Real Variables*, Yaroslavl', Yaroslavl' State University, 1988, 128-144 (in Russian).

[19] L.G. Hanin, Closed ideals in algebras of functions satisfying generalized Zygmund condition, *Geometrical Problems of the Theory of Functions and Sets*, Kalinin, Kalinin State University, 1989, 74-83 (in Russian).

[20] L.G. Hanin, Spectral synthesis of ideals in algebras of smooth functions, Preprint, Department of Mathematics, Technion, Haifa, Israel, 1994.

[21] L.I. Hedberg, Spectral synthesis in Sobolev spaces and uniqueness of solutions of the Dirichlet problem, *Acta Math.* **147** (1981), 237-264.

[22] L.I. Hedberg and T.H. Wolff, Thin sets in nonlinear potential theory, *Ann. Inst. Fourier* **33, 4** (1983), 161-187.

[23] E. Hewitt and K.A. Ross, *Abstract Harmonic Analysis*, Volume 1, Springer-Verlag, Berlin-Heidelberg-New York 1963.

[24] E. Hewitt and K.A. Ross, *Abstract Harmonic Analysis*, Volume 2, Springer-Verlag, Berlin-Heidelberg-New York 1970.

[25] J.-P. Kahane, *Séries de Fourier Absolument Convergentes*, Springer-Verlag, Berlin-Heidelberg-New York 1970.

[26] E. Lasker, Zur Theorie der Moduln und Ideale, *Math. Ann.* **60** (1905), 20-116.

[27] B. Malgrange, *Ideals of Differentiable Functions*, Oxford University Press 1966.

[28] P. Malliavin, Sur l'impossibilité de la synthèse spectrale dans une algèbre de fonctions presque périodique, *C. R. Acad. Sci. Paris* **248** (1959), 1756-1759.

[29] P. Malliavin, Sur l'impossibilité de la synthèse spectrale sur la droite, *C. R. Acad. Sci. Paris* **248** (1959), 2155-2157.

[30] P. Malliavin, Impossibilité de la synthèse spectrale sur le groupes abéliens non compacts, *Publications Mathématiques de l'Institut des Hautes Études Scientifiques Paris* **2** (1959), 61-68.

[31] V.G. Maz'ya and T.O. Shaposhnikova, Theory of multipliers in spaces of differentiable functions, *Russian Math. Surveys* **38, 3** (1983), 23-86.

[32] E.J. McShane, Extension of range of functions, *Bull. Amer. Math. Soc.* **40** (1934), 837-842.

[33] E. Noether, Idealtheorie in Ringbereichen, *Math. Ann.* **83** (1921), 24-66.

[34] N.M. Osadchij, Algebras $L_n^2(\Gamma)$ and the structure of closed ideals in these algebras, *Ukrainian Math. J.* **26, 5** (1974), 669-670 (in Russian).

[35] L. Schwartz, Théorie génerale des fonctions moyenne-périodique, *Ann. of Math.* **48** (1947), 857-929.

[36] L. Schwartz, Sur une proprieté de synthèse spectrale dans les groupes non compacts, *C. R. Acad. Sci. Paris* **227** (1948), 424-426.

[37] D.R. Sherbert, The structure of ideals and point derivations in Banach algebras of Lipschitz functions, *Trans. Amer. Math. Soc.* **111, 2** (1964), 240-272.

[38] G.E. Shilov, *On Regular Normed Rings, Trudy Mat. Inst. Steklova* **21** (1947), 1-118 (in Russian).

[39] E.E. Shnol', Closed ideals in the ring of continuously differentiable functions, *Mat. Sbornik* **27** (1950), 281-284 (in Russian).

[40] P.A. Shvartsman, On the trace of functions of two variables that satisfy Zygmund condition, *Studies in the Theory of Functions of Several Real Variables*, Yaroslavl', Yaroslavl' State University, 1982, 145-168 (in Russian).

[41] S.L. Sobolev, On a theorem in functional analysis, *Mat. Sb.* **4, 3** (1938), 471-497 (in Russian).

[42] M.H. Stone, Applications of the theory of Boolean rings to general topology, *Trans. Amer. Math. Soc.* **41** (1937), 375-481.

[43] R.S. Strichartz, Multipliers on fractional Sobolev spaces, *J. Math. Mech.* **16, 9** (1967), 1031-1060.

[44] L. Waelbroeck, Closed ideals of Lipschitz functions, in: *Proceedings Int. Symp. "Function Algebras"*, F. Birtel (ed.), Scott-Foresmann, Chicago 1966, 322-325.

[45] H. Whitney, Analytic extensions of differentiable functions defined in closed sets, *Trans. Amer. Math. Soc.* **36** (1934), 63-89.

[46] H. Whitney, On ideals of differentiable functions, *Amer. J. Math.* **70** (1948), 635-658.

[47] A. Zygmund, Smooth functions, *Duke Math. J.* **12, 1** (1945), 47-76.

A Characterization of Lacunary Sets and Spectral Properties of Fourier Multipliers

OSAMU HATORI* Department of Mathematics, Tokyo Medical College, 6 - 1 - 1 Shinjuku, Shinjuku-ku Tokyo 160, Japan

ABSTRACT

For a wide class of spaces of continuous functions including certain Besov spaces the functions which operate on them are Lipschitz. As a consequence of our results we obtain a characterization of $\Lambda(p)$ sets in a discrete abelian group. We also determine the maximal ideal space of the algebra of p-q multipliers on an infinite compact abelian group for $1 < p < q < \infty$.

1 INTRODUCTION

Let G be an infinite compact abelian group, $1 \leq p \leq q \leq \infty$ and let $M(p,q)$ be

*Supported in part by the Grants in Aid for Scientific Reseafrch, the Ministry of Education, Science and Culture, Japan

the Banach algebra of all the translation invariant bounded operators from $L^p(G)$ to $L^q(G)$. We say that such an operator on $M(p,q)$ is a p-q multiplier. Igari and Sato [13] show that certain non-Lipschitz but continuous functions operate on the Fourier transforms $\widehat{M(p,q)}$ of $M(p,q)$ if $p \neq q$. On the other hand for various classes of spaces of functions the operating functions are Lipschitzian. It is well-known that functions on $[-1,1]$ which operate on $\widehat{M(p,p)}$ for $1 \leq p < \infty$, $p \neq 2$ are entire functions (cf. [12]). Bourdaud and Kateb [5] and Bourdaud [4] characterize the operating functions on the Besov space $B^s_{p,q}(\mathbb{R}^n)$ in the case of $0 < s < 1$, $1 \leq p < \infty$, $1 \leq q \leq \infty$. In short they show that non-Lipschitz functions never operate on $B^s_{p,q}(\mathbb{R}^n)$.

In this paper we show that only Lipschitz functions operate on a space of continuous functions which satisfies a very natural condition which is common to various classes of spaces. As a consequence we show that results similar to those of Bourdaud and Kateb [5] and Bourdaud [4] concerning operating functions on $B^s_{p,q}(\mathbb{R}^n)$ still hold for the case that $B^s_{p,q}(\mathbb{R}^n)$ consists of continuous functions, that is, $n < p \leq \infty$, $0 < q \leq \infty$ and $n/p < s < 1$ or $n < p < \infty$, $0 < q \leq 1$ and $n/p = s$.

Figà-Talamanca [7] anounced a characterization of $\Lambda(p)$ sets via $\widehat{M(p,p)}$ on the dual group $\widehat{G}$ of G for $2 < p < \infty$. We show a characterization of $\Lambda(\max\{p',q\})$ set via $\widehat{M(p,q)}$ on $\widehat{G}$ for $1 < p < q < \infty$. The proof depends on the above result of operating functions and a theorem of Igari and Sato [13].

In the third section we determine the maximal ideal space Y of $M(p,q)$ on an infinite compact abelian group G. By the result we see that the dual group Γ is dense in Y. We also see that Y is homeomorphic to an open subset of the maximal ideal space of $M(r,r)$ for $r = \max\{p',q\}$. We also show that $M(p,q)$ is a regular and self-adjoint commutative Banach algebra. We also study the spectral properties of Fourier multipliers. In particular we show that for every $T \in M(q,q)$ there exist T_1 and T_2 in $M(q,q)$ such that $T = T_1 + T_2$ and each T_i has the natural spectrum for $i = 1,2$.

2 Symbolic calculus

In this section we show that only locally Lipschitz functions operate on a wide class of spaces of continuous functions. Let Y be a locally compact Hausdorff space. We denote $C^b(Y)$ (resp. $C^b_R(Y)$) the space of all complex-valued bounded functions on Y. If Y is compact, then we simply write $C(Y)$ (resp. $C_R(Y)$) instead of $C^b(Y)$ (resp. $C^b_R(Y)$). The uniform norm of a complex-valued bounded function f on a topological space T is denoted by $\|f\|_{\infty(T)}$.

Lemma 1. *Let T be a topological space, $\mathfrak{A}$ a complex algebra which consists of complex-valued bounded functions on T, and D a plane domain which contains 0.*

Suppose that for every compact subset K of T there is f in $\mathfrak{A}$ such that $f = 1$ on K and that $\|f\|_{\infty(T)} = 1$. Suppose also that a complex-valued function φ defined on D operates on $\mathfrak{A}$. Then, for every $w_0 \in D$ and for every compact subset K of T there is a positive number ε such that $\{w + w_0 : w \in \mathbb{C}, |w| < \varepsilon\} \subset D$ and $\Phi \circ f|K \in \mathfrak{A}|K$ for every f in $\mathfrak{A}$ with $f(T) \subset \{w \in \mathbb{C} : |w| < \varepsilon\}$, where $\Phi(w) = \varphi(w + w_0) - \varphi(w_0)$ on $\{w \in \mathbb{C} : |w| < \varepsilon\}$.

Proof. The case $w_0 = 0$ is trivial, so we assume $w_0 \neq 0$. Let D' be a simply connected bounded domain such that $\{0, w_0\} \subset D' \subset D$ and π the inverse Riemann map from $\{z \in \mathbb{C} : |z| < 1\}$ onto D' with $\pi(0) = 0$. Let $z_0 = \pi^{-1}(w_0)$ and take a positive real number ε_1 so that $\varepsilon_1 < 1 - |z_0|$. Let $\Delta = \{z \in \mathbb{C} : |z| \leq 1 - \varepsilon_1\}$. Then $\varepsilon_3 = \inf\{|w - \pi(z)| : w \in \mathbb{C} \setminus D', z \in \Delta\}$ is positive since $\pi(\Delta)$ is a compact subset of D'. Let ε_2 be a positive number less than $\varepsilon_3/2$. Then there is an analytic polynomial P_{ε_2} such that $P_{\varepsilon_2}(0) = 0$ and $\|P_{\varepsilon_2} - \pi\|_{\infty(\Delta)} < \varepsilon_2 |z_0|$. Put

$$P(z) = P_{\varepsilon_2}(z) + z(\pi(z_0) - P_{\varepsilon_2}(z_0))/z_0.$$

Then $P(0) = 0$ and $P(z_0) = w_0$ and $P(\Delta) \subset D'$. Choose a positive number ε such that

$$\inf\{|w - P(z)| : w \in \mathbb{C} \setminus D', |z| \leq 1 - \varepsilon_1\} > \varepsilon.$$

Let f_K be a function in $\mathfrak{A}$ such that $f_K = z_0$ on K, $\|f_K\|_{\infty(T)} = |z_0|$. Then $P(f_K) \in \mathfrak{A}$ and $(f + P(f_K))(T) \subset D'$ for every $f \in \mathfrak{A}$ with $f(T) \subset \{z \in \mathbb{C} : |z| < \varepsilon\}$. It follows that $\varphi(f + P(f_K)) - \varphi(P(f_K)) \in \mathfrak{A}$ for every $f \in \mathfrak{A}$ with $f(T) \subset \{z \in :|z| < \varepsilon\}$. Since $P(f_K)|K = w_0$ we see that $\Phi(f)|K \in \mathfrak{A}|K$ for every $f \in \mathfrak{A}$ with $f(T) \subset \{z \in \mathbb{C} : |z| < \varepsilon\}$. □

Lemma 2. *Let X be a compact Hausdorff space and E a real quasi-Banach space with $1 \in E$, continuously embedded in $C_R(X)$. Suppose that there is a positive real number ε such that for every pair of compact subsets K_1 and K_2 of X there is a function $f \in E$ with $\|f\|_E \leq 1$ such that $f \geq \varepsilon$ on K_1 and $f \leq -\varepsilon$ on K_2. Then $E = C_R(X)$.*

Proof. E is quasi-Banach means that there is a finite constant c that satisfies $\|f+g\|_E \leq c(\|f\|_E + \|g\|_E)$ for all f and g in E, where $\|\cdot\|_E$ has all the properties of a norm except (possibly) subadditivity. By multiplying $\|\cdot\|_E$ by a sufficiently large positive constant without loss of generality, we may suppose that $\|f\|_{\infty(X)} \leq \|f\|_E$ for every f in E. Let $\widetilde{E}$ be the space of all bounded (with respect to the norm $\|\cdot\|_E$) sequences in E. We may identify a sequence $\langle f_n \rangle \in \widetilde{E}$ with the function $\mathfrak{f}$ on $X \times \mathbb{N}$ such that $\mathfrak{f}(x, n) = f_n(x)$. Since $\|\cdot\|_{\infty(X)} \leq \|\cdot\|_E$ the function $\mathfrak{f}$ is uniquely extended to the function $\tilde{f}$ on the Stone-Čech compactification $\widetilde{X}$ of the product space $X \times \mathbb{N}$, where $\mathbb{N}$ is the discrete space of positive integers. Thus we may suppose that $\widetilde{E}$ is a quasi-Banach space of continuous functions on $\widetilde{X}$ with respect to the norm $\|\cdot\|_{\widetilde{E}}$ defined by $\|\tilde{f}\|_{\widetilde{E}} = \sup_n \|f_n\|_E$ for $\tilde{f} \in \widetilde{E}$ (cf.[2]). Suppose that $\widetilde{K}_1$ and $\widetilde{K}_2$ are disjoint compact subsets in $\widetilde{X}$. Let $\widetilde{G}_i$ be a compact neighborhood of $\widetilde{K}_i$ for each $i = 1, 2$ such that $\widetilde{G}_1 \cap \widetilde{G}_2 = \emptyset$. Then we see that $\widetilde{K}_i \subset \overline{\widetilde{G}_i \cap (X \times \mathbb{N})}$, where $\bar{\cdot}$ denotes the closure in $\widetilde{X}$. Let P_n be the projection from $X \times \{n\}$ onto X such that $P_n((x, n)) = x$. Then for every positive integer n there is $u_n \in E$ with $\|u_n\|_E \leq 1$ such that $u_n \geq \varepsilon$ on $P_n(\widetilde{G}_1 \cap (X \times \{n\}))$ and $u_n \leq -\varepsilon$ on $P_n(\widetilde{G}_2 \cap (X \times \{n\}))$. Then we see that $\tilde{u} = \langle u_n \rangle \in \widetilde{E}$, $\|\tilde{u}\|_{\infty(\widetilde{X})} \leq 1$, $\tilde{u} \geq \varepsilon$ on $\widetilde{K}_1$ and $\tilde{u} \leq -\varepsilon$ on $\widetilde{K}_2$. It follows that $\widetilde{E}$ is uniformly dense in $C_R(\widetilde{X})$. (Suppose that $\tilde{f} \in C_R(\widetilde{X})$. We may suppose that $\|\tilde{f}\|_{\infty(\widetilde{X})} \leq 1$. Put $\widetilde{K}_1 = \{p \in \widetilde{X} : \tilde{f}(p) \geq 1/2\}$ and $\widetilde{K}_2 = \{p \in \widetilde{X} : \tilde{f}(p) \leq -1/2\}$. Then there is $\tilde{u} \in \widetilde{E}$. with $\|\tilde{u}\|_{\infty(\widetilde{X})} \leq 1$, $\tilde{u} \geq \varepsilon$ on $\widetilde{K}_1$ and $\tilde{u} \leq -\varepsilon$ on $\widetilde{K}_2$. Put $\tilde{g} = \tilde{f} - \tilde{u}/3$. Then $\|\tilde{g}\|_{\infty(X)} \leq \max\{1 - \varepsilon/3, 5/6\}$. By a successive approximation argument we see that $\widetilde{E}$ is uniformly dense in $C_R(\widetilde{X})$.) Thus there is a positive constant M such that for every $u \in C_R(X)$ with $\|u\|_{\infty(X)} \leq 1$ there is $u_1 \in E$ with $\|u_1\|_E < M$ such that $\|u - u_1\|_{\infty(X)} \leq (2c)^{-1}$. By induction we can choose a sequence $\{u_n\}$ in E such that

$$\|u_n\|_E < M(2c)^{-n+1}, \ \|u - \sum_{k=1}^{n} u_k\|_{\infty(X)} \leq (2c)^{-n}.$$

It follows that $\sum_{k=1}^{\infty} u_n$ converges in E and coincides with u since E is continuously embedded in $C_R(X)$. We conclude that $E = C_R(X)$. □

It should be noted that Lemma 2 was proved for Banach spaces of continuous functions by Ellis [6] and Bernard [3]. Similar result was proved for Banach spaces of complex valued continuous functions by Bade and Curtis [1]. We say that B is a multiplication algebra if B is a quasi-Banach space and is also an algebra such that there is a constant c with $\|fg\|_B \leq c\|f\|_B\|g\|_B$ for every pair $f, g \in B$.

Lemma 3. *Let B be a complex quasi-Banach space with $1 \in B$ which is continuously embedded in $C(X)$. Suppose that B is a multiplication algebra and that there is a positive real number ε such that for every pair of disjoint compact subsets K_1 and K_2 of X there is $f \in B$ with $\|f\|_B \leq 1$ such that $Ref \geq \varepsilon$ on K_1 and $Ref \leq -\varepsilon$ on K_2. Then $B = C(X)$.*

Proof. By simple calculation we see that $\mathrm{Re}B$ is a real quasi-Banach space with respect to the quotient norm $\|\cdot\|_{\mathrm{Re}B}$ defined by $\|u\|_{\mathrm{Re}B} = \inf\{\|f\|_B : f \in B, \mathrm{Re}f = u\}$ for $u \in \mathrm{Re}B$. So by Lemma 2 we have $\mathrm{Re}B = C_R(X)$. Let $\widetilde{B}$ be the space of all bounded sequences in B. Then we may suppose that $\widetilde{B}$ is a space of continuous functions on $\widetilde{X}$. We see that $\widetilde{B}$ is an algebra since B is a multiplication algebra. We also see that $\widetilde{B}$ separates the different points in $\widetilde{X}$ since $\mathrm{Re}\widetilde{B} = \widetilde{\mathrm{Re}B} = C_R(\widetilde{X})$. It follows by a theorem of Hoffman and Wermer [11] that $\widetilde{B}$ is dense in $C(\widetilde{X})$. In the same way as in the proof of Lemma 2 we see that $B = C(X)$. □

Proposition 4. *Let Y be a locally compact Hausdorff space and A a complex quasi-Banach space which is continuously embedded in $C^b(Y)$ and D a plane domain which contains 0. Suppose that A is a multiplication algebra. Suppose also that A satisfies the following two conditions: (1) for every pair of disjoint compact subsets K_1 and K_2 of Y there is $f \in A$ with $\|f\|_{\infty(Y)} = 1$ such that $f = 1$ on K_1 and $f = 0$ on K_2; (2) there exists a sequence $\{y_n\}$ of distinct points in Y which has a cluster point in Y such that $A|G \neq C(G)$ for every compact neighborhood G of each y_n. Suppose that φ is a complex-valued continuous function on D which operates on A. Then φ is locally Lipschitz.*

Proof. By multiplying $\|\cdot\|_A$ by a sufficiently large number we may assume without loss of generality that $\|\cdot\|_{\infty(Y)} \leq \|\cdot\|_A$ since A is continuously embedded in $C^b(Y)$. We may also assume that $\|fg\|_A \leq \|f\|_A\|g\|_A$ since A is a multiplication algebra. Note that $\|1\|_A$ need not equal to 1 even if $1 \in A$. We may assume that there is a compact subset K of Y whose interior $\mathrm{int}K$ contains every y_n. Suppose that φ is not locally Lipschitzian at w_0, that is, for every $\varepsilon > 0$ there are $w_1, w_2 \in D$ with $|w_0 - w_1| + |w_0 - w_2| < \varepsilon$ such that $\varepsilon|\varphi(w_1) - \varphi(w_2)| > |w_1 - w_2|$. By Lemma 1 there is a positive real number ε_K such that $\Phi(w) = \varphi(w + w_0) - \varphi(w_0)$ is well-defined on $\{w \in \mathbb{C} : |w| < \varepsilon_K\}$ and $\Phi(f)|K \in A|K$ for every f in A with $f(Y) \subset \{w \in \mathbb{C} : |w| < \varepsilon_K\}$. We shall show that there is a finite subset S of $\mathrm{int}K$ such that for every $x \in \mathrm{int}K \setminus S$ there are two positive real numbers $\delta_x < \varepsilon_K$ and ε_x, and a compact neighborhood G_x of x such that $\Phi(u) \in A|G_x$ and $\|\Phi(u)\|_{A|G_x} < \varepsilon_x$ for every $u \in A|G_x$ with $\|u\|_{A|G_x} < \delta_x$, where

$$\|u\|_{A|G_x} = \inf\{\|U\|_A : U \in A, U|G_x = u\}.$$

Suppose not. Then there is a sequence $\{x_n\}$ in $\mathrm{int}K$ such that for any compact neighborhood G of any x_n and for any real numbers δ and ε there is $u \in A|G$ such that $\|u\|_{A|G} < \delta$, $\Phi(u) \in A|G$ and $\|\Phi(u)\|_{A|G} > \varepsilon$. Without loss of generality we may suppose that there is a compact neighborhood $G_n \subset \mathrm{int}K$ of each x_n such that $G_n \cap \overline{(\bigcup_{m\neq n} G_m)} = \emptyset$ for every n. Then by condition (1) there is $f_n \in A$ with $\|f_n\|_{\infty(Y)} = 1$ such that $f_n = 1$ on G_n and $f_n = 0$ on $\overline{(\bigcup_{m\neq n} G_m)}$ for every n. We see that for every n there is $u_n \in A|G_n$ such that

$$\|u_n\|_{A|G_n} < 2^{-n-1}(\|f_n\|_A)^{-1}c^{-n}\varepsilon_K,$$

$\Phi(u_n) \in A|G_n$ and $\|\Phi(u_n)\|_{A|G_n} > n$, where c is a constant greater than or equal to 1 such that $\|f + g\|_A \leq c(\|f\|_A + \|g\|_A)$ for every f, $g \in A$. Then for each n there is $\tilde{u}_n \in A$ such that $\tilde{u}_n|G_n = u_n$ and $\|\tilde{u}_n\|_A < 2^{-n-1}(\|f_n\|_A)^{-1}c^{-n}\varepsilon_K$. Since the inequality $\|\sum_{n=1}^m h_n\|_A \leq \sum_{n=1}^m c^n\|h_n\|_A$ holds for $h_n \in A$ we see that $g = \sum_{n=1}^\infty f_n\tilde{u}_n$ converges in A and $\sum_{n=1}^\infty f_n\tilde{u}_n$ also converges uniformly to g since $\|\cdot\|_{\infty(Y)} \leq \|\cdot\|_A$. We also see that $g(Y) \subset \{w \in \mathbb{C} : |w| < \varepsilon_K\}$ so that $\Phi(g)|K \in A|K$. Since $G_n \subset K$ we have $\|\Phi(g)|G_n\|_{A|G_n} \leq \|\Phi(g)|K\|_{A|K}$. On the other hand we have

$$n < \|\Phi(u_n)\|_{A|G_n} = \|\Phi(g)|G_n\|_{A|G_n}$$

since $\Phi(g)|G_n = \Phi(u_n)$, which is a contradiction for a sufficiently large n. We conclude that a finite subset S of int$K$ with the required properties exists.

Choose a point $x \in \{y_n\} \setminus S$. Then there are two positive real numbers $\delta_x < \varepsilon_K$ and ε_x and a compact neighborhood G_x of x such that $\Phi(u) \in A|G_x$ and $\|\Phi(u)\|_{A|G_x} < \varepsilon_x$ for every $u \in A|G_x$ with $\|u\|_{A|G_x} < \delta_x$. By multiplying Φ by $\delta_x/(4c\varepsilon_x)$ we may suppose that $\|\Phi(u)\|_{A|G_x} < \delta_x/(4c)$ for every $u \in A|G_x$ with $\|u\|_{A|G_x} < \delta_x$. Put $\alpha = \delta_x/(20c^2\|1\|_{A|G_x})$. Note that $\alpha < \varepsilon_K/(2c)$ since $c \geq 1$ and $\|1\|_{A|G_x} \geq 1$. Since Φ is not locally Lipschitzian at 0 we see by a simple calculation that there are complex numbers w_1, θ, t_0 and η with $|w_1| < \alpha$, $0 \leq \theta < 2\pi$, $0 \leq t_0 < \alpha$ and $t_0 < \eta$ such that

$$|\Phi(t_0e^{i\theta} + w_1) - \Phi(w_1)| = 10t_0c, \quad |\Phi(te^{i\theta} + w_1) - \Phi(w_1)| > 10tc$$

for every t with $t_0 < t < \eta$. If $t_0 \neq 0$, then put

$$\Phi_0(w) = \{\Phi(we^{i\theta} + w_1) - \Phi(w_1)\}|\Phi(t_0e^{i\theta} + w_1) - \Phi(w_1)|/(\Phi(t_0e^{i\theta} + w_1) - \Phi(w_1)),$$

if $t_0 = 0$, then put

$$\Phi_0(w) = \Phi(we^{i\theta} + w_1) - \Phi(w_1)$$

on $\{w \in \mathbb{C} : |w| < \varepsilon_K/(2c)\}$. Then the function Φ_0 is well-defined since Φ is defined on $\{w \in \mathbb{C} : |w| < \varepsilon_K\}$ and $c \geq 1$. Let $u \in A|G_x$ with $\|u\|_{A|G_x} < \delta_x/(2c)$. Then $\Phi_0(u) \in A|G_x$ and $\|\Phi_0(u)\|_{A|G_x} < \delta_x/(2c)$ since

$$\|e^{i\theta}u + w_1\|_{A|G_x} \leq c(\|u\|_{A|G_x} + |w_1|\|1\|_{A|G_x}) < \delta_x.$$

We show that $A|G_x = C(G_x)$. Suppose not. By Lemma 3 we see that there is a sequence $\{(G_1^{(n)}, G_2^{(n)})\}$ of pairs of disjoint compact subsets of G_x such that

$$M_n = \inf\{\|u\|_{A|G_x} : u \in A|G_x, u(G_1^{(n)}) = 0, u(G_2^{(n)}) = 1\} \to \infty$$

as $n \to \infty$. Note that $M_n < \infty$ by the condition (1). (Suppose that there is not such a sequence. Then there is a positive number M such that for every pair of disjoint compact sets G_1 and G_2 of G_x there exists $u, v \in A|G_x$ such that $\|u\|_{A|G_x} < M$, $\|v\|_{A|G_x} < M$, $u = 0$ on G_1, $v = 1$ on G_1, $u = 1$ on G_2, $v = 0$ on G_2. Put $\varepsilon = 1/(2M)$. Then we see that $(v-u)/(2M) \in A|G_x$, $\|(v-u)/(2M)\|_{A|G_x} < 1$, $(v-u)/(2M) = \varepsilon$ on G_1 and $(v-u)/(2M) = -\varepsilon$ on G_2. It follows that $A|G_x = C(G_x)$ by Lemma 3.) For each n choose $u_n \in A$ with $\|u_n\|_A < 2M_n$ such that $u_n(G_1^{(n)}) = 0$ and $u_n(G_2^{(n)}) = 1$. Put

$$v_n = (\delta_x u_n/(8M_nc) + t_0e_x),$$

where e_x is a function in A such that $e_x = 1$ on G_x and $\|e_x\|_A < 2\|1\|_{A|G_x}$. Then we have $\Phi_0(v_n|G_x) \in A|G_x$ and $\|\Phi_0(v_n|G_x)\|_{A|G_x} < \delta_x/(2c)$ since $\|v_n|G_x\|_{A|G_x} \leq \|v_n\|_A < \delta_x/(2c)$. Put

$$f_n = 4\{\Phi_0(v_n|G_x) - \Phi_0(t_0)\}M_n/(5\delta_x).$$

Then $\|f_n\|_{A|G_x} < 4M_n/5$, $f_n = 0$ on $G_1^{(n)}$ and

$$f_n = 4\{\Phi_0(\delta_x/(8M_nc)+t_0) - \Phi_0(t_0)\}M_n/(5\delta_x)$$

on $G_2^{(n)}$. Since $M_n \to \infty$ as $n \to \infty$ we have that there is n_0 such that $\delta_x/(8M_{n_0}c)+t_0 < \eta$. Thus we see that f_{n_0} is constant and $|f_{n_0}| \geq 1$ on $G_2^{(n_0)}$, while $\|f_{n_0}\|_{A|G_x} < 4M_{n_0}/5$ which contradicts the definition of M_{n_0}. It follows that φ is locally Lipschitzian on D. □

Proposition 5. *Let Y be a locally compact Hausdorff space with a non-empty derived set. Let B be a complex quasi-Banach space which is continuously embedded in $C^b(Y)$, and D a plane domain which contains the origin. Suppose that $fg \in B$ for every f and g in B. Suppose also that B satisfies that for every pair of disjoint compact subsets K_1 and K_2 there is $f \in B$ with $\|f\|_{\infty(Y)} = 1$ such that $f = 1$ on K_1 and $f = 0$ on K_2. Then every complex-valued function φ defined on D which operates on B is continuous on D.*

Proof. Suppose that φ is discontinuous at $w_0 \in D$. Let x be an accumulation point in Y and G a compact neighborhood of x. Then there is a sequence $\{x_n\} \subset G$ such that $x_n \notin \overline{\{x_k\}_{k\neq n}}$ for every n. We see by the condition on B that for every n there is $f_n \in B$ with $f_n(x_n) = 1$ and $f_n = 0$ on $\overline{\{x_k\}_{k\neq n}}$. In the same way as in the proof of Lemma 1 we see that there is $f_0 \in B$ such that $f_0 = w_0$ on G and

$$\inf\{|w - f_0(y)| : w \in \mathbb{C}\setminus D, y \in Y\} = \varepsilon > 0.$$

Since φ is discontinuous at w_0 there is a sequence $\{w_n\} \subset D$ such that $|w_n - w_0| \leq 2^{-n-1}(\|f_n\|_A)^{-1}c^{-n}\varepsilon$ and $\inf_n\{|\varphi(w_n) - \varphi(w_0)|\} > 0$, where c is a constant with $c \geq 1$ such that $\|f+g\|_A \leq c(\|f\|_A + \|g\|_A)$ for every f and g in A. Then

$$f = f_0 + \sum_{n=1}^{\infty}(w_n - w_0)f_n$$

converges in A and $f(Y) \subset D$, so we have $\varphi \circ f \in A$. Choose $x_0 \in \overline{\{x_n\}} \setminus \{x_n\}$. Note that $\overline{\{x_n\}} \setminus \{x_n\} \neq \emptyset$ since $\{x_n\} \subset G$ and $x_n \notin \overline{\{x_k\}_{k\neq n}}$ for every n. Since $f(x_n) = w_n$ we have $f(x_0) = w_0$ and $\varphi \circ f(x_n) = \varphi(w_n)$. Thus we see for every m that

$$|\varphi \circ f(x_0) - \varphi \circ f(x_m)| = |\varphi(w_0) - \varphi(w_m)| \geq \inf\{|\varphi(w_n) - \varphi(w_0)|\} > 0,$$

which is a contradiction since $\varphi \circ f$ is continuous. □

Note that the condition that Y has a non-empty derived set is essential. For example, let G be a compact abelian group and $1 \leq p < q \leq \infty$. Then the Fourier transform $\widehat{M(p,q)}$ of the algebra $M(p,q)$ of p-q multipliers is a Banach algebra of bounded functions on the discrete group $\widehat{G}$ and a discontinuous function operates on the algebra $\widehat{M(p,q)}$ [13].

Proposition 6. *Let Y be a locally compact Hausdorff space and A a complex quasi-Banach space which is continuously embedded in $C^b(Y)$. Suppose that A is a multiplication algebra, separates the distinct points in Y and $\{y \in Y : f(y) = 0$ for $\forall f \in A\} = \emptyset$. Suppose also that there is a sequence $\{y_n\}$ of distinct points in Y which has a cluster point in Y such that $A|G \neq C(G)$ for every compact neighborhood G of y_n for every n. Let D be a plane domain which contains 0 and ε a positive real number such that $\{z \in \mathbb{C} : |z| \leq 2\varepsilon\} \subset D$. Suppose that φ is a complex-valued continuous function defined on D such that $|\varphi| \leq 1$ on D, $\varphi = 0$ on $\{z \in \mathbb{C} : |z| \leq \varepsilon\}$ and $\varphi = 1$ on $\{z \in \mathbb{C} : |z - 3\varepsilon/2| \leq \varepsilon/3\}$. Suppose that φ operates on A. Then every function defined on D which operates on A is locally Lipschitzian on D.*

Proof. Let βY be the Stone-Čech compactification of Y. Then we may suppose that each function $f \in A$ is defined on βY. Let X be the quotient space of βY by identifying the points in βY which are not separated by functions in A. Put $\mathrm{Ker}_X A = \{x \in X : f(x) = 0, \forall f \in A\}$. Then we may suppose that $A \subset C_0(X \setminus \mathrm{Ker}_X A)$, where $C_0(X \setminus \mathrm{Ker}_X A)$ is the space of all the complex-valued continuous functions on $X \setminus \mathrm{Ker}_X A$ which vanish at infinity. Since φ operates on A defined on Y and φ is a continuous function we see that φ operates on A defined on $X \setminus \mathrm{Ker}_X A$. Thus by [9] we see that A is dense in $C_0(X \setminus \mathrm{Ker}_X A)$ since φ is not analytic. We may suppose that a compact subset in Y is a compact subset in $X \setminus \mathrm{Ker}_X A$ since A separates points in Y and $\{y \in Y : f(y) = 0$ for $\forall f \in A\} = \emptyset$. It follows that for every pair of disjoint compact subsets K_1 and K_2 of Y there is $f \in A$ such that $f(Y) \subset D$, $f(K_1) \subset \{z \in \mathbb{C} : |z| \leq \varepsilon\}$ and $f(K_2) \subset \{z \in \mathbb{C} : |z - 3\varepsilon/2| \leq \varepsilon/3\}$. So we have $\varphi \circ f \in A$, $\|\varphi \circ f\|_{\infty(Y)} \leq 1$, $\varphi \circ f = 1$ on K_1 and $\varphi \circ f = 0$ on K_2. We have just shown that A satisfies the condition (1) in Proposition 4. Thus by Propositions 4 and 5 we see that every function defined on D which operates on A is locally Lipschitzian on D. □

Theorem 7. *Let Y be a locally compact Hausdorff space and A a complex quasi-Banach space which is continuously embedded in $C^b(Y)$. Suppose that A is a multiplication algebra which separates points in Y and $\{y \in Y : f(y) = 0$ for $\forall f \in A\} = \emptyset$. Suppose also that there is a sequence $\{y_n\}$ of distinct points in Y which has a cluster point in Y such that $A|G \neq C(G)$ for every compact neighborhood G of y_n for every n. Let D be a plane domain which contains the origin. Then we have the following.*

(i) *If $A \ni 1$, then the following* (a) *and* (b) *are equivalent.*

(a) *Every complex-valued locally Lipschitz function defined on D operates on A.*

(b) *The set of all the complex-valued functions defined on D which operate on A coincides with the set of all the complex-valued locally Lipschitz functions on D.*

(ii) *If $A \not\ni 1$, then the following* (a') *and* (b') *are equivalent.*

(a') *Every complex-valued locally Lipschitz function φ defined on D with $\varphi(0) = 0$ operates on A.*

(b') *The set of all the complex-valued functions defined on D which operate on A*

coincides with the set of all the complex-valued locally Lipschitz function φ defined on D with $\varphi(0) = 0$.

Proof. It is trivial that the condition (b) implies the condition (a). We also see that (b′) implies (a′). We show the inverse implications. Suppose that (a) holds. Then a locally Lipschitz function φ on D such that $|\varphi| \leq 1$ on D, $\varphi = 0$ on $\{z \in \mathbb{C} : |z| \leq \varepsilon\}$ and $\varphi = 1$ on $\{z \in \mathbb{C} : |z - 3\varepsilon/2| \leq \varepsilon/3\}$ operates on A, where ε is a positive real number such that $\{z \in \mathbb{C} : |z| \leq 2\varepsilon\} \subset D$. Then by Proposition 6 we see that a function defined on D which operates on A is locally Lipschitzian. We conclude that (b) holds.

Suppose (a′) holds. In the same way as above we see that a function defined on D which operates on A is locally Lipschitzian. We also see that $\varphi(0) \in A$ since $0 \in A$, so $\varphi(0) = 0$ since $1 \notin A$ in this case. We conclude that (b′) holds. □

3 APPLICATIONS

Bourdaud and Kateb [5] and Bourdaud [4] characterize the operating functions on the Besov space $B^s_{p,q}(\mathbb{R}^n)$ defined on $\mathbb{R}^n$ in the case that $0 < s < 1$, $1 \leq p < \infty$ and $1 \leq q \leq \infty$. The situation is different in the following two cases: (1) $0 < s < \min\{1, n/p\}$ and $1 \leq q \leq \infty$ or $s = n/p < 1$ and $1 < q \leq \infty$; (2) $n/p < s < 1$ and $1 \leq q \leq \infty$ or $s = n/p < 1$ and $q = 1$. In the case (1) the operating functions defined on $\mathbb{C}$ are globally Lipschitz and in the case of (2) the operating functions defined on $\mathbb{C}$ are locally Lipschitz. The reason why the situations are different from each other is that $B^s_{p,q}(\mathbb{R}^n)$ consists of continuous functions in the case (2) and there are certain discontinuous functions in $B^s_{p,q}(\mathbb{R}^n)$ in the case (1). We show that the similar results hold in the case that $0 < s < 1$, $0 < p \leq \infty$ and $0 < q \leq \infty$ and $B^s_{p,q}(\mathbb{R}^n)$ consists of continuous functions.

Corollary 8. *Suppose that $n < p \leq \infty$, $0 < q \leq \infty$ and $n/p < s < 1$ or $n < p < \infty$, $0 < q \leq 1$ and $n/p = s$. Let D be a plane domain which contains the origin. Then we see that the set of all the functions defined on D which operate on the Besov space $B^s_{p,q}(\mathbb{R}^n)$ coincides with the set of all the complex-valued locally Lipschitz functions φ defined on D such that $\varphi(0) = 0$.*

Proof. By Theorem 2.8.3 in [17] we see that $B^s_{p,q}(\mathbb{R}^n)$ is a complex quasi-Banach space and a multiplication algebra which consists of complex-valued bounded and uniformly continuous functions on $\mathbb{R}^n$. We also see by Theorems 2.5.12 and 2.8.3 in [17] that every locally Lipschitz function defined on D operates on $B^s_{p,q}(\mathbb{R}^n)$. We also see that $\{y \in \mathbb{R}^n : f(y) = 0 \text{ for } \forall f \in B^s_{p,q}(\mathbb{R}^n)\} = \emptyset$ and $B^s_{p,q}(\mathbb{R}^n)|G \neq C(G)$ for every compact subset G of $\mathbb{R}^n$. It follows by Theorem 7 that the conclusion holds. □

In the same way as above we can characterize the operaiting functions on Besov spaces and Triebel-Lizorkin spaces defined on domains in $\mathbb{R}^n$.

Next we consider a characterization of certain lacunary sets in a discrete abelian group. Let G be an infinite compact abelian group, $\widehat{G}$ the dual group of G and let $1 \le p, q \le \infty$. We denote p' the conjugate of p, i.e., $1/p + 1/p' = 1$. We say that a bounded operator T from $L^p(G)$ to $L^q(G)$ is a p-q multiplier if T is translation invariant. The algebra of all the p-q multipliers is denoted by $M(p,q)$ and $\widehat{M(p,q)}$ is the set of the Fourier transforms of $T \in M(p,q)$. Since G is compact we see that $\widehat{M(p,q)}$ is an algebra of bounded functions on the discrete group $\widehat{G}$. Let E be a subset of $\widehat{G}$ and $2 < r < \infty$. We denote the space of all bounded functions on E by $\ell^\infty(E)$. Trig$G$ denotes the set of all trigonometric polynomials on G and Trig$_E G$ stands for the set of all $f \in$ Trig$G$ such that the Fourier transform $\hat{f}$ of f vanishes off E. We say that E is a $\Lambda(r)$ set if for some constant c, $\|f\|_r \le c\|f\|_2$ for all $f \in$ Trig$_E G$. Figà-Talamanca [7] announced that E is a $\Lambda(r)$ set if and only if E is an interpolation set for $\widehat{M(r,r)}$ (resp. $\widehat{M(r',r')}$), i.e., $\widehat{M(r,r)}|E = \ell^\infty(E)$. An analogous result for the Sidon sets is in [16, Theorem 5.7.3]. By using the results in section 1 we shall obtain a characterization of $\Lambda(r)$ sets in terms of $\widehat{M(p,q)}$ for certain p and q.

Lemma 9. *Let $\mathfrak{Z}$ be a discrete space and $\beta\mathfrak{Z}$ the Stone-Čech compactification of $\mathfrak{Z}$. Suppose that A is a complex Banach algebra which consists of complex-valued continuous functions on $\beta\mathfrak{Z}$ which satisfies the following three conditions: (1) the algebra A separates points in $\beta\mathfrak{Z} \setminus \mathrm{Ker}A$; (2) for every pair of disjoint compact subset K_1 and K_2 of $\beta\mathfrak{Z} \setminus \mathrm{Ker}A$ there is a function f in A such that $f = 1$ on K_1 and $f = 0$ on K_2; (3) there is a discrete subset S of $\beta\mathfrak{Z} \setminus \mathrm{Ker}A$ such that $A|K = C(K)$ for every compact subset K of $\beta\mathfrak{Z} \setminus ((\mathrm{Ker}A) \cup S)$. Then $A|K = C(K)$ for every compact subset K of $\beta\mathfrak{Z} \setminus \mathrm{Ker}A$.*

Proof. We show that for every $x \in S$ there is a compact neighborhood G_x of x such that $A|G_x = C(G_x)$ holds. It follows that the conclusion holds by using a decomposition of unity argument. Let x be a point of S. Suppose that $A|G \neq C(G)$ for every compact neighborhood G of x. Let G_0 be a compact neighborhood of x such that $G_0 \cap ((S \setminus \{x\}) \cup \mathrm{Ker}A) = \emptyset$. By induction there are two sequence $\{X_n\}$ and $\{Y_n\}$ of compact subset of $G_0 \setminus \{x\}$ such that

$$X_n \cap \overline{(\cup_{m\neq n} X_m) \cup (\cup_{k=1}^\infty Y_k)} = \emptyset, \quad Y_n \cap \overline{(\cup_{m\neq n} Y_m) \cup (\cup_{k=1}^\infty X_k)} = \emptyset$$

for every positive integer n, $\overline{X_n \cap \mathfrak{Z}} = X_n$, $\overline{Y_n \cap \mathfrak{Z}} = Y_n$ for every n,

$$M_n = \inf\{\|f\|_A : f \in A,\ f|X_n = 1,\ f|Y_n = 0\} \to \infty$$

as $n \to \infty$. First choose a pair of disjoint non-empty compact subsets X_1 and Y_1 of $G \setminus \{x\}$ with $\overline{\mathrm{int}X_1} = X_1$, $\overline{\mathrm{int}Y_1} = Y_1$, so $\overline{X_1 \cap \mathfrak{Z}} = X_1$ and $\overline{Y_1 \cap \mathfrak{Z}} = Y_1$ since $\overline{U \cap \mathfrak{Z}} \supset U$

holds for every open set U. Suppose that we have chosen $X_1, \dots, X_n$ and $Y_1, \dots, Y_n$ with the required properties. Put

$$M_n = \inf\{\|f\|_A : f \in A,\ f|X_n = 1,\ f|Y_n = 0\}.$$

Let G_n be a compact neighborhood of x such that $G_n \subset G_0$ and

$$G_n \cap ((\cup_{k=1}^n X_k) \cup (\cup_{k=1}^n Y_k)) = \emptyset.$$

Then $A|G_n$ is a Banach algebra of complex-valued continuous functions on G_n with respect to the quotient norm. Note that $\|1\|_{A|G_n}$ need not be equal to 1. By Bernard's lemma [1] we see that $\mathrm{cl}(\widetilde{A|G_n}) \neq C(\widetilde{G_n})$, where $\widetilde{G_n}$ is the Stone-Čech compactification of the direct product $G_n \times N$ and cl· denotes the uniform closure. Thus by a theorem of Hoffman and Wermer [11] we see that $\mathrm{Re}\,\mathrm{cl}(\widetilde{A|G_n}) \neq C_R(\widetilde{G_n})$. It follows that $\mathrm{Re}A|G_n \neq C_R(G_n)$. So by Lemma 5 in [10] that there is a pair of disjoint compact subsets K_1 and K_2 of $G_n \setminus \{x\}$ such that

$$\inf\{\|u\|_{\mathrm{Re}A|G_n} : u \in \mathrm{Re}A|G_n,\quad u = 1 \text{ on } K_1,\ u = 0 \text{ on } K_2\} \geq M_n + 1,$$

where $\|u\|_{\mathrm{Re}A|G_n} = \inf\{\|F\|_A : F \in A,\ \mathrm{Re}F|G_n = u\} = \inf\{\|f\|_{A|G_n} : f \in A|G_n,\ \mathrm{Re}f = u\}$. Note that we may suppose that $K_1 \cup K_2 \subset \mathrm{int}G_n$. It follows by the definition of the quotient norm that

$$\inf\{\|f\|_A : f \in A,\quad f = 1 \text{ on } K_1,\ f = 0 \text{ on } K_2\} \geq M_n + 1.$$

Let O_1 and O_2 be open neighborhoods of K_1 and K_2 respectively such that $\bar{O}_1 \cup \bar{O}_2 \subset \mathrm{int}G_n \setminus \{x\}$ and $\bar{O}_1 \cap \bar{O}_2 = \emptyset$. Put $X_{n+1} = \bar{O}_1$ and $Y_{n+1} = \bar{O}_2$. Then we have

$$\begin{aligned} M_{n+1} &= \inf\{\|f\|_A : f \in A,\quad f = 1 \text{ on } X_{n+1},\ f = 0 \text{ on } Y_{n+1}\} \\ &\geq \inf\{\|f\|_A : f \in A,\quad f = 1 \text{ on } K_1,\ f = 0 \text{ on } K_2\}. \end{aligned}$$

We also see that $\overline{\mathrm{int}X_{n+1} \cap \mathfrak{Z}} = X_{n+1}$ and $\overline{Y_{n+1} \cap \mathfrak{Z}} = Y_{n+1}$ since $\overline{\mathrm{int}X_{n+1}} = X_{n+1}$ and $\overline{\mathrm{int}Y_{n+1}} = Y_{n+1}$. By induction we can choose a sequence $\{X_n\}$ and $\{Y_n\}$ with the required properties. Put

$$\mathfrak{Z}_o = \bigcup_{n=1}^{\infty} \{(X_{2n-1} \cap \mathfrak{Z}) \cup (Y_{2n-1} \cap \mathfrak{Z})\},$$

$$\mathfrak{Z}_e = \bigcup_{n=1}^{\infty} \{(X_{2n} \cap \mathfrak{Z}) \cup (Y_{2n} \cap \mathfrak{Z})\}.$$

Then we have $A|\bar{\mathfrak{Z}}_o \neq C(\bar{\mathfrak{Z}}_o)$ since

$$\inf\{\|f\|_{A|\bar{\mathfrak{Z}}_o} : f \in A|\bar{\mathfrak{Z}}_o,\ f|X_{2n-1} = 1,\ f|Y_{2n-1} = 0\} = M_{2n-1}$$

holds for every positive integer n by the definition of the quotient norm and $M_{2n-1} \to \infty$ as $n \to \infty$. In the same way we see that $A|\bar{\mathfrak{Z}}_e \neq C(\bar{\mathfrak{Z}}_e)$. We also see that $\bar{\mathfrak{Z}}_o \cap \bar{\mathfrak{Z}}_e = \emptyset$ since $\mathfrak{Z}_o \cap \mathfrak{Z}_e = \emptyset$ and $\bar{\cdot}$ denotes the closure in $\beta\mathfrak{Z}$, so $x \notin \bar{\mathfrak{Z}}_o$ or $x \notin \bar{\mathfrak{Z}}_e$. Suppose that $x \notin \bar{\mathfrak{Z}}_o$. The set $\bar{\mathfrak{Z}}_o$ is a compact subset of $\beta\mathfrak{Z} \setminus ((\mathrm{Ker}A) \cup S)$ since $\bar{\mathfrak{Z}}_o \subset G_0$ and $G_0 \cap ((\mathrm{Ker}A) \cup (S \setminus \{x\})) = \emptyset$. It follows that $A|\bar{\mathfrak{Z}}_o = C(\bar{\mathfrak{Z}}_o)$ by the condition on A. We also see that $A|\bar{\mathfrak{Z}}_e = C(\bar{\mathfrak{Z}}_e)$ if $x \notin \bar{\mathfrak{Z}}_e$. In any case we have a contradiction. Consequently we see that there is a compact neighborhood G_x of x such that $A|G_x = C(G_x)$. □

Proposition 10. *Let G be an infinite compact abelian group and Γ the dual group of G. Suppose that p and q are real numbers such that $1 < p < q < \infty$. Suppose that E is a subset of Γ. Then the following are equivalent.*

(1) *E is a $\Lambda(\max\{p', q\})$ set.*

(2) $\chi_E \in \widehat{M(p,q)}$.

(3) *There is $T \in M(p,q)$ such that* $\inf\{|\widehat{T}(\gamma)| : \gamma \in E\} > 0$.

Proof. First we show that (1) and (2) are equivalent. We consider in the following three cases: (i) $1 < p < q \leq 2$; (ii) $1 < p < 2 < q < \infty$; (iii) $2 \leq p < q < \infty$.

Case (iii). Suppose that E is a $\Lambda(q)$ set. By a theorem of Hare [8, Corollary 1.9] we see that $\chi_E \in \widehat{M(2,q)}$. Since $2 \leq p$ we see that $\widehat{M(2,q)} \subset \widehat{M(p,q)}$, so $\chi_E \in \widehat{M(p,q)}$. Suppose conversely that $\chi_E \in \widehat{M(p,q)}$. Let T_E be an operator in $M(p,q)$ such that $\widehat{T_E} = \chi_E$. It is well-known that $M(p,q) = M(q',p')$. By a proof of a theorem of Igari and Sato [13, Lemma 1 and Remark 1] there is a positive integer n such that $T_E^n \in M(q',2)$. (In fact, let n_0 be the smallest integer such that $n_0 \geq (1/p'-1/2)/(1/q'-1/p')$. Then $T^{n_0+1} \in M(q',r)$, where $1/r = 1/q' - (n_0+1)(1/q' - 1/p')$. Note that $r \geq 2$. Thus we have that $T^{n_0+1} \in M(q',2)$). Thus $\chi_E = \widehat{T_E}^n = \widehat{T_E^n} \in \widehat{M(q',2)} = \widehat{M(2,q)}$. It follows by a theorem of Hare [8, Corollary 1.9] that E is a $\Lambda(q)$ set.

Case (ii). Without loss of generality we may suppose that $p' \leq q$ since $M(p,q) = M(q',p')$. Suppose that E is a $\Lambda(q)$ set. Then there is a constant c such that $\|T_E f\|_q \leq c\|T_E f\|_2$ for every $f \in \mathrm{Trig}G$, where $\widehat{T_E f} = \chi_E \hat{f}$. By, for example, Theorem 5.3 in [15] we see that E is a $\Lambda(p')$ set since $2 < p' \leq q$, so by a theorem of Hare we may suppose that $T_E \in M(2,p') = M(p,2)$, thus there is a constant c' such that $\|T_E g\|_2 \leq c'\|g\|_p$ for every $g \in L^p(G)$. It follows that the inequality $\|T_E f\|_q \leq cc'\|f\|_p$ holds for every $f \in \mathrm{Trig}G$, so $T_E \in M(p,q)$. Thus we conclude that $\chi_E \in \widehat{M(p,q)}$.

Suppose conversely that $\chi_E \in \widehat{M(p,q)}$. Since $1 < p < 2$ we see that $\widehat{M(p,q)} \subset \widehat{M(2,q)}$, so that $\chi_E \in M(2,q)$. Thus E is a $\Lambda(q)$ set by a theorem of Hare [8, Corollary 1.9].

Case (i) follows form case (iii) since $M(p,q) = M(q',p')$ and $2 \leq q' < p' < \infty$.

Next we show that (3) implies (1). Consider the complex-valued function φ defined on the complex plane by

$$\varphi(z) = \begin{cases} 0, & |z| < \inf\{|\widehat{T}(\gamma)| : \gamma \in E\} \\ 1, & |z| \geq \inf\{|\widehat{T}(\gamma)| : \gamma \in E\}. \end{cases}$$

Then by a theorem of Igari and Sato [13, Theorem 1 and Remark 1] we see that φ operates on $\widehat{M(p,q)}$, so we have $\varphi(\widehat{T}) \in \widehat{M(p,q)}$ and $\varphi(\widehat{T}) = \chi_F$, where

$$F = \{\gamma' \in \Gamma : |T(\gamma')| \geq \inf\{|\widehat{T}(\gamma)| : \gamma \in E\}\}.$$

Thus F is a $\Lambda(\max\{p', q\})$ set since (1) and (2) are equivalent. It follows that E is a $\Lambda(\max\{p', q\})$ set since $E \subset F$.

It is trivial that (2) implies (3). We conclude that (1), (2) and (3) are equivalent. □

Proposition 11. *Let G be an infinite compact abelian group and Γ the dual group of G. Suppose that p and q are real numbers such that $1 < p < q < \infty$. Suppose that $\beta\Gamma$ is the Stone-Čech compactification of Γ and B is the algebra of complex-valued continuous functions on $\beta\Gamma$ which are extensions of functions in $\widehat{M(p,q)}$ on Γ to $\beta\Gamma$. Let $\mathrm{Ker}B = \{x \in \beta\Gamma : f(x) = 0 \text{ for } \forall f \in B\}$. Then $B|K = C(K)$ for every compact subset K of $\beta\Gamma \setminus \mathrm{Ker}B$.*

Proof. We show that B on $\beta\Gamma$ satisfies the three properties of A in Lemma 9. Let x and y be a pair of distinct points in $\beta\Gamma \setminus \mathrm{Ker}B$. Suppose that $x \in \Gamma$. Then $\chi_{\{x\}} \in \widehat{M(p,q)}$ so we may suppose that $\chi_{\{x\}} \in B$ and $\chi_{\{x\}}(x) = 1$, $\chi_{\{x\}}(y) = 0$. If $y \in \Gamma$, then we also see that $\chi_{\{y\}}$ is in B and separates x and y. Suppose that $x, y \in \beta\Gamma \setminus \Gamma$. Then there is $f \in B$ such that $f(x) = 1$ since $x \notin \mathrm{Ker}B$. If $f(y) \neq 1$, then f separates x and y. Suppose that $f(y) = 1$. Put $E = \{\gamma \in \Gamma : |f(\gamma)| \geq 1/2\}$. Then by Proposition 10 we see that E is a $\Lambda(\max\{p', q\})$ set. Choose a compact neighborhood U of x with $U \subset \beta\Gamma \setminus (\{y\} \cup \mathrm{Ker}B)$. Then by simple calculation we have that $x \in \overline{U \cap E}$ and $y \notin \overline{U \cap E}$. Since E is a $\Lambda(\max\{p', q\})$ set, the set $U \cap E$ is also a $\Lambda(\max\{p', q\})$ set. So by Proposition 10 we have that $\chi_{U \cap E} \in \widehat{M(p,q)}$. Thus we see that the extended function (also denoted by) $\chi_{U\cap E}$ on $\beta\Gamma$ separates x and y. In any case we conclude that B separates the different points in $\beta\Gamma \setminus \mathrm{Ker}B$. Thus we see that (1) of Lemma 9 holds.

By a theorem of Igari and Sato [13, Theorem 1 and Remark 1] $\varphi(z) = \bar{z}$ on $\mathbb{C}$ operates on B, thus we see by the Stone-Weierstrass theorem that B is uniformly dense in $\{f \in C(\beta\Gamma) : f = 0 \text{ on } \mathrm{Ker}B\}$. Let φ_1 be a complex-valued continuous function defined on $\mathbb{C}$ such that $|\varphi_1(w)| \leq |w|$ for every $w \in \mathbb{C}$,

$$\varphi_1(w) = \begin{cases} 0, & |w| \leq 1 \\ 1, & |w - 3/2| \leq 1/3. \end{cases}$$

Then φ_1 operates on B by a theorem of Igari and Sato [13, Theorem 1 and Remark 1]. Let K_1 and K_2 be a pair of disjoint compact subsets of $\beta\Gamma \setminus \mathrm{Ker}B$. Then there is f in B with $|f(x) - 3/2| \leq 1/3$ for every $x \in K_1$ and $|f(y)| \leq 1$ for every $y \in K_2$, so $\varphi_1 \circ f \in B$. Thus we see that (2) of Lemma 9 holds.

By a theorem of Igari and Sato [13, Theorem 1 and Remark 1] we see that there is a continuous function defined on $\mathbb{C}$ which is not locally Lipschitz and operates on $B|(\beta\Gamma \setminus \mathrm{Ker}B)$ and we also see that φ_1 operates on $B|(\beta\Gamma \setminus \mathrm{Ker}B)$. It follows by Proposition 6 that

$$S = \{y \in \beta\Gamma \setminus \mathrm{Ker}B : A|G \neq C(G) \quad \text{for every compact neighborhood } G \text{ of } y\}$$

is a discrete subset of $\beta\Gamma\setminus\mathrm{Ker}B$. Suppose that K is a compact subset of $\beta\Gamma\setminus((\mathrm{Ker}B)\cup S)$. Then for every $x \in K$ there is a compact neighborhood G_x of x such that $A|G_x = C(G_x)$, so we see that $A|K = C(K)$. Thus we see that (3) of Lemma 9 holds.

It follows by Lemma 9 that $B|K = C(K)$ for every compact subset K of $\beta\Gamma \setminus \mathrm{Ker}B$. □

Theorem 12. *Let G be an infinite compact abelian group and Γ the dual group of G. Suppose that $1 < p < q < \infty$ and E is a subset of Γ. Then the following are equivalent.*

(1) *E is a $\Lambda(\max\{p', q\})$ set.*

(2) $\chi_E \in \widehat{M(p,q)}$.

(3) *There is $T \in M(p,q)$ such that* $\inf\{|\widehat{T}(\gamma)| : \gamma \in E\} > 0$.

(4) $\widehat{M(p,q)}|E = \ell^\infty(E)$.

Proof. We have already shown in Proposition 10 that (1), (2) and (3) are equivalent. It is trivial that (4) implies (3). We show that (2) imples (4). Let $\bar{E}$ be the closure of E in $\beta\Gamma$. Then we see by Proposition 11 that $B|\bar{E} = C(\bar{E})$, where B is the algebra of extensions of functions in $\widehat{M(p,q)}$ on Γ to $\beta\Gamma$. It follows that $\widehat{M(p,q)}|E = \ell^\infty(E)$. □

4 SPECTRAL PROPERTIES OF MULTIPLIERS

In this section we show that $Y = \beta\Gamma\setminus\mathrm{Ker}B$ is homeomorphic to the maximal ideal space of $M(p,q)$ for $1 < p < q < \infty$. We also show that Y is homeomorphic to a subset τY of the maximal ideal space of $M(r,r)$, where $r = \max\{p', q\}$. Since $M(p,q)$ is isometrically isomorphic to $M(q', p')$ we may suppose that

$$M(p,q)|L^r(G) \subset M(r,r),$$

for short, $M(p,q) \subset M(r,r)$. We show that the Gelfand transformation of $T \in M(p,q)$ as an element of $M(r,r)$ vanishes off τY. As a consequence we see that $T \in M(p,q)$ has a natural spectrum as an operator in $M(r,r)$ (cf. [13]).

For a locally compact Hausdorff space Y we denote by $C_0(Y)$ the algebra of all complex-valued continuous functions on Y which vanish at infinity. For a commutative Banach algebra A we denote by $\check{f}$ the Gelfand transformation of $f \in A$. The maximal ideal space of A is denoted by Φ_A. The closure of $S \subset \mathbf{C}$ is denoted by cl(S).

We can prove the following standard lemma by a routine argument.

Lemma 13. *Let X be a compact Hausdorff space and B a commutative Banach algebra included in $C(X)$ which separates the points of X and contains constants. Then the maximal ideal space Φ_B coincides with X if and only if for every finite number of functions $f_1, \ldots, f_n \in B$ such that $\sum_{i=1}^n |f_i| > 0$ on X there are $g_1, \ldots, g_n \in B$ such that $\sum_{i=1}^n f_i g_i = 1$.*

Proposition 14. *Let Y be a locally compact Hausdorff space and A a commutative Banach algebra included in $C_0(Y)$ which separates the points in Y such that $kerA = \{y \in Y : f(y) = 0, \forall f \in A\} = \emptyset$. Let $D = \{z \in \mathbf{C} : |z| < 1\}$. Suppose that A is conjugate closed, i.e., the conjugate function $\bar{f}$ is in A for every $f \in A$. Then the following are equivalent.*

(1) *The maximal ideal space Φ_A of A coincides with Y.*

(2) cl$(f(Y)) =$ cl$(\check{f}(\Phi_A))$ *for every $f \in A$.*

(3) $\|f\|_{\infty(Y)} = \|\check{f}\|_{\infty(\Phi_A)}$ *for every $f \in A$.*

(4) *The function $\sum_{n=1}^{\infty} z^n$ defined on D operates on A.*

(5) *Some rational function φ on D with poles off D, such that $\varphi(0) = 0$ and φ is not a polynomial, operates on A.*

(6) *Every analytic function φ defined on D such that $\varphi(0) = 0$ operates on A.*

Proof. By the definition (1) implies (2) and (2) implies (3). The condition (6) clearly implies (5). Since $\sum_{n=1}^{\infty} z^n = z/(1-z)$ we see that (4) implies (5). By the well-known result on analytic functional calculus we see that (1) implies (6). We show that (3) implies (4) and that (5) implies (1).

Suppose that (3) holds. Suppose that $f \in A$ with $f(Y) \subset D$. Then $\|f\|_{\infty(Y)} < 1$ since cl$(f(Y)) \subset D$ for f vanishes at infinity. Since

$$\|\check{f}\|_{\infty(\Phi_A)} = \lim_{n\to\infty} \|f^n\|_A^{1/n}$$

we see that $g_m = \sum_{k=1}^{m} f^k$ converges to a function $g \in A$. By the elementary fact that $\|\cdot\|_{\infty(Y)} \leq \|\cdot\|_A$ we see that g_m also converges uniformly to g on Y. Thus $\varphi(f) = g$. We have proved that φ operates on A.

Suppose that (5) holds. Let $A[e]$ be the unitalization for A. To show that $Y = \Phi_A$ it is sufficient to show $\Phi_{A[e]} = Y \cup \{\infty\}$, the one point compactification of Φ_A, since $\Phi_{A[e]}$ is the one point compactification of Φ_A [14, Theorem 3.2.2]. So we show that $\Phi_{A[e]} = Y \cup \{\infty\}$. By Lemma 13 it is enough to show that for $f_1, \dots, f_n \in A[e]$ such that $\sum_{i=1}^n |f_i| > 0$ on $Y \cup \{\infty\}$ the ideal $I = f_1 A[e] + \cdots + f_n A[e]$ coincides with $A[e]$. Suppose that $I \neq A[e]$. Then there is a maximal ideal M of $A[e]$ such that $I \subset M$. The algebra $A[e]$ is conjugate closed since A is. Thus $\overline{f_1}, \dots, \overline{f_n} \in A$. So $F = \sum_{i=1}^n |f_i|^2 \in M$ and $F > 0$ on $Y \cup \{\infty\}$. Put

$$\varphi(z) = z^s P(z)(a - z)^{-m} \prod_{k=1}^{l} (a_k - z)^{-1},$$

where $P(z)$ is a polynomial of z with $P(0) \neq 0$, $P(a) \neq 0$, $P(a_k) \neq 0$ and $a \neq a_k$, where a, $a_k \in \mathbb{C} \setminus D$. By the Weierstrass approximation theorem there is a polynomial $Q(t)$ such that $Q(0) = 0$, $Q(F(\infty)) = a$ and $(a - Q(F))(Y \cup \{\infty\}) \subset D$. Put $f = Q(F)$. Then $f \in M$ since $Q(0) = 0$. Since $(a - f)(\infty) = 0$ we see that $a - f \in A$, hence $\varphi(a - f) \in A \subset A[e]$. It follows that $(a - f)^s P(a - f) \in M$ since

$$(a - f)^s P(a - f) = \varphi(a - f) f^m \prod_{k=1}^{l} (a_k - a + f).$$

Since $\sum_{k=1}^{s} \binom{s}{k} a^{s-k} (-f)^k P(a - f) \in M$ we see that $a^s P(a - f) \in M$. Put $P(a - f) = \sum_{k=1}^{N} b_k f^k + c$. Then $c \neq 0$ since $P(a) = c$ and $P(a) \neq 0$. So $a^s P(a - f) = a^s \sum_{k=1}^{N} b_k f^k + a^s c$ and $a^s c \in M$. Hence M contains constants, which is a contradiction. We conclude that $I = A$, thus $\Phi_{A[e]} = Y \cup \{\infty\}$. □

Let G be a compact abelian group and $1 < p < q < \infty$. Suppose that $M(p, q)$ is the Banach algebra of p-q multipliers and $\widehat{M(p,q)}$ denotes the Fourier transforms of $M(p, q)$. For $f \in \widehat{M(p,q)}$ we denote by $\tilde{f}$ the continuous functions on $\beta\Gamma$ which is the extension of f. We say that a commutative Banach algebra is self-adjoint if the Gelfand transform $\check{A}$ of A is conjugate closed on Φ_A. A commutative Banach algebra A is regular if and only if for a point x in Φ_A and a closed subset K with $x \notin K$ there exists a function $\check{f} \in \check{A}$ such that $\check{f}(x) = 1$ and $\check{f} = 0$ on K.(cf. [14])

Theorem 15. *Let G be an infinite compact abelian group and $1 < p < q < \infty$. Put*

$$Y = \beta\Gamma \setminus \{x \in \beta\Gamma : \tilde{f}(x) = 0 \quad \forall f \in \widehat{M(p,q)}\}.$$

Then Y is an open subset of $\beta\Gamma$ and

$$Y = \cup \overline{E},$$

where E varies over all $\Lambda(\max\{p', q\})$ set for Γ and $^-$ denotes the closure in $\beta\Gamma$. The maximal ideal space for $M(p,q)$ is homeomorphic to Y, where the topology on Y is the relative topology induced by $\beta\Gamma$. The Banach algerba $M(p,q)$ is regular and self-adjoint.

Proof. Since $\{x \in \beta\Gamma : \tilde{f}(x) = 0, \forall f \in \widehat{M(p,q)}\}$ is closed the set Y is an open subset of $\beta\Gamma$. Thus Y is a locally compact Hausdorff space with respect to the relative topology induced by $\beta\Gamma$. Let $A = \{\tilde{f}|Y : f \in \widehat{M(p,q)}\}$. Then A is a Banach algebra with the norm $\|\tilde{f}|Y\|_A = \|f\|_{\widehat{M(p,q)}}$, which is included in $C_0(Y)$. Note that $\|\widehat{T}\|_{\widehat{M(p,q)}}$ is defined by $\|T\|_{M(p,q)}$ for $T \in M(p,q)$. Let $Y^1 = \cup\overline{E}$, where E varies over all $\Lambda(\max\{p', q\})$ set. We show that $Y = Y^1$. Suppose that $y \in Y^1$. Then there is a $\Lambda(\max\{p', q\})$ set E such that $y \in \overline{E}$. By Theorem 12 we see that $\chi_E \in \widehat{M(p,q)}$. Thus $\widetilde{\chi_E}(y) = 1$ since $y \in \overline{E}$. We conclude that $y \in Y$. Suppose that $y \in Y$. Then there is $f \in \widehat{M(p,q)}$ such that $\tilde{f}(y) = 1$. Put $E = \{\gamma \in \Gamma : |f(\gamma)| \geq 1/2\}$. Then by Theorem 12 we see that E is a $\Lambda(\max\{p', q\})$ set. Suppose that U is an open neighborhood of y. Since Γ is dense in Y and $U \cap \{y \in Y : |\tilde{f}(y) - 1| < 1/2\}$ is an open neighborhood of y we see that

$$U \cap E \supset U \cap \{y \in Y : |\tilde{f}(y) - 1| < 1/2\} \cap \Gamma \neq \emptyset.$$

It follows that $y \in \overline{E}$, hence $y \in Y^1$. Next we show that A separates the points in Y. Let y_1 and y_2 be distinct points in Y. Then there is $f \in A$ such that $f(y_1) = 1$. Choose open neighborhoods G_1 and G_2 of y_1 and y_2 respectively such that $\overline{G}_1 \cap \overline{G}_2 = \emptyset$. Put $U = \{y \in Y : |f(y)| > 1/2\}$. Then $U \cap \Gamma$ is a $\Lambda(\max\{p', q\})$ set by Theorem 12. Since a subset of a $\Lambda(\max\{p', q\})$ set is also a $\Lambda(\max\{p', q\})$ set we see that $E_1 = G_1 \cap U \cap \Gamma$ and $E_2 = G_2 \cap U \cap \Gamma$ are $\Lambda(\max\{p', q\})$ set such that $y_1 \in \overline{E}_1$, $y_2 \in \overline{E}_2$ and $\overline{E}_1 \cap \overline{E}_2 = \emptyset$. It follows that $\widetilde{\chi_{E_1}}|Y \in A$, $\widetilde{\chi_{E_1}}(y_1) = 1$ and $\widetilde{\chi_{E_1}}(y_2) = 0$. Thus we see that A separates the different points in A. By a theorem of Igari and Sato [13] we see that A is conjugate closed and there is a positive real number k_0 such that $z^{k_0}\varphi(z)$ defined on $D = \{z \in \mathbb{C} : |z| < 1\}$ operates on A, where φ is a bounded analytic function on D which is not a polynomial. Hence by Proposition 14 we conclude that $Y = \Phi_A$. Thus A is self-adjoint. By a theorem of Igari and Sato [13] we see that a complex-valued continuous function φ defined on D with $0 \leq |\varphi| \leq 1$ such that

$$\varphi(z) = \begin{cases} 0, & |z| < 1/5 \\ 1, & |z - 1/2| < 1/5 \end{cases}$$

operates on A. Since A is uniformly dense in $C_0(Y)$ by the Stone-Weierstrass theorem, we see that for every $x \in Y$ and closed subset K of Y such that $x \notin K$ there is $f \in A$ such that $f(x) = 1/2$, $f(K) \subset \{z \in \mathbb{C} : |z| < 1/10\}$ and $f(Y) \subset D$. It follows that $\varphi \circ f \in A$ and $\varphi \circ f(x) = 1$ and $\varphi \circ f(K) = 0$. Since $Y = \Phi_A$ we conclude that A is a regular Banach algebra. Since A is isometrically isomorphic to $M(p,q)$ the conclusion holds. $\square$

Let G be an infinite compact abelian group and Γ the dual group. Suppose that $1 < r < \infty$. For $T \in M(r,r)$ we denote by $\sigma_r(T)$ the spectrum of T as an element of $M(r,r)$. Hare [8, Proposition 1.15] showed $\sigma_s(T) = \overline{\widehat{T}(\Gamma)}$ for $T \in M(2,p) \cap M(q,q)$, $2 < p \leq q$ and $2 \leq s \leq q$. Igari and Sato [13] showed that every $T \in M(p,q)$ has the natural spectrum in $M(p,q)$, that is, the spectrum of T in $M(p,q)$ coincides with $\overline{\widehat{T}(\Gamma)}$ for every $T \in M(p,q)$. As a corollary of Theorem 15 we can prove a theorem of Igari and Sato.

Corollary 16. *Let G be an infinite compact abelian group, Γ the dual group and $1 < p < q < \infty$. Put $r = \max\{p', q\}$. Then every $T \in M(p,q)$ has the natural spectrum in $M(p,q)$, thus $\sigma_s(T) = \overline{\widehat{T}(\Gamma)}$ for $2 \leq s \leq r$.*

Proof. Since Γ is dense in the maximal ideal space of $M(p,q)$ by Theorem 15, we see that the spectrum of $T \in M(p,q)$ coincides with $\overline{\widehat{T}(\Gamma)}$. Hence $\sigma_s(T) = \overline{\widehat{T}(\Gamma)}$ holds for $2 \leq s \leq r$ since $M(p,q) \subset M(r,r) \subset M(s,s)$. □

Lemma 17. *Let Y be the open subset of $\beta\Gamma$ defined by*

$$Y = \beta\Gamma \setminus \{x \in \beta\Gamma : \tilde{f}(x) = 0 \quad \forall f \in \widehat{M(p,q)}\}.$$

Then the topology Y inherits from $\beta\Gamma$ is identical to the weak topologies induced by $\{f \in C(\beta\Gamma) : f = 0 \quad on \quad \beta\Gamma \setminus Y\}$ and by $\widehat{M(p,q)}$.

In the following the topology on Y is the one in Lemma 17. Put $r = \max\{p', q\}$. Let $y \in Y$. We define $\tau_y : M(r,r) \to \mathbb{C}$ by $\tau_y T = \tilde{\widehat{T}}(y)$ for $T \in M(r,r)$, where $\widehat{T}$ is the Fourier transform of T and $\tilde{\widehat{T}}$ is the continuous extension of $\widehat{T}$ on $\beta\Gamma$. Then the operator τ_y is linear and multiplicative on $M(r,r)$. Thus $\tau_y \in \Phi_{M(r,r)}$.

Lemma 18. *Let $\tau Y = \{x \in \Phi_{M(r,r)} : x = \tau_y$ for some $y \in Y\}$. The mapping $\tau : Y \to \tau Y$ defined by $\tau(y) = \tau_y$ is a homeomorphism such that $\tau(y) = y$ for every $y \in \Gamma$.*

Proof. Since $(\widehat{M(p,q)})\tilde{}$ separates points in Y and $\widehat{M(p,q)} \subset \widehat{M(r,r)}$ the mapping τ is injective, hence τ is bijection. It is trivial that $\tau(y) = y$ for $y \in \Gamma$ by the definition.

Choose $\tau_y \in \tau Y$, $T \in M(r,r)$ and $\varepsilon > 0$. Then

$$\{\tau_x \in \tau Y : |\tau_x T - \tau_y T| < \varepsilon\}$$

is an open subset of τY and the family of such an open set is a subbase for the topology on τY. Since

$$\tau^{-1}(\{\tau_x \in \tau Y : |\tau_x T - \tau_y T| < \varepsilon\} = \{x \in Y : |\tilde{\widehat{T}}(x) - \tilde{\widehat{T}}(y)| < \varepsilon\}$$

is an open set in Y_1 we see that τ is continuous.

On the other hand for $y \in Y$, $T \in M(p,q)$ and $\varepsilon > 0$ the set

$$\{x \in Y : |\tilde{\hat{T}}(x) - \tilde{\hat{T}}(y)| < \varepsilon\}$$

is an open set in Y and the family of such open sets is a subbase for the topology on Y. Since

$$\tau(\{x \in Y : |\tilde{\hat{T}}(x) - \tilde{\hat{T}}(y)| < \varepsilon\}) = \{\tau_x : \tau Y : |\tau_x T - \tau_y T| < \varepsilon\}$$

is an open set in τY and since τ is a bijection we see that τ is an open mapping. We conclude that τ is a homeomorphism. □

Lemma 19. *Let E be a $\Lambda(r)$ set in Γ. Then τ(the closure of E in Y) =the closure of E in $\Phi_{M(r,r)}$.*

Proof. We denote by $\overline{E}$ the closure of E in Y and by $\overline{\overline{E}}$ the closure of E in $\Phi_{M(r,r)}$. Since Y is the union of the closure in $\beta\Gamma$ of $\Lambda(r)$ set by Theorem 15 we see that $\overline{E}$ is compact. So $\tau(\overline{E})$ is compact in τY. Thus we see that $\tau(\overline{E})$ is closed in $\Phi_{M(r,r)}$. Hence $E \subset \overline{\overline{E}} \subset \tau(\overline{E}) \subset \tau Y$. So $\overline{\overline{E}}$ is a closed subset in τY. It follows that $\tau(\overline{E}) = \overline{\overline{E}}$ since τ is a homeomorphism. □

Theorem 20. *Let G be an infinite compact abelian group and $1 < p < q < \infty$. Put $r = \max\{p', q\}$. Then*

$$(M(p,q)|L^r(G))^{\vee(r)} \subset \{f \in M(r,r)^{\vee(r)} : f(x) = 0 \quad \forall x \in \Phi_{M(r,r)} \setminus \tau Y\},$$

where $S^{\vee(r)}$ denotes the Gelfand transform of $S \in M(r,r)$ in $M(r,r)$.

Proof. Note that we may suppose that $M(p,q) = M(p,q)|L^r(G)$. Let $x \in \Phi_{M(r,r)} \setminus \tau Y$. Suppose that there is $T \in M(p,q)$ such that $T^{\vee(r)}(x) \neq 0$. Then $I = \{T \in M(p,q) : T^{\vee(r)}(x) = 0\}$ is a proper ideal of $M(p,q)$. There exists $T_x \in M(p,q)$ such that $T_x^{\vee(r)}(x) = 1$, so $ST - S \in I$ for every $S \in M(p,q)$. Hence I is a proper regular ideal of $M(p,q)$. Thus there is a maximal regular ideal M of $M(p,q)$ such that $I \subset M$. Since $Y = \Phi_{M(p,q)}$ there exists $y \in Y$ such that

$$M = \{T \in M(p,q) : \tilde{\hat{T}}(y) = 0\}.$$

Since

$$T^{\vee(r)}(\tau_y) = \tau_y(T) = \tilde{\hat{T}}(y)$$

for $T \in M(p,q)$ we see that

$$M = \{T \in M(p,q) : T^{\vee(r)}(\tau_y) = 0\}.$$

There exists a $\Lambda(r)$ set E of Γ such that $\tau_y \in \overline{\overline{E}}$. By Theorem 12 we see that $\chi_E \in \widehat{M(p,q)}$, so there is $T_E \in M(p,q)$ such that $\widehat{T_E} = \chi_E$. Since $\tau_y \in \overline{\overline{E}}$ we see that $T_E^{\vee(r)}(\tau_y) = 1$. Since $M(r,r)^{\vee(r)}$ separates the different points in $\Phi_{M(r,r)}$ there exists $T \in M(r,r)$ such that $T^{\vee(r)}(\tau_y) = 1$ and $T^{\vee(r)}(x) = 0$. It follows that $TT_E \in M(p,q)$ since $M(p,q)$ is an ideal of $M(r,r)$ and $(TT_E)^{\vee(r)}(x) = T^{\vee(r)}(x)T_E^{\vee(r)}(x) = 0$, so $TT_E \in I$, hence $(TT_E)^{\vee(r)}(\tau_y) = 0$ for $I \subset M$. On the other hand $(TT_E)^{\vee(r)}(\tau_y) = T^{\vee(r)}(\tau_y)T_E^{\vee(r)}(\tau_y) = 1$, which is a contradiction, proving the theorem. □

Igari [12] showed that for every $1 < q < \infty$ with $q \neq 2$ there exists a finite regular Borel measure μ on G such that $\overline{\widehat{\mu}(\Gamma)}$ is properly included in $\sigma_q(T_\mu)$ for every nondiscrete locally compact abelian group G, where T_μ is the convolution operator on $L^q(G)$ defined by $T_\mu f = \mu * f$ for $f \in L^q(G)$ and $\widehat{\mu}$ is the Fourier-Stieltjes transform of μ. On the other hand we see the following.

Corollary 21. *Let G be an infinite compact abelian group, Γ the dual group of G and $1 < q < \infty$, $q \neq 2$. For every $T \in M(q,q)$ there exist $T_1, T_2 \in M(q,q)$ such that $T = T_1 + T_2$ and $\sigma_q(T_i) = \overline{\widehat{T_i}(\Gamma)}$ $(i = 1,2)$. If $2 < q < \infty$ (resp. $1 < q < 2$), then T_1 can be choosen from $M(2,q)$ (resp. $M(q,2)$).*

Proof. Without loss of generality we may assume that $2 < q$ since $M(r,r)$ is isometrically isomorphic to $M(r',r')$ for $1 < r < \infty$. Choose an infinite $\Lambda(q)$ set E for Γ. By Theorem 12 we see that $\widehat{M(2,q)}|E = \ell^\infty(E)$, so there is $S_1 \in M(2,q)$ such that $\widehat{S_1} = \widehat{T}$ on E. Since E is infinite and $\widehat{M(2,q)}|E = \ell^\infty(E)$ we see that there is $S_2 \in M(2,q)$ such that

$$\overline{\widehat{S_2}(E)} = (T - S_1)^{\vee(q)}(\Phi_{M(q,q)}).$$

It is easy to see that $T - S_1 + S_2$ has the natural spectrum. Put $T_1 = S_1 - S_2$ and $T_2 = T - S_1 + S_2$. The required conditions are satisfied. □

Suppose that G is an I-group. Zafran [18] showed that there exist measure μ and ν in $M(G)$, the measure algebra on G, such that $\sigma(\mu) = \overline{\widehat{\mu}(\Gamma)}$ and $\sigma(\nu) = \overline{\widehat{\nu}(\Gamma)}$ while $\sigma(\mu + \nu) \neq \overline{(\mu + \nu)\widehat{}(\Gamma)}$, where σ denotes the spectrum of a measure as an element of $M(G)$. By Corollary 21 we see that for every infinite compact abelian group G and $1 < q < \infty$, $q \neq 2$ there exist $T, S \in M(q,q)$ such that $\sigma_q(T) = \overline{\widehat{T}(\Gamma)}$ and $\sigma_q(S) = \overline{\widehat{S}(\Gamma)}$ while $\sigma_q(T + S) \neq \overline{(T + S)\widehat{}(\Gamma)}$.

References

1. W. Bade and P. Curtis, *Embedding theorem for commutative Banach algebras*, Pacific J. Math. **18** (1966), 391–409.

2. A. Bernard, *Espaces des parties réelles des éléments d'une algèbre de fonctions*, J. Funct. Anal. **10** (1972), 387–409.

3. A. Bernard, *Une fonction Lipschitzienne peut-elle opérer sur un espace de Banach de fonctions non trivial?*, J. Funct. Anal. **122** (1994), 451–477.

4. G. Bourdaud, *Le calcul fonctionel dans l'espace de Besov critique*, Proc. Amer. Math. Soc. **116** (1992), 983–986.

5. G. Bourdaud et D. Kateb, *Fonctions qui operent sur les espaces de Besov*, Proc. Amer. Math. Soc. **112** (1991), 1067–1076.

6. A. J. Ellis, *Separation and ultraseparation properties for continuous function spaces*, J. London Math. Soc. **29** (1984), 521–532.

7. A. Figà -Talamanca, *Multipliers of p-integrable functions*, Bull. Amer. Math. Soc. **70** (1964), 666–669.

8. K. Hare, *Properties and examples of* (L^p, L^q) *multipliers*, Indiana Univ. Math. J. **38** (1989), 211–227.

9. O. Hatori, *Range transformations on a Banach function algebra. II*, Pacific J. Math. **138** (1989), 89–118.

10. O. Hatori, *Separation properties and operating functions on a space of continuous functions*, Internat. J. Math. **4** (1993), 551–600.

11. K. Hoffman and J. Wermer, *A characterization of* $C(X)$, Pacific J. Math. **12** (1962), 941–944.

12. S. Igari, *Functions of* L^p *multipliers*, Tôhoku Math. J. **21** (1969), 304–320.

13. S. Igari and E. Sato, *Operating functions on Fourier multipliers*, Tôhoku Math. J. **46** (1994), 357–366.

14. R. Larsen, *Banach algebras*, Pure and Appl. Math. 24, Marcel Dekker, New York, 1973.

15. L. R. López and K. A. Ross, *Sidon Sets*, Lecture Notes in Pure and Appl. Math. 13, Marcel Dekker, New York, 1975.

16. W. Rudin, *Fourier analysis on groups*, Interscience, New York, 1962.

17. H. Triebel, *Theory of function spaces*, Birkhäuser, Basel, 1983.

18. M. Zafran, *On the spectra of multipliers*, Pacific J. Math. **47** (1973), 609–629.

On the Extremal Structure of the Unit Ball of the Space $C(K,X)^*$

ZHIBAO HU Department of Mathematics, Miami University, Oxford, Ohio 45056

MARK A. SMITH Department of Mathematics, Miami University, Oxford, Ohio 45056

ABSTRACT

A review of the various known types of extreme points in the unit ball of the Banach space $C(K,X)^*$ is given. Next, the strongly extreme points of this unit ball are characterized in a manner consistent with these known results. Then the weak* points of continuity and weak* points of sequentially continuity of this unit ball are identified. Finally, the points of continuity and points of sequential continuity of this unit ball are discussed.

1 INTRODUCTION

In 1936, J.A. Clarkson [C] invented the abstract notions of strict convexity and uniform convexity within the study of the geometry of Banach spaces. Since then, several other geometric notions have been defined and studied; some of these are global concepts that either lie in strength between strict convexity and uniform convexity or are otherwise related to these properties, while others are localizations of one of these global concepts. For example, recall that a point x in the closed unit ball B_E of a real Banach space E is called an *extreme point* of B_E provided x is not the midpoint of any nontrivial line segment lying in B_E. This is the localized notion of strict convexity in the sense that, by

The second author was supported in part by a Miami University Research Appointment.

definition, a Banach space E is said to be *strictly convex* if and only if every norm-one element of B_E is an extreme point of B_E. Additionally, recall that a point x in B_E is called a *point of continuity* of B_E provided that whenever $\{x_\lambda\}_{\lambda\in\Lambda}$ is a net in B_E which converges weakly to x, it follows that $\{x_\lambda\}_{\lambda\in\Lambda}$ converges in norm to x. This localizes the Kadec property in that, by definition, a Banach space E is said to have the *Kadec property* if and only if every norm-one element of B_E is a point of continuity of B_E. The subject matter of this paper is a collection of these localized notions.

2 DEFINITIONS AND PRELIMINARIES

For a Banach space E, let S_E denote the unit sphere of E, let E^* denote the dual space of E, and let the value of x^* in E^* at x in E be denoted by $\langle x^*, x\rangle$. The following is a list of geometrical, localized notions, in addition to those given above, with which this paper is concerned; it is included here for completeness sake. The notation $\pm$ in, for example, $\|x \pm z\| \leq 1+\delta$, is to be read as "$\|x+z\| \leq 1+\delta$ and $\|x-z\| \leq 1+\delta$."

(i) A point x in B_E is called a *strongly extreme point* of B_E provided that for every $\varepsilon > 0$ there exists a $\delta > 0$ such that $\|x \pm z\| \leq 1+\delta$ for z in E implies $\|z\| \leq \varepsilon$.

(ii) A point x in B_E is called a *point of sequential continuity* of B_E provided that whenever $\{x_n\}_{n=1}^{\infty}$ is a sequence in B_E which converges weakly to x, it follows that $\{x_n\}_{n=1}^{\infty}$ converges in norm to x.

(iii) A point x in B_E is called a *denting point* of B_E provided x is not an element of the closed convex hull of $\{y \in B_E : \|y - x\| > \varepsilon\}$ for each $\varepsilon > 0$.

(iv) A point x in B_E is called a *strongly exposed point* of B_E provided that there exists x^* in B_{E^*} such that $\langle x^*, x\rangle = \|x^*\| = \|x\| = 1$ and whenever $\{x_n\}_{n=1}^{\infty}$ is a sequence in B_E such that $\lim_{n\to\infty}\langle x^*, x_n\rangle = 1$, it follows that $\lim_{n\to\infty}\|x - x_n\| = 0$; in this case, the functional x^* is said to *strongly expose* B_E at x.

It is known, from [**LLT**], that a point in B_E is a denting point of B_E if and only if it is a point of continuity and an extreme point of B_E.

If E is a dual space, the notions of a *weak* point of continuity* and a *weak* point of sequential continuity* of B_E are defined as in the definitions of point of continuity and point of sequential continuity above replacing weak convergence by weak* convergence. The notion of a *weak* denting point* of B_E is defined as in (iii) above replacing the closed convex hull by the weak* closed convex hull of the set given there, and the notion of *weak* strongly exposed point* of B_E is defined as in (iv) above insisting that the strongly exposing functional belongs to the predual of E. Note that a point in B_E is a weak* denting point of B_E if and only if it is a weak* point of continuity and an extreme point of B_E.

With the obvious abbreviations, the diagram below gives the relative strengths of all these notions.

$$\begin{array}{ccccccc}
w^*\text{-str-exp} & \rightarrow & w^*\text{-dent} & \rightarrow & w^*\text{-pc} & \rightarrow & w^*\text{-psc} \\
\downarrow & & \downarrow & & \downarrow & & \downarrow \\
\text{str-exp} & \rightarrow & \text{dent} & \rightarrow & \text{pc} & \rightarrow & \text{psc} \\
 & & \downarrow & & & & \\
 & & \text{str-ext} & \rightarrow & \text{ext.} & &
\end{array}$$

Let K be a compact Hausdorff topological space and let X be a real Banach space. The symbol $C(K, X)$ denotes the real Banach space of all continuous functions from K into X with the usual supremum norm. The dual space $C(K, X)^*$ can be identified with the space of all countably additive X^*-valued Borel measures of bounded variation on K equipped with the variation norm, where the action of μ in $C(K, X)^*$ on f in $C(K, X)$ is given by $\langle \mu, f \rangle = \int_K f \, d\mu$.

The goal of this paper is to completely identify all (#) points of the closed unit ball of $C(K, X)^*$, where (#) denotes for any one of the ten localized, geometric notions given in the diagram above. Exact characterizations are known for five of the ten notions; the other five are the topic of this paper.

For a point k in K, let δ_k denote the element of $C(K)^*$, here $C(K) = C(K, \mathbb{R})$, that is point mass measure at k. For k in K and x^* in X^*, let L_{k,x^*} denote the element of $C(K, X)^*$ given by

$$\langle L_{k,x^*}, f \rangle \quad = \quad \langle x^*, f(k) \rangle \quad \text{for all} \quad f \quad \text{in} \quad C(K, X);$$

the element L_{k,x^*} is sometimes denoted by $\delta_k \otimes x^*$.

It is well-known that the set of extreme points of $B_{C(K)^*}$ is the set $\{\delta_k : k \in K\}$; in fact, it is straightforward to show that, for each k in K, the measure δ_k is a strongly exposed point of $B_{C(K)^*}$.Thus an element of $B_{C(K)^*}$ is an extreme point of $B_{C(K)^*}$ if and only if it is a strongly exposed point of $B_{C(K)^*}$ if and only if it is δ_k for some k in K. Similarly, it can be shown that an element of $B_{C(K)^*}$ is a weak* denting point of $B_{C(K)^*}$ if and only if it is a weak* strongly exposed point of $B_{C(K)^*}$ if and only if it is δ_k for some isolated point k in K. So whenever $X = \mathbb{R}$, the sets of strongly exposed points, denting points, strongly extreme points and extreme points all coincide as do the sets of weak* strongly exposed points and weak* denting points. Such need not be the case whenever the dimension of X is greater than one, as will be seen in Example 5.

3 THE RESULTS

The first theorem below is, by now, classical. The second and third theorems, stated here in the context of $C(K,X)^*$, are special cases of several results in **[RS1]** and **[RS2]**. The fourth theorem completes the picture relative to the five notions which are stronger than that of an extreme point in the diagram above. Even though Theorem 4 is natural and expected, its proof requires a substantial amount of work and it does not follow from known results.

Theorem 1 (Singer [Si]). An element of $C(K,X)^*$ is an extreme point of the unit ball of $C(K,X)^*$ if and only if it has the form L_{k,x^*} where k is in K and x^* is an extreme point of the unit ball of X^*.

Theorem 2 (Reuss and Stegall [RS1] and [RS2]). An element of $C(K,X)^*$ is a weak* strongly exposed point (respectively, weak* denting point) of the unit ball of $C(K,X)^*$ if and only if it has the form L_{k,x^*} where k is an isolated point of K and x^* is a weak* strongly exposed point (respectively, weak* denting point) of the unit ball of X^*.

Theorem 3 (Reuss and Stegall [RS1]). An element of $C(K,X)^*$ is a strongly exposed point (respectively, denting point) of the unit ball of $C(K,X)^*$ if and only if it has the form L_{k,x^*} where k is in K and x^* is a strongly exposed point (respectively, denting point) of the unit ball of X^*.

Theorem 4. An element of $C(K,X)^*$ is a strongly extreme point of the unit ball of $C(K,X)^*$ if and only if it has the form L_{k,x^*} where k is in K and x^* is a strongly extreme point of the unit ball of X^*.

Proof. Suppose μ is a strongly extreme point of $B_{C(K,X)^*}$. In particular, μ is an extreme point of $B_{C(K,X)^*}$ and hence, by Theorem 1, there exist k in K and x^* in B_{X^*} such that $\mu = L_{k,x^*}$. Since the mapping which sends z^* in X^* to L_{k,z^*} in $C(K,X)^*$ is a linear isometry (here k is fixed), it follows that x^* is a strongly extreme point of B_{X^*}.

Conversely, suppose k is in K and x^* is a strongly extreme point of B_{X^*}. To show L_{k,x^*} is a strongly extreme point of $B_{C(K,X)^*}$, let $\varepsilon > 0$ be given. By the hypothesis on x^*, there exists $\delta > 0$ with $\delta < \varepsilon$ such that if z^* is in X^* with $\|x^* \pm z^*\| \le 1 + \delta$, then $\|z^*\| < \varepsilon$. Suppose μ is an element of $C(K,X)^*$ such that $\|L_{k,x^*} \pm \mu\| < 1 + \frac{1}{2}\delta$.

Claim 1. If U is an open subset of K that contains k, then $\|\mu(U)\| < \varepsilon$.

To establish this claim, let U be an open set containing k. Choose an open set V containing k such that $\overline{V} \subseteq U$ and $|\mu|(U \setminus \overline{V}) < \frac{1}{4}\delta$. Then choose a continuous function $g : K \to [0,1]$ such that $g(K \setminus U) = 0$ and $g(\overline{V}) = 1$. Let x be in B_X and define $f : K \to X$ by $f(t) = g(t)x$ for each t in K. Then f is in $B_{C(K,X)}$ and

$$\left|\langle L_{k,x^*} \pm \mu, f\rangle\right| \le \|L_{k,x^*} \pm \mu\| < 1 + \frac{\delta}{2}.$$

From this and the fact that $\text{supp}(f) \subseteq U$, it follows that

$$\begin{aligned}
\left|\langle x^* \pm \mu(U), x\rangle\right| &\leq \left|\langle L_{k,x^*} \pm \mu, f\rangle\right| + \left|\langle x^* \pm \mu(U), x\rangle - \langle L_{k,x^*} \pm \mu, f\rangle\right| \\
&< 1 + \frac{\delta}{2} + \left|\langle x^*, x\rangle \pm \langle \mu(U), x\rangle - \left(\langle x^*, x\rangle \pm \int_K f\, d\mu\right)\right| \\
&= 1 + \frac{\delta}{2} + \left|\langle \mu(U), x\rangle - \int_U f\, d\mu\right| \\
&= 1 + \frac{\delta}{2} + \left|\langle \mu(U), x\rangle - \int_{\overline{V}} x\, d\mu - \int_{U\setminus\overline{V}} f\, d\mu\right| \\
&\leq 1 + \frac{\delta}{2} + \left|\langle \mu(U), x\rangle - \langle \mu(\overline{V}), x\rangle\right| + |\mu|(U \setminus \overline{V}) \\
&< 1 + \frac{\delta}{2} + \|\mu(U) - \mu(\overline{V})\| + \frac{\delta}{4} \\
&\leq 1 + \frac{3\delta}{4} + |\mu|(U \setminus \overline{V}) \\
&< 1 + \delta.
\end{aligned}$$

Since x in B_X was arbitrary, $\|x^* \pm \mu(U)\| \leq 1 + \delta$ and so it follows, from definition of δ, that $\|\mu(U)\| < \varepsilon$.

Claim 2. If U is an open subset of K that contains k, then $|\mu|(K \setminus U) \leq 2\varepsilon$.

To establish this claim, let U be an open set containing k and let $\eta > 0$. Choose an open set V containing k such that $\overline{V} \subseteq U$ and $|\mu|(U \setminus V) < \eta$. Next, choose open sets V_1 and V_2 such that $\overline{V} \subseteq V_1 \subseteq \overline{V}_1 \subseteq V_2 \subseteq \overline{V}_2 \subseteq U$. Then choose a continuous function $g : K \to [0,1]$ such that $g(\overline{V}) = g(K \setminus U) = 1$ and $g(\overline{V}_2 \setminus V_1) = 0$. Finally, choose x in B_X such that $\langle x^*, x\rangle > 1 - \eta$ and choose f in $B_{C(K \setminus V, X)}$ such that $\int_{K\setminus V} f d\mu > |\mu|(K \setminus V) - \eta$. Now, define $h : K \to X$ by

$$h(t) = \begin{cases} g(t)\ f(t) & \text{for } t \text{ in } K \setminus \overline{V}_1 \\ g(t)\ x & \text{for } t \text{ in } V_2. \end{cases}$$

Note that h is well-defined since $g(V_2 \setminus \overline{V}_1) = 0$ and note that h is in $B_{C(K,X)}$. It follows that

$$1 + \varepsilon > 1 + \delta$$

$$
\begin{aligned}
&> \langle L_{k,x^*} \pm \mu, h\rangle \\
&= \langle x^*, h(k)\rangle \pm \int_K h \, d\mu \\
&= \langle x^*, x\rangle \pm \left(\int_V h \, d\mu + \int_{U\setminus V} h \, d\mu + \int_{K\setminus U} h \, d\mu\right) \\
&= \langle x^*, x\rangle \pm \left(\int_V x \, d\mu + \int_{U\setminus V} h \, d\mu + \int_{K\setminus U} f \, d\mu\right) \\
&> 1 - \eta \pm \langle \mu(V), x\rangle - |\mu|(U \setminus V) \pm \int_{K\setminus U} f \, d\mu \\
&> 1 - \eta \pm \langle \mu(V), x\rangle - \eta \pm \int_{K\setminus U} f \, d\mu.
\end{aligned}
$$

Now, $|\langle \mu(V), x\rangle| \leq \|\mu(V)\| < \varepsilon$, by Claim 1, and hence, from the last inequality above,

$$1 + \varepsilon > 1 - 2\eta - \varepsilon \pm \int_{K\setminus U} f d\mu.$$

Thus $\int_{K\setminus U} f d\mu < 2\varepsilon + 2\eta$ and hence

$$
\begin{aligned}
|\mu|(K \setminus U) &\leq |\mu|(K \setminus V) \\
&< \int_{K\setminus V} f \, d\mu + \eta \\
&= \int_{U\setminus V} f \, d\mu + \int_{K\setminus U} f \, d\mu + \eta \\
&< |\mu|(U \setminus V) + 2\varepsilon + 2\eta + \eta \\
&< 4\eta + 2\varepsilon.
\end{aligned}
$$

Since η was arbitrary, the claim is established.

Claim 3. For μ as above, $\|\mu\| \leq 3\varepsilon$.

To establish this claim, let $\eta > 0$ be given. Choose f in $B_{C(K,X)}$ such that $\int_K f d\mu > \|\mu\| - \eta$ and then choose an open set U containing k such that diam $f(U) < \eta$. As in the proof of Claim 2, choose open sets V, V_1 and V_2 containing k such that $\overline{V} \subseteq V_1 \subseteq \overline{V}_1 \subseteq V_2 \subseteq \overline{V}_2 \subseteq U$ and $|\mu|(U \setminus V) < \eta$. Then choose a continuous function $g : K \to [0,1]$ such that $g(\overline{V}) = g(K \setminus U) = 1$ and $g(\overline{V}_2 \setminus V_1) = 0$. Let $x = f(k)$ and define $h : K \to X$ by

$$h(t) = \begin{cases} g(t)\, f(t) & \text{for } t \text{ in } K \setminus \overline{V}_1 \\ g(t)\, x & \text{for } t \text{ in } V_2. \end{cases}$$

Then, as in the proof of the previous claim, h is in $B_{C(K,X)}$. Now,

$$\begin{aligned} \int_K h \, d\mu &= \int_V x \, d\mu + \int_{U\setminus V} h \, d\mu + \int_{K\setminus U} f \, d\mu \\ &\leq \langle \mu(V), x\rangle + |\mu|(U \setminus V) + |\mu|(K \setminus U). \end{aligned}$$

Note that $\langle \mu(V), x\rangle \leq \|\mu(V)\| < \varepsilon$, by Claim 1, and $|\mu|(K \setminus U) \leq 2\varepsilon$, by Claim 2. Combining these with the last inequality above and using the fact that $|\mu|(U \setminus V) < \eta$, it follows that

$$\int_K h \, d\mu < \varepsilon + \eta + 2\varepsilon.$$

On the other hand,

$$\begin{aligned} \int_K h \, d\mu &= \int_K f \, d\mu + \int_K (h - f) \, d\mu \\ &= \int_K f \, d\mu + \int_V (x - f(t)) \, d\mu(t) + \int_{U\setminus V} (h - f) \, d\mu \\ &> \|\mu\| - \eta - |\mu|(V)\eta - |\mu|(U \setminus V)\|h - f\| \\ &\geq \|\mu\| - \eta - |\mu|(K)\eta - 2\eta \\ &= (1 - \eta)\|\mu\| - 3\eta. \end{aligned}$$

Thus $(1 - \eta)\|\mu\| - 3\eta < 3\varepsilon + \eta$ and so

$$\|\mu\| < \frac{3\varepsilon + 4\eta}{1 - \eta}.$$

Since η was arbitrary, the claim is established.

It now follows, from Claim 3, that L_{k,x^*} is a strongly extreme point of $B_{C(K,X)^*}$ and the proof of the theorem is complete.

As mentioned above, with respect to the notions given in the diagram, there are only two distinct classes of extreme points in $B_{C(K)^*}$ and, actually, only one whenever K has no isolated points. If X is finite dimensional, then the sets of extreme points, strongly extreme points and denting points in B_{X^*} coincide and hence so do the corresponding sets in $B_{C(K,X)^*}$. However, even when X is two dimensional, it can be the case that the other three types of extreme points form distinct sets as illustrated in the next example.

Example 5. Let E be $\mathbb{R}^2$ equipped with the norm whose unit sphere is the "bathtub" with line segments from $(-1,1)$ to $(1,1)$ and from $(-1,-1)$ to $(1,-1)$, and with semicircular arcs from $(1,1)$ to $(1,-1)$ centered at $(1,0)$ and from $(-1,1)$ to $(-1,-1)$ centered at $(-1,0)$. Let X be E^* (so that X^* is E) and let $K = [0,1] \cup \{2\}$ with the topology inherited from $\mathbb{R}$. Let $x^* = (0,1), y^* = (1,1)$ and $z^* = (2,0)$ in X^* and let $k = 2$ and $m = \frac{1}{2}$ in K. Then, from the theorems above, it follows that L_{k,y^*} is a weak* denting point but is not a weak* strongly exposed point of $B_{C(K,X)^*}$. The functional L_{m,y^*} is a denting point but is neither a strongly exposed point nor a weak* denting point of $B_{C(K,X)^*}$. The functional L_{m,z^*} is a strongly exposed point but is not a weak* strongly exposed point of $B_{C(K,X)^*}$. Finally, note L_{k,z^*} is a weak* strongly exposed point of $B_{C(K,X)^*}$ but L_{k,x^*} and L_{m,x^*} are not even extreme points of $B_{C(K,X)^*}$. This completes Example 5.

To provide an example where the sets of denting points, strongly extreme points and extreme points are distinct in $B_{C(K,X)^*}$, it suffices to provide a space X where these sets of points are distinct in B_{X^*}. As mentioned above, this requires X to be infinite dimensional but such an X does exist within the class of reflexive Banach spaces. For example, let $E = M \oplus_2 W$ where M denotes the space $(\ell^2, \|\cdot\|_M)$ given in [**DHS**] and W denotes the space $(\ell^2, \|\cdot\|_W)$ given in [**Sm**]. Then let $X = E^*$ (so that $X^* = E$). The space X has the desired properties since B_M has a strongly extreme point that is not a denting point and B_W has an extreme point that is not a strongly extreme point.

The remainder of this paper focuses on the four notions listed in the diagram that are not necessarily extreme points. The next theorem identifies precisely the strongest of these four in the unit ball of the space $C(K,X)^*$.

Theorem 6. An element μ of $C(K,X)^*$ is a weak* point of continuity of the unit ball of $C(K,X)^*$ if and only if it has the form $\mu = \sum_{k\in I} L_{k,x_k^*}$ where $I = \{k \in K : k \text{ is an isolated point of } K\}$ and, for each k in I, either $x_k^* = 0$ or $x_k^*/\|x_k^*\|$ is a weak* point of continuity of B_{X^*} and $\sum_{k\in I}\|x_k^*\| = 1$.

Proof. Suppose $\mu = \sum_{k\in I} L_{k,x_k^*}$ is as described in the statement of the theorem. Then the set $I_0 = \{k \in I : x_k^* \neq 0\}$ is countable and

$$\|\mu\| \;=\; |\mu|(I_0) \;=\; \sum_{k\in I_0} \|\mu(\{k\})\| \;=\; \sum_{k\in I_0} \|x_k^*\| \;=\; 1.$$

Now, suppose $\{\mu_\lambda\}_{\lambda\in\Lambda}$ is a net in $B_{C(K,X)^*}$ that converges weak* to μ in $C(K,X)^*$. For each k in I and each x in X, observe that the function which sends k to x and all other points of K to 0 is in $C(K,X)$. Applying each μ_λ to this function yields that the net $\{\mu_\lambda(\{k\})\}_{\lambda\in\Lambda}$ converges weak* to the point $\mu(\{k\}) = x_k^*$ in X^* for each k in I. Thus

$$\liminf_{\lambda\in\Lambda} \|\mu_\lambda(\{k\})\| \quad \geq \quad \|x_k^*\| \quad \text{for each } k \text{ in } I.$$

Since $\|\mu_\lambda\| \leq 1$, for each λ in Λ, it follows that

$$\begin{aligned} 1 &= \sum_{k\in I_0} \|x_k^*\| \\ &\leq \sum_{k\in I_0} \liminf_{\lambda\in\Lambda} \|\mu_\lambda(\{k\})\| \\ &\leq \liminf_{\lambda\in\Lambda} \sum_{k\in I_0} \|\mu_\lambda(\{k\})\| \\ &\leq \liminf_{\lambda\in\Lambda} \sum_{k\in I_0} |\mu_\lambda|(\{k\}) \\ &\leq \liminf_{\lambda\in\Lambda} |\mu_\lambda|(K) \\ &\leq 1. \end{aligned}$$

Combining these last two inequalities yields

$$\liminf_{\lambda\in\Lambda} \|\mu_\lambda(\{k\})\| \quad = \quad \|x_k^*\| \quad \text{for each } k \text{ in } I_0.$$

Now, for each k in I_0, the preceding argument can be applied to an arbitrary subnet of $\{\mu_\lambda(\{k\})\}_{\lambda\in\Lambda}$ and hence, by the last equation, that subnet has in turn a subnet that converges to $\|x_k^*\|$. From this, it follows that

$$\lim_{\lambda\in\Lambda} \|\mu_\lambda(\{k\})\| \quad = \quad \|x_k^*\| \quad \text{for each } k \text{ in } I_0.$$

Since $x_k^*/\|x_k^*\|$ is a weak* point of continuity of B_{X^*} for each k in I_0, it follows that

$$\lim_{\lambda\in\Lambda} \|\mu_\lambda(\{k\}) - x_k^*\| \quad = \quad 0 \quad \text{for each } k \text{ in } I_0.$$

Let $\varepsilon > 0$ be given. Choose $\{k_1, ..., k_m\} \subseteq I_0$ such that $\sum_{i=1}^{m} \|x_{k_i}^*\| > 1 - \varepsilon$. Now, there exists λ_0 in Λ such that, for all $\lambda \geq \lambda_0$, it is the case that

$$\sum_{i=1}^{m} \|\mu_\lambda(\{k_i\}) - x_{k_i}^*\| \ < \ \varepsilon.$$

Thus, for $\lambda \geq \lambda_0$, it follows that

$$\begin{aligned}
\|\mu_\lambda - \mu\| &= |\,\mu_\lambda - \mu\,|(\{k_1, ..., k_m\}) + |\,\mu_\lambda - \mu\,|(K \setminus \{k_1, ..., k_m\}) \\
&\leq \sum_{i=1}^{m} \|(\mu_\lambda - \mu)(\{k_i\})\| + (|\,\mu_\lambda\,| + |\,\mu\,|)(K \setminus \{k_1, ..., k_m\}) \\
&< \varepsilon + \|\mu_\lambda\| - |\,\mu_\lambda\,|(\{k_1, ..., k_m\}) + \|\mu\| - |\,\mu\,|(\{k_1, ..., k_m\}) \\
&\leq \varepsilon + 2 - \sum_{i=1}^{m} \|\mu_\lambda(\{k_i\})\| - \sum_{i=1}^{m} \|\mu(\{k_i\})\| \\
&\leq \varepsilon + 2 + \sum_{i=1}^{m} \|\mu_\lambda(\{k_i\}) - \mu(\{k_i\})\| - 2\sum_{i=1}^{m} \|\mu(\{k_i\})\| \\
&= \varepsilon + 2 + \sum_{i=1}^{m} \|\mu_\lambda(\{k_i\}) - x_{k_i}^*\| - 2\sum_{i=1}^{m} \|x_{k_i}^*\| \\
&< \varepsilon + 2 + \varepsilon - 2(1 - \varepsilon) \\
&= 4\varepsilon.
\end{aligned}$$

This shows that $\{\mu_\lambda\}_{\lambda \in \Lambda}$ converges in norm to μ and thus μ is a weak* point of continuity of $B_{C(K,X)^*}$.

To establish the converse, suppose μ is a weak* point of continuity of $B_{C(K,X)^*}$.

Claim 1. For every nonisolated point k in K, it is the case that $\mu(\{k\}) = 0$.

To establish this claim, let k be in $K \setminus I$ and suppose $\mu(\{k\}) \neq 0$. Let V be an arbitrary open subset of K that contains k. Since k is a nonisolated point, V must be infinite and so there exists a point k_V in V such that $k_V \neq k$. Then let $\mu_V = \mu|_{K \setminus V} + L_{k_V, \mu(V)}$; that is, for each f in $C(K, X)$,

$$\langle \mu_V, f \rangle = \int_{K \setminus V} f \, d\mu + \langle \mu(V), f(k_V) \rangle.$$

It follows that μ_V is in $C(K,X)^*$. Note

$$\|\mu_V\| \;=\; |\mu|(K \setminus V) \;+\; \|\mu(V)\| \;\leq\; \|\mu\| \;\leq\; 1$$

and $\mu_V(\{k\}) = 0$ since k is in V and $k_V \neq k$. Now, $\{\mu_V\}_{V \in \mathcal{V}}$, where $\mathcal{V}$ is the collection of open subsets of K containing k partially ordered by set inclusion, forms a net in $B_{C(K,X)^*}$. To show this net converges weak* to μ , let f be in $C(K,X)$ and let $\varepsilon > 0$ be given. Choose an open subset V_ε of K that contains k and such that $\operatorname{diam} f(V_\varepsilon) < \varepsilon$. Then, for any V in $\mathcal{V}$ with $V \subseteq V_\varepsilon$, it follows that

$$\begin{aligned}\left|\langle \mu_V - \mu, f\rangle\right| &= \left|\langle \mu(V), f(k_V)\rangle - \int_V f \, d\mu\right| \\ &= \left|\int_V (f(k_V) - f(t)) \, d\mu(t)\right| \\ &\leq \int_V \|f(k_V) - f(t)\| d|\mu|(t) \\ &< \varepsilon.\end{aligned}$$

Thus $\{\mu_V\}_{V \in \mathcal{V}}$ converges weak* to μ and hence, by the hypothesis on μ , it follows that $\{\mu_V\}_{V \in \mathcal{V}}$ converges in norm to μ. But this contradicts the fact that, for all V in $\mathcal{V}$,

$$\|\mu_V - \mu\| \;\geq\; \|\mu_V(\{k\}) - \mu(\{k\})\| \;=\; \|\mu(\{k\})\| \;>\; 0$$

and hence the claim is established.

Let Λ be the collection of all finite open coverings of K partially ordered as follows: for λ_1 and λ_2 in Λ, write $\lambda_1 \leq \lambda_2$ provided, for every open set U in λ_2, there exists an open set V in λ_1 such that $U \subseteq V$. For each λ in Λ , there exists a finite partition Π_λ of K consisting of Borel sets such that, for every E in Π_λ, there exists V in λ with $E \subseteq V$. Now, for each E in Π_λ, choose a point $k_{\lambda,E}$ in E. Then define μ_λ by

$$\mu_\lambda \;=\; \sum_{E \in \Pi_\lambda} L_{k_{\lambda,E},\mu(E)}.$$

It follows that μ_λ is in $C(K,X)^*$ and

$$\|\mu_\lambda\| \;\leq\; \sum_{E \in \Pi_\lambda} \|\mu(E)\| \;\leq\; \sum_{E \in \Pi_\lambda} |\mu|(E) \;\leq\; \|\mu\| \;\leq\; 1.$$

To show that the net $\{\mu_\lambda\}_{\lambda \in \Lambda}$ converges weak* to μ, let f be in $C(K,X)$ and let $\varepsilon > 0$ be given. Choose λ_ε in Λ such that, for all V in λ_ε, it is the case that $\operatorname{diam} f(V) < \varepsilon$. Then, for any λ in Λ with $\lambda \geq \lambda_\varepsilon$, it follows, for every E in Π_λ, that $\operatorname{diam} f(E) < \varepsilon$ and hence

$$\left|\langle \mu_\lambda - \mu, f\rangle\right| \;=\; \left|\sum_{E \in \Pi_\lambda} \left[\langle \mu(E), f(k_{\lambda,E})\rangle \;-\; \int_E f \, d\mu\right]\right|$$

$$= \Big| \sum_{E\in\Pi_\lambda} \Big[\int_E f(k_{\lambda,E})\, d\mu \; - \; \int_E f\, d\mu\Big]\Big|$$

$$\leq \sum_{E\in\Pi_\lambda} \Big|\int_E (f(k_{\lambda,E}) - f(t))\, d\mu(t)\Big|$$

$$\leq \sum_{E\in\Pi_\lambda} \int_E \|f(k_{\lambda,E}) - f\|\, d|\mu|$$

$$\leq \sum_{E\in\Pi_\lambda} \varepsilon|\mu|(E)$$

$$\leq \varepsilon.$$

Thus $\{\mu_\lambda\}_{\lambda\in\Lambda}$ converges weak* to μ and hence, by the hypothesis on μ, it follows that $\{\mu_\lambda\}_{\lambda\in\Lambda}$ converges in norm to μ.

Claim 2. If $I_0 = \{k \in I : \mu(\{k\}) \neq 0\}$, then I_0 is countable, $|\mu|(K \setminus I_0) = 0$ and (hence) $\mu = \sum_{k\in I_0} L_{k,x_k^*}$ where $x_k^* = \mu(\{k\})$.

To establish this claim, first note that I_0 is countable since μ is in $C(K, X)^*$. Now, let $\varepsilon > 0$ be given. Since $\sum_{k\in I_0} \|\mu(\{k\})\| = |\mu|(I_0) \leq \|\mu\|$, there exists $\{k_1, ..., k_m\} \subseteq I_0$ such that $\sum_{k\in I_0'} \|\mu(\{k\})\| < \varepsilon$ where $I_0' = I_0 \setminus \{k_1, ..., k_m\}$. Let

$$\lambda_0 = \Big\{\{k_1\}, \{k_2\}, ..., \{k_m\}, K \setminus \{k_1, ..., k_m\}\Big\}.$$

Then λ_0 is in Λ and note, for all λ in Λ with $\lambda \geq \lambda_0$, it follows that $\{k_i\}$ is in Π_λ for each $1 \leq i \leq m$. Since $\{\mu_\lambda\}_{\lambda\in\Lambda}$ converges in norm to μ, there exists λ in Λ with $\lambda \geq \lambda_0$ and such that $\|\mu_\lambda - \mu\| < \varepsilon$. Thus, letting $\Pi_\lambda' = \Pi_\lambda \setminus \{\{k_1\}, ..., \{k_m\}\}$, it follows that

$$\varepsilon > \|\mu_\lambda - \mu\|$$

$$\geq \sum_{E\in\Pi_\lambda'} |\mu_\lambda - \mu|(E)$$

$$\geq \sum_{E\in\Pi_\lambda'} |\mu_\lambda - \mu|(E \setminus \{k_{\lambda,E}\})$$

$$= \sum_{E\in\Pi'_\lambda} |\mu|(E\setminus\{k_{\lambda,E}\})$$

$$= |\mu|(K\setminus\{k_1,\dots,k_m\}) - \sum_{E\in\Pi'_\lambda} \|\mu(\{k_{\lambda,E}\})\|.$$

By Claim 1 and the definition of I_0, it is the case that $\mu(\{k\}) = 0$ for all k in $K\setminus I_0$. Using this and the fact that $k_{\lambda,E}$ is in E for each E in Π'_λ, it follows that

$$\sum_{E\in\Pi'_\lambda} \|\mu(\{k_{\lambda,E}\})\| \leq \sum_{k\in I'_0} \|\mu(\{k\})\| < \varepsilon.$$

Combining the last two inequalities, it follows that

$$|\mu|(K\setminus I_0) \leq |\mu|(K\setminus\{k_1,\dots,k_m\}) < 2\varepsilon.$$

Since ε was arbitrary, $|\mu|(K\setminus I_0) = 0$. Finally, letting $x_k^* = \mu(\{k\})$ for k in I_0, it immediately follows that $\mu = \sum_{k\in I_0} L_{k,x_k^*}$ and so the claim is established.

Let k in I be fixed, let $x_k^* = \mu(\{k\})$ and suppose $x_k^* \neq 0$. To show $x_k^*/\|x_k^*\|$ is a weak* point of continuity of B_{X^*}, suppose $\{y_\alpha^*\}_{\alpha\in A}$ is a net in X^* which converges weak* to x_k^* and such that $\|y_\alpha^*\| \leq \|x_k^*\|$ for all α in A. For each α in A, let $\nu_\alpha = \mu|_{K\setminus\{k\}} + L_{k,y_\alpha^*}$; that is, for each f in $C(K,X)$,

$$\langle \nu_\alpha, f\rangle = \int_{K\setminus\{k\}} f\, du + \langle y_\alpha^*, f(k)\rangle.$$

It follows that ν_α is in $C(K,X)^*$, for each α in A, and

$$\|\nu_\alpha\| = |\mu|(K\setminus\{k\}) + \|y_\alpha^*\| \leq \|\mu\|.$$

It is straightforward, from the definition of ν_α, to verify that $\{\nu_\alpha\}_{\alpha\in A}$ converges weak* in $C(K,X)^*$ to μ and hence, by the hypothesis of μ, it follows that $\{\nu_\alpha\}_{\alpha\in A}$ converges in norm to μ. But $\|\nu_\alpha - \mu\| = \|y_\alpha^* - x_k^*\|$, for each α in A, and hence it follows that $\{y_\alpha^*\}_{\alpha\in A}$ converges in norm to x_k^*. This shows that $x_k^*/\|x_k^*\|$ is a weak* point of continuity of B_{X^*}.

Finally, since $C(K,X)$ is infinite dimensional, it follows that $\|\mu\| = 1$ and hence it is the case, by Claim 2, that

$$\sum_{k\in I}\|x_k^*\| = \sum_{k\in I_0}\|x_k^*\| = \sum_{k\in I_0}|\mu|(\{k\}) = |\mu|(K) = 1.$$

This completes the proof of the theorem.

The first of the next two corollaries follows immediately from Theorem 6 and the second follows by combining Theorem 1 and Theorem 6 (note this corollary is one of the two results stated in Theorem 2).

Corollary 7. If K has no isolated points, then the unit ball of $C(K,X)^*$ has no weak* points of continuity (and hence has neither weak* denting points nor weak* strongly exposed points).

Corollary 8 (Reuss and Stegall [RS2]). An element of $C(K,X)^*$ is a weak* denting point of the unit ball of $C(K,X)^*$ if and only if it has the form L_{k,x^*} where k is an isolated point of K and x^* is a weak* denting point of the unit ball of X^*.

In the next theorem, a restriction is made on K in order to obtain a complete characterization of the weak* points of sequential continuity of $B_{C(K,X)^*}$. Note that in the proof this restriction is employed for only one of the implications.

Theorem 9. Suppose K is metrizable and let $I = \{k \in K : k \text{ is an isolated point of } K\}$. An element μ in $C(K,X)^*$ is a weak* point of sequential continuity of the unit ball of $C(K,X)^*$ if and only if it has the form $\mu = \sum_{k\in I} L_{k,x^*}$ where, for each k in I, either $x_k^* = 0$ or $x_k^* /\|x_k^*\|$ is a weak* point of sequential continuity of B_{X^*} and $\sum_{k\in I}\|x_k^*\| = 1$.

Proof. If μ has the stated form, then the proof that μ is a weak* point of sequential continuity of $B_{C(K,X)^*}$ follows that given in Theorem 6 for weak* points of continuity; note that no restriction on K is necessary.

Conversely, suppose μ is a weak* point of sequential continuity of $B_{C(K,X)^*}$. Since $C(K,X)$ is infinite dimensional, it follows that $\|\mu\| = 1$. Let d be a metric on K which induces the topology on K. For each k in $K \setminus I$, it can be shown (see Claim 1 in the proof of Theorem 6) that $\mu(\{k\}) = 0$. There exists a countable subset I_0 of I such that $\mu(\{k\}) = 0$ for each k in $I \setminus I_0$. Choose a countable, dense subset D of K such that $I_0 \subseteq D$. Then let

$$\Lambda = \{\lambda : \lambda \text{ is a finite open covering of } K \text{ such that each } V \in \lambda \text{ is of the form } U(k,n) \text{ for some } k \in D \text{ and } n \text{ in } \mathbb{N}\},$$

where $U(k,n) = \{t \in K : d(t,k) < \frac{1}{n}\}$. Since Λ is countable, it may be written as $\Lambda = \{\lambda_n\}_{n\in\mathbb{N}}$. For each n in $\mathbb{N}$, let

$$\gamma_n = \{V_1 \cap V_2 \cap \cdots \cap V_n : V_i \in \lambda_i \quad \text{for} \quad 1 \le i \le n\}$$

and note that γ_n is a finite open covering of K which refines λ_i for each $1 \le i \le n$. For each n in $\mathbb{N}$, choose a finite partition Π_n of K consisting of Borel sets such that, for each

E in Π_n, there exists W in γ_n with $E \subseteq W$. Now, for each E in Π_n, choose a point $k_{n,E}$ in E. Then define μ_n by

$$\mu_n = \sum_{E\in\Pi_n} L_{k_{n,E},\mu(E)}.$$

Note μ_n is in $C(K,X)^*$ and

$$\|\mu_n\| \leq \sum_{E\in\Pi_n} \|\mu(E)\| \leq \sum_{E\in\Pi_n} |\mu|(E) \leq \|\mu\| = 1.$$

To show that the sequence $\{\mu_n\}_{n=1}^{\infty}$ converges weak* to μ, let f be in $C(K,X)$ and let $\varepsilon > 0$ be given. There exists m in $\mathbb{N}$ such that $\operatorname{diam} f(V) < \varepsilon$ for all V in λ_m. Thus, for $n \geq m$ and every E in Π_n, it follows that $\operatorname{diam} f(E) < \varepsilon$ and hence (as in the proof of Theorem 6)

$$\begin{aligned}\Big|\langle \mu_n - \mu, f \rangle\Big| &= \Big| \sum_{E\in\Pi_n} \Big[\langle \mu(E), f(k_{n,E})\rangle - \int_E f\, d\mu \Big]\Big| \\ &\leq \sum_{E\in\Pi_n} \Big| \int_E (f(k_{n,E}) - f(t))d\mu(t) \Big| \\ &\leq \sum_{E\in\Pi_n} \int_E \|f(k_{n,E}) - f\|\, d|\mu| \\ &\leq \varepsilon.\end{aligned}$$

Thus $\{\mu_n\}_{n=1}^{\infty}$ converges weak* to μ and hence, by invoking the hypothesis on μ, it follows that $\{\mu_n\}_{n=1}^{\infty}$ converges in norm to μ.

Claim. It is the case that $|\mu|(K\setminus I_0) = 0$ and (hence) $\mu = \sum_{k\in I_0} L_{k,x_k^*}$ where $x_k^* = \mu(\{k\})$.

To establish this claim, let $\varepsilon > 0$ be given. As in the proof of Claim 2 of Theorem 6, there exists $\{k_1, ..., k_m\} \subseteq I_0$ such that $\sum_{k\in I_0'} \|\mu(\{k\})\| < \varepsilon$ where $I_0' = I_0 \setminus \{k_1, ..., k_m\}$. Since $\{\mu_n\}_{n=1}^{\infty}$ converges in norm to μ, there exists N in $\mathbb{N}$ such that $\|\mu_n - \mu\| < \varepsilon$ whenever $n \geq N$. Choose M in $\mathbb{N}$ so large that $M \geq N$ and

$$\{k_i\} \text{ is in } \lambda_1 \cup \lambda_2 \cup \cdots \cup \lambda_M \text{ for all } 1 \leq i \leq m.$$

Note that $\{k_i\}$ is in Π_M for $1 \leq i \leq M$. Letting $\Pi_M' = \Pi_m \setminus \{\{k_1\}, ..., \{k_m\}\}$, it now follows, as in the proof of Claim 2 of Theorem 6, that

$$\varepsilon > \|\mu_M - \mu\|$$

$$\geq \sum_{E\in\Pi'_M} |\mu_M - \mu|(E)$$

$$\geq \sum_{E\in\Pi'_M} |\mu_M - \mu|(E \setminus \{k_{n,E}\})$$

$$= \sum_{E\in\Pi'_M} |\mu|(E \setminus \{k_{n,E}\})$$

$$= |\mu|(K \setminus \{k_1, ..., k_m\}) - \sum_{E\in\Pi'_M} \|\mu(\{k_{n,E}\})\|.$$

As before, it is the case that

$$\sum_{E\in\Pi'_M} \|\mu(\{k_{n,E}\})\| \leq \sum_{k\in I'_0} \|\mu(\{k\})\| < \varepsilon.$$

Combining these yields that

$$|\mu|(K \setminus I_0) \leq |\mu|(K \setminus \{k_1, ..., k_m\}) < 2\varepsilon.$$

Since ε was arbitrary, $|\mu|(K \setminus I_0) = 0$. Letting $x_k^* = \mu(\{k\})$, for k in I_0, as before completes the proof of the Claim.

The remainder of the proof continues to follow along the same lines as the proof of Theorem 6. Let k in I be fixed with $x_k^* = \mu(\{k\})$ and suppose $x_k^* \neq 0$. As before, it can be shown that $x_k^*/\|x_k^*\|$ is a weak* point of sequential continuity of B_{X^*} (see the proof of Theorem 6). Finally, as before, since $\|\mu\| = 1$,

$$\sum_{k\in I} \|x_k^*\| = \sum_{k\in I_0} \|x_k^*\| = \sum_{k\in I_0} |\mu|(\{k\}) = |\mu|(K) = 1.$$

This completes the proof of the theorem.

Corollary 10. If K is metrizable and has no isolated points, then the unit ball of $C(K, X)^*$ has no weak* points of sequential continuity.

Of the ten types of points given in the diagram, only two (points of continuity and points of sequential continuity) have yet to be discussed in the unit ball of $C(K, X)^*$. Since both these types involve the weak topology and since there is no known concrete representation (in general) for the space $C(K, X)^{**}$, it is essentially hopeless to find (in general) a complete characterization of these types of points in the unit ball of $C(K, X)^*$.

It is possible to identify some of these points which arise in a very natural manner.

Proposition 11. If $\mu = \sum_{k\in K} L_{k,x_k^*}$ where, for each k in K, either $x_k^* = 0$ or $x_k^*/\|x_k^*\|$ is a point of continuity (respectively, point of sequential continuity) of B_{X^*} and $\sum_{k\in K}\|x_k^*\| = 1$, then μ is a point of continuity (respectively, point of sequential continuity) of the unit ball of $C(K,X)^*$.

Proof. The proof of this proposition follows exactly along the lines of the first part of the proof of Theorem 6; in fact, the only changes to make in the line-by-line argument there are the following (in the case of points of continuity): The set $I_0 = \{k \in K : x_k^* \neq 0\}$ is countable. Suppose $\{\mu_\lambda\}_{\lambda\in\Lambda}$ is a net in $B_{C(K,X)^*}$ that converges weakly to μ in $C(K,X)^*$. For each k in K and x^{**} in X^{**}, let $F_{k,x^{**}}$ denote the element of $C(K,X)^{**}$ given by $\langle F_{k,x^{**}}, \nu\rangle = \langle x^{**}, \nu(\{k\})\rangle$ for ν in $C(K,X)^*$. Then apply $F_{k,x^{**}}$ to each μ_λ to conclude that $\{\mu_\lambda(\{k\})\}_{\lambda\in\Lambda}$ converges weakly to the point $\mu(\{k\}) = x_k^*$ in X^* for each k in K. Now, this proof is completed in the same manner as the proof of the first part of Theorem 6.

4 CONCLUDING REMARKS

A natural question is that under what circumstance (if any) is the converse of Proposition 11 true for either points of continuity or points of sequential continuity, thereby establishing the analogue of Theorem 6 or Theorem 9. Suppose, for example, that μ is a point of continuity of the unit ball of $C(K,X)^*$. Let $I_0 = \{k \in K : \mu(\{k\}) \neq 0\}$ and note that I_0 is countable since $|\mu|(K) = 1$. To establish that μ has the form given in Proposition 11, it suffices to show (as in the proof of Theorem 6) that $|\mu|(K \setminus I_0) = 0$. Discovering conditions on K and/or X that yield $|\mu|(K \setminus I_0) = 0$ is left as an open problem. One avenue (not pursued here) may be to consider the circumstance when either K is dispersed or X^* has the Radon-Nikodým property; for, in either of these situations, a representation of $C(K,X)^{**}$ is known (see [CG]).

ACKNOWLEDGEMENTS

Each of the authors wishes to thank Professor K. Jarosz and all the supporting faculty and staff at SIUE for four productive and pleasant days in May in Edwardsville.

REFERENCES

[CG] M. Cambern and P. Greim, The bidual of $C(X, E)$, *Proc. Amer. Math. Soc.* **85** (1982), 53–58 .

[C] J.A. Clarkson, Uniformly convex spaces, *Trans. Amer. Math. Soc.* **40** (1936), 396–414.

[DHS] P.N. Dowling, Z. Hu and M.A. Smith, MLUR renormings of Banach spaces, *Pac. J. Math.* (to appear).

[LLT] B.-L. Lin, P.-K. Lin and S.L. Troyanski, Characterizations of denting points, *Proc. Amer. Math. Soc.* **102** (1988), 526–528.

[RS1] W.M. Ruess and C.P. Stegall, Exposed and denting points in duals of operator spaces, *Israel J. Math.* **53** (1986), 163–190.

[RS2] W.M. Ruess and C.P. Stegall, Weak*-denting points in duals of operator spaces, *"Banach Spaces," Lec. Notes in Math.* **1166** (1984), 158–168.

[Si] I. Singer, Linear funtionals on the space of continuous mappings of a compact Hausdorff space into a Banach space, *Rev. Math. Pures Appl.* **2** (1957), 301–315.

[Sm] M.A. Smith, Some examples concerning rotundity in Banach spaces, *Math. Ann.* **233** (1976), 155–161.

On Certain Banach Algebras of Vector-Valued Functions

ROBERT KANTROWITZ Department of Mathematics, Hamilton College, Clinton, New York 13323

MICHAEL M. NEUMANN [1] Department of Mathematics and Statistics, Mississippi State University, Mississippi State, Mississippi 39762

1 INTRODUCTION

In this article, we are concerned with the classical algebras of continuous, Lipschitz, differentiable, and integrable functions with values in a complex Banach algebra. We introduce the relevant notations and review briefly the evolution of the theory of Banach algebras of vector-valued functions in Section 2. Although the basics of this theory were developed quite some time ago, it seems that some natural problems have been addressed only very recently. Here, our emphasis will be on three topics, namely the automatic continuity theory, the description of the greatest regular subalgebra, and the local spectral theory for Banach algebras of vector-valued functions.

In Section 3, we discuss the automatic continuity results for such Banach algebras in the spirit of our recent paper [18]. Under fairly mild conditions, it turns out that the classical Banach algebras of vector-valued continuous functions carry a unique complete norm topology. Moreover, all epimorphisms onto such algebras and all derivations on such algebras are automatically continuous. Note

[1]Research supported by Grant SNF 11-1015 from the Danish Science Research Council

that these Banach algebras need not be semi-simple, so that standard automatic continuity theory is not available in this context.

We then proceed to the description of the greatest regular subalgebra for Banach algebras of vector-valued functions. The material of Section 4 is based on our paper [17] and contains a brief discussion of the greatest regular subalgebra *Reg* A of an arbitrary commutative Banach algebra A. A typical result in this context states that the greatest regular subalgebra of the Banach algebra of all continuous A-valued functions on a compact Hausdorff space consists precisely of those continuous functions which map into *Reg* A. A similar characterization holds for the classical Banach algebras of vector-valued Lipschitz and differentiable functions and also for group algebras of Bochner integrable functions.

The main intention of the present article is to derive the companion results for the Apostol algebra *Dec* A of a commutative Banach algebra A with a bounded approximate identity. The subalgebra *DecA* dates back to classical work of Apostol [5] and consists of those elements of A for which the corresponding multiplication operators on A are decomposable in the sense of Foiaş. The relevant notions from the theory of decomposable operators and local spectral theory are collected at the end of Section 4.

Recent results from [20], [22], [23], [24] show that the Apostol algebra is closely related to the greatest regular subalgebra and that it plays a natural role in the theory of Banach algebras and harmonic analysis. In fact, if A is semi-simple, then *Dec* A contains *Reg* A and consists exactly of those elements whose Gelfand transforms are continuous with respect to the hull-kernel topology on the spectrum $\Delta(A)$ of A. In light of the indicated results for the greatest regular subalgebra, one may therefore expect that the Apostol algebra of a Banach algebra of A-valued functions consists exactly of those functions which map into *Dec* A.

In the main theorem of this paper, we show that this is indeed the case for each of the classical algebras of vector-valued functions considered above. This theorem is contained in Section 6 and requires a number of general results on decomposable multipliers. These results are developed in Section 5 and should be of independent interest. In particular, we establish that, for multipliers on semi-prime commutative Banach algebras with an approximate identity, decomposability in the sense of Foiaş is equivalent to the decomposition property (δ) from [3], which characterizes, up to similarity, the quotients of decomposable operators.

The authors are indebted to Kjeld B. Laursen (Copenhagen) for several stimulating discussions on decomposable multipliers. In particular, the first part of Theorem 5 and the simple proof of Theorem 6 have been obtained jointly by Kjeld B. Laursen and the second-named author.

2 THE ALGEBRAS $C(\Omega, A)$ AND $L^1(G, A)$

Banach algebras of vector-valued functions date back to the early days of the theory of Banach algebras and play a natural role in functional analysis. Probably the most important example is the space $C(\Omega, A)$ of all continuous functions from a compact Hausdorff space Ω into a complex Banach algebra A. Endowed with pointwise operations and the usual supremum norm $\| \cdot \|_\infty$, the space $C(\Omega, A)$ becomes a complex Banach algebra. Evidently, $C(\Omega, A)$ is commutative precisely when A is commutative. Further information about $C(\Omega, A)$ becomes available in connection with topological tensor products.

We shall use the elementary theory of cross-norms on the algebraic tensor product $A \otimes B$ of two Banach spaces A and B, as developed in [25]. The most important examples of cross-norms on $A \otimes B$ are the projective tensor norm γ and the weak tensor norm λ, given by

$$\gamma(c) := \inf \left\{ \sum_{k=1}^{n} \|u_k\| \, \|v_k\| : u_k \in A, v_k \in B \text{ such that } c = \sum_{k=1}^{n} u_k \otimes v_k \right\}$$

$$\lambda(c) := \sup \left\{ \left| \sum_{k=1}^{n} \sigma(a_k)\, \tau(b_k) \right| : \sigma \in A^*, \tau \in B^* \text{ such that } \|\sigma\|, \|\tau\| \leq 1 \right\}$$

for all $c = \sum_{k=1}^{n} a_k \otimes b_k \in A \otimes B$. If both A and B are Banach algebras, then the algebraic tensor product $A \otimes B$ carries a natural multiplication so that

$$(a_1 \otimes b_1)\,(a_2 \otimes b_2) = (a_1 a_2) \otimes (b_1 b_2) \qquad \text{for all } a_1, a_2 \in A \text{ and } b_1, b_2 \in B.$$

A cross-norm $\| \cdot \|_\alpha$ on $A \otimes B$ need not be submultiplicative with respect to this multiplication, but if it is, then the completed tensor product $A \widehat{\otimes}_\alpha B$ becomes a Banach algebra. This applies, for instance, to the projective tensor norm γ. We refer to Section 42 of [6] for further information on tensor products of Banach algebras.

A classical result due to Grothendieck [10] shows that $C(\Omega, A)$ is, as a Banach algebra, isometrically isomorphic to the completed tensor product $C(\Omega) \widehat{\otimes}_\lambda A$, where $C(\Omega)$ denotes the usual Banach algebra of continuous complex-valued functions on Ω and λ is the weak tensor norm on $C(\Omega) \otimes A$. For the case of a commutative Banach algebra A, this result implies that the spectrum of the Banach algebra $C(\Omega, A)$ may be canonically identified with the cartesian product of Ω and $\Delta(A)$. Here $\Delta(A)$ denotes the spectrum of A, *i.e.* the set of all non-trivial multiplicative linear functionals on A, equipped with the relative weak-star topology inherited from the topological dual space A^* of A.

More precisely, the indicated characterization of $\Delta(C(\Omega), A))$ means that every non-trivial multiplicative linear functional ϕ on $C(\Omega, A)$ may be represented by a unique pair $(\omega, \varphi) \in \Omega \times \Delta(A)$, with $\phi(f) = \varphi(f(\omega))$ for all $f \in C(\Omega, A)$.

This description of the spectrum of $C(\Omega, A)$ was obtained by Hausner [11] in 1957 and has interesting consequences. For instance, it follows immediately that $C(\Omega, A)$ is semi-simple exactly when A is semi-simple. Furthermore, the regularity of the Banach algebra $C(\Omega, A)$ turns out to be equivalent to the regularity of A. Recall that a commutative complex Banach algebra A is said to be *regular* if, for every closed subset of the spectrum $\Delta(A)$ and every $\varphi \in \Delta(A)\backslash F$, there exists some $a \in A$ for which $\varphi(a) = 1$ and $\psi(a) = 0$ for all $\psi \in F$.

For vector-valued group algebras, the situation is remarkably similar. Indeed, let $L^1(G, A)$ denote the Banach space of all equivalence classes modulo Haar measure of the Bochner integrable functions from a locally compact abelian group G into a commutative complex Banach algebra A. With multiplication given by A-valued convolution, $L^1(G, A)$ becomes a commutative Banach algebra. Note that, again by a result of Grothendieck [10, Théorème 2.2.2], the Banach algebra $L^1(G, A)$ is isometrically isomorphic to the completed tensor product $L^1(G) \widehat{\otimes}_\gamma A$, where $L^1(G)$ denotes the scalar-valued group algebra of G and γ is the projective tensor norm on the algebraic tensor product $L^1(G) \otimes A$. It is then not too surprising that the spectrum of $L^1(G, A)$ may be identified with the product of the dual group $\widehat{G}$ and $\Delta(A)$. As in the case of continuous functions, it follows that $L^1(G, A)$ is semi-simple if and only if A is semi-simple and that the regularity of $L^1(G, A)$ is equivalent to that of A. For further information, we refer to the work of Johnson [16] from 1959.

The striking similarities between the Banach algebras $C(\Omega, A)$ and $L^1(G, A)$ received immediate attention. In 1960, Tomiyama [28] developed a unified approach, based on the theory of tensor products of Banach algebras. Let A and B be commutative Banach algebras, and consider a submultiplicative cross-norm $\| \cdot \|_\alpha$ on $A \otimes B$, which dominates the weak tensor norm λ. In this setting, Tomiyama [28] proved that the completed tensor product $A \widehat{\otimes}_\alpha B$ becomes a commutative Banach algebra, whose spectrum may be canonically identified with the cartesian product $\Delta(A) \times \Delta(B)$ of the spectra of A and B. Moreover, the algebra $A \widehat{\otimes}_\alpha B$ turns out to be regular precisely when both A and B are regular. Tensor products of Banach algebras will also play an important role in the present paper. We shall see that canonical results may be obtained also in the case of non-regular algebras.

We close this section with some further examples of Banach algebras of vector valued-functions which are covered by the theory of topological tensor products. Given an arbitrary complex Banach algebra A and a locally compact Hausdorff space Ω, let $C_0(\Omega, A)$ denote the space of all continuous A-valued functions on Ω which vanish at infinity. With pointwise operations, $C_0(\Omega, A)$ is a Banach algebra with respect to the supremum norm. Similarly, for every integer $n \geq 0$, let $C^n(I, A)$ be the Banach algebra of all n-times continuously differentiable functions from a compact interval I into the Banach algebra A, as usual endowed with the norm $\|\cdot\|_n$ given by

$$\|f\|_n := \sum_{k=0}^{n} \frac{1}{k!} \|f^{(k)}\|_\infty \qquad \text{for all } f \in C^n(I, A).$$

Finally, for any $0 < \alpha < 1$, let $Lip_\alpha(I, A)$ consist of all A-valued functions f on a compact interval I with the property that $s_\alpha(f) < \infty$, where

$$s_\alpha(f) := \sup\left\{\frac{\|f(u) - f(v)\|}{| u - v |^\alpha} : u, v \in I \text{ with } u \neq v\right\}.$$

As in the well-known scalar-valued case, the vector-valued Lipschitz algebra $Lip_\alpha(I, A)$ of order α becomes a complex Banach algebra with respect to pointwise operations, when endowed with the norm $\|\cdot\|_\alpha$ given by $\|f\|_\alpha := \|f\|_\infty + s_\alpha(f)$ for all $f \in Lip_\alpha(I, A)$. Let $lip_\alpha(I, A)$ consist of all $f \in Lip_\alpha(I, A)$ for which $m_\alpha(f, \delta) \to 0$ as $\delta \to 0$, where

$$m_\alpha(f, \delta) := \sup\left\{\frac{\|f(u) - f(v)\|}{| u - v |^\alpha} : u, v \in I \text{ with } 0 <| u - v |\leq \delta\right\}$$

for all $\delta > 0$. It is easily seen that $lip_\alpha(I, A)$ is a closed subalgebra of $Lip_\alpha(I, A)$. For the basic theory of these algebras in the scalar-valued case, we refer to [26]. We note that each of the Banach algebras $C_0(\Omega, A)$, $C^n(I, A)$ and $lip_\alpha(I, A)$ may be represented as a completed tensor product with respect to the canonical norms; for details we refer to Section 3 of [17].

3 AUTOMATIC CONTINUITY

Throughout this section, let Ω be a compact Hausdorff space, and consider a complex Banach algebra A with identity. Then $C(\Omega, A)$ is a unital Banach algebra, which is, in general, neither commutative nor semi-simple. Thus the classical results due to Johnson and Sinclair [13] , [15] on the uniqueness of norm and the automatic continuity of derivations on semi-simple Banach algebras are not available in this context. Nevertheless, as we shall see in this section, the algebra $C(\Omega, A)$ allows an extensive automatic continuity theory without the assumption of semi-simplicity. The following results have recently been obtained in [18].

Let B denote an arbitrary complex Fréchet algebra, *i.e.* a complete metrizable topological algebra over $\mathbf{C}$, not assumed to be locally convex. Given an algebra homomorphism $\phi : B \to C(\Omega, A)$ and a point $\omega \in \Omega$, let $\phi_\omega : B \to A$ denote the homomorphism given by $\phi_\omega(u) := \phi(u)(\omega)$ for all $u \in B$. An elementary application of the principle of uniform boundedness shows that ϕ is continuous if and only if ϕ_ω is continuous for every $\omega \in \Omega$. In particular, it follows that all homomorphisms from B into $C(\Omega, A)$ are continuous exactly when all homomorphisms from B into A are continuous. But this observation does not preclude the possibility that all epimorphisms from B onto $C(\Omega, A)$ are automatically continuous, even if A does not enjoy a unique Banach algebra norm.

Theorem 1 *Assume that Ω is a compact Hausdorff space without isolated points, and consider an arbitrary Fréchet algebra B and a unital Banach algebra A. Then the Banach algebra $C(\Omega, A)$ has a unique Fréchet algebra topology, and all surjective algebra homomorphisms $\phi : B \to C(\Omega, A)$ and all derivations $\delta : C(\Omega, A) \to C(\Omega, A)$ are automatically continuous.*

For the sake of illustration, we sketch the idea of the proof for the continuity of an epimorphism $\phi : B \to C(\Omega, A)$. Let Δ_ϕ denote the set of all $\omega \in \Omega$ for which the homomorphism ϕ_ω is discontinuous. It is easily seen from the uniform boundedness principle that Δ_ϕ is always an open subset of Ω. On the other hand, a gliding hump argument shows that the set Δ_ϕ is finite. Since Ω has no isolated points, it follows that Δ_ϕ is empty. But this implies the continuity of ϕ by the application of the uniform boundedness principle mentioned earlier.

In particular, it follows from Theorem 1 that every unital complex Banach algebra may be embedded into a Banach algebra with unique complete norm topology. Simple examples show that the absence of isolated points is essential for Theorem 1. If Ω does have isolated points, then appropriate assumptions on A are needed to ensure automatic continuity results for $C(\Omega, A)$. For instance, it has been shown in [18] that, for any compact Hausdorff space Ω, the algebra $C(\Omega, A)$ has a unique Banach algebra topology whenever A does and that every derivation on $C(\Omega, A)$ is continuous whenever every derivation on A is continuous.

The theory developed in [18] is general enough to apply to a variety of other cases. For instance, it is shown that, for any unital complex Banach algebra A, all the statements of Theorem 1 remain valid for the Banach algebras $C^n(I, A)$, $Lip_\alpha(I, A)$ and $lip_\alpha(I, A)$.

4 THE GREATEST REGULAR SUBALGEBRA

In this section, we shall discuss the greatest regular subalgebra for certain algebras of vector-valued functions. Recall from [23, Theorem 2.8] that every commutative Banach algebra A contains a greatest regular closed subalgebra, denoted by $Reg\, A$. The existence of this subalgebra means that the closed subalgebra generated by the union over all regular closed subalgebras of A is again regular. Since the notion of this subalgebra is both simple and natural, it is somewhat surprising that $Reg\, A$ was first discovered as an application of the spectral theory of decomposable operators in [2, Theorem 2.4]. The original approach due to Albrecht [2] was in the context of semi-simple commutative Banach algebras with identity, while the case of arbitrary commutative Banach algebras was settled only recently in [12] and, independently, in [23]. As one might expect, these two proofs for the general case work entirely within the framework of elementary Banach algebra theory.

Of course, $Reg\,A$ may well be trivial. For instance, by a simple application of the identity theorem for analytic functions, it is seen that $Reg\,A$ consists only of the complex constants, whenever A is a unital Banach algebra of analytic functions on a domain in the complex plane. On the other hand, the greatest regular subalgebra of the measure algebra $M(G)$ for a locally compact group G is large enough to contain all complex Borel measures on G, whose continuous parts are absolutely continuous with respect to Haar measure on G, and also certain singular measures; see [23, Example 2.7]. Note, however, that $Reg\,M(G)$ is, in general, strictly contained in $M(G)$. In fact, if the group G is non-discrete, then $M(G)$ fails to be regular, since the Shilov boundary of $M(G)$ is strictly smaller than $\Delta(M(G))$; see for instance [9, Corollary 8.2.4]. It remains an interesting open problem to characterize $Reg\,M(G)$ in measure-theoretic terms. For the Banach algebra $M_0(G)$ of those measures on G whose Fourier-Stieltjes transforms on the dual group $\widehat{G}$ vanish at infinity, the greatest regular subalgebra has been determined in [20, Theorem 3.1]: $Reg\,M_0(G)$ consists precisely of those measures on G whose Fourier-Stieltjes transforms on the spectrum $\Delta(M(G))$ of the measure algebra vanish outside the dual group $\widehat{G}$. Thus $Reg\,M_0(G) = Rad\,L^1(G)$ in the terminology of [9]. For further information and applications to commutative harmonic analysis, we refer to [20], [22], [23], [24].

In the following result from [17], the greatest regular subalgebra is determined for some typical Banach algebras of vector-valued functions. Note that the result shows, in particular, that each of these algebras is regular precisely when the range algebra A is regular. The proof of Theorem 2 in [17] is based on the theory of tensor products of Banach algebras from [28].

Theorem 2 *For every commutative complex Banach algebra A, the following identities hold:*

(a) $Reg\,C_0(\Omega, A) = C_0(\Omega, Reg\,A)$ *for any locally compact Hausdorff space* Ω.
(b) $Reg\,C^n(I, A) = C^n(I, Reg\,A)$ *for any compact interval* I *and* $n \geq 0$.
(c) $Reg\,lip_\alpha(I, A) = lip_\alpha(I, Reg\,A)$ *for any compact interval* I *and* $0 < \alpha < 1$.
(d) $Reg\,L^1(G, A) = L^1(G, Reg\,A)$ *for any locally compact abelian group* G.

In the following, we shall establish a variant of Theorem 2 for a certain subalgebra of A which arises naturally in local spectral theory. Recall from [29] that a continuous linear operator $T \in L(X)$ on a complex Banach space X is said to be *decomposable* if, for every open cover $\{U, V\}$ of the complex plane $\mathbf{C}$, there exist T-invariant closed linear subspaces Y and Z of X such that $Y + Z = X$, $\sigma(T\,|\,Y) \subseteq U$, and $\sigma(T\,|\,Z) \subseteq V$, where σ denotes the spectrum. Obviously, all normal operators on Hilbert spaces and, more generally, all spectral operators in the sense of Dunford on arbitrary Banach spaces are decomposable. Moreover, the class of decomposable operators contains all operators with a non-analytic functional calculus and, in particular, all generalized scalar and all compact operators. We refer to the monographs by Colojoară-Foiaş [7] and Vasilescu [29] for a thorough discussion of local spectral theory and the theory of decomposable operators.

The investigation of decomposable operators in the context of Banach algebras has a long history. For a commutative complex Banach algebra A, let $Dec\,A$ denote the set of all $a \in A$ for which the corresponding multiplication operators $L_a : A \to A$, given by $L_a u := au$ for all $u \in A$, are decomposable. If A has an identity, then a classical result due to Apostol [5, Theorem 3.6] implies that $Dec\,A$ is a closed subalgebra of A. Therefore, $Dec\,A$ is sometimes called the *Apostol algebra* of A. We include a brief discussion of this algebra, which we believe to play an interesting role in the theory of Banach algebras and commutative harmonic analysis.

Let A be a commutative complex Banach algebra with spectrum $\Delta(A)$ and, for each $a \in A$, let $\hat{a} : \Delta(A) \to \mathbf{C}$ denote the corresponding Gelfand transform given by $\hat{a}(\varphi) := \varphi(a)$ for all $\varphi \in \Delta(A)$. On the spectrum $\Delta(A)$, there are two natural topologies of interest to us. The first one is the usual *Gelfand topology,* which is the relative weak-star topology and hence the coarsest topology on $\Delta(A)$ for which all the Gelfand transforms $\hat{a}$ are continuous. The second one is the *hull-kernel topology,* which is generated by the Kuratowski closure operation $cl(E) := \{\psi \in \Delta(A) : \psi(u) = 0 \text{ for all } u \in A \text{ with } \varphi(u) = 0 \text{ for all } \varphi \in E\}$ for each subset E of $\Delta(A)$. It is well-known that the hull-kernel topology is always coarser then the Gelfand topology and that the two topologies coincide precisely when A is regular. Thus, for non-regular A, there will always exist some $a \in A$ for which the Gelfand transform $\hat{a}$ is not hull-kernel continuous on $\Delta(A)$. For further information, we refer to Section 23 of [6].

A classical result of Colojoară and Foiaş states that all multiplication operators on a regular semi-simple commutative Banach algebra are decomposable; see [7, Theorem 6.2.6]. Thus $Dec\,A = A$ whenever A is semisimple and regular, but much more turns out to be true. As mentioned in the introduction, for a semi-simple commutative Banach algebra A, it is shown in [23, Theorem 2.5] that $Reg\,A$ is always contained in $Dec\,A$. Moreover, again for a semi-simple commutative Banach algebra A, it has been established in [23, Theorem 1.2] that $Dec\,A$ consists precisely of all $a \in A$ for which the Gelfand transform $\hat{a}$ is hull-kernel continuous on $\Delta(A)$. This characterization of $Dec\,A$ follows from the Shilov idempotent theorem and implies that, in the semi-simple case, $Dec\,A$ is always a closed subalgebra of A, even if A has no identity element. Moreover, it follows that A is regular if and only if A is generated by a system of elements for which the corresponding multiplication operators are all decomposable on A. For further results and applications of this theory, we refer to [2], [8], [20], [22], [23], [24].

Note, however, that some challenging problems in this field are still open. For instance, it is not known for which algebras A the identity $Reg\, A = Dec\, A$ holds. In particular, this problem remains unsettled in the important case of the measure algebra $M(G)$ for a locally compact abelian group G. The theory developed in [20] and [22] shows that the Apostol algebra $Dec\, M(G)$ is a full and symmetric closed *-subalgebra of the measure algebra $M(G)$, which is stable with respect to absolute continuity, but a measure-theoretic characterization of this subalgebra and a better insight into the hull-kernel topology of the spectrum of the measure algebra are still missing. On the other hand, in the case of the subalgebra $M_0(G)$, the Apostol algebra is completely understood, since we know from [20, Theorem 4.5] that $Reg\, M_0(G) = Dec\, M_0(G) = Rad\, L^1(G)$.

To prove the analog of Theorem 2 for the Apostol algebra, we need to recall a few basic notions from local spectral theory; see [7] and [29] for further information. Given an arbitrary continuous linear operator $T \in L(X)$ on a complex Banach space X, let $\rho_T(x)$ denote the *local resolvent set* of T at $x \in X$, *i.e.* the set of all $\lambda \in \mathbf{C}$ for which there exists an analytic function $f : U \to X$ on some open neighborhood U of λ so that $(T-\mu)f(\mu) = x$ holds for all $\mu \in U$. One may think of the function $f : U \to X$ as a local extension of the resolvent function, since obviously $f(\lambda) = (T-\lambda)^{-1}x$ for all those $\lambda \in U$ which belong to the resolvent set $\rho(T)$. The complement $\sigma_T(x) := \mathbf{C}\backslash\rho_T(x)$ is called the *local spectrum* of T at x. An operator $T \in L(X)$ is said to have *Dunford's property* (C), if the space $X_T(F) := \{x \in X : \sigma_T(x) \subseteq F\}$ is closed for every closed $F \subseteq \mathbf{C}$. By [21, Proposition 1.2], property (C) for an operator $T \in L(X)$ always implies that T has the *single valued extension property* (SVEP), which means that, for any open set $U \subseteq \mathbf{C}$, the only X-valued analytic solution of the equation $(T-\lambda)f(\lambda) = 0$ for all $\lambda \in U$ is the constant function $f \equiv 0$. We mention in passing that both SVEP and property (C) play an important role in the theory of spectral operators.

Finally, an operator $T \in L(X)$ is said to have the *decomposition property* (δ), if the representation $X = \mathcal{X}_T(\overline{U}) + \mathcal{X}_T(\overline{V})$ holds for every open cover $\{U, V\}$ of $\mathbf{C}$. Here, for each closed set $F \subseteq \mathbf{C}$, the space $\mathcal{X}_T(F)$ is defined to consist of all $x \in X$ for which there exists an analytic function $f : \mathbf{C}\backslash F \to X$ with the property that $(T-\lambda)f(\lambda) = x$ for all $\lambda \in \mathbf{C}\backslash F$. If T has SVEP, then clearly $\mathcal{X}_T(F) = X_T(F)$ for all closed $F \subseteq \mathbf{C}$, in which case property (δ) means precisely a decomposition of local spectra. It is well-known and easily seen that an operator $T \in L(X)$ is decomposable if and only if T has both (C) and (δ). Moreover, by a recent result due to Albrecht and Eschmeier [3], property (δ) characterizes those operators which are similar to quotients of decomposable operators. By [29, Theorem 4.4.28], it follows that property (δ) for an operator $T \in L(X)$ implies that the representation $X = \mathcal{X}_T(\overline{U}_1) + \dots + \mathcal{X}_T(\overline{U}_n)$ holds for every finite open cover $\{U_1, \dots, U_n\}$ of $\mathbf{C}$.

5 MULTIPLIERS AND LOCAL SPECTRAL THEORY

Throughout this section, let A denote a commutative complex Banach algebra, which is assumed to be *without order* in the sense that $au = 0$ for all $a \in A$ implies that $u = 0$. Obviously, all unital Banach algebras and, more generally, all Banach algebras with an approximate identity are without order. Moreover, if A is semi-simple or at least semi-prime, then A is easily seen to be without order. In this connection, recall that a commutative Banach algebra is semi-simple, precisely when it contains no non-zero quasi-nilpotent element, and semi-prime, precisely when it contains no non-zero nilpotent element; cf. [6].

We shall use the elementary theory of multipliers, as developed, for instance, in Chapter 1 of [19]. A mapping $S : A \to A$ is called a *multiplier* on A, if $uSv = (Su)v$ holds for all $u, v \in A$. Since A is without order, every multiplier S on A satisfies $S(uv) = (Su)v$ for all $u, v \in A$ and is automatically linear and continuous. Moreover, the set $M(A)$ of all multipliers on A is a closed commutative subalgebra of $L(A)$. Obviously, $M(A)$ contains the identity operator I_A on A and the multiplication operators L_a for all $a \in A$. Again, since A is without order, A may be viewed as an ideal in $M(A)$ via the left regular representation, but observe that the norms of A and $M(A)$ need not be equivalent on A.

Perhaps the most important example is the group algebra $A = L^1(G)$ for a locally compact abelian group G, in which case the multiplier algebra $M(A)$ may be identified, via convolution, with the measure algebra $M(G)$ of all regular complex Borel measures on G. As another natural example, we mention the Banach algebra $A = C_0(\Omega)$ for a locally compact Hausdorff space Ω, in which case the multiplier algebra $M(A)$ may be canonically identified with the Banach algebra $C_b(\Omega)$ of all bounded continuous complex-valued functions on Ω.

Now consider a complex Banach space X and an algebra homomorphism $\phi : A \to L(X)$. Thus X becomes a left A-module with the module action given by $a \cdot x := \phi(a)x$ for all $a \in A$ and $x \in X$, but since we do not assume the continuity of the homomorphism ϕ, the module action need not be continuous. This level of generality will be relevant in the proof of Proposition 10. The left A-module X is said to be *without order,* if $a \cdot x = 0$ for all $a \in A$ implies that $x = 0$. In terms of the homomorphism ϕ, this condition means precisely that $\bigcap_{a\in A} \ker \phi(a) = \{0\}$. For arbitrary subsets $B \subseteq A$ and $Y \subseteq X$, let $B \cdot Y$ denote the linear span of the products $b \cdot y$ for all $b \in B$ and $y \in Y$. In the following, we shall require that $A \cdot X = X$, which is fulfilled whenever $\bigcup_{a\in A} \operatorname{ran} \phi(a) = X$.

If A has an identity $e \in A$ and if $\phi(e) = I_X$, the identity operator on X, then X is clearly without order and satisfies $A \cdot X = X$. Note, however, that these conditions will also be fulfilled under much weaker assumptions. For instance, if the homomorphism ϕ is continuous and if A has a bounded approximate identity $(e_\lambda)_{\lambda\in\Lambda}$ for which $\phi(e_\lambda)x \to x$ for each $x \in X$, then the A-module X is without order, and the module version of the Cohen factorization theorem implies that

$A \cdot X = X$; see [6, Theorem 11.10]. We next show that, in this case, X becomes a left module over the multiplier algebra $M(A)$. For related techniques in the cohomology theory of Banach algebras, we refer to Johnson [14].

Lemma 3 *Consider a commutative Banach algebra A without order, a Banach space X, and an algebra homomorphism $\phi : A \to L(X)$. Assume that the induced left A-module X is without order and satisfies $A \cdot X = X$. Then, for $S \in M(A)$ and $x \in X$, the definition*

$$\Phi(S)x := \sum_{k=1}^{n} \phi(Sa_k)y_k \qquad \text{for} \qquad x = \sum_{k=1}^{n} \phi(a_k)y_k,$$

with $a_k \in A$ and $y_k \in X$ for $k = 1, \ldots, n$, yields a unital algebra homomorphism $\Phi : M(A) \to L(X)$, which extends ϕ in the sense that $\Phi(L_a) = \phi(a)$ for all $a \in A$. Moreover, Φ is continuous whenever ϕ is continuous.

Proof. First observe that the condition $A \cdot X = X$ ensures that every $x \in X$ admits a representation of the type considered above. Now, given an arbitrary $S \in M(A)$, let $a_k \in A$ and $y_k \in X$ for $k = 1, \ldots, n$ such that $\sum_{k=1}^{n} \phi(a_k)y_k = 0$. Then, for every $u \in A$, we obtain that

$$\phi(u)(\sum_{k=1}^{n} \phi(Sa_k)y_k) = \sum_{k=1}^{n} \phi(uSa_k)y_k = \phi(Su)(\sum_{k=1}^{n} \phi(a_k)y_k) = 0$$

and hence that $\sum_{k=1}^{n} \phi(Sa_k)y_k = 0$, since the A-module X is without order. This observation shows that $\Phi(S) : X \to X$ is well-defined, and the linearity of this mapping is immediate. To prove the continuity of $\Phi(S)$, consider a null sequence $(x_m)_{m\in\mathbf{N}}$ in X so that the sequence $(\Phi(S)x_m)_{m\in\mathbf{N}}$ converges to some $z \in X$. Given an arbitrary $u \in A$, it follows from the continuity of $\phi(u)$ that $\phi(u)\Phi(S)x_m \to \phi(u)z$ as $m \to \infty$. On the other hand, the same computation as above shows that $\phi(u)\Phi(S)x_m = \phi(Su)x_m$ for all $m \in \mathbf{N}$, which implies that $\phi(u)\Phi(S)x_m \to 0$ as $m \to \infty$ by the continuity of $\phi(Su)$. We conclude that $\phi(u)z = 0$ for all $u \in A$ and hence that $z = 0$, since X is without order. Thus $\Phi(S)$ is continuous by the closed graph theorem. It is now routine to check that Φ is a unital homomorphism which extends ϕ. Finally, if ϕ is continuous, it is immediate that, for each $x \in X$, the mapping $S \to \Phi(S)x$ is continuous on $M(A)$, and this implies the continuity of Φ by the principle of uniform boundedness. ■

We shall need the fact that decomposability is preserved under algebra homomorphisms. This is contained in the following lemma, which is related to corresponding results due to Albrecht [1], [2] and Eschmeier [8].

Lemma 4 *Let A be a commutative Banach algebra without order, and consider a Banach space X and an algebra homomorphism $\phi : A \to L(X)$, for which the* *induced left A-module X is without order and satisfies $A \cdot X = X$. In addition, assume that $A \cdot Y_B = Y_B$ holds for every $B \subseteq A$, where $Y_B := \bigcap_{u \in B} \ker \phi(u)$. Then, if the multiplier $S \in M(A)$ has SVEP and property (δ), the operator $T := \Phi(S) \in L(X)$ is decomposable and satisfies $X_T(F) = E_T(F)$ for all closed $F \subseteq \mathbf{C}$, where*

$$E_T(F) := \{x \in X : \phi(a)x = 0 \textit{ for all } a \in A \textit{ with } \sigma_S(a) \cap F = \emptyset\}.$$

Proof. We shall use the spaces $E_T(F)$ to verify the definition of decomposability. First, for each closed $F \subseteq \mathbf{C}$, it is easily seen that $E_T(F)$ is a closed linear subspace of X with the property that $\Phi(R)\, E_T(F) \subseteq E_T(F)$ for every $R \in M(A)$. In particular, the space $E_T(F)$ is invariant under the operator T.

Moreover, for every open cover $\{U, V\}$ of $\mathbf{C}$, we have that $X = E_T(\overline{U}) + E_T(\overline{V})$. Indeed, given an arbitrary $x \in X$, we employ the condition $A \cdot X = X$ to get a representation $x = \sum_{k=1}^{n} \phi(a_k) y_k$ with $a_k \in A$ and $y_k \in X$ for $k = 1, \dots, n$. Since S has property (δ), we then obtain $u_k, v_k \in A$ such that $\sigma_S(u_k) \subseteq \overline{U}$, $\sigma_S(v_k) \subseteq \overline{V}$, and $a_k = u_k + v_k$ for $k = 1, \dots, n$. Thus $x = u + v$, where $u := \sum_{k=1}^{n} \phi(u_k) y_k$ and $v := \sum_{k=1}^{n} \phi(v_k) y_k$. To show that u belongs to $E_T(\overline{U})$, consider an arbitrary $a \in A$ for which $\sigma_S(a) \cap \overline{U} = \emptyset$. For $k = 1, \dots, n$, it follows that $\sigma_S(a u_k) \subseteq \sigma_S(a) \cap \sigma_S(u_k) \subseteq \sigma_S(a) \cap \overline{U} = \emptyset$ and therefore $a u_k = 0$, since S has SVEP. This implies that $\phi(a) u = 0$ and hence that $u \in E_T(\overline{U})$. The same argument shows that $v \in E_T(\overline{V})$, which completes the proof of the identity $X = E_T(\overline{U}) + E_T(\overline{V})$.

We next claim that $\sigma(T|E_T(F)) \subseteq F$ for every closed $F \subseteq \mathbf{C}$. To establish this inclusion, we fix an arbitrary open neighborhood U of F and observe that $A \cdot E_T(F) = E_T(F)$ by the additional assumption of the lemma, since $E_T(F) = Y_B$ with the choice $B := \{a \in A : \sigma_S(a) \cap F = \emptyset\}$. Hence, given any $x \in E_T(F)$, we obtain a representation $x = \sum_{k=1}^{n} \phi(a_k) z_k$ with $a_k \in A$ and $z_k \in E_T(F)$ for $k = 1, \dots, n$. Since S has property (δ) and $\mathbf{C} = U \cup (\mathbf{C} \setminus F)$, we may write $a_k = u_k + v_k$ for suitable $u_k, v_k \in A$ such that $\sigma_S(u_k) \subseteq U$ and $\sigma_S(v_k) \cap F = \emptyset$ for $k = 1, \dots, n$. Because $z_k \in E_T(F)$, we conclude that $\phi(v_k) z_k = 0$ for $k = 1, \dots, n$ and hence that $x = \sum_{k=1}^{n} \phi(u_k) z_k$. Next, since S has SVEP and $\sigma_S(u_k) \subseteq U$ for $k = 1, \dots, n$, there exist analytic functions $f_1, \dots, f_n : \mathbf{C} \setminus \overline{U} \to A$ such that $(S - \lambda) f_k(\lambda) = u_k$ for all $\lambda \in \mathbf{C} \setminus \overline{U}$ and $k = 1, \dots, n$. An application of the homomorphism Φ gives that $(T - \lambda)\phi(f_k(\lambda)) = \phi(u_k)$ for $k = 1, \dots, n$ and therefore $(T - \lambda) y_\lambda = x$, where $y_\lambda := \sum_{k=1}^{n} \phi(f_k(\lambda)) z_k \in E_T(F)$ for all $\lambda \in \mathbf{C} \setminus \overline{U}$. Consequently, $T - \lambda$ maps $E_T(F)$ onto itself for each $\lambda \in \mathbf{C} \setminus \overline{U}$. To establish the injectivity, let $\lambda \in \mathbf{C} \setminus \overline{U}$ and $z \in E_T(F)$ so that $(T - \lambda) z = 0$. Since the A-module X is without order, it suffices to show that $\phi(a) z = 0$ for each $a \in A$. Clearly $x := \phi(a) z$ belongs to $E_T(F)$ so that the preceding argument for the special case $n = 1$ supplies an analytic function $f : \mathbf{C} \setminus \overline{U} \to A$ for which $(T - \lambda)\phi(f(\lambda)) z = x$. Thus $x = \phi(f(\lambda))(T - \lambda) z = 0$, which proves that the restriction $(T - \lambda)|E_T(F)$ is injective. It follows that $\sigma(T|E_T(F)) \subseteq \overline{U}$ for every open neighborhood U of F and therefore $\sigma(T|E_T(F)) \subseteq F$.

We conclude that T is decomposable and that $E_T(F) \subseteq X_T(F)$ for every closed $F \subseteq \mathbf{C}$. The opposite inclusion follows easily whenever the homomorphism ϕ is continuous. In fact, given any $x \in X_T(F)$ and $a \in A$ for which $\sigma_S(a) \cap F = \emptyset$, we obtain from the continuity of ϕ that $\rho_S(a) \subseteq \rho_T(\phi(a)x)$ and hence that $\sigma_T(\phi(a)x) \subseteq \sigma_S(a) \cap \sigma_T(x) \subseteq \sigma_S(a) \cap F = \emptyset$. Since T has SVEP, this implies that $\phi(a)x = 0$ and therefore $x \in E_T(F)$. The general case, however, seems to require a deeper argument. The identity $E_T(F) = X_T(F)$ for all closed $F \subseteq \mathbf{C}$ will be a consequence of the uniqueness theorem for 2-spectral capacities [29, Theorem 4.1.9], once the spaces $E_T(F)$ are known to preserve intersections. To verify this condition, consider a family $\{F_\alpha : \alpha \in J\}$ of closed subsets of the complex plane, and let $F := \cap_{\alpha \in J} F_\alpha$. Since obviously $E_T(F) \subseteq \cap_{\alpha\in J} E_T(F_\alpha)$, it remains to prove the reverse inclusion. Let $x \in \cap_{\alpha \in J} E_T(F_\alpha)$ and consider an arbitrary $a \in A$ with $\sigma_S(a) \cap F = \emptyset$. Then, by the compactness of $\sigma_S(a)$, there exist finitely many $\alpha_1, \dots, \alpha_n \in J$ such that $\sigma_S(a) \subseteq \cup_{k=1}^n (\mathbf{C}\backslash F_{\alpha_k})$. Since S has property (δ), every $b \in A$ admits a representation $b = u + u_1 + \dots + u_n$, where $\sigma_S(u) \subseteq \rho_S(a)$ and $\sigma_S(u_k) \subseteq \mathbf{C}\backslash F_{\alpha_k}$ for $k = 1, \dots, n$. We obtain that $\sigma_S(ua) \subseteq \sigma_S(u) \cap \sigma_S(a) \subseteq \rho_S(a) \cap \sigma_S(a) = \emptyset$ and therefore $ua = 0$ by SVEP for S. Moreover, from $\sigma_S(u_k a) \subseteq \sigma_S(u_k) \subseteq \mathbf{C}\backslash F_{\alpha_k}$ and $x \in E_T(F_{\alpha_k})$, we conclude that $\phi(u_k a)x = 0$ for $k = 1, \dots, n$. It follows that $\phi(b)\phi(a)x = \phi(ba)x = \sum_{k=1}^n \phi(u_k a)x = 0$ for all $b \in A$ and hence that $\phi(a)x = 0$, since the A-module X is without order. Thus $x \in E_T(F)$, as desired. ∎

Theorem 5 *Let A be a commutative Banach algebra with a bounded approximate identity $(e_\lambda)_{\lambda\in\Lambda}$. Then a multiplier on A is decomposable if and only if it has SVEP and property (δ). Moreover, if X is a Banach space and $\phi : A \to L(X)$ is a continuous algebra homomorphism such that $\phi(e_\lambda)x \to x$ for all $x \in X$, then, for every decomposable multiplier $S \in M(A)$, the operator $T := \Phi(S) \in L(X)$ is decomposable and satisfies $X_T(F) = E_T(F)$ for all closed $F \subseteq \mathbf{C}$.*

Proof. First observe that the left regular representation of A induces the identity mapping on $M(A)$. Also note that, in the setting of the preceding lemma, the set Y_B is always a closed submodule of X in the A-module structure induced by the homomorphism ϕ. Hence the assertions follow immediately from Lemma 4 and the module version of the Cohen factorization theorem [6, Theorem 11.10]. ∎

It is well-known and easily seen that every multiplier on a semi-simple commutative Banach algebra has SVEP; see [7, Proposition 6.2.3]. The first part of the following result extends this to the case of semi-prime algebras. Theorem 6 has been obtained earlier in [22, Theorem 3.3], but, for completeness, we include an independent short proof.

Theorem 6 *Every multiplier on a semi-prime commutative Banach algebra has SVEP. Moreover, if A is a semi-prime commutative Banach algebra with* *a bounded approximate identity, then a multiplier on A is decomposable if and only if it has property (δ).*

Proof. By Theorem 5, it suffices to establish the first statement. Given any multiplier $T \in M(A)$, we shall need the fact that $\ker T \cap \operatorname{ran} T = \{0\}$. To see this, let $u \in \ker T \cap \operatorname{ran} T$, so that $Tu = 0$ and $u = Tv$ for some $v \in A$. Hence $0 = 0v = T(Tv)v = (Tv)^2$ and therefore $u = Tv = 0$, since A contains no non-zero nilpotent element.

Now, let $f : U \to A$ be an analytic function on an open set $U \subseteq \mathbf{C}$ such that $(T-\mu)f(\mu) = 0$ for all $\mu \in U$. Fix $\lambda \in U$ and note that $T - \lambda \in M(A)$ and $f(\lambda) \in \ker(T-\lambda)$. Consequently, the result of the preceding paragraph will establish SVEP for the operator T, once we have seen that $f(\lambda) \in \operatorname{ran}(T-\lambda)$. Define $g : U \to A$ by $g(\lambda) := f'(\lambda)$ and $g(\mu) := (\lambda-\mu)^{-1}(f(\lambda) - f(\mu))$ for all $\mu \in U \backslash \{\lambda\}$. Then the function $g : U \to A$ is analytic and satisfies $(T-\mu)g(\mu) = (\lambda-\mu)^{-1}(T-\mu)f(\lambda) = f(\lambda)$ for all $\mu \in U\backslash\{\lambda\}$. Passing to the limit as $\mu \to \lambda$, we obtain that $(T-\lambda)g(\lambda) = f(\lambda)$ and hence that $f(\lambda) \in \operatorname{ran}(T-\lambda)$. ■

Another essential tool for our investigation of decomposable multiplication operators will be the following result on algebraic properties of decomposable multipliers.

Theorem 7 *If A is a commutative Banach algebra with a bounded approximate identity, then both the decomposable multipliers and the multipliers with property (δ) form a closed subalgebra of the multiplier algebra $M(A)$.*

Proof. The more elementary case of multipliers with property (δ) has been settled in [22, Corollary 3.2]. The result for decomposable multipliers is due to Albrecht [2, Theorem 2.6], but since the original proof involves the theory of several commuting operators, it is perhaps interesting to note that the decomposable case may be reduced to that of property (δ). In fact, if $S, T \in M(A)$ are decomposable multipliers, then $S+T$ and ST have property (δ), since the multipliers with property (δ) form an algebra. Moreover, $S+T$ and ST have SVEP by a recent result of Sun [27, Theorem 5], which states that sums and products of commuting operators with Dunford's property (C) always have SVEP. By Theorem 5, it then follows that $S+T$ and ST are decomposable. Finally, an immediate application of a classical perturbation result due to Apostol shows that the algebra of decomposable multipliers is closed in $M(A)$; see [4, Corollary 2.8] and also [22, Theorem 2.7]. ■

6 DECOMPOSABLE MULTIPLICATION OPERATORS

We finally return to the discussion of the set $Dec\, A$ for a commutative complex Banach algebra A. Recall that $Dec\, A$ is defined to consist of all $a \in A$ for which the corresponding multiplication operators $L_a : A \to A$ are decomposable. As mentioned earlier, it follows from a result of Apostol [5, Theorem 3.6] that $Dec\, A$ is a closed subalgebra of A whenever A has an identity. Theorem 7 immediately implies the following stronger result.

Proposition 8 *For every commutative Banach algebra A with a bounded approximate identity, the Apostol algebra $Dec\, A$ is a closed subalgebra of A.*

In this section, we shall show that the Apostol algebra behaves canonically with respect to certain Banach algebras of vector-valued functions. Our main results will be contained in Theorem 11. These results are completely parallel to the case of the greatest regular subalgebra in Theorem 2, although the technicalities are rather different. Here, the theory of tensor products of Banach algebras has to be combined with the theory of decomposable multipliers.

Proposition 9 *Let A and B be commutative Banach algebras with bounded approximate identities, and consider a submultiplicative cross-norm $\|\cdot\|_\alpha$ on $A\otimes B$. Then the completed tensor product $A\widehat{\otimes}_\alpha B$ is a commutative Banach algebra with a bounded approximate identity and satisfies $Dec\, A \otimes Dec\, B \subseteq Dec(A\widehat{\otimes}_\alpha B)$.*

Proof. Clearly, $A\widehat{\otimes}_\alpha B$ is a commutative Banach algebra. Now, let $(e_\lambda)_{\lambda\in\Lambda}$ and $(f_\mu)_{\mu\in M}$ be bounded approximate identities in A and B, respectively, and choose a constant $\kappa > 0$ such that $\|e_\lambda\|, \|f_\mu\| \le \kappa$ for all $\lambda \in \Lambda$ and $\mu \in M$. Then we obtain that $\|e_\lambda \otimes f_\mu\|_\alpha \le \kappa^2$ and

$$\|(e_\lambda \otimes f_\mu)(a \otimes b) - a \otimes b\|_\alpha \le \|e_\lambda\|\, \|a\|\, \|f_\mu b - b\| + \|e_\lambda a - a\|\, \|b\|$$

for all $(a, b) \in A \times B$ and $(\lambda, \mu) \in \Lambda \times M$. It follows that $(e_\lambda \otimes f_\mu)\, c \to c$ for each $c \in A \otimes B$ and consequently for each $c \in A\widehat{\otimes}_\alpha B$, since the norms $\|e_\lambda \otimes f_\mu\|_\alpha$ are uniformly bounded. Thus $A\widehat{\otimes}_\alpha B$ has a bounded approximate identity. By Proposition 8, we conclude that $Dec(A\widehat{\otimes}_\alpha B)$ is a closed subalgebra of $A\widehat{\otimes}_\alpha B$. Hence it remains to show that $u \otimes v \in Dec(A\widehat{\otimes}_\alpha B)$ whenever $u \in Dec\, A$ and $v \in Dec\, B$.

To this end, choose an arbitrary $w \in B$ and consider the left multiplication operators $S_\lambda := L_{e_\lambda \otimes w} \in L(A\widehat{\otimes}_\alpha B)$ for all $\lambda \in \Lambda$. Clearly, $\|S_\lambda\| \le \|e_\lambda \otimes w\|_\alpha = \|e_\lambda\|\, \|w\| \le \kappa \|w\|$ for all $\lambda \in \Lambda$ and $S_\lambda(a\otimes b) = (e_\lambda a)\otimes(wb) \to a\otimes(wb)$ for all $a \in A$ and $b \in B$. Hence the net of operators $(S_\lambda)_{\lambda\in\Lambda}$ converges pointwise on $A \otimes B$. Since the operator norms $\|S_\lambda\|$ are uniformly bounded, for arbitrary $c \in A\widehat{\otimes}_\alpha B$ it follows that $(S_\lambda c)_{\lambda\in\Lambda}$ is a Cauchy net and hence convergent in $A\widehat{\otimes}_\alpha B$. Denoting this limit by $T_w c$ for each $c \in A\widehat{\otimes}_\alpha B$, we obtain an operator $T_w \in L(A\widehat{\otimes}_\alpha B)$ such that $\|T_w\| \le \kappa \|w\|$ and $T_w(a \otimes b) = a \otimes (wb)$ for all $a \in A$ and $b \in B$. Moreover, since T_w is the pointwise limit of a net of multiplication operators, it is clear that T_w is a multiplier on $A\widehat{\otimes}_\alpha B$. Now, let $\phi : B \to L(A\widehat{\otimes}_\alpha B)$ be given by $\phi(w) := T_w$ for all $w \in B$. Then it is easily seen that ϕ is a continuous algebra homomorphism with $\|\phi\| \le \kappa$ and $\phi(f_\mu)c \to c$ for all $c \in A\widehat{\otimes}_\alpha B$. Because $v \in Dec\, B$, we conclude from Theorem 5 that the operator T_v is decomposable on $A\widehat{\otimes}_\alpha B$.

The same argument leads to a decomposable multiplier R_u on $A\widehat{\otimes}_\alpha B$ with the property that $R_u(a \otimes b) = (ua) \otimes b$ for all $a \in A$ and $b \in B$. It follows that $L_{u\otimes v}(a \otimes b) = (ua) \otimes (vb) = (R_u T_v)(a \otimes b)$ for all $a \in A$, $b \in B$ and consequently, by linearity and continuity, that $L_{u\otimes v} = R_u T_v$ on $A\widehat{\otimes}_\alpha B$. This shows that the multiplication operator $L_{u\otimes v}$ is the product of two decomposable multipliers on a Banach algebra with a bounded approximate identity and hence decomposable by Theorem 7. Thus $u \otimes v \in Dec(A\widehat{\otimes}_\alpha B)$. ■

Proposition 10 *Let $F(\Omega)$ be a regular Banach algebra of complex-valued functions on a non-empty set Ω, with pointwise operations. Assume that $F(\Omega)$ has a bounded approximate identity and that, for every commutative Banach algebra A, there exists a Banach algebra $F(\Omega, A)$ of A-valued functions on Ω, with pointwise operations, such that the following conditions are fulfilled:*

(i) $F(\Omega, A)$ contains $F(\Omega) \otimes A$ as a dense subalgebra.
(ii) The norm of $F(\Omega, A)$ is a cross-norm on $F(\Omega) \otimes A$.
(iii) For each closed subalgebra B of A, $F(\Omega, B) = \{f \in F(\Omega, A) : f(\Omega) \subseteq B\}$, and the norm of $F(\Omega, A)$ is an extension of the norm of $F(\Omega, B)$.

Then the identity $Dec\, F(\Omega, A) = F(\Omega, Dec\, A)$ holds for every commutative Banach algebra A with a bounded approximate identity.

Proof. Since the Banach algebra $F(\Omega)$ is regular and semi-simple, we conclude from a classical result of Colojoară and Foiaş [7, Theorem 6.2.6] that $Dec\, F(\Omega) = F(\Omega)$; see also [23, Theorem 2.1] for a different approach. Furthermore, by Proposition 8, we know that $Dec\, A$ is a closed subalgebra of A. Hence we may apply the conditions (i) and (ii) to the Banach algebra $Dec\, A$ to obtain the identity $F(\Omega, Dec\, A) = F(\Omega)\widehat{\otimes}_\alpha Dec\, A$ for a suitable submultiplicative cross-norm $\|\cdot\|_\alpha$ on $F(\Omega) \otimes Dec\, A$. By conditions (i)-(iii), this norm is the restriction of the submultiplicative cross-norm for $F(\Omega) \otimes A$ induced by the norm of $F(\Omega, A)$, for simplicity also denoted by $\|\cdot\|_\alpha$. Since the Apostol algebra is closed, Proposition 9 now implies that $F(\Omega, Dec\, A) = F(\Omega)\widehat{\otimes}_\alpha Dec\, A \subseteq Dec(F(\Omega)\widehat{\otimes}_\alpha A) = Dec\, F(\Omega, A)$.

To show the opposite inclusion, we first note that, by Proposition 9, the Banach algebra $F(\Omega, A) = F(\Omega)\widehat{\otimes}_\alpha A$ has a bounded approximate identity and hence is without order. Now, choose an arbitrary $\omega \in \Omega$ and consider the algebra homomorphism $\phi_\omega : F(\Omega, A) \to L(A)$ given by $\phi_\omega(f) := L_{f(\omega)}$ for all $f \in F(\Omega, A)$. Since we do not assume that convergence in $F(\Omega, A)$ implies pointwise convergence, the homomorphism ϕ_ω need not be continuous, but fortunately Lemma 4 works without a continuity assumption. To verify that A is without order in the

$F(\Omega, A)$-module structure induced by ϕ_ω, let $a \in A$ be such that $\phi_\omega(f)a = 0$ for all $f \in F(\Omega, A)$. By the regularity of $F(\Omega)$, there exists some $g \in F(\Omega)$ such that $g(\omega) = 1$. Hence, if $(e_\lambda)_{\lambda \in \Lambda}$ denotes a bounded approximate identity for A, then $e_\lambda a = g(\omega)e_\lambda a = \phi_\omega(g \otimes e_\lambda)a = 0$ for all $\lambda \in \Lambda$ and therefore $a = 0$, as desired. Also, given any $a \in A$, we obtain from the Cohen factorization theorem [6, Theorem 11.10] that $a = uv$ for suitable $u, v \in A$ and therefore $a = \phi_\omega(g \otimes u)v$, which shows that $F(\Omega, A) \cdot A = A$. The last assumption of Lemma 4 follows from a similar argument, since, for each subset B of $F(\Omega, A)$, the space $Y_B := \bigcap_{f \in B} \ker \phi_\omega(f)$ is a closed ideal in A. Hence, again by the Cohen factorization theorem, every $a \in Y_B$ may be written as $a = uv$ with $u \in A$ and $v \in Y_B$, which implies that $a = \phi_\omega(g \otimes u)v$ and therefore $F(\Omega, A) \cdot Y_B = Y_B$. Now Lemma 4 shows that, for every $f \in Dec\, F(\Omega, A)$, the multiplication operator $L_{f(\omega)}$ is decomposable on A and hence that $f(\omega) \in Dec\, A$. Thus $Dec\, F(\Omega, A) \subseteq F(\Omega, Dec\, A)$. ∎

Theorem 11 *For every commutative complex Banach algebra A with a bounded approximate identity, the following identities hold:*

(a) $Dec\, C_0(\Omega, A) = C_0(\Omega, Dec\, A)$ *for any locally compact Hausdorff space* Ω.
(b) $Dec\, C^n(I, A) = C^n(I, Dec\, A)$ *for any compact interval* I *and* $n \geq 0$.
(c) $Dec\, lip_\alpha(I, A) = lip_\alpha(I, Dec\, A)$ *for any compact interval* I *and* $0 < \alpha < 1$.
(d) $Dec\, L^1(G, A) = L^1(G, Dec\, A)$ *for any locally compact abelian group* G.

Proof. The assertions (a)-(c) follow immediately from Proposition 10, since each of the Banach algebras $C_0(\Omega, A)$, $C^n(I, A)$ and $lip_\alpha(I, A)$ satisfies the assumptions (i)-(iii) of this result. The main point is, of course, that these Banach algebras may be represented as completed tensor products with respect to the canonical norms. For $C_0(\Omega, A)$, this is a classical fact noted, for instance, in [10]. The case of the Banach algebras $C^n(I, A)$ and $lip_\alpha(I, A)$ has been treated in Section 3 of [17].

Assertion (d) requires a slightly different argument, since the vector-valued group algebra $L^1(G, A)$ consists, strictly speaking, not of functions, but of equivalence classes of functions modulo Haar measure m on G, and since the multiplication of $L^1(G, A)$ is not given pointwise, but via convolution. First note that $L^1(G)$ is a regular semi-simple Banach algebra with a bounded approximate identity. It follows again from [7, Theorem 6.2.6] that $Dec\, L^1(G) = L^1(G)$. Moreover, as already noted in Section 2, for each commutative Banach algebra A, the Banach algebras $L^1(G, A)$ and $L^1(G) \widehat{\otimes}_\gamma A$ are isometrically isomorphic by a result of Grothendieck [10, Théorème 2.2.2]. Hence, if A has a bounded approximate identity, then the inclusion $L^1(G, Dec\, A) \subseteq Dec\, L^1(G, A)$ is immediate from Proposition 9.

To prove the opposite inclusion, let $\gamma \in \widehat{G}$ be an arbitrary continuous character on G, and consider the corresponding mapping $\phi_\gamma : L^1(G, A) \to L(A)$ given by the multiplication operators $\phi_\gamma(f) := L_{\widehat{f}(\gamma)}$ for all $f \in L^1(G, A)$, where $\widehat{f}(\gamma)$

denotes the vector-valued Fourier transform

$$\widehat{f}(\gamma) := \int_G \overline{\gamma(t)}\, f(t)\, dm(t) \qquad \text{for all } f \in L^1(G, A).$$

By [16, Theorem 2.1], we know that ϕ_γ is a continuous algebra homomorphism. Moreover, consider a bounded approximate identity $(a_\mu)_{\mu \in M}$ for A, and let $(e_\lambda)_{\lambda \in \Lambda}$ denote the usual bounded approximate identity for $L^1(G)$, as in [6, Example 11.7]. Thus Λ is the directed set of all compact neighborhoods of the identity element of G, and we have that $e_\lambda \geq 0$, $\operatorname{supp} e_\lambda \subseteq \lambda$, and $\int_G e_\lambda(t)\, dm(t) = 1$ for all $\lambda \in \Lambda$. By the proof of Proposition 9, we know that $(e_\lambda \otimes a_\mu)_{(\lambda,\mu) \in \Lambda \times M}$ is a bounded approximate identity for $L^1(G) \widehat{\otimes}_\gamma A = L^1(G, A)$, and it is easily seen that $\hat{e}_\lambda(\gamma) \to 1$ and hence that $\phi_\gamma(e_\lambda \otimes a_\mu)x \to x$ for each $x \in A$. Thus Theorem 5 applies to the homomorphism $\phi_\gamma : L^1(G, A) \to L(A)$ and ensures that $\widehat{f}(\gamma) \in Dec\, A$ for all $f \in Dec\, L^1(G, A)$ and $\gamma \in \widehat{G}$. Now, let $(Dec\, A)^\perp$ denote the annihilator of $Dec\, A$ in the topological dual space of A, and consider an arbitrary $f \in Dec\, L^1(G, A)$. It follows that, for each $\varphi \in (Dec\, A)^\perp$, the composition $\varphi \circ f$ belongs to $L^1(G)$ and satisfies $\widehat{\varphi \circ f}(\gamma) = \varphi(\widehat{f}(\gamma)) = 0$ for all $\gamma \in \widehat{G}$. Since $\widehat{\varphi \circ f} = 0$ on the dual group $\widehat{G}$, we conclude from the uniqueness theorem for the scalar-valued Fourier transform that $\varphi \circ f = 0$ for all $\varphi \in (Dec\, A)^\perp$. By [17, Lemma 4], this means precisely that $f \in L^1(G, Dec\, A)$. Consequently, $Dec\, L^1(G, A)$ is contained in $L^1(G, Dec\, A)$, which completes the proof. ■

References

[1] E. Albrecht, Spectral decompositions for systems of commuting operators. *Proc. Royal Irish Acad.* **81A** (1981), 81-98.

[2] E. Albrecht, Decomposable systems of operators in harmonic analysis. In *Toeplitz Centennial* (I. Gohberg, ed.), Birkhäuser, Basel 1982, pp. 19-35.

[3] E. Albrecht and J. Eschmeier, Analytic functional models and local spectral theory. Preprint, Saarbrücken 1992.

[4] C. Apostol, Remarks on the perturbation and a topology for operators. *J. Funct. Analysis* **2** (1968), 395-409.

[5] C. Apostol, Decomposable multiplication operators. *Rev. Roum. Math. Pures Appl.* **17** (1972), 323-333.

[6] F. F. Bonsall and J. Duncan, *Complete normed algebras.* Springer-Verlag, New York 1973.

[7] I. Colojoară and C. Foiaş, *Theory of generalized spectral operators.* Gordon and Breach, New York 1968.

[8] J. Eschmeier, Spectral decompositions and decomposable multipliers. *Manuscripta math.* **51** (1985), 201-224.

[9] C. C. Graham and O. C. McGehee, *Essays in commutative harmonic analysis.* Springer-Verlag, New York 1979.

[10] A. Grothendieck, *Produits tensoriels topologiques et espaces nucléaires.* Mem. Amer. Math. Soc. **16**, Providence, RI 1955.
[11] A. Hausner, Ideals in a certain Banach algebra. *Proc. Amer. Math. Soc.* **8** (1957), 246-249.
[12] J. Inoue and S.-E. Takahasi, A note on the largest regular subalgebra of a Banach algebra. *Proc. Amer. Math. Soc.* **116** (1992), 961-962.
[13] B. E. Johnson, The uniqueness of the (complete) norm topology. *Bull. Amer. Math. Soc.* **73** (1967), 537-539.
[14] B. E. Johnson, *Cohomology in Banach algebras.* Mem. Amer. Math. Soc. **127**, Providence, RI 1972.
[15] B. E. Johnson and A. M. Sinclair, Continuity of derivations and a problem of Kaplansky. *Amer. J. Math.* **90** (1968), 1067-1073.
[16] G. P. Johnson, Spaces of functions with values in a Banach algebra. *Trans. Amer. Math. Soc.* **92** (1959), 411-429.
[17] R. Kantrowitz and M. M. Neumann, The greatest regular subalgebra of certain Banach algebras of vector-valued functions. *Rend. Circ. Mat. Palermo (2)* (to appear).
[18] R. Kantrowitz and M. M. Neumann, Automatic continuity of homomorphisms and derivations on algebras of continuous vector-valued functions. *Czech. Math. J.* (to appear).
[19] R. Larsen, *An introduction to the theory of multipliers.* Springer-Verlag, New York 1971.
[20] K. B. Laursen and M. M. Neumann, Decomposable multipliers and applications to harmonic analysis. *Studia Math.* **101** (1992), 193-214.
[21] K. B. Laursen and M. M. Neumann, Asymptotic intertwining and spectral inclusions on Banach spaces. *Czech Math. J.* **43** (1993), 483-497.
[22] V. G. Miller and M. M. Neumann, Local spectral theory for multipliers and convolution operators. In *Algebraic methods in operator theory* (R. E. Curto and P. E. T. Jørgensen, eds.), Birkhäuser, Boston 1994, pp. 25-36.
[23] M. M. Neumann, Commutative Banach algebras and decomposable operators. *Mh. Math.* **113** (1992), 227-243.
[24] M. M. Neumann, Banach algebras, decomposable convolution operators, and a spectral mapping property. In *Function spaces* (K. Jarosz, ed.), Marcel Dekker, New York 1992, pp. 307-323.
[25] R. Schatten, *A theory of cross-spaces.* Annals of Math. Studies **26**, Princeton University Press, Princeton, NJ 1950.
[26] D. R. Sherbert, The structure of ideals and point derivations in Banach algebras of Lipschitz functions. *Trans. Amer. Math. Soc.* **111** (1964), 240-272.
[27] S. L. Sun, The sum and product of decomposable operators. *Northeastern Math. J.* **5** (1989), 105-117. In Chinese.
[28] J. Tomiyama, Tensor products of commutative Banach algebras. *Tôhoku*

Math. J. **12** (1960), 147-154.
[29] F.-H. Vasilescu, *Analytic functional calculus and spectral decompositions.* Editura Academiei and D. Reidel Publishing Company, Bucharest and Dordrecht 1982.

Weighted Spaces of Holomorphic Functions and Analytic Functionals

Le Hai Khoi Uppsala University, S-751 06 Uppsala, Sweden;
E-mail: khoi@math.uu.se

1. In many problems of function theory, such as existence of bases, representation of functions by means of series, solvability of functional equations and so on, a description of dual spaces, i.e. a description of analytic functionals, plays an important role.

Most of the results concern either spaces of holomorphic functions or weighted spaces of holomorphic functions in convex domains of $\mathbf{C}^n$ having prescribed behaviour near the boundary.

What about weighted spaces of holomorphic functions in convex domains having given boundary smoothness? In this paper we consider this problem.

I would like to thank Prof. C. O. Kiselman for valuable discussions.

2. We use some basic notation.

$H(G)$ (G being a domain in $\mathbf{C}^n$) denotes the space of functions holomorphic in G, with the topology of uniform convergence on compact subsets of G.

$H(K)$, respectively $C^\infty(K)$ (K being a compact set in $\mathbf{C}^n$) denotes the space of germs of functions holomorphic on K, endowed with the topology of inductive limit: $H(K) = \lim \operatorname{ind} H(\omega)$, ω are open neighbourhoods of K, respectively the space of functions infinitely differentiable on K.

If $\alpha = (\alpha_1, \ldots, \alpha_n)$ is a multi-index from $\mathbf{N}^n$, then we denote $\|\alpha\| = \alpha_1 + \cdots + \alpha_n$, $\alpha! = \alpha_1! \cdots \alpha_n!$ and $D^\alpha = \frac{\partial^{\|\alpha\|}}{\partial^{\alpha_1} \cdots \partial^{\alpha_n}}$. For all $z \in \mathbf{C}^n$ we put $z^\alpha = z_1^{\alpha_1} \cdots z_n^{\alpha_n}$. If $z, \zeta \in \mathbf{C}^n$ then $|z| = (z_1\bar{z}_1 + \cdots + z_n\bar{z}_n)^{1/2}$, $\langle z, \zeta \rangle = z_1\zeta_1 + \cdots + z_n\zeta_n$.

Let $\rho(z) \in C^2$ in $\mathbf{C}^n$. We put

$$\nabla_z \rho = \left(\frac{\partial \rho}{\partial z_1}, \ldots, \frac{\partial \rho}{\partial z_n}\right);$$

Supported in part by the State Program for Fundamental Research in Natural Sciences and the Swedish Institute.

$$R_j(z) = \det \begin{pmatrix} \frac{\partial \rho}{\partial z_1} & \cdots & \frac{\partial \rho}{\partial z_n} \\ \frac{\partial^2 \rho}{\partial \bar{z}_1 \partial z_1} & \cdots & \frac{\partial^2 \rho}{\partial \bar{z}_1 \partial z_n} \\ \cdots & [j] & \cdots \\ \frac{\partial^2 \rho}{\partial \bar{z}_n \partial z_1} & \cdots & \frac{\partial^2 \rho}{\partial \bar{z}_n \partial z_n} \end{pmatrix};$$

$$\bar{\omega}(z, \nabla_z \rho) = \langle z, \nabla_z \rho \rangle^{-n} \sum_{j=1}^{n} R_j(z) \partial \bar{z}_1 \wedge ...[j]... \wedge \partial \bar{z}_n \wedge \partial z_1 \wedge ... \wedge \partial z_n,$$

where $\wedge$ is the symbol of exterior multiplication;

$$u(z) = \langle z, \nabla_z \rho \rangle^{-1} \nabla_z \rho.$$

The supporting function of a convex set M in $\mathbf{C}^n$ is

$$h_M(\xi) = \sup_{z \in M} \operatorname{Re} \langle z, \xi \rangle, \quad \xi \in \mathbf{C}^n,$$

(see, e.g., [2]).

For a set E in $\mathbf{C}^n$ $(0 \in E)$ we denote by $\widetilde{E}$ the conjugate set of E, i.e.

$$\widetilde{E} = \{\xi \in \mathbf{C}^n : \langle z, \xi \rangle \neq 1 \ \forall z \in E\}.$$

3. Let E be a compact set in $\mathbf{C}^n$. A complex-valued function f on E is said to be p-smooth on E (in the sense of Whitney) if there exists a family of functions $(f^\alpha)_{\|\alpha\| \leqslant p}$, continuous on E, such that $f^0 = f$ and

$$\forall \zeta, z \in E, \ \forall k = (k_1, \dots, k_n), \ \|k\| \leqslant p$$

$$f^k(\zeta) = \sum_{\|\alpha\| \leqslant p - \|k\|} \frac{f^{k+\alpha}(z)}{\alpha!} (\zeta - z)^\alpha + R_p^k(\zeta, z),$$

where $R_p^k(\zeta, z) = o(|\zeta - z|^{p - \|k\|})$ uniformly for all $\zeta, z \in E$.

If f is p-smooth for all p then we say that f is infinitely smooth.

Let $M = (M_m)$ be an increasing sequence of positive numbers. An infinitely smooth function f on E belongs to the class $A_E^0(M)$ if

$$\forall p \geqslant 1, \ \forall k = (k_1, \dots, k_n), \ \|k\| \leqslant p$$

$$|R_p^k(\zeta, z)| \leqslant C_f \frac{M_{p+1}}{(p - \|k\| + 1)!} |\zeta - z|^{p - \|k\| + 1}, \ \forall \zeta, z \in E.$$

We call $A_E(M) = \bigcup_{Q>0} A_E^0(M_Q)$, where $M_Q = (Q^m M_m)$, the Carleman class on E.

Note that if the set E satisfies $\overline{\operatorname{int} E} = E$, then the functions f^α are uniquely determined by the function f.

Everywhere in what follows the sequence (M_m) is supposed to be regular, i.e., to be an increasing sequence of positive numbers satisfying the following conditions; cf. [3]:

$$\left(\frac{M_m}{m!}\right)^2 \leqslant \frac{M_{m-1}}{(m-1)!} \cdot \frac{M_{m+1}}{(m+1)!};$$

$$\sup_m \left(\frac{M_{m+1}}{M_m}\right)^{1/m} < +\infty;$$

$$\lim_{m\to\infty} \left(\frac{M_m}{m!}\right)^{1/m} = +\infty.$$

The corresponding Carleman class $A_E(M)$ also is called regular.

We see that the first condition is the most important: it is the logarithmic convexity of the sequence $(M_m/m!)$. The second condition is the condition of differentiability of the class, i.e. coincidence of $A_E(M)$ and $A_E(M')$, where $M'_m = M_{m+1}$. Finally, the third condition means that the class $A_E(M)$ contains all holomorphic functions.

We define an associated weight to (M_m) as follows

$$h(r) = \inf_m \frac{M_m}{m!} r^{m-1}, \ r > 0.$$

A regular class can be reconstructed from its associated weight, namely

$$M_m = m! \sup_{r>0} \frac{h(r)}{r^{m-1}}.$$

Note that if we change (M_m) to $(Q^m M_m)$, then $h(r)$ is replaced by $h(Qr)$.

If h_1 and h_2 are two functions such that

$$A h_1(\alpha r) \leqslant h_2(r) \leqslant B h_1(\beta r), \ r > 0,$$

where $A,\ B,\ \alpha,\ \beta$ are constants, we shall write $h_1 \sim h_2$.

The most typical example of the regular sequence $M = (M_m)$ are the Gevrey classes

$$M_m = (m!)^{\alpha+1}, \ \alpha > 0.$$

Then

$$h(r) \sim \exp\left\{-\left(\frac{1}{r}\right)^{1/\alpha}\right\}.$$

Other examples can be given (see, e.g., [4]).

4. We need the following result on pseudoanalytic extension of smooth functions [1].

Lemma. *Let $A_E(M)$ be a regular Carleman class on a compact set $E \subset \mathbf{C}^n$, $f \in A_E(M)$. Then f can be extended to a function $\widetilde{f} \in C_0^\infty(\mathbf{C}^n)$ such that $\widetilde{f} = f$ on E and*

$$\left|\frac{\partial \widetilde{f}}{\partial \bar{z}_j}(z)\right| \leqslant C_1\, h(C_2 \rho(z,E)), \ \forall j = 1,\dots,n, \ \forall z \in \mathbf{C}^n \setminus E,$$

where $\rho(z,E)$ is the Euclidean distance from z to E and $C_1,\ C_2$ are constants.

5. Now let G be a bounded convex C^2-smooth domain in $\mathbf{C}^n$, $0 \in G$. Let further (M_m) be a regular sequence.

We put

$$\rho(z) = \begin{cases} -\inf_{\zeta\in\partial G} |\zeta - z|, & z \in G \\ \inf_{\zeta\in\partial G} |\zeta - z|, & z \notin G. \end{cases}$$

Then $\rho(z) \in C^2$ in some neighbourhood of ∂G and it is a defining function of the domain G.

We suppose that an associated weight $h(r)$ of the sequence (M_m) satisfies the following condition

$$\forall k \in \mathbf{N}\ \exists p \in \mathbf{N}: \ \sup_{r>0} \frac{h(\frac{r}{k+p})}{r^2 h(\frac{r}{k})} = N_{k,p} < +\infty.$$

It should be noted that for many concrete examples of (M_m) (see, e.g., [4]) this condition always holds.

Consider the following space

$$I_M(G) = \lim\operatorname{proj} I_k(G),$$

where

$$I_k(G) = \left\{ f \in H(G) \cap C^\infty(\overline{G}) : \|f\|_k = \sup_m \sup_{\|\alpha\|\leqslant m} \sup_{z\in G} \frac{|D^\alpha f(z)| k^m}{M_m} < \infty \right\}.$$

We shall describe analytic functionals on the space $I_M(G)$. For this purpose we put

$$\psi(r) = \sup_m \frac{r^m}{M_m}, \ r > 0 \text{ (as a central term of power series)}$$

and form the following space

$$E_M(G) = \lim\operatorname{ind} E_k(G),$$

where

$$E_k(G) = \left\{ g(\xi) \in H(\mathbf{C}^n) : \ |g|_k = \sup_{\mathbf{C}^n} \frac{|g(\xi)|}{e^{h_G(\xi) + \ln \psi(k|\xi|)}} < +\infty \right\}.$$

(Recall that h_G is the supporting function of the domain G.)

6. Let G be a bounded convex domain in $\mathbf{C}^n$ with C^2 boundary and (M_m) be a regular sequence. The main result of this note is the follwing

Theorem. *The Laplace transformation establishes a topological isomorphism between the strong dual of $I_M(G)$, the space $I_M(G)^*$, and the space $E_M(G)$.*

Proof. Let $T \in I^*_M$. Then $\exists k_0 \geqslant 1\ \forall k \geqslant k_0 : \ T \in I_k(G)^*$. Consider the Laplace transform of this functional

$$\hat{T}(\xi) = \langle T_z, e^{\langle z,\xi\rangle}\rangle, \ \xi \in \mathbf{C}^n.$$

In this case

$$|\hat{T}(\xi)| \leqslant \|T\|^*_k \cdot \|e^{\langle z,\xi\rangle}\|_k,$$

where $\|T\|^*_k$ is a norm of T in $I_k(G)^*$.

We estimate

$$\|e^{\langle z,\xi\rangle}\|_k = \sup_m \sup_{\|\alpha\|\leqslant m} \sup_{z\in G} \frac{|D^\alpha e^{\langle z,\xi\rangle}|k^m}{M_m}$$

$$= \sup_m \sup_{\|\alpha\|\leqslant m} \sup_{z\in G} \frac{|\xi^\alpha|e^{Re\langle z,\xi\rangle}k^m}{M_m}$$

$$\leqslant Const \,.\, e^{h_G(\xi) + \ln\psi(k|\xi|)}$$

Thus, $T \to \hat{T}$ is a linear continuous mapping from I^*_M into E_M.

Furthermore, as well-known, exponents form a complete system in $H(\overline{G})$. Moreover, we can check that $H(\overline{G})$ is dense in I_M and convergence in $H(\overline{G})$ is stronger than convergence in I_M. This means that $T \to \hat{T}$ is injective.

The most difficult part of the proof is to show the surjectivity of $T \to \hat{T}$.

Denote by G_u the image of $\overline{G}$ by the map $z \in \mathbf{C}^n \mapsto \langle u, z\rangle \in \mathbf{C}$, $u \in \mathbf{C}^n$, i.e. $G_u = \{\xi \in \mathbf{C} : \ \xi = \langle u, z\rangle, \ z \in \overline{G}\}$.

For every $u \in \tilde{\overline{G}}$ we take $\theta_u \in [0, 2\pi)$ so that

$$Re(e^{i\theta_u}) = h_{G_u}(e^{i\theta_u}) + \tilde{\rho}(u),$$

where $\tilde{\rho}(u) = \inf_{\xi\in G}|1 - \langle u, \xi\rangle|$, $u \in \tilde{\overline{G}}$.

Fix $g(\xi) \in E_M(G)$. We consider the following function

$$L_g(u) = \frac{1}{(n-1)!}\int_0^{\infty e^{i\theta_u}} g(ut)\, t^{n-1}\, e^{-t} dt,$$

here as the radius of the integration we take $t = re^{i\theta_u}$, $0 < r < +\infty$.

We shall estimate $|L_g(u)|$.

First step. Estimate $\tilde{\rho}(u) = \inf_{\xi\in G}|1 - \langle u, \xi\rangle|$, $u \in \tilde{\overline{G}}$.

We keep the notation from section 2: $u(z) = \langle z, \nabla_z\rho\rangle^{-1}\nabla_z\rho$. Then for $z \in G$ we have

$$\tilde{\rho}(u(z)) = \inf_{\xi\in G}|1 - \langle u(z), \xi\rangle| = \inf_{\xi\in G}|1 - \langle z, \nabla_z\rho\rangle^{-1} \,.\, \langle\nabla_z\rho, \xi\rangle|$$

$$= |\langle z, \nabla_z \rho\rangle^{-1}| \cdot \inf_{\xi \in G} |\langle \nabla_z \rho, z - \xi\rangle|.$$

On the other hand, as noted in [4],

$$\inf_{\xi \in G} |\langle \nabla_z \rho, z - \xi\rangle| = \frac{1}{2} |\rho(z)|, \ z \in G.$$

Hence,

$$\widetilde{\rho}(u(z)) = \frac{1}{2} |\langle z, \nabla_z \rho\rangle^{-1}| \, |\rho(z)|, \ \forall z \in G^\delta \setminus G,$$

where $G^\delta = \{z \in \mathbf{C}^n : \rho(z) < \delta\}$ (in this case $|\rho(z)| = \rho(z), \ \forall z \in G^\delta \setminus G$).

From this fact it follows that for $\delta > 0$ sufficiently small there exist constants $A(\delta)$, $B(\delta) > 0$ such that

$$A|\rho(z)| \leqslant \widetilde{\rho}(u(z)) \leqslant B|\rho(z)| \ , \ \forall z \in G^\delta \setminus G.$$

Second step. Let k be the smallest integer number such that $g \in E_k(G)$. Then, by definition, we have

$$|g(\xi)| \leqslant |g|_k \ e^{h_G(\xi) + \ln \psi(k|\xi|)}, \ \forall \xi \in \mathbf{C}^n.$$

Hence,

$$|L_g(u)| \leqslant Const. \ |g|_k \cdot \int_0^{\infty e^{i\theta_u}} e^{h_G(ut) - Ret} \ \psi(|ut|k)|t|^{n-1} d|t|$$

$$= Const. \ |g|_k \cdot \int_0^{\infty e^{i\theta_u}} \psi(|ut|k)|t|^{n-1} e^{-\widetilde{\rho}(u)|t|} d|t|$$

$$= Const. \ |g|_k \cdot \int_0^{\infty} \psi(|u|xk) x^{n-1} e^{-\widetilde{\rho}(u)x} dx$$

$$\leqslant Const. |g|_k \cdot \sup_{x>0} \{\psi(|u|xk) x^{n-1} e^{-\widetilde{\rho}(u)\frac{x}{2}}\} \frac{2}{\widetilde{\rho}(u)}.$$

On the other hand,

$$\sup_{x>0} \{\psi(|u|xk) x^{n-1} e^{-\widetilde{\rho}(u)\frac{x}{2}}\} = \sup_{x>0} \sup_{m} \frac{(k|u|x)^m}{M_m} x^{n-1} e^{-\widetilde{\rho}(u)\frac{x}{2}}$$

$$= \sup_m \frac{k^m |u|^m}{M_m} \sup_{x>0} \{x^{m+n-1} e^{-\widetilde{\rho}(u)\frac{x}{2}}\} = \sup_m \frac{k^m |u|^m}{M_m} \left(\frac{2(m+n-1)}{e\widetilde{\rho}(u)}\right)^{m+n-1}$$

$$= \sup_m \frac{k^m |u|^m}{M_m} \left(\frac{m+n-1}{e}\right)^{m+n-1} \left(\frac{2}{\widetilde{\rho}(u)}\right)^{m+n-1}$$

$$\leqslant \sup_m \frac{k^m|u|^m}{M_m}(m+n-1)!\left(\frac{2}{\widetilde{\rho}(u)}\right)^{m+n-1}.$$

Since (M_m) is a regular sequence there exists $q > 1$ such that $M_{m+n-1} \leqslant q^m M_m, \forall m = 1, 2, \ldots$ Then we can write

$$\sup_{x>0}\{\psi(|u|xk)x^{n-1}e^{-\widetilde{\rho}(u)\frac{x}{2}}\} \leqslant \sup_m \frac{k^m|u|^m q^m}{M_{m+n-1}}(m+n-1)!\left(\frac{2}{\widetilde{\rho}(u)}\right)^{m+n-1}$$

$$= \sup_m \frac{(m+n-1)!}{M_{m+n-1}\left(\frac{\tilde{\rho}(u)}{2q|u|k}\right)^{m+n-2}} \frac{2}{\widetilde{\rho}(u)} \frac{1}{(q|u|k)^{n-2}}$$

$$= \frac{2}{\widetilde{\rho}(u)}(q|u|k)^{-(n-2)} \Big/ h\left(\frac{\tilde{\rho}(u)}{2q|u|k}\right).$$

So, we have proved that

$$|L_g(u)| \leqslant |g|_k \cdot \frac{4}{(\tilde{\rho}(u))^2}(q|u|k)^{-(n-2)} \Big/ h\left(\frac{\tilde{\rho}(u)}{2q|u|k}\right).$$

Furthermore, note that for $\delta > 0$ there exist constants $\tilde{A}(\delta),\ \tilde{B}(\delta) > 0$ such that

$$\tilde{A} \leqslant |u(z)| \leqslant \tilde{B},\ \forall z \in G^\delta \setminus G.$$

Consequently, for some constants $C_3(\delta),\ C_4(\delta) > 0$ one gets

$$|L_g(u(z))| \leqslant |g|_k \cdot \frac{C_3}{(\rho(z))^2} \cdot \frac{1}{h\left(\frac{C_4|\rho(z)|}{k}\right)},\ \forall z \in G^\delta \setminus G.$$

We mention that for k given there exist $p \in \mathbf{N}$ such that

$$\sup_{x>0} \frac{h(\frac{x}{k+p})}{x^2 h(\frac{x}{k})} = N_{k,p} < +\infty.$$

Now let $f \in I_M(G)$. Then $f \in A_{\overline{G}}(M')$, where $M'_m = \frac{M_m}{(k+p)^m}$, the regular Carleman class on the compact set $\overline{G}$, for some (for all) $p \geqslant 1$, in particular for the p indicated above. By the Lemma on pseudoanalytic extension of smooth functions f can be extended to a function $\widetilde{f} \in C_0^\infty(G^\delta)$ such that $\widetilde{f} = f$ on $\overline{G}$ and

$$\left|\frac{\partial \widetilde{f}}{\partial \bar{z}_j}(z)\right| \leqslant C_1\, h\left(\frac{C_2|\rho(z)|}{k+p}\right),\ \forall j = 1, \ldots, n,\ \forall z \in G^\delta \setminus G.$$

We put

$$\langle f, g\rangle = \frac{(n-1)!}{(2\pi i)^n}\int_{G^\delta\setminus G}\sum_{j=1}^n (-1)^{j-1} R_j(z)\frac{\partial \widetilde{f}}{\partial \bar{z}_j}(z)\, L_g\big(u(z)\big)\langle z, \nabla_z \rho\rangle^{-n}\, d\bar{z} \wedge dz.$$

We can check that this integral is defined correctly. Indeed, one has

$$\left|\frac{\partial \widetilde{f}}{\partial \bar{z}_j}(z)L_g(u(z))\right| \leqslant C_5\ |g|_k \frac{h\left(\frac{C_2|\rho(z)|}{k+p}\right)}{\rho(z)^2\ h\left(\frac{C_4\ |\rho(z)|}{k}\right)}$$

$$\leqslant C_6 \sup_{x>0} \frac{h(\frac{x}{k+p})}{x^2 h(\frac{x}{k})} = C_7\ N_{k,p} < +\infty.$$

Furthermore, from Stokes's formula it follows that this integral is independent of the choice of the extension of f.

Thus, with fixed $g \in E_M(G)$ the formula $\langle f, g\rangle$ defines a linear continuous functional T_g on $I_M(G)$.

Our *final step* is to prove that the Laplace transform of T_g coincides with g. Let $\varphi \in H(\overline{G})$. Then $\varphi \in H(\overline{G^\varepsilon})$ for some $\varepsilon > 0$ and

$$\langle \varphi, g\rangle = \frac{(n-1)!}{(2\pi i)^n} \int_{\partial G^\epsilon} \varphi(z)\ L_g(u(z))\ \bar{\omega}(z, \nabla_z \rho).$$

By Martineau formula on Fourier-Borel transform [5] we obtain

$$\left\langle e^{\langle z,\xi\rangle}, g\right\rangle = \frac{(n-1)!}{(2\pi i)^n} \int_{\partial G^\epsilon} e^{\langle z,\xi\rangle}\ L_g(u(z))\ \bar{\omega}(z, \nabla_z \rho) = g(\xi).$$

The theorem is proved completely.

REFERENCES

[1] Dyn'kin E. M., Pseudoanalytic extension of smooth functions. The uniform scale (Russian), *Mathematical programming and related questions (Proc. Seventh Winter School, Drogobych,* 1974). *Theory of functions and functional analysis. Central Ekonom.-Mat. Inst. Akad. Nauk SSSR, Moscow,* 1976, 40–73. English translation in *AMS Transl.*, **115** (1980), 33–58 MR 58 # 28536.

[2] Hörmander L., *An introduction to complex analysis in several variables*, 3rd ed., North-Holland Publ. Co. (1990), xii + 254 pages.

[3] Komatsu H., Ultradistributions. I, Structure theorems and a characterization, *J. Fac. Sci. Tokyo Sec. IA*, **20** (1973), 25–105.

[4] Lê Hai Khôi, Espaces conjugués, ensembles faiblement suffisants discrets et systèmes de représentation exponentielle, *Bull. Sc. Math.*, (2) **113**(1989), N3, 309-347.

[5] Martineau A., Equations différentielles d'ordre infini, *Bull. Soc. Math. France*, **95**(1967) 109-154.

Isometries of L_p-Spaces of Solutions of Homogeneous Partial Differential Equations

ALEXANDER KOLDOBSKY Division of Mathematics, Computer Science, and Statistics, University of Texas at San Antonio, San Antonio, TX 78249, U.S.A.

ABSTRACT

Let $n \geq 2, A = (a_{ij})_{i,j=1}^n$ be a real symmetric matrix, $a = (a_i)_{i=1}^n \in R^n$. Consider the differential operator $D_A = \sum_{i,j=1}^n a_{ij} \frac{\partial^2}{\partial x_i \partial x_j} + \sum_{i=1}^n a_i \frac{\partial}{\partial x_i}$. Let E be a bounded domain in R^n, $p > 0$. Denote by $L_{D_A}^p(E)$ the space of solutions of the equation $D_A f = 0$ in the domain E provided with the L_p-norm.

We prove that, for matrices A, B, vectors a, b, bounded domains E, F, and every $p > 0$ which is not an even integer, the space $L_{D_A}^p(E)$ is isometric to a subspace of $L_{D_B}^p(F)$ if and only if the matrices A and B have equal signatures, and the domains E and F coincide up to a natural mapping which in the most cases is affine. We use the extension method for L_p-isometries which reduces the problem to the question of which weighted composition operators carry solutions of the equation $D_A f = 0$ in E to solutions of the equation $D_B f = 0$ in F.

1. INTRODUCTION

Let $n \geq 2, A = (a_{ij})_{i,j=1}^n$ be a real symmetric matrix, $a = (a_i)_{i=1}^n \in R^n$. Consider the differential operator

$$D_A = \sum_{i,j=1}^n a_{ij} \frac{\partial^2}{\partial x_i \partial x_j} + \sum_{i=1}^n a_i \frac{\partial}{\partial x_i}$$

Let E be a bounded domain in $R^n, p > 0$. Denote by $L_{D_A}^p(E)$ the space of real functions

$f \in C^2(E)$ for which $D_A f = 0$ and

$$\|f\| = (\int_E |f(x)|^p dm(x))^{1/p} < \infty$$

where m is Lebesgue measure in R^n.

Suppose that, for different matrices A, B, vectors a, b, and bounded domains E, F, the space $L^p_{D_A}(E)$ is isometric to a subspace of $L^p_{D_B}(F)$. Does the similarity of geometric structures of the spaces imply any equivalence of differential operators and domains ?

We shall answer this question in positive and show that, for every $p > 0$ which is not an even integer, there is a close connection between the geometry of the space $L^p_{D_A}(E)$ and properties of D_A and E.

The case of the spaces of harmonic functions was considered by A.Plotkin (1972). He proved that, for $n \geq 3$, $p \neq 2k, k \in N$ and $p \neq 2n/(n-2)$, the space $L^p_\Delta(E)$ is isometric to a subspace of $L^p_\Delta(F)$(Δ is the Laplace operator) if and only if the domains E and F are similar (coincide up to the composition of a translation, rotation, reflection and homothety). If $p = 2n/(n-2)$ one can add an inversion to the composition. For $n = 2$, E and F must be similar. In Koldobsky (1980) , Plotkin's result was extended to the case of elliptic operators D_A and D_B.

The result of this paper generalizes Plotkin's theorem to the case of arbitrary differential operators D_A and D_B. Our main tool is the following extension theorem for L_p-isometries:

Extension Theorem. (A.Plotkin (1974, 1976), C.Hardin (1981)) Let $p > 0$, where p is not an even integer, (X_1, σ_1) and (X_2, σ_2) be finite measure spaces, Y a subspace of $L_p(X_1)$ containing the constant function $1(x) \equiv 1$, and let T be an arbitrary linear isometry from Y to $L_p(X_2)$. Then there exists a linear isometry $\tilde{T} : L^p(X_1, \Omega_0, \sigma_1) \mapsto L^p(X_2)$ such that $\tilde{T}|_Y = T$. Here Ω_0 is the minimal σ-algebra making the functions from Y measurable.

By the Extension Theorem, every isometry $T : L^p_{D_A}(E_1) \mapsto L^p_{D_B}(E_2)$ can be extended to the whole space $L_p(E_1)$. By the classical characterizations of the isometries of L_p-spaces due to S.Banach (1932) and J.Lamperti (1958), the extension is a weighted composition operator. Therefore, our problem can be reduced to the following question: Which weighted composition operators carry functions from $L^p_{D_A}(E_1)$ to functions from $L^p_{D_B}(E_2)$?

In Koldobsky (1991), one can find references to other applications and generalizations of the extension method.

2. THE MAIN RESULT

We start with necessary definitions and notation.

Let $A = (a_{ij})^n_{i,j=1}, detA \neq 0$ be a real symmetric matrix. There exists a matrix M diagonalizing the matrix A in the sense that $M^*AM = I_\ell$ for some integer $\ell, 0 \leq \ell \leq n$ where $I_\ell = (\ell_{ij})^n_{i,j=1}$ stands for the matrix with $\ell_{ij} = 0, i \neq j$, $\ell_{ii} = 1, 1 \leq i \leq \ell$ and $\ell_{ii} = -1, \ell+1 \leq i \leq n$. We call the number $2\ell - n$ a signature of the matrix A.

We denote by D_ℓ the differential operator corresponding to the matrix I_ℓ :

$$D_\ell = \sum_{i=1}^{\ell} \frac{\partial^2}{\partial x_i^2} - \sum_{i=\ell+1}^{n} \frac{\partial^2}{\partial x_i^2}.$$

For two bounded domains E_1 and E_2 in R^n with $int(clE_1) = E_1$ and a mapping $\tau : E_2 \mapsto cl(E_1)$ of the class C^1, we say that E_1 and E_2 coincide up to τ if $m(E_1 \backslash \tau(E_2)) = 0$.

If $x \in E_2$, $J(x)$ stands for the Jacobi matrix of τ at the point x, and $\tau'(x) = detJ(x)$ is the Jacobian of τ at x.

We say that τ is ℓ-conformal at a point x if $J^*(x)I_\ell J(x) = C(x)I_\ell$ where $C(x) \in R$. A mapping is ℓ-conformal in a domain if it is ℓ-conformal at every point of the domain.

For $z_0 \in R^n$, the ℓ-inversion with center z_0 is the mapping

$$z \mapsto \frac{z - z_0}{\|z - z_0\|_l^2} + z_0, z \in R^n$$

where $\|z\|_\ell^2 = \sum_{i=1}^{\ell} z_i^2 - \sum_{i=\ell+1}^{n} z_i^2$. A homothety with center z_0 and coefficient $t \in R$ is the mapping $z \mapsto t(z - z_0) + z_0, z \in R^n$. We call ℓ-similarity a mapping which is the composition of a homothety and a mapping preserving the metric $\|z\|_\ell^2$ (all such mappings are affine, Dubrovin et al (1992)).

The following characterization of ℓ-conformal mappings was given by Liouville in 1850 for the mappings of the class C^3. This result was extended by Hartman (1958) to the C^2-mappings and, finally, Reshetnyak (1967) formulated and proved it without any smoothness assumptions.

Liouville's Theorem. Let D be a domain in $R^n, n \geq 3$ and τ a ℓ-conformal mapping from D to R^n where $\ell \in N, 1 \leq \ell \leq n$. Then τ is the composition of a ℓ-similarity and a ℓ-inversion.

We are ready to formulate the main result of the paper.

Theorem 1. Let $n \geq 3$, p be a positive number which is not an even integer, E_1, E_2 bounded domains in R^n with $int(clE_1) = E_1$, A, B are real symmetric matrices with non-zero determinants, $a, b \in R^n$, M, N be the matrices diagonalizing A and B and $2\ell - n, 2m - n$ be the signatures of A and B.
Let $a, b \in R^n$, and D_A, D_B be the differential operators corresponding to (A, a) and (B, b), respectively. Then:
(i) If either $\ell \neq m$ or $\ell = m$ and one of the vectors a, b is zero and another is non-zero, then the space $L^p_{D_A}(E_1)$ is not isometric to a subspace of $L^p_{D_B}(E_2)$.
(ii) If $\ell = m$ and $a = b = 0$ then, for $p \neq 2n/(n-2)$, the space $L^p_{D_A}(E_1)$ is isometric to a subspace of $L^p_{D_B}(E_2)$ if and only if the domains ME_1 and NE_2 coincide up to a ℓ-similarity τ. For $p = 2n/(n-2)$, the domains ME_1 and NE_2 may coincide up to the composition of a ℓ-similarity and a ℓ-inversion.
(iii) If $\ell = m$ and $a \neq 0, b \neq 0$ then the space $L^p_{D_A}(E_1)$ is isometric to a subspace of $L^p_{D_B}(E_2)$ if and only if the domains ME_1 and NE_2 coincide up to a ℓ-similarity τ such that $JNb = |\tau'|^{2/n}Ma$ (since τ is an affine mapping the Jacobi matrix J does not depend on the choice of a point.)
Finally, in all the cases where an isometric embedding T exists it has the form $Tf = \pm|det(M^{-1}JN)|^{1/p}f(M^{-1}\tau N)$, $f \in L^p_{D_A}(E_1)$.

In the case $n = 2$ the result is different. The reason is that the Liouville Theorem is not valid in this case and the class of ℓ-conformal mappings is larger.

Theorem 2. Let $n = 2$ and $p, A, B, a, b, E_1, E_2, \ell, m$ be as in Theorem 1. Then: (i) If either $\ell \neq m$, or $\ell = m = 2$ and one of the vectors a, b is zero and another is non-zero, or $\ell = m = 1$ and one of the numbers $\|Ma\|_1^2, \|Nb\|_1^2$ is zero and another is non-zero, then the space $L^p_{D_A}(E_1)$ is not isometric to a subspace of $L^p_{D_B}(E_2)$.
(ii) If $\ell = m$ and $a = b = 0$ then, for every p, the answer is the same as for $p \neq 2n/(n-2)$ in Theorem 1. If $\ell = m = 2$ and $a \neq 0, b \neq 0$ or $\ell = m = 1$ and $\|Ma\|_1^2 \neq 0, \|Nb\|_1^2 \neq 0$ the answer is the same as in the part (iii) of Theorem 1.
(iii) In the case $\ell = m = 1$ and $\|Ma\|_1^2 = \|Nb\|_1^2 = 0$ the class of mappings τ generating isometric embeddings is different. The answer depends on the coordinates of the vectors $Ma = (c_1, c_2)$ and $Nb = (d_1, d_2)$.
For $c_1 = \pm c_2 = c \neq 0$ and $d_1 = \pm d_2 = d \neq 0$, the coordinates u_1, u_2 of the mapping τ are as follows:

$$\begin{cases} u_1(x_1,x_2) = (-1/pc)\ln|\gamma\exp(-pdx_1/2 \pm pdx_2/2) - 1| + kx_1 \pm kx_2 + \alpha \\ u_2(x_1,x_2) = (\pm 1/pc)\ln|\gamma\exp(-pdx_1/2 \pm pdx_2/2) - 1| \pm kx_1 + kx_2 + \beta \end{cases}$$

or

$$\begin{cases} u_1(x_1,x_2) = (-1/pd)\gamma\exp(-pdx_1/2 \pm pdx_2/2) + (1/pc)\ln|pcx_1 \pm pcx_2 + \delta| + \alpha \\ u_2(x_1,x_2) = (\mp 1/pd)\gamma\exp(-pdx_1/2 \pm pdx_2/2) \mp (1/pc)\ln|pcx_1 \pm pcx_2 + \delta| + \beta \end{cases}$$

If $c_1 = \pm c_2 = c \neq 0, d_1 = \mp d_2 = d \neq 0$ then

$$\begin{cases} u_1(x_1,x_2) = (-1/pd)\gamma\exp(-pdx_1/2 \mp pdx_2/2) - (1/pc)\ln|\mp pcx_1 + pcx_2 + \delta| + \alpha \\ u_2(x_1,x_2) = (\mp 1/pd)\gamma\exp(-pdx_1/2 \mp pdx_2/2) \pm (1/pc)\ln|\mp pcx_1 + pcx_2 + \delta| + \beta \end{cases}$$

or

$$\begin{cases} u_1(x_1,x_2) = (1/pc)\ln|\gamma\exp(-pdx_1/2 \mp pdx_2/2) - 1| + kx_1 \mp kx_2 + \alpha \\ u_2(x_1,x_2) = (\mp 1/pc)\ln|\gamma\exp(-pdx_1/2 \mp pdx_2/2) - 1| \pm kx_1 - kx_2 + \beta \end{cases}$$

For $c_1 = c_2 = 0, d_1 = \pm d_2 = d \neq 0$,

$$\begin{cases} u_1(x_1,x_2) = (-1/pd)\gamma\exp(-pdx_1/2 \pm pdx_2/2) + kx_1 \pm kx_2 + \alpha \\ u_2(x_1,x_2) = (\pm 1/pd)\gamma\exp(-pdx_1/2 \pm pdx_2/2) \pm kx_1 + kx_2 + \beta \end{cases}$$

or

$$\begin{cases} u_1(x_1,x_2) = (-1/pd)\gamma\exp(-pdx_1/2 \pm pdx_2/2) + kx_1 \pm kx_2 + \alpha \\ u_2(x_1,x_2) = (\mp 1/pd)\gamma\exp(-pdx_1/2 \pm pdx_2/2) \mp kx_1 - kx_2 + \beta \end{cases}$$

Finally, for $c_1 = \pm c_2 = c \neq 0, d_1 = d_2 = 0$,

$$\begin{cases} u_1(x_1,x_2) = -(1/pc)\ln|\mp pcx_1 + pcx_2 + \delta| + kx_1 \pm kx_2 + \alpha \\ u_2(x_1,x_2) = \pm(1/pc)\ln|\mp pcx_1 + pcx_2 + \delta| \pm kx_1 + kx_2 + \beta \end{cases}$$

or

$$\begin{cases} u_1(x_1,x_2) = \mp(1/pc)\ln|pcx_1 \pm pcx_2 + \delta| + kx_1 \mp kx_2 + \alpha \\ u_2(x_1,x_2) = (1/pc)\ln|pcx_1 \pm pcx_2 + \delta| \pm kx_1 - kx_2 + \beta \end{cases}$$

In these formulas $\alpha, \beta, \gamma, \delta, k$ are real numbers (if γ and k are both present in a formula, one of them must be non-zero). We use $\pm$ and $\mp$ as follows: first read the text with the upper signs everywhere, and then read it for the second time with the lower signs.

3. WEIGHTED COMPOSITION OPERATORS

We start with two facts whose simple proofs we leave to the reader. The first one shows that, for all subspaces of L_p we are going to deal with, the σ-algebra Ω_0 appearing in the Extension Theorem is, in fact, the σ-algebra of all Borel sets.

Lemma 1. Let E be a bounded open set in R^n, H a family of continuous functions on E containing the function $1(x) \equiv 1$ and separating the points of E (for every $x, y \in E$, there exists $f \in H$ such that $f(x) \neq f(y)$.) Then the minimal σ-algebra of subsets of E making functions from H measurable is the σ-algebra of all Borel subsets of E.

The second fact reduces the main question of this paper to the case where the matrices A and B are equal to I_ℓ and I_m, respectively.

Lemma 2. Let $n \geq 2$, E_1, p, A, a, M, ℓ be as in Theorem 1, and

$$H = D_\ell + \sum_{i=1}^{n} \alpha_i \frac{\partial}{\partial x_i} \tag{1}$$

where $\alpha = (\alpha_1, ..., \alpha_n) = Ma \in R^n$. Then the operator T defined by

$$Tf(x) = |\det M|^{-1/p} f(M^{-1}x), f \in L^p_{D_A}(E_1), x \in ME_1$$

is a linear isometry from $L^p_{D_A}(E_1)$ onto $L^p_H(ME_1)$.

Now we can apply the Extension Theorem to the space $Y = L^p_H(E_1)$.

Theorem 3. Let $n \geq 2$, E_1, E_2 be bounded domains in R^n, $int(clE_1) = E_1$, $p > 0$ and p is not an even integer, $\ell \in N, 1 \leq \ell \leq n$, $\alpha \in R^n$, and define H by (1). Then, for every isometry $T : L^p_H(E_1) \mapsto L^p(E_2) \cap C^2(E_2)$, there exists a mapping $\tau : E_2 \mapsto cl(E_1)$ such that:
(i) τ is of the class C^2 on $E_2 \setminus \{x \in E_2 : T1(x) = 0\}$ and $|\tau'(x)| = |T1(x)|^p$ on E_2,
(ii) E_1 and E_2 coincide up to τ,
(iii) for every $f \in L^p_H(E_1)$, $Tf(x) = T1(x)f(\tau(x))$ on E_2.

Proof: Clearly, the function $1(x) \equiv 1$ belongs to the space $L^p_H(E_1)$. Besides, the space $L^p_H(E_1)$ separates the points of E_1. To see that, take two different points $y, z \in R^n$. There exists k such that $y_k \neq z_k$. If $\alpha_k = 0$ then the function $f_k(x) \equiv x_k$ belongs to $L^p_H(E_1)$ and separates y and z. If $\alpha_k \neq 0$ then one of the functions $u(x) = \exp(-\alpha_k x_k)$ and $v(x) = \exp(\alpha_k x_k)$ belongs to $L^p_H(E_1)$ and separates the points.

By the Extension Theorem and Lemma 2, the isometry T can be extended to an isometry $\tilde{T}$ from the whole space $L^p(E_1)$ to $L^p(E_2)$. By the classical result of J.Lamperti (1958), there exists an isometric homomorphism ϕ from the algebra $L^\infty(E_1)$ to $L^\infty(E_2)$ such that, for every $f \in L^\infty(E_1)$, $\tilde{T}f = F\phi(f)$ where $F = T1$.

For the functions $f_k(x) = x_k$, put $\phi f_k = u_k$ and consider a mapping $\tau : E_2 \mapsto R^n$ defined by $\tau(x) = (u_1(x), ..., u_n(x))$. We are going to prove that τ satisfies the conditions (i)-(iii).

Since ϕ is a homomorphism of algebras, for every polynomial $P(x_1, ..., x_n)$, $\tilde{T}P = FP(\tau)$. Polynomials form a dense subset in $L_p(E_1)$ and, therefore, we have (iii).

The function $F = T1$ belongs to the class $C^2(E_2)$. Since one of the functions f_k, $u(x) = \exp(-\alpha_k x_k)$, $v(x) = \exp(\alpha_k x_k)$ belongs to $L^p_H(E_1)$, one of the functions Fu_k, $F\exp(-\alpha_k u_k)$, $F\exp(\alpha_k u_k)$ belongs to $C^2(E_2)$, and it follows that τ is a mapping of the class C^2 on $E_2 \setminus \{x \in E_2 : F(x) = 0\}$.

Let us prove that $\tau(E_2) \subset cl(E_1)$. Suppose that there exists $x_0 \in E_2$ for which $\tau(x_0) \notin cl(E_1)$. Consider a polynomial $P(x) = A - \sum_{i=1}^{n}(x_i - u_i(x_0))^2$ where we choose $A > 0$ so that P is positive on E_1. Then,

$$A = \sup_{E_2} P(\tau) = \|\phi(P)\|_{L^\infty(E_2)} = \|P\|_{L^\infty(E_1)} = \sup_{E_1} P < A,$$

and we get a contradiction.

Let χ be the characteristic function of the set $E_1 \setminus \tau(E_2)$. Then $\tilde{T}\chi = F\chi(\tau) = 0$. Since $\tilde{T}$ is an isometry we get $\chi = 0$ which means that $m(E_1 \setminus \tau(E_2)) = 0$, and the domains E_1, E_2 coincide up to τ. We have proved (ii).

To finish the proof of (i), note that, for every function $f \in L_p(E_1)$,

$$\|f\|^p_{L_p(E_1)} = \int_{E_1} |f(y)|^p dm(y) =$$

$$\int_{E_2} |f(\tau(x))|^p |\tau'(x)| dm(x) =$$

$$\|\tilde{T}f\|^p_{L_p(E_2)} = \int_{E_2} |f(\tau(x))|^p |F(x)|^p dm(x).$$

(We made a change of variables $y = \tau(x)$.) Since f is an arbitrary function and $m(E_1 \setminus \tau(E_2)) = 0$ we get $|\tau'| = |F|^p$, which completes the proof of the theorem.

Part (iii) of Theorem 3 shows that every isometry from $L^p_H(E_1)$ to $L^p(E_2) \cap C^2(E_2)$ is generated by a mapping τ. Now we are going to choose those mappings for which the images of functions from $L^p_H(E_1)$ are solutions of another differential equation.

We need the following elementary fact.

Lemma 3. Let $n, \ell \in N, 1 \leq \ell \leq n$. Put $\epsilon_i = 1, 1 \leq i \leq \ell$ and $\epsilon_i = -1, \ell < i \leq n$. Suppose that a real symmetric matrix $B = (b_{ij})_{i,j=1}^n$, a vector $a \in R^n$, and a number $c \in R$ satisfy the following: for any choice of complex numbers $s_1, ..., s_n$, the equality $\sum_{i=1}^n \epsilon_i s_i^2 + a_i s_i = 0$ implies $\sum_{i,j=1}^n b_{ij} s_i s_j + \sum_{i=1}^n c a_i s_i = 0$. Then the matrices B and I_ℓ differ by a constant multiple only.

Theorem 4. Let $n \geq 2$, E_1, E_2 be bounded domains in R^n, $\tau : E_2 \mapsto cl(E_1)$ a mapping of the class C^2, $\alpha \in R^n$, and H the differential operator defined by (1). Consider any real functions $b_{ij}(x), b_i(x), i,j = 1, ..., n$ on E_2 and denote by D_B the differential operator

$$D_B = \sum_{i,j=1}^{n} b_{ij}(x) \frac{\partial^2}{\partial x_i \partial x_j} + \sum_{i=1}^{n} b_i(x) \frac{\partial}{\partial x_i}.$$

Suppose that there exists a function $F \in L_{D_B}(E_2)$ such that, for every $f \in L_H^\infty(E_1)$, the function $Ff(\tau)$ belongs to $L_{D_B}(E_2)$. Then:
(i) for every $x \in E_2 \setminus \{x \in E_2 : F(x) = 0\}$, the matrix $B(x) = (b_{ij}(x))_{i,j=1}^n$ has the signature $2l - n$.
(ii) there exists a real function $C : E_2 \mapsto R$ such that, for every $x \in E_2 \setminus \{x \in E_2 : F(x) = 0\}$,

$$J^*(x)B(x)J(x) = \frac{C(x)}{F(x)} I_\ell.$$

Proof: Let $\tau = (u_1, ..., u_n)$. For any $a \in R^n$,

$$\sum_{k=1}^{n} a_k x_k \in L_H^\infty(E_1) \Longleftrightarrow \sum_{k=1}^{n} \alpha_k a_k = 0. \tag{2}$$

On the other hand, if $\sum_{k=1}^n a_k x_k \in L_H^\infty(E_1)$ then $F \sum_{k=1}^n a_k u_k \in L_{D_B}(E_2)$. Since $F \in L_{D_B}(E_2)$ the latter condition gives

$$\sum_{k=1}^{n} a_k \Big(\sum_{i,j=1}^{n} b_{ij}(x) \big(2 \frac{\partial F}{\partial x_i} \frac{\partial u_k}{\partial x_j} + F \frac{\partial^2 u_k}{\partial x_i \partial x_j}\big) + \sum_{i=1}^{n} b_i(x) F \frac{\partial u_k}{\partial x_i} \Big) = 0. \tag{3}$$

Since (2) implies (3) for every vector a, the coefficients at a_k's must be proportional. It means that , for every $x \in E_2$, there exists $c(x) \in R$ such that, for each $k = 1, ..., n$,

$$\sum_{i,j=1}^{n} b_{ij}(x) \big(2 \frac{\partial F}{\partial x_i}(x) \frac{\partial u_k}{\partial x_j}(x) + F(x) \frac{\partial^2 u_k}{\partial x_i \partial x_j}(x)\big) + \sum_{i=1}^{n} b_i(x) F(x) \frac{\partial u_k}{\partial x_i}(x) = c(x) \alpha_k. \tag{4}$$

Consider the function $\exp(x, s)$ where $s = (s_1, ... s_n)$ is a n-tuple of complex numbers and (x, s) stands for the scalar product. Clearly, $H(\exp(x, s)) = 0$ if and only if

$$\sum_{i=1}^{n} \epsilon_i s_i^2 + \alpha_i s_i = 0 \tag{5}$$

where the numbers ϵ_i are the same as in Lemma 3.

On the other hand, $D_B(F \exp(\tau(x), s)) = 0$ and using (4) and the fact that $F \in L_{D_B}(E_2)$ we get

$$F(x)\sum_{k,m=1}^{n} s_k s_m \sum_{i,j=1}^{n} b_{ij}(x)\frac{\partial u_k}{\partial x_i}(x)\frac{\partial u_m}{\partial x_j}(x) + c(x)\sum_{k=1}^{n}\alpha_k s_k = 0. \tag{6}$$

Thus, for any choice of complex numbers $s_1, ..., s_n$, (5) implies (6). It means that, for every $x \in E_2 \setminus \{x \in E_2 : F(x) = 0\}$, the matrix $J^*(x)B(x)J(x)$ satisfies the conditions of Lemma 3. Therefore, there exists a function $C : E_2 \mapsto R$ such that

$$J^*(x)B(x)J(x) = \frac{C(x)}{F(x)} I_\ell,$$

and we get (ii). Part (i) follows from the uniqueness of the diagonalization.

4. PROOF OF THE MAIN RESULT

We are ready to prove Theorems 1 and 2. The first part of the proof applies to both of the cases $n > 2$ and $n = 2$.

Using Lemma 2 one can reduce the problem to the case of diagonal matrices. Let T be an isometry from $L^p_{D_A}(E_1)$ to $L^p_{D_B}(E_2)$ and define differential operators H and G by

$$H = D_\ell + \sum_{i=1}^{n} c_i \frac{\partial}{\partial x_i}, G = D_m + \sum_{i=1}^{n} d_i \frac{\partial}{\partial x_i}$$

where $c = Ma, d = Nb$.

By Lemma 2, the operators

$$T_1 f(x) = |\det M|^{-1/p} f(M^{-1}x), T_2 g(x) = |\det N|^{-1/p} g(N^{-1}x)$$

are isometries from $L^p_{D_A}(E_1)$ and $L^p_{D_B}(E_2)$ onto $L^p_H(ME_1)$ and $L^p_G(NE_2)$, respectively. Therefore, $S = T_2 T T_1^{-1}$ is an isometry from $L^p_H(ME_1)$ to $L^p_G(NE_2)$.

Put $F = S1$ and $E = \{x \in NE_2 : F(x) = 0\}$. By Theorem 3, there exists a mapping $\tau : NE_2 \mapsto ME_1$ of the class C^2 on $NE_2 \setminus E$ such that the domains ME_1 and NE_2 coincide up to τ, $|\tau'| \equiv |F|^p$ and, for every $f \in L^p_H(ME_1)$, $Sf = Ff(\tau)$.

Clearly, the mapping τ satisfies the conditions of Theorem 4 with $B(x) = G$ for every x, so the matrices H and G have equal signatures which means that $\ell = m$. Besides, there exists a real function $C : NE_2 \mapsto R$ such that

$$J^*(x) I_\ell J(x) = \frac{C(x)}{F(x)} I_\ell \tag{7}$$

for every $x \in NE_2 \setminus E$.

Thus, τ is a ℓ-conformal mapping on $NE_2 \setminus E$.

Calculating the determinants in both sides of (7) we get

$$J^*(x) I_\ell J(x) = |\tau'(x)|^{2/n} I_\ell \tag{8}$$

Let $\epsilon_i = 1, 1 \leq i \leq \ell$ and $\epsilon_i = -1, \ell < i \leq n$. If $\tau = (u_1, ..., u_n)$ then, for every function $f \in L^p_H(ME_1)$,

$$(9) \qquad 0 = G(Ff(\tau)) = \sum_{k=1}^{n} \frac{\partial f}{\partial x_k}(\tau) \sum_{i=1}^{n} \Big((\epsilon_i (2 \frac{\partial F}{\partial x_i} \frac{\partial u_k}{\partial x_i} + F \frac{\partial^2 u_k}{\partial x_i^2}) + F \frac{\partial u_k}{\partial x_i} d_i - c_k F |\tau'(x)|^{2/n} \Big)$$

We used (8) and the fact that $F \in L^p_G(NE_2)$.

Starting from this point we consider the cases $n \geq 3$ and $n = 2$ separately.

The case $n \geq 3$.

It follows from (8) that τ is a ℓ-conformal mapping on $NE_2 \setminus E$. By Liouville's theorem, the mapping τ is either a ℓ-similarity or the composition of a ℓ-similarity and a ℓ-inversion on every connected subset U of $NE_2 \setminus E$. The Jacobian of the ℓ-inversion with center x_0 is equal to $\|x - x_0\|_\ell^{-2n}$ (see Dubrovin et al (1992)) and the Jacobian of any ℓ-similarity is a constant, so we have $\tau'(x) = k\|x - x_0\|_\ell^{-2n}$ on U.

By Theorem 4, $|\tau'| \equiv |F|^p$, and we have $|F| = k^{1/p}\|x - x_0\|_\ell^{-2n/p}$ on U. Clearly, $k \neq 0$ because $U \cap E = \emptyset$. Therefore, there exists a constant $\alpha > 0$ such that $|F(x)| > \alpha$ on U. The function F is continuous, so $cl(U) \cap E = \emptyset$ for every connected subset U of $NE_2 \setminus E$. This is possible only if $E = \emptyset$. Thus, the mapping τ is either a ℓ-similarity or the composition of a ℓ-similarity and a ℓ-inversion on the whole set NE_2.

Suppose that τ is the composition of a ℓ-similarity and a ℓ-inversion on NE_2. Since the function $F(x) = k^{1/p}\|x - x_0\|_\ell^{-2n/p}$ belongs to the space $L^p_G(NE_2)$ we have

$$(10) \qquad 0 = G(F) = (-nk/p) \sum_{i=1}^{n} d_i \|x - x_0\|_\ell^{-2n/p-2} 2(x_i - (x_0)_i)\epsilon_i + (nk/p)(n/p+1) \sum_{i=1}^{n} \|x - x_0\|_\ell^{-2n/p-4} 4(x_i - (x_0)_i)^2 \epsilon_i + (2kn^2/p)\|x - x_0\|_\ell^{-2n/p-2}$$

for every $x \in NE_2$. It is easy to see that (10) implies $p = 2n/(n-2)$ and $d = 0$.

Simple calculations show that, for $F(x) = k\|x - x_0\|_\ell^{2-n}$ and $u_m(x) = (x_0)_m + (x_m - (x_0)_m)/\|x - x_0\|_\ell^2$ (these are the coordinate functions for ℓ-inversion),

$$\sum_{k=1}^{n} \epsilon_i (2 \frac{\partial F}{\partial x_i} \frac{\partial u_k}{\partial x_i} + F \frac{\partial^2 u_k}{\partial x_i^2}) \equiv 0$$

Now (9) implies $c = 0$.

Thus, the mapping τ can contain a ℓ-inversion only if $p = 2n/(n-2)$ and $c = d = 0$. On the other hand, if $p = 2n/(n-2)$, $c = d = 0$ and τ is the composition of a ℓ-similarity and a ℓ-inversion then, by (10), $|\tau'|^{1/p} \in L^p_G(NE_2)$ and , by (9), $|\tau'|^{1/p} f(\tau) \in L^p_G(NE_2)$ for every $f \in L^p_H(ME_1)$. Therefore, $Sf = |\tau'|^{1/p} f(\tau)$ is an isometry from $L^p_H(ME_1)$ to $L^p_G(NE_2)$. We have proved part (ii) of the theorem.

If $p \neq 2n/(n-2)$ or one of the vectors c, d is non-zero the mapping τ is a ℓ-similarity. Therefore, $F = |\tau'|^{1/p}$ is a constant function, and the coordinate functions u_k of the mapping τ are affine. The equality (9) implies

$$\sum_{i=1}^{n} F \frac{\partial u_k}{\partial x_i} d_i = F|\tau'|^{2/n} c_k$$

for every $k = 1, ..., n$. It means that $Jd = |\tau'|^{2/n} c$ which proves (iii) and, besides, shows that an isometric embedding does not exist if one of the vectors c, d is zero and another is non-zero. We have proved Theorem 1.

The case $n = 2$.

In this case, Liouville's theorem is no longer valid. We consider the cases $\ell = 2$ and $\ell = 1$ separately.

First, let $\ell = 2$. It follows from (8) that the mapping τ is either holomorphic or antiholomorphic, and

$$\frac{\partial u_1}{\partial x_1} = \pm \frac{\partial u_2}{\partial x_2}, \frac{\partial u_1}{\partial x_2} = \mp \frac{\partial u_2}{\partial x_1} \tag{11}$$

(Read this formula and the following text with the upper signs first, and then read it for the second time with the lower signs.)

The equality (9) shows that, for $k = 1, 2$,

$$(2\frac{\partial F}{\partial x_1} + d_1 F)\frac{\partial u_k}{\partial x_1} + (2\frac{\partial F}{\partial x_2} + d_2 F)\frac{\partial u_k}{\partial x_2} = F|\tau'|c_k \tag{12}$$

We get from (11) and (12) that , for $k = 1, 2$,

$$\frac{\frac{\partial F}{\partial x_k}}{F} = \pm(1/2)(c_1 \frac{\partial u_1}{\partial x_k} + c_2 \frac{\partial u_2}{\partial x_k} - d_k) \tag{13}$$

Solving this system of equations with respect to F we get

$$F(x) = \exp(\pm(1/2)(c_1 u_1(x) + c_2 u_2(x) - d_1 x_1 - d_2 x_2)) \tag{14}$$

Since $F \in L_G^p(NE_2)$ we get using (11), (13) and (14) that

$$\begin{aligned} G(F) = (1/4)F((c_1 \frac{\partial u_1}{\partial x_1} + c_2 \frac{\partial u_2}{\partial x_1} - d_1)^2 + (c_1 \frac{\partial u_1}{\partial x_2} + c_2 \frac{\partial u_2}{\partial x_2} - d_2)^2 + \\ 2(c_1 \frac{\partial^2 u_1}{\partial x_1^2} + c_2 \frac{\partial^2 u_2}{\partial x_1^2}) \pm 2(c_1 \frac{\partial^2 u_1}{\partial x_2^2} + c_2 \frac{\partial^2 u_2}{\partial x_2^2}) + \\ 2d_1(c_1 \frac{\partial u_1}{\partial x_1} + c_2 \frac{\partial u_2}{\partial x_1} - d_1) + 2d_2(c_1 \frac{\partial u_1}{\partial x_2} + c_2 \frac{\partial u_2}{\partial x_2} - d_2) = \\ (1/4)F(\pm(c_1^2 + c_2^2)|\tau'| - (d_1^2 + d_2^2)) = 0 \end{aligned} \tag{15}$$

If $c \neq 0, d \neq 0$ we get from the latter equality that $|\tau'|$ is constant.(Note that, by (13), F is non-zero at every point.) Clearly, τ is a similarity, and we get $Jd = |\tau'|c$ in the same way as in the case $n \geq 3$. If one of the vectors c, d is zero and another is non-zero, (15) is impossible, so an isometry does not exist. If $c = d = 0$, (9) implies $\frac{\partial F}{\partial x_1} \equiv 0$ and $\frac{\partial F}{\partial x_2} \equiv 0$. Therefore, $|\tau'| = |F|^p$ is constant and τ is a similarity. This finishes the proof in the case $\ell = 2$.

Let $\ell = 1$. Then, instead of (11), we get

$$\frac{\partial u_1}{\partial x_1} = \pm\frac{\partial u_2}{\partial x_2}, \frac{\partial u_1}{\partial x_2} = \pm\frac{\partial u_2}{\partial x_1} \tag{16}$$

and $\tau' = \pm((\frac{\partial u_1}{\partial x_1})^2 - (\frac{\partial u_1}{\partial x_2})^2)$.

We get from (9) that, for $k = 1, 2$,

$$(2\frac{\partial F}{\partial x_1} + d_1 F)\frac{\partial u_k}{\partial x_1} + (-2\frac{\partial F}{\partial x_2} + d_2 F)\frac{\partial u_k}{\partial x_2} = F|\tau'|c_k \tag{17}$$

It follows from (16) and (17) that

$$\frac{\frac{\partial F}{\partial x_k}}{F} = \pm(1/2)(c_1\frac{\partial u_1}{\partial x_k} - c_2\frac{\partial u_2}{\partial x_k} - (-1)^k d_k) \tag{18}$$

$k = 1, 2$, and we can calculate F :

$$F(x) = \exp(\pm(1/2)(c_1 u_1(x) - c_2 u_2(x) - d_1 x_1 + d_2 x_2)) \tag{19}$$

Similarly to the case $\ell = 2$ we get $\pm(c_1^2 - c_2^2)|\tau'| - (d_1^2 - d_2^2)) = 0$. We can finish the proof in the same way as for $\ell = 2$ if either $\|c\|_1^2 \neq 0, \|d\|_1^2 \neq 0$ or one of the numbers $\|c\|_1^2, \|d\|_1^2$ is zero and another is non-zero.

Finally, consider the case where $c_1^2 = c_2^2$ and $d_1^2 = d_2^2$. Suppose we have (16) with +. Since $|F| = |\tau'|^{1/p} = ((\frac{\partial u_1}{\partial x_1})^2 - (\frac{\partial u_1}{\partial x_2})^2)^{1/p}$ we can write (18) in the following form:

$$\begin{cases} \frac{\partial u_1}{\partial x_1}\frac{\partial^2 u_1}{\partial x_1^2} - \frac{\partial u_1}{\partial x_2}\frac{\partial^2 u_1}{\partial x_1 \partial x_2} = (p/4)((\frac{\partial u_1}{\partial x_1})^2 - (\frac{\partial u_1}{\partial x_2})^2)(c_1\frac{\partial u_1}{\partial x_1} - c_2\frac{\partial u_1}{\partial x_2} - d_1) \\ \frac{\partial u_1}{\partial x_1}\frac{\partial^2 u_1}{\partial x_1 \partial x_2} - \frac{\partial u_1}{\partial x_2}\frac{\partial^2 u_1}{\partial x_1^2} = (p/4)((\frac{\partial u_1}{\partial x_1})^2 - (\frac{\partial u_1}{\partial x_2})^2)(c_1\frac{\partial u_1}{\partial x_2} - c_2\frac{\partial u_1}{\partial x_1} + d_2) \end{cases} \tag{20}$$

Adding and subtracting the equations (20) we get

$$\begin{cases} \frac{\partial^2 u_1}{\partial x_1^2} + \frac{\partial^2 u_1}{\partial x_1 \partial x_2} = (p/4)(\frac{\partial u_1}{\partial x_1} + \frac{\partial u_1}{\partial x_2})((c_1 - c_2)(\frac{\partial u_1}{\partial x_1} + \frac{\partial u_1}{\partial x_2}) - (d_1 - d_2)) \\ \frac{\partial^2 u_1}{\partial x_1^2} - \frac{\partial^2 u_1}{\partial x_1 \partial x_2} = (p/4)(\frac{\partial u_1}{\partial x_1} - \frac{\partial u_1}{\partial x_2})((c_1 + c_2)(\frac{\partial u_1}{\partial x_1} - \frac{\partial u_1}{\partial x_2}) - (d_1 + d_2)) \end{cases} \tag{21}$$

If we found a function u_1 satisfying (21) and defined u_2 so that (16) holds then the mapping $\tau = (u_1, u_2)$ would generate an isometry from $L^p_H(ME_1)$ to $L^p_G(NE_2)$ because $|\tau'|^{1/p} \in L^p_G(NE_2)$ and $|\tau'|^{1/p} f(\tau) \in L^p_G(NE_2)$ for every $f \in L^p_H(ME_1)$.

First, assume that $c_1 = c_2 = c \neq 0$ and $d_1 = d_2 = d \neq 0$. Then (21) implies

$$\begin{cases} \frac{\partial^2 u_1}{\partial x_1^2} + \frac{\partial^2 u_1}{\partial x_1 \partial x_2} = \frac{\partial}{\partial x_1}(\frac{\partial u_1}{\partial x_1} + \frac{\partial u_1}{\partial x_2}) = 0 \\ \frac{\partial}{\partial x_1}(\frac{\partial u_1}{\partial x_1} - \frac{\partial u_1}{\partial x_2}) = (p/2)(\frac{\partial u_1}{\partial x_1} - \frac{\partial u_1}{\partial x_2})(c(\frac{\partial u_1}{\partial x_1} - \frac{\partial u_1}{\partial x_2}) - d) \end{cases} \tag{22}$$

Modify (22) using (16) to get

$$(23) \qquad \begin{cases} \frac{\partial^2 u_1}{\partial x_1^2} + \frac{\partial^2 u_1}{\partial x_1 \partial x_2} = \frac{\partial}{\partial x_2}(\frac{\partial u_1}{\partial x_1} + \frac{\partial u_1}{\partial x_2}) = 0 \\ -\frac{\partial}{\partial x_1}(\frac{\partial u_1}{\partial x_1} - \frac{\partial u_1}{\partial x_2}) = (p/2)(\frac{\partial u_1}{\partial x_1} - \frac{\partial u_1}{\partial x_2})(c(\frac{\partial u_1}{\partial x_1} - \frac{\partial u_1}{\partial x_2}) - d) \end{cases}$$

It follows from (22) and (23) that

$$(24) \qquad \frac{\partial u_1}{\partial x_1} + \frac{\partial u_1}{\partial x_2} = K = const$$

Integrating the second equalities in (22) and (23) and using (16) we get

$$(25) \qquad \frac{\partial u_1}{\partial x_1} - \frac{\partial u_1}{\partial x_2} = \frac{d\gamma \exp((-pd/2)x_1 + (pd/2)x_2)}{c(\gamma \exp((-pd/2)x_1 + (pd/2)x_2) - 1)}$$

for some $\gamma \in R$. By (24) and (25),

$$u_1 = (-1/cp)\ln|\gamma \exp((-pd/2)x_1 + (pd/2)x_2) - 1| + (K/2)x_1 + (K/2)x_2 + \alpha$$

where $\alpha \in R$. Now we can use (16) to find u_2 :

$$u_2 = (1/cp)\ln|\gamma \exp((-pd/2)x_1 + (pd/2)x_2) - 1| + (K/2)x_1 + (K/2)x_2 + \beta.$$

Similarly, one can calculate u_1 and u_2 for all other cases considered in the third part of Theorem 2. Note that, in every case, one gets the first solution if (16) holds with positive signs and the second solution appears if we have (16) with negative signs. We have proved Theorem 2.

REFERENCES

1. S. Banach, *Theorie des operations lineaires*, Monografie Matematycne, Warsaw, (1932).
2. B. A. Dubrovin, A. T. Fomenko, S. P. Novikov, *Modern geometry - methods and applications. Part 1. The geometry of surfaces, transformation groups, and fields*, Springer-Verlag New York (1992).
3. C. D. Hardin, Isometries of subspaces of L_p, *Indiana Univ. Math. J.* *30* (1981), 449–465.
4. P. Hartman, On isometries and on a theorem of Liouville, *Math. Zeitschrift* *69* (1958), 202–210.
5. A. Koldobsky, Isometric classification of L_p-spaces of solutions of homogeneous elliptic differential equations, *Contemporary questions of function theory and functional analysis*, Karaganda Univ., Karaganda (1990), 90–99.
6. A. Koldobsky, Isometries of $L_p(X; L_q)$ and equimeasurability, *Indiana Univ. Math. J.* *40* (1991), 677–705.
7. J. Lamperti, On the isometries of certain function spaces, *Pacific J. Math.* *8* (1958), 459–466.
8. A. I. Plotkin, Isometric operators in L_p-spaces of analytic and harmonic functions, *Zap. Nauchn. Semin. Leningr. Otd. Mat. Inst. Steklov* *30* (1972), 130–145.

9. A. I. Plotkin, Continuation of L_p-isometries, *J. Soviet Math.* *2* (1974), 143–165.
10. A. I. Plotkin, An algebra generated by translation operators and L_p-norms, *Functional Analysis* *6* (1976), 112–121.
11. Yu. G. Reshetnyak, On stability of conformal mappings in multidimensional spaces, *Siberian Math. J.* *8* (1967), 91–114.
12. W. Rudin, L_p-isometries and equimeasurability, *Indiana Univ. Math. J.* *25* (1976), 215–228.

L-Projections on Banach Lattices

Pei-Kee Lin Department of Mathematics, University of Memphis, Memphis, TN 38152

1. Introduction

Let X be a Banach space. A projection P on X is said to be *L-projection* (respectively, *M-projection*) if

$$\|x\| = \|P(x)\| + \|(I-P)(x)\|,$$
$$\text{(respectively, } \|x\| = \max\{\|P(x)\|, \|(I-P)(x)\|\}).$$

Recall a Banach lattice E is said to be ***order complete*** ***(σ-order complete)*** if every order bounded set (sequence) in E has a least upper bound (l.u.b.). It is known that $C(0,1)$ is not σ-order complete. On the other hand, the dual space of any Banach lattice is order complete (see [5] p. 85 Proposition 5.5).

A sublattice F of E is said to be an ***ideal*** if $x \in E$, $y \in F$, and $|x| \le |y|$ implies $x \in F$. An ideal F of E is said to be a ***band*** if $A \subseteq F$ and $\sup A = x \in E$ implies $x \in F$. Recall $x,\ y \in E$ is said to be ***disjoint*** if $|x| \wedge |y| = 0$. For a nonempty subset A of E, $A^{\perp}$ denotes the set of $x \in E$ disjoint from each $y \in A$. A band F of E is called a ***projection band*** if $E = F + F^{\perp}$. The associated projection $P : E \to F$ with kernel $F^{\perp}$ is called a ***band projection***. It is known that every band of an order complete Banach lattice is a projection band (see [4] p.10 Proposition 1.a.10, and [5] p. 63 Proposition 2.11).

Let E be a (real or complex) Banach lattice. We say that the norm on E is ***p additive*** (for $1 \le p \le \infty$) if $|x| \wedge |y| = 0$ implies

$$\|x+y\| = \begin{cases} (\|x\|^p + \|y\|^p)^{1/p} & \text{when } p < \infty, \\ \max\{\|x\|, \|y\|\} & \text{when } p = \infty. \end{cases}$$

A Banach lattice E with a p additive norm is said to be an *abstract L_p space*, (respectively, an *abstract M space*) if $1 \leq p < \infty$ (respectively, $p = \infty$). It is known that

(i) E is an abstract L_p space if and only if it is linearly isometric and lattice isomorphic to $L_p(\Omega)$ for some measure (Ω, μ) (see [3] p. 135 Theorem 3);

(ii) E is an abstract M space, if and only if it is isometric and lattice isomorphic to a closed sublattice of $C(T, \mathbb{R})$ (respectively, $C(T, \mathbb{C})$) for some compact Hausdorff space (see [3] p 23 Corollary of Theorem 5).

Suppose that E is a σ-order complete Banach space. For every $x \in E$, the associated projection P_x is defined in the following way (see [4] p. 8). For $z \geq 0$,

$$P_x(z) = \bigvee_{n=1}^{\infty} (n|x| \wedge z),$$

and for a general $y = y_1 - y_2 + iy_3 - iy_4 \in E$, $y_1, y_2, y_3, y_4 \geq 0$,

$$P_x(y) = P_x(y_1) - P_x(y_2) + iP_x(y_3) - iP_x(y_4).$$

In [2], L. Drewnowski, A. Kamińska and the author studied the L-projections and M-projections on complex Banach lattices. They proved that any L-projection (respectively, M-projection) on complex Banach lattices is a band projection. They also showed that if E is a real Banach lattice with a strictly monotone norm, then every L-projection on E is a band projection. In this article, we continue studying the L-projections on real σ-order complete Banach lattices and we prove the following theorem.

THEOREM 1. Let P be an L-projection on a σ-order complete real Banach lattice E, and let x, y be two elements in E such that $(P(x))^+ \wedge ((I - P)(x))^+ \geq y > 0$. If $Q = P_y$, and $R = P_{P(y)}$, then

(a) $Q(E)$ is an abstract L_1 space, and $z = 2Q(P(z))$ for every $z \in Q(E)$;

(b) $R(E)$ and $(I - R)(E)$ are invariant subspaces of P, and R is an L-projection.

Using this results, we get a characterization of L-projections on real Banach lattices.

THEOREM 2. Let P be an L-projection on a real Banach lattice E. Then there exist two projection bands F_1, F_2 of E such that

(a) F_1 and F_2 are two invariant subspaces of P and $E = (F_1 \oplus F_2)_1$;

(b) the restriction of P to F_2 is a band projection;

(c) there is a measure (Ω, μ) such that F_1 is lattice isometrically isomorphic to the Bochner L_1-space $L_1(\Omega, \ell_\infty^2)$ (so we may identify F_1 with $L_1(\Omega, \ell_\infty^2)$);

(d) for every $f \in L_1(\Omega, \ell_\infty^2)$ and $t \in \Omega$,

$$P(f)(t) = \frac{\langle e_1^* + e_2^*, f(t)\rangle}{2}(e_1 + e_2) \text{ or } P(f)(t) = \frac{\langle e_1^* - e_2^*, f(t)\rangle}{2}(e_1 - e_2).$$

2. Some Lemmas

In this section, we prove several lemmas that we use to prove our theorems.

LEMMA 3. Let P be an L-projection on a (real or complex) Banach lattice, and let x and y be two elements in E such that $\|P(x)\| \geq \|P(x)-y\|$ and $\|(I-P)(x)\| \geq \|(I-P)(x)-y\|$. Then

(i) $\|P(x)-y\| = \|P(x)\|$;
(ii) $\|(I-P)(x)-y\| = \|(I-P)(x)\|$;
(iii) $\|P(y)\| = \|(I-P)(y)\| = \frac{1}{2}\|y\|$.

Proof: Since

$$\begin{aligned}\|P(x)\| + \|(I-P)(x)\| =&\|P(x) - (I-P)(x)\| \\ =&\|(P(x)-y) - ((I-P)(x)-y)\| \\ \leq&\|P(x)-y\| + \|(I-P)(x)-y\| \\ \leq&\|P(x)\| + \|(I-P)(x)\|,\end{aligned}$$

$\|P(x)\| = \|P(x)-y\|$ and $\|(I-P)(x)-y\| = \|(I-P)(x)\|$. So

$$\begin{aligned}\|P(x)\| =&\|P(x)-y\| = \|P(x)-P(y)\| + \|(I-P)(y)\| \\ \|(I-P)(x)\| =&\|(I-P)(x)-y\| = \|(I-P)(x)-(I-P)(y)\| + \|P(y)\|.\end{aligned} \tag{1}$$

On the other hand,

$$\begin{aligned}&\|P(x)\| + \|(I-P)(x)\| \\ =&\|P(x)-y\| + \|(I-P)(x)-y\| \\ =&\|P(x)-P(y)\| + \|(I-P)(y)\| + \|(I-P)(x)-(I-P)(y)\| + \|P(y)\| \\ \geq&\|P(x)\| + \|(I-P)(x)\|.\end{aligned}$$

This implies

$$\begin{aligned}\|P(x)\| =&\|P(x-y)\| + \|P(y)\| \\ \|(I-P)(x)\| =&\|(I-P)(x-y)\| + \|(I-P)(y)\|.\end{aligned} \tag{2}$$

By (1) and (2), we have $\|(I-P)(y)\| = \|P(y)\| = \frac{1}{2}(\|P(y)\| + \|(I-P)(y)\|) = \frac{1}{2}\|y\|$. □

Let E be a complex lattice. It is known that if x, y are any two positive elements in E such that $x \leq y$, then for any real element z in E,

$$|x+iz| \leq |y+iz| \quad \text{(see [4] section 1.d.).}$$

So we have the following corollary.

COROLLARY 4. Let E be a σ-order complete (real or complex) Banach lattice. For any four non-negative disjoint elements x_1, x_2, x_3, x_4 in E ($x_3 = 0 = x_4$ when E is a real

Banach lattice), let

$$\begin{aligned} Q_j &= P_{x_j} && \text{for } 1 < j \le 4 \\ Q_1 &= I - Q_2 - Q_3 - Q_4 \\ R &= Q_1 - Q_2 - iQ_3 + iQ_4 \\ y &= x_1 - x_2 + ix_3 - ix_4 \\ z &= x_1 + x_2 + x_3 + x_4. \end{aligned}$$

If P is an L-projection on E, and if x is any element in E which satisfies the following condition

$$(\Re(R(P(x))))^+ \wedge (\Re(R((I-P)(x))))^+ \ge z$$
$$\text{(respectively, } (\Re(R(P(x))))^+ \wedge (\Re(R((I-P)(x))))^- \ge z),$$

then

(i) $\|P(x) - y\| = \|P(x)\|$;
(ii) $\|(I-P)(x) - y\| = \|(I-P)(x)\|$, (respectively, $\|(I-P)(x)+y\| = \|(I-P)(x)\|$);
(iii) $\|P(y)\| = \|(I-P)(y)\| = \frac{1}{2}\|y\|$;
(iv) $\|P(x)\| = \|P(x-y)\| + \|(I-P)(y)\| = \|P(x-y)\| + \|P(y)\|$.

Proof: We only prove the corollary if $(\Re(R(P(x))))^+ \wedge (\Re(R((I-P)(x))))^+ \ge z$. The remainder is left to the reader.

Since R is an isometry on E, $RPR^{-1} = Q$ is an L-projection. The assumption implies

$$\Big(\Re(Q(R(x)))\Big)^+ \wedge \Big(\Re((I-Q)(R(x)))\Big)^+ \ge z = R(y).$$

So

$$\|QR(x)\| \ge \|QR(x) - z\| \quad \text{and} \quad \|(I-Q)R(x)\| \ge \|(I-Q)R(x) - z\|.$$

By Lemma 3, we have

$$\begin{aligned} \|P(x) - y\| &= \|RP(x) - R(y)\| = \|QR(x) - z\| = \|QR(x)\| = \|RP(x)\| = \|P(x)\|, \\ \|(I-P)(x) - y\| &= \|R(I-P)(x) - R(y)\| = \|(I-Q)R(x) - z\| \\ &= \|(I-Q)R(x)\| = \|R(I-P)(x)\| = \|(I-P)(x)\|, \\ \|P(y)\| &= \|Q(z)\| = \|(I-Q)(z)\| = \|(I-P)(y)\| = \frac{1}{2}\|z\| = \frac{1}{2}\|y\|. \end{aligned}$$

Hence,

$$\|P(x)\| = \|P(x) - y\| = \|P(x-y)\| + \|(I-P)(y)\| = \|P(x-y)\| + \|P(y)\|. \quad \square$$

LEMMA 5. Let P be any L-projection on a σ-order complete (real or complex) Banach lattice. Let x, x_1 and x_2 be three elements in E such that $x_1 \ge 0$, $x_2 \ge 0$, and

$$\Big(\Re(P(x))\Big)^+ \wedge \Big(\Re((I-P)(x))\Big)^+ \ge x_1 + x_2$$
$$\text{(respectively, } \Big(\Re(P(x))\Big)^+ \wedge \Big(\Re((I-P)(x))\Big)^- \ge x_1 + x_2).$$

Then $\|x_1 + x_2\| = \|x_1\| + \|x_2\|$.

Proof: We only prove the lemma when $\Big(\Re(P(x))\Big)^+ \wedge \Big(\Re((I-P)(x))\Big)^+ \geq x_1 + x_2$. The remainder is left to reader.

If $x_1 = 0$ or $x_2 = 0$, then it is trivial. So we may assume that $x_1 > 0$, and $x_2 > 0$.

Step 1. We claim that $\|x_1 + x_2\| > \max\{\|x_1\|, \|x_2\|\}$. Let $a = \min\{\frac{1}{4}, \frac{\|x_2\|}{4\|x_1\|}\}$ and $x_3 = ax_1$. Then

$$\begin{aligned}
&\Big(\Re(P(x - x_3))\Big)^+ \wedge \Big(\Re((I-P)(x - x_3))\Big)^+ \\
&\geq \Big(x_2 - (\Re(P(x_3)))^+\Big)^+ \wedge \Big(x_2 - (\Re((I-P)(x_3)))^+\Big)^+ \\
&\geq (x_2 - |P(x_3)|)^+ \wedge (x_2 - |(I-P)(x_3)|)^+ \\
&\geq x_2 - \Big((|P(x_3)| + |(I-P)(x_3)|) \wedge x_2\Big) > 0 \qquad \text{because its norm is bigger than 0.}
\end{aligned}$$

Let $x_4 = ax_2 \wedge \bigg(x_2 - \Big((|P(x_3)| + |(I-P)(x_3)|) \wedge x_2\Big)\bigg)$. Then we have

$$\begin{aligned}
&0 \leq x_3 \leq x_3 + x_4 \leq a x_1 + a x_2 \leq (\Re(P(x)))^+ \wedge (\Re((I-P)(x)))^+, \\
&0 \leq x_4 \leq (\Re(P(x - x_3)))^+ \wedge (\Re((I-P)(x - x_3)))^+.
\end{aligned}$$

By Corollary 4,

$$\begin{aligned}
\|P(x)\| &= \|P(x - x_3 - x_4)\| + \|P(x_3 + x_4)\| = \|P(x - x_3)\| + \|P(x_3)\|, \\
\|P(x - x_3)\| &= \|P(x - x_3 - x_4)\| + \|P(x_4)\|, \\
\|x_3\| &= 2\|P(x_3)\|, \\
\|x_3 + x_4\| &= 2\|P(x_3 + x_4)\|.
\end{aligned}$$

This implies

$$\begin{aligned}
a\|x_1\| = \|x_3\| &= 2\|P(x_3)\| < 2(\|P(x_3)\| + \|P(x_4)\|) \\
&= 2\|P(x_3 + x_4)\| = \|x_3 + x_4\| \leq a\|x_1 + x_2\|.
\end{aligned}$$

Similarly, $\|x_1 + x_2\| > \|x_2\|$.

Step 2. We claim that if $\big((P(x)) \wedge (I-P)(x)\big) \geq z_1 \geq y > 0$ and if $P_{z_1}(z_2) > y$, then $\|z_2\| > \|y\|$.

Note: in step 1, we proved that $\Big((P(2x))^+ \wedge ((I-P)(2x))^+\Big) \geq z_2 > y > 0$, then $\|z_2\| > \|y\|$.

Since $P_{z_1}(z_2) > y$,

$$(P_{z_1}(z_2) - y) \wedge z_1 > 0.$$

This implies

$$P_{z_1}(z_2) \wedge \big((P(2x))^+ \wedge ((I-P)(2x))^+\big) \geq y + \big((P_{z_1}(z_2) - y) \wedge z_1\big) > y.$$

By step 1,

$$\|z_2\| \geq \|P_{z_1}(z_2)\| > \|y\|.$$

Step 3. Let y be any positive element ($y > 0$) such that $(P(x))^+ \wedge ((I-P)(x))^+ \geq y$. Let $Q = P_y$. We claim that $y \geq QP(y) > 0$ and $y \geq Q(I-P)(y) > 0$.

First, we show that $QP(y) \geq 0$ and $Q(I-P)(y) \geq 0$. Suppose it is not true. Then either $|QP(y)| > \Re(QP(y))$ or $|Q(I-P)(y)| > \Re(Q(I-P)(y))$. So we have

$$Q\big(|P(y)| + |(I-P)(y)|\big) = |QP(y)| + |Q(I-P)(y)| > y.$$

By step 2,

$$\begin{aligned}\|y\| &< \| |QP(y)| + |Q(I-P)(y)| \| \\ &\leq \|QP(y)\| + \|Q(I-P)(x)\| \\ &\leq \|P(y)\| + \|(I-P)(x)\| = \|y\|.\end{aligned}$$

This is impossible. So we must have $QP(y) \geq 0$ and $Q(I-P)(y) \geq 0$.

On the other hand,

$$\begin{aligned}y &= Q(P(y) + (I-P)y) = QP(y) + Q(I-P)(y) \\ \frac{\|y\|}{2} &= \max\{\|P(y)\|, \|(I-P)(y)\|\} \geq \max\{\|QP(y)\|, \|Q(I-P)(y)\|\}.\end{aligned}$$

This implies $\|QP(y)\| = \frac{\|y\|}{2} = \|Q(I-P)(y)\|$. So $QP(y) > 0$ and $Q(I-P)(y) > 0$.

Step 4. Let $Q = P_{x_1}$ and $R = P_{x_1+x_2}$. Then $Q \circ R = Q = R \circ Q$. We note that

$$\begin{aligned}|R(P(x_1))| &\geq |Q(P(x_1))|, \\ |R((I-P)(x_1))| &\geq |Q((I-P)(x_1))|, \\ \|Q(P(x_1))\| &= \|P(x_1)\| \geq \|R(P(x_1))\| \\ \|Q((I-P)(x_1))\| &= \|(I-P)(x_1)\| \geq \|R((I-P)(x_1))\|.\end{aligned}$$

By step 2 again, we have

$$\begin{aligned}Q(P(x_1)) &= R(P(x_1)), \\ Q((I-P)(x_1)) &= R((I-P)(x_1)).\end{aligned}$$

Hence,

$$\begin{aligned}&\Big(\Re(P(x-x_1))\Big)^+ \wedge \Big(\Re((I-P)(x-x_1))\Big)^+ \\ \geq& R\Big((\Re(P(x-x_1)))^+ \wedge (\Re((I-P)(x-x_1)))^+\Big) \\ \geq& R\Big((\Re(P(x)))^+ \wedge (\Re((I-P)(x)))^+\Big) - \Big(|R((P(x_1))| + |R((I-P)(x_1))|\Big) \\ =& R\Big((\Re(P(x)))^+ \wedge (\Re((I-P)(x)))^+\Big) - \Big(R((P(x_1)) + R((I-P)(x_1))\Big) \\ \geq& (x_1+x_2) - x_1 = x_2.\end{aligned}$$

By Corollary 4,

$$\begin{aligned}\|P(x)\| - \|P(x_1 + x_2)\| =& \|P(x - x_1 - x_2)\| \\ =& \|P(x - x_1)\| - \|P(x_2)\| \\ =& \|P(x)\| - \|P(x_1)\| - \|P(x_2)\|.\end{aligned}$$

This implies $\|x_1 + x_2\| = \|x_1\| + \|x_2\|$. $\square$

REMARK 1. Let P be an L-projection on a σ-order complete Banach lattice. Let x, y be two elements in E such that $(\Re(P(x)))^+ \wedge (\Re((I-P)(x)))^+ \geq y > 0$. Let $Q = P_y$. By the proof of Lemma 5, if R is a band projection such that $Q \circ R = R$, then

$$Q(P(R(y))) = R(P(y)), \quad \text{and} \quad \|R(y)\| = 2\|R(P(y))\|.$$

3. Proof of Theorem 1

Now, we prove Theorem 1.

Proof of THEOREM 1 (a): Let z_1 be any positive element in $Q(E)$. By Lemma 5, for any $m \in \mathbb{N}$,

$$\|(m+1)y \wedge z_1\| = \|my \wedge z_1\| + \|((m+1)y \wedge z_1) - (my \wedge z_1))\|.$$

By induction, for any $n > m$,

$$\|ny \wedge z_1\| = \|my \wedge z_1\| + \sum_{k=m+1}^{n} \|(ky \wedge z_1) - ((k-1)y \wedge z_1).\|$$

This implies $\{ny \wedge z_1\}$ converges to z_1. Let z_2 be another positive element in $Q(E)$. Then

$$\begin{aligned}\|z_1 + z_2\| &= \lim_{n\to\infty} \|(z_1 + z_2) \wedge 2ny\| \\ &\geq \lim_{n\to\infty} \|(z_1 \wedge ny) + (z_2 \wedge ny)\| \\ &= \lim_{n\to\infty} (\|z_1 \wedge ny\| + \|z_2 \wedge ny\|) = \|z_1\| + \|z_2\|.\end{aligned}$$

So $Q(E)$ is an abstract L_1 space.

For any $0 \leq z \in Q(E)$, $\lim_{n\to\infty} ny \wedge z = z$. We only need to show that if $0 < z < y$, then $z = 2Q(P(z))$.

Since $\|z\| = 2\|Q(P(z))\|$ and $Q(E)$ is an abstract L_1 space, it is enough to show that $z' = \left(z - 2Q(P(z))\right)^+ = 0$. Suppose that it is not true. Let $R = P_{z'}$. Since $Q(E)$ is an abstract L_1 space,

$$\|R(z)\| > 2\|RQP(z)\| = 2\|RP(z)\| = \|R(z)\|.$$

This is impossible. So z must be equal to $2Q(P(z))$. □

Now, we give a direct proof that every L-projection on a complex Banach lattice is a band projection.

THEOREM 6. ([2] Theorem 1.5) Let E be a complex Banach lattice. Then every L-projection on E is a band projection.
Proof: First, we assume that E is σ-order complete. Let P be an L-projection on E. It is enough to show that if $P(x_1) = x_1$ and $(I-P)(x_2) = x_2$, then x_1 and x_2 are disjoint.

Suppose it is not true. Without loss of generality, we may assume that $(\Re(x_1))^+ \wedge (\Re(x_2))^+ \geq y > 0$. (So $(\Re(P(x_1+x_2)))^+ \wedge (\Re((I-P)(x_1+x_2)))^+ \geq y$.) Let $Q = P_y$, and $z = 2(I-Q)P(y)$. By Theorem 1(a), we have

$$P(y) = \frac{y+z}{2} = P(z)$$
$$(I-P)(y) = \frac{y-z}{2} = -(I-P)(z)$$

Hence,

$$\begin{aligned}\|y+z\| &= \|y+iz\| \\ &= \|P(y+iz)\| + \|(I-P)(y+iz)\| \\ &= \|\frac{(1+i)(y+z)}{2}\| + \|\frac{(1-i)(y-z)}{2}\| \\ &= \sqrt{2}\|y+z\|.\end{aligned}$$

This is impossible. So P must be a band projection.

Now, consider the general case. Since P^{**} is an L-projection and E^{**} is order complete, by above proof, P^{**} is a band projection. Hence, $P = P^{**}|_E$ is a band projection. □

From now on, we only deal with real Banach lattices.

REMARK 2. Let E be a real σ-order complete Banach lattice, and let P be any L-projection on E. Suppose that x, y be two elements in E such that $(P(x))^+ \wedge ((I-P)(x))^+ \geq y > 0$. Let $Q_1 = P_y$, $y_1 = (I-P_y)(P(y))$, and $Q_2 = P_{y_1}$.

(1) Then
$$y_1 = Q_2(P(y)) = -Q_2((I-P)(y)).$$
Let $y_2 = (y_1)^-$, and $Q_3 = P_{y_2}$. Then $I - 2Q_3$ is an isometry, $(I-2Q_3)\circ P \circ (I-2Q_3)$ is an L-projection, and

$$Q_2 \circ (I-2Q_3) \circ P \circ (I-2Q_3)(y) = Q_2 \circ (I-2Q_3) \circ P(y) = (y_1)^+ + (y_1)^- = |y_1|.$$

Hence, without loss of generality, we may assume that $y_1 \geq 0$. By Theorem 1 (a), $Q_2(E)$ is an abstract L_1 space.

(2) For any $z \in Q_1(E)$, let $z' = 2(I-P_z)P(z) = 2(I-Q_1)P(z)$. Then

$$P(z) = \frac{z+z'}{2} = P(z')$$
$$(I-P)(z) = \frac{z-z'}{2} = -(I-P)(z').$$

So

$$2\|(I-Q_1)P(z)\| = \|z'\| = \|P(z')\| + \|(I-P)(z')\| = \|P(z)\| + \|(I-P)(z)\| = \|z\|.$$

(3) Let z_1, z_2 be two disjoint positive elements in $Q_1(E)$. We claim that $P(z_1), P(z_2)$ are disjoint. Let $z_3 = (I-Q_1)P(z_1+z_2)$, and let $R = P_{z_3}$. Since $Q_1(P(z_1)) = \frac{z_1}{2}$ and $Q_1(P(z_2)) = \frac{z_2}{2}$, we only need to show that $|(I-Q_1)(P(z_1))| \wedge |(I-Q_1)((P(z_2))| = 0$.

Subclaim: $|R(P(z_1))| \wedge |R((P(z_2))| = 0$. Suppose the subclaim is not true. Without loss of generality, we may assume that

$$(R(P(z_1)))^+ \wedge (R((P(z_2)))^+ = z_4 > 0.$$

Then

$$\begin{aligned}
&\|R(P(z_1))\| + \|R(P(z_2))\| \\
=&\|Q_1(P(z_1))\| + \|Q_1(P(z_2))\| \\
=&\|Q_1(P(z_1 - z_2))\| \\
=&\|R(P(z_1 - z_2))\| \\
=&\|R(P(z_1)) - z_4 - (R(P(z_2)) - z_4)\| \\
\leq&\|R(P(z_1)) - z_4\| + \|R(P(z_2)) - z_4\| \\
=&\|R(P(z_1))\| - \|z_4\| + \|R(P(z_2))\| - \|z_4\|.
\end{aligned}$$

This is impossible. So $|R(P(z_1))| \wedge |R((P(z_2))| = 0$.

Let $z_5 = (I-Q_1)(z_1)$, $z_6 = (I-Q_1)(z_2)$, $R_1 = P_{z_5}$, and $R_2 = P_{z_6}$. Note:

$$\begin{aligned}
|R(P(z_1))| &\leq |R_1(P(z_1))| \\
|R(P(z_2))| &\leq |R_2(P(z_2))| \\
2\|R(P(z_1))\| &\leq \|z_1\| = 2\|R_1(P(z_1))\| \\
2\|R(P(z_2))\| &\leq \|z_2\| = 2\|R_2(P(z_2))\| \\
2\|R(P(z_1+z_2))\| &= \|z_1+z_2\| = \|z_1\| + \|z_2\|.
\end{aligned}$$

This implies $2\|R(P(z_1))\| = \|z_1\|$ and $2\|R(P(z_2))\| = \|z_2\|$. By the proof of Lemma 5, we have $R(P(z_1)) = R_1(P(z_1))$ and $R(P(z_2)) = R_2(P(z_2))$. And we proved our claim.

Proof of THEOREM 1 (b): Step 1. We claim that $R(E)$ is an invariant subspace of P. It is enough to show that if $0 \leq z \in Q(E)$ or if $0 \leq z \in (R-Q)(E)$, then $R(P(z)) = P(z)$. Without loss of generality, we may assume that $z \in Q(E)$. Since $Q(E)$ is an abstract L_1 space, we may assume that $0 \leq z \leq y$. Then

$$\begin{aligned}
&\|y - z\| \\
=&2\|(I-Q)P(y-z)\| \\
\geq&2\|(R-Q)P(y-z)\| \\
\geq&2(\|(R-Q)P(y)\| - \|(R-Q)P(z)\|) \\
=&2\|(R-Q)P(y)\| - 2\|(R-Q)P(z)\| \\
=&\|y\| - 2(\|(I-Q)P(z)\| - \|(I-R)P(z)\|) \quad \text{since } P_{(I-Q)(z)}(E) \text{ is an abstract } L_1 \text{ space} \\
=&\|y\| - \|z\| + 2\|(I-R)P(z)\| \\
=&\|y - z\| + 2\|(I-R)P(z)\|.
\end{aligned}$$

So $(I-R)P(z)=0$, and we proved our claim.

Step 2. We claim that if $z_1 = P(z_1) \in (I-R)(E)$ and $z_2 = P(z_2) \in R(E)$ (respectively, $z_1 = (I-P)(z_1) \in (I-R)(E)$ and $z_2 = (I-P)(z_2) \in R(E)$), then $\|z_1+z_2\| = \|z_1\|+\|z_2\|$.

Without loss of generality, we assume that $z_1 = P(z_1) \in (I-R)(E)$ and $z_2 = P(z_2) \in R(E)$. Let $z_3 = Q(z_2)$ and let $z_4 = (I-Q)(z_2)$. Then

$$\begin{aligned} P(2z_3) &= z_2 = P(2z_4) \\ (I-P)(2z_3) &= z_3 - z_4 = (I-P)(2z_4). \end{aligned}$$

So

$$\begin{aligned} \|z_1+z_2\| &= \|z_1+z_3-z_4\| \\ &= \|P(z_1+z_3-z_4)\| + \|(I-P)(z_1+z_3-z_4)\| \\ &= \|z_1\| + \|z_3-z_4\| \\ &= \|z_1\| + \|z_2\|. \end{aligned}$$

Step 3. We claim $(I-R)(E)$ is an invariant subspace of P. For any $z \in (I-R)(E)$, let $z_1 = RP(z)$ and $z_2 = (I-R)P(z)$. Then $P(z) = z_1 + z_2$, and

$$\begin{aligned} z_2 &= (I-R)P(z) \\ &= (I-R)P(z_1+z_2) \\ &= (I-R)P(z_1) + (I-R)P(z_2) \\ &= (I-R)P(z_2) && \text{since } P(z_1) \in R(E). \end{aligned}$$

So $\|z_2\| \leq \|P(z_2)\|$. But $\|z_2\| = \|P(z_2)\| + \|(I-P)(z_2)\|$. We must have

$$\begin{aligned} (I-P)(z_1) = -(I-P)(z_2) = 0, \\ P(z) = P(z_1+z_2) = z_1+z_2. \end{aligned}$$

Hence,

$$\begin{aligned} \|z\| &= \|P(z)\| + \|(I-P)(z)\| \\ &= \|z_1\| + \|z_2\| + \|(I-P)(z)\| && \text{By step 2} \\ &\geq \|z_2 + (I-P)(z)\| + \|z_1\| \\ &\geq \|(I-R)(z_2 + (I-P)(z))\| + \|z_1\| \\ &= \|(I-R)(z_2) + (I-R)(z) - (I-R)P(z)\| + \|z_1\| \\ &= \|(I-R)(z)\| + \|z_1\| \\ &= \|z\| + \|z_1\|. \end{aligned}$$

This implies $z_1 = 0$ and $P(z) \in (I-R)(E)$.

Step 4. Let z be any element in $(I-R)(E)$ and let w be any element in $R(E)$. By step 1 and 3, $P(z) \in (I-R)(E)$ and $P(w) \in R(E)$. By step 2, we have

$$\begin{aligned} \|z+w\| &= \|P(z+w)\| + \|(I-P)(z+w)\| \\ &= \|P(z)\| + \|P(w)\| + \|(I-P)(z)\| + \|(I-P)(w)\| \\ &= \|w\| + \|z\|. \end{aligned}$$

So R is an L-projection. □

4. Proof of Theorem 2

Now, we give the proof of Theorem 2.

Proof of THEOREM 2: First, we assume that E is σ-order complete

A positive element y in E is said to have ***property*** $(*)$ if

(i) there is $x \in E$ such that $(P(x))^+ \wedge ((I-P)(x))^+ \geq y > 0$;

(ii) $(I-P_y)P(y)$ is either positive or negative.

Step 1. We claim that if P is not a band projection, then there is $y > 0$ which has property $(*)$.

If P is not a band projection, then there are $x \in E$ and $z > 0$ such that $(P(x))^+ \wedge ((I-P)(x))^+ \geq z$. Let $y_1 = ((I-P_z)P(z))^+ \neq 0$ and $y_2 = ((I-P_z)P(z))^- \neq 0$. By Remark 2, $P(y_1)$ and $P(y_2)$ are disjoint. Suppose that $y_1 \neq 0$. Let $Q = P_{y_1}$ and $R = P_{P(y_1)}$. Then

$$Q(z) = 2y_1$$
$$(I-Q)(P(2y_1)) = (I-Q)(P(Q(z))) = \frac{1}{2}(R-Q)(z).$$

So y_1 has property $(*)$. Similarly, if $y_2 \neq 0$, then y_2 has property $(*)$.

Step 2. For each $z = y_\lambda$ with property $(*)$, let

$$Q_\lambda = P_z \quad \text{and} \quad R_\lambda = P_{P(z)}.$$

By the Hausdorff Maximal Principle, there is a maximal family

$$\{y_\lambda : \lambda \in \Lambda\} = \{y_\lambda \in E : y_\lambda \text{ has property } (*), \text{ and } R_\lambda R_{\lambda'} = 0 \text{ whenever } \lambda \neq \lambda'\}.$$

Since for each $\lambda \in \Lambda$, R_λ is an L-projection,

$$\sum_{\lambda \in \Lambda} \|Q_\lambda(z)\| = \sum_{\lambda \in \Lambda} \|R_\lambda(Q_\lambda(z))\| \leq \sum_{\lambda \in \Lambda} \|R_\lambda(z)\| \leq \|z\|$$

for any $z \in E$. Hence, for any $z \in E$, there are countable λ such that $R_\lambda(z) \neq 0$. let $R(z)$ and $Q(z)$ be defined by

$$R(z) = \sum_{\lambda \in \Lambda} R_\lambda(z),$$
$$Q(z) = \sum_{\lambda \in \Lambda} Q_\lambda(z).$$

If $z \in Q(E)$, then

$$\|z\| = \sum_{\lambda \in \Lambda} \|Q_\lambda(z)\|.$$

So $Q(E)$ is an abstract L_1 space. We claim that

(iii) $(I-R)(E)$, and $R(E)$ are invariant subspaces of P;

(iv) if $x \in (I-R)(E)$, then $P(x)$ and $(I-P)(x)$ are disjoint.

Note: (iii) and (iv) imply that the restriction of P to $(I-R)(E)$ is a band projection.

Proof of (iii). For any $\lambda \in \Lambda$, $R_\lambda P = PR_\lambda$. Hence, for every $x \in E$,

$$P(R(x)) = P(\sum_{\lambda\in\Lambda} R_\lambda(x)) = \sum_{\lambda\in\Lambda} PR_\lambda(x) = \sum_{\lambda\in\Lambda} R_\lambda P(x) = R(P(x)),$$

and

$$\begin{aligned} P((I-R)(x)) =& P(x - \sum_{\lambda\in\Lambda} R_\lambda(x)) \\ =& P(x) - \sum_{\lambda\in\Lambda} PR_\lambda(x) \\ =& P(x) - \sum_{\lambda\in\Lambda} R_\lambda P(x) \\ =& P(x) - R(P(x)) = (I-R)(P(x)). \end{aligned}$$

So $R(E)$ and $(I-R)(E)$ are invariant subspaces of P.

Proof of (iv). Suppose it is not true. Then there exist $x, y_\alpha \in (I-R)(E)$ such that $(P(x))^+ \wedge ((I-P)(x))^+ \geq y_\alpha > 0$. By the proof of step 1, we may assume that either $(I-P_{y_\alpha})(P(y_\alpha)) > 0$ or $(I-P_{y_\alpha})(P(y_\alpha)) < 0$. But $(I-R)(E)$ is an invariant subspace of P. This implies for any $\lambda \in \Lambda$,

$$R_\lambda P_{P(y_\alpha)} = 0.$$

This is a contradiction. So $P(x)$ and $(I-P)(x)$ are disjoint for every $x \in (I-R)(E)$.

Step 3. We claim that R is an L-projection. For a fixed $x \in E$, let

$$\{R_i : i \in \mathbb{N}\} = \{R_\lambda : R_\lambda(x) \neq 0\}.$$

Then

$$\begin{aligned} \|x\| =& \|R(x) + (I-R)(x)\| \\ =& \lim_{n\to\infty} \|\sum_{i=1}^{n} R_i(x) + (I-R)(x)\| \\ =& \lim_{n\to\infty} \|\sum_{i=1}^{n} R_i(x)\| + \|(I-R)(x)\| = \|R(x)\| + \|(I-R)(x)\|. \end{aligned}$$

So P is an L-projection.

Step 4. For each y_λ, let

$$a_\lambda = \begin{cases} 1 & \text{if } (I-Q_\lambda)(y_\lambda) > 0, \\ -1 & \text{if } (I-Q_\lambda)(y_\lambda) < 0. \end{cases}$$

For each $z \in Q_\lambda(E)$, define

$$T_\lambda(z) = 2(I-Q_\lambda)P(z).$$

By Theorem 1 and Remark 2, T_λ is an isometric isomorphism from $Q_\lambda(E)$ onto $(R_\lambda - Q_\lambda)(E)$.

We claim that T_λ is a positive (respectively, negative) operator if $a_\lambda = 1$ (respectively, $a_\lambda = -1$). Since $Q_\lambda(E)$ is an abstract L_1 space, it is enough to show that if $0 \le z \le y_\lambda$, then $a_\lambda T_\lambda(z) \ge 0$. But $(R_\lambda - Q_\lambda)(E)$ is an abstract L_1 space, $a_\lambda T_\lambda(y_\lambda) \ge 0$, and

$$\|T_\lambda(y_\lambda - z)\| = \|y_\lambda - z\| = \|y_\lambda\| - \|z\|.$$

So $a_\lambda T_\lambda(z)$ must be positive, and $T = \sum_{\lambda\in\Lambda} a_\lambda T_\lambda$ is a lattice isometric isomorphism from $Q(E)$ onto $(R-Q)(E)$. Let $T' = \sum_{\lambda\in\Lambda} T_\lambda$. Then for $x \in Q(E)$ and $y \in (R-Q)(E)$,

$$P(x+y) = \frac{1}{2}(I+T')(x+(T')^{-1}(y)).$$

Since $Q(E)$ is an abstract L_1 space, there is a measure (Ω, μ) such that $Q(E)$ is lattice isometrically isomorphic to $L_1(\Omega,\mu)$. Let S_1 be the lattice isometric isomorphism from L_1 onto $Q(E)$. Let $\{e_1, e_2\}$ be the natural basis of ℓ^2_∞. Define $S : L_1(\Omega, \ell^2_\infty) \to R(E)$ by

$$S(f\,e_1 + g\,e_2) = S_1(f) + TS_1(g) \qquad (\text{for } f, g \in L_1(\Omega,\mu)).$$

By Theorem 1 and Lemma 3, $\|S(f\,e_1 + f\,e_2)\| = \|S(f\,e_1)\| = \|f\|_1$.

The verification of S being a linear isometric and lattice isomorphism form $L_1(\Omega, \ell^2_\infty)$ onto $R(E)$ is left to the reader.

Now, we prove the general case. Let P be an L-projection on a real Banach lattice E. It is enough to show that if x, y are two elements in E such that $(P(x))^+ \wedge ((I-P)(x))^+ \ge y > 0$, then P_y and $P_{P(y)}$ are well-defined.

Since E^{**} is order complete and P^{**} is an L-projection on E^{**}, the above proof showed there exist a measure (Ω, μ) and a projection band F of E^{**} such that

$$E^{**} = (L_1(\Omega, \ell^2_\infty) \oplus F)_1.$$

Since $L_1(\Omega, \ell^2_\infty)$ does not contain any subspace which is isomorphic to c_o, every sublattice of $L_1(\Omega, \ell^2_\infty)$ is σ-order complete. But $y, P(y) \in L_1(\Omega, \ell^2_\infty)$. So P_y and $P_{P(y)}$ are well-defined. □

COROLLARY 7. Let E be a real Banach lattice. If E is not linear isometric and lattice isomorphic to ℓ^2_∞, and if there is a nontrivial L-projection on E, then there is a nontrivial L-band projection on E.

Proof: Suppose P is a nontrivial L-projection on E. By Theorem 2, there exist two projection bands F_1 and F_2 of E which satisfy (a), (b) and (c) in Theorem 2. If $F_1 = \{0\}$, then by Theorem 2.b, P is a L-band projection. If $F_1 \ne \{0\} \ne F_2$, then the band projection from E onto F_1 is a nontrivial L-band projection. Hence, we may assume $F_2 = \{0\}$. But E is not isomorphic to ℓ^2_∞. Ω is not singleton. So there exists a nontrivial L-band projection on E. □

Note: in Theorem 2, F_1 and F_2 are dependent to P. It is natural to ask whether there are two projection bands in E such that

(i) for any L-projection P, F_1, F_2 are two invariant subspaces of P; moreover, the restriction of P to F_2 is a band projection;

(ii) F_1 is linear isometric and lattice isomorphic to $L_1(\Omega, \ell^2_\infty)$ for some measure (Ω, μ).

The following theorem shows the answer is affirmative.

THEOREM 8. Let E be a real Banach lattice. There exist two projection bands F_1 and F_2 of E such that $(F_1 \oplus F_2)_1 = E$, and for any L-projection P on E

(i) F_1 and F_2 are invariant subspaces of P;
(ii) the restriction of P to F_2 is a band projection;
(iii) there is a measure (Ω, μ) such that F_1 is linear isometric and lattice isomorphic to $L_1(\Omega, \ell^2_\infty)$.

Proof: Let $\{y_\lambda \in E : \lambda \in \Lambda\}$ be a maximal family such that

(1) for each $\lambda \in \Lambda$, there is an L-projection P_λ such that $y_\lambda = 2\,P_{y_\lambda}(P_\lambda(y_\lambda))$;
(2) if $\lambda \neq \lambda'$, then $|P_\lambda(y_\lambda)| \wedge |P_{\lambda'}(y_{\lambda'})| = 0$.

Let $R_\lambda = P_{P_\lambda(y_\lambda)}$. Then

(3) for each λ, R_λ is an L-projection;
(4) if $\lambda \neq \lambda'$, then $R_\lambda \circ R_{\lambda'} = 0$;
(5) for each $\lambda \in \Lambda$, $R_\lambda(E)$ is Banach lattice isometrically isomorphic to $L_1(\mu_\lambda, \ell^2_\infty)$.

As the proof of Theorem 2, the supremum R of the $\{R_\lambda : \lambda \in \Lambda\}$ exists. Let $F_1 = R(E)$ and $F_2 = (I - R)(E)$. Then F_1 is linear isometric and lattice isomorphic to $L_1(\Omega, \mu, \ell^2_\infty)$ for some measure space (Ω, μ).

Let P be any L-projection on E, and let $\{y_\nu \in E : \lambda \in \Lambda'\}$ be a maximal family such that

(1) for each $\nu \in \Lambda'$, $y_\nu = 2\,P_{y_\nu}(P(y_\nu))$;
(2) if $\nu \neq \nu'$, then $|P(y_\nu)| \wedge |P(y_{\nu'})| = 0$.

The supremum R' of the $\{R_\nu = P_{P(y_\nu)} : \nu \in \Lambda'\}$ exists. We claim that $R' \circ R = R'$.

Suppose the claim were proved. We have

(6) $R'(E) \subseteq F_1$ and $F_2 \subseteq (I - R')(E)$;
(7) the restriction of P to $(I - R')(E)$ is a band projection;
(8) $R'(E)$ and $(I - R')(E)$ are invariant subspaces of P.

Note: (7) implies that every projection band of $(I - R')(E)$ is an invariant subspace of P. So F_1 and F_2 are invariant subspaces of P and the restriction of P to F_2 is a band projection. And we proved (i) and (ii).

Proof of the claim. If the claim is not true, then there is $\nu \in \Lambda'$ such that $(I-R) \circ R_\nu \neq 0$ (i.e. either $(I - R)(y_\nu) \neq 0$ or $(I - R)(I - P_{y_\nu})(P(y_\nu)) \neq 0$). Replace y by $2(I - P_{y_\nu})(y_\nu)$ if necessary, we may assume that $(I - R)(y_\nu) = z \neq 0$. Let $Q = P_{P(z)}$. By the proof of Theorem 2, $z = 2\,P_z(P(z))$. Since $\{y_\lambda : \lambda \in \Lambda\}$ is a maximal family satisfies (1) and (2), there is $\lambda \in \Lambda$ such that $R_\lambda \circ Q \neq 0$. So either $P_{y_\lambda}(P(z)) \neq 0$ or $(Q_\lambda - P_{y_\lambda})(P(z)) \neq 0$. Without loss of generality, we assume that $P_{y_\lambda}(P(z)) = w > 0$. Let $x_1 = 2\,(I - P_w)(P(w))$ and $x_2 = 2\,(I - P_w)(P_\lambda(w))$. Then $|x_1| \wedge w = 0 = |x_2| \wedge w$. So we have

$$(P(x_1))^+ \wedge ((P - I)(x_1))^+ \geq \frac{w}{2} > 0$$
$$(P_\lambda(x_2))^+ \wedge ((P_\lambda - I)(x_2))^+ \geq \frac{w}{2} > 0.$$

Let $R_1 = P_{P(w)}$ and $R_2 = P_{P_\lambda(w)}$. Note:

(9) The proof of Theorem 2 showed that $P_z(P(w)) \neq 0$. So $(I - R) \circ P(w) \neq 0$.
(10) Both R_1 and R_2 are L-projection.
(11) $2\,P_w(P(w)) = w$.

(9) implies that $|P(w) - R_2(P(w))| \geq (I - R)(|P(w)|) > 0$. So

$$\frac{\|w\|}{2} = \|P(w)\| = \|P(w) - R_2(P(w))\| + \|R_2(P(w))\|$$
$$\geq \|P(w) - R_2(P(w))\| + \|\frac{w}{2}\| > \frac{\|w\|}{2}.$$

This is impossible. Hence, the claim must be true. The proof is complete. □

REFERENCES

1. E. Behrends, "*M*-Structure an the Banach-Stone Theorem," Springer-Verlag (1979).
2. Lech Drewnowski, Anna Kamińska, and P. K. Lin, On multipliers and *L*- and *M*-projections in Banach lattices and Kõthe function spaces, preprint.
3. H. Elton Lacey, "The Isometric Theory of Classical Banach Spaces," Springer-Verlag (1974).
4. J. Lindenstrauss and L. Tzafriri, "Classical Banach Spaces, II: Function Spaces," Springer-Verlag (1979).
5. H. H. Schaefer, "Banach Lattices and Positive Operators," Springer-Verlag (1974).

A Survey of Mean Convergence of Orthogonal Polynomial Expansions

D.S. LUBINSKY Department of Mathematics, University of the Witwatersrand, P.O. Wits 2050, Rep. of South Africa; e-mail: 036DSL@COSMOS.WITS.AC.ZA
Fax: 27-11-403-1926; 27-11-403-2017

ABSTRACT

We survey in part the mean convergence of orthogonal polynomial expansions associated with weights on $(-1, 1)$ and $\mathbb{R}$. We first present some of the ideas in the context of Fourier series and Legendre expansions. Then we survey some of the results for weights on $(-1, 1)$ and on $\mathbb{R}$.

1 FOURIER SERIES

Orthogonal expansions are L_2 tools, and yet some of their most interesting technical features occur in L_p, $p \neq 2$. Marcel Riesz discovered much of it for Fourier series in the 1920's. Let's look at some of the ideas that initiated our subject. If Γ denotes the unit circle, and $f \in L_1(\Gamma)$, we can form its Fourier series

$$f(z) \simeq \sum_{n=-\infty}^{\infty} c_n z^n$$

where $\forall n$,

$$c_n = \frac{1}{2\pi i} \int_{\Gamma} f(t) t^{-n-1} dt.$$

The non-symmetric partial sum

$$S_{m,n}[f](z) := \sum_{j=-m}^{n} c_j z^j$$

is easily seen to have the compact representation

$$S_{m,n}[f](z) = \frac{1}{2\pi i} \int_{\Gamma} f(t) \left(\frac{t}{z}\right)^m \frac{1-(z/t)^{m+n+1}}{t-z} dt. \tag{1.1}$$

It takes talent or perversity to express this respectable integral as a difference of principal value integrals: Let $B(z;\epsilon) := \{t\colon |t-z| < \epsilon\}$ be the ball centre z, radius ϵ, and define the Hilbert transform on the unit circle

$$H[g](z) := \frac{1}{2\pi i} \lim_{\epsilon \to 0+} \int_{\Gamma \setminus B(z;\epsilon)} \frac{g(t)}{t-z} dt, \tag{1.2}$$

whenever it exists.

This is of course a Cauchy principal value integral, and with hindsight, we know that it exists a.e. if $g \in L_1(\Gamma)$ [48]. So

$$S_{m,n}[f](z) = z^{-m} H\left[f(t)t^m\right](z) - z^{n+1} H\left[f(t)t^{-n-1}\right](z). \tag{1.3}$$

Of course, now comes the famous piece, namely: The Hilbert transform is a bounded operator from $L_p(\Gamma)$ to $L_p(\Gamma)$, $1 < p < \infty$, announced in 1924, but with a proof first appearing in 1927 [45]:

$$\left\|H[g]\right\|_{L_p(\Gamma)} \le C \|g\|_{L_p(\Gamma)}, \tag{1.4}$$

with C independent of g. Applying this in (1.3) (recall $|t| = |z| = 1$) gives

$$\left\|S_{m,n}[f]\right\|_{L_p(\Gamma)} \le 2C \|f\|_{L_p(\Gamma)},\ m,\ n \ge 0. \tag{1.5}$$

Thus the partial sums of the Fourier series are uniformly bounded operators from L_p to L_p. But if

$$R(z) = \sum_{j=-k}^{\ell} c_j z^j$$

is a trigonometric polynomial, we have

$$S_{m,n}[R] = R, \ m \geq k, \ n \geq \ell,$$

and Weierstrass' theorem tells us that such R are dense in $L_p(\Gamma)$. So we obtain Riesz' theorem of 1924/7: If $1 < p < \infty$,

$$\lim_{m,\, n \to \infty} \|f - S_{m,n}[f]\|_{L_p(\Gamma)} = 0 \, \forall \, f \in L_p(\Gamma). \tag{1.6}$$

These (perhaps boringly familiar) details set the stage for all subsequent investigations into mean convergence of orthogonal polynomial expansions.

2 ORTHOGONAL POLYNOMIAL EXPANSIONS: THE BEGINNING

Let $d\alpha$ be a positive measure on the real line with infinitely many points in its support, and with all power moments

$$\int t^j d\alpha(t), \ j = 0, 1, 2, \ldots,$$

finite. Then we can use the Gram–Schmidt process to generate orthonormal polynomials

$$p_n(x) = p_n(d\alpha, x) = \gamma_n x^n + \cdots, \ \gamma_n > 0, \ n = 0, 1, 2, \ldots,$$

satisfying

$$\int p_n p_m d\alpha = \gamma_{mn}. \tag{2.1}$$

The orthonormal polynomials $\{p_n\}_{n=0}^{\infty}$ satisfy a *three term recurrence relation*

$$x p_{n-1}(x) = \alpha_n p_n(x) + \beta_n p_{n-1}(x) + \alpha_{n-1} p_{n-2}(x), \tag{2.2}$$

where

$$\alpha_n := \gamma_{n-1}/\gamma_n \text{ and } \beta_n := \int x p_{n-1}^2(x) d\alpha(x). \tag{2.3}$$

When $d\alpha(x)$ is absolutely continuous, we write $d\alpha(x) = w(x)dx$, and call w a *weight*. Where necessary, we write $p_n(w, x)$ rather than $p_n(d\alpha, x)$, and so on.

For functions $f: \mathbb{R} \to \mathbb{R}$ for which $f(t)t^j$ is integrable with respect to $d\alpha$, $j = 0, 1, 2, \ldots$, we can form the formal orthonormal *orthonormal polynomial expansion*

$$f \simeq \sum_{j=0}^{\infty} c_j p_j, \tag{2.4}$$

where

$$c_j := \int f p_j d\alpha, \; j \geq 0.$$

We denote the partial sums of the orthonormal expansion by

$$S_m[f] := \sum_{j=0}^{m-1} c_j p_j. \tag{2.5}$$

If we are to imitate what was done for Fourier series, we need a compact expression for $S_m[f]$ like (1.1). Using the definition of the c_j, we see that

$$S_m[f](x) = \int K_m(x,t) f(t) d\alpha(t), \tag{2.6}$$

where

$$K_m(x,t) := \sum_{j=0}^{m-1} p_j(x) p_j(t). \tag{2.7}$$

For Fourier series, this reproducing kernel would be a finite geometric series involving terms $(z/t)^j$, which leads to (1.1). Unfortunately, here things are not so simple, but at least there is a *Christoffel-Darboux formula:*

$$K_m(x,t) = \alpha_m \frac{p_m(x)p_{m-1}(t) - p_m(t)p_{m-1}(x)}{x-t}. \tag{2.8}$$

(For those unfamiliar with orthogonal polynomials, all this can be found in [6], [12], or [49]).

The presence of $x - t$ in the denominator suggests using the Hilbert transform on the real line: For $g \in L_1(\mathbb{R})$, define

$$H[g](x) := \lim_{\epsilon \to 0+} \int_{|x-t| \geq \epsilon} \frac{g(t)}{x-t} dt, \tag{2.9}$$

whenever the limit exists. We see that if $d\alpha(x) = w(x)dx$

$$S_m[f](x) = \alpha_m \Big\{ p_m(x) H[fp_{m-1}w](x) - p_{m-1}(x) H[fp_m w](x) \Big\}. \tag{2.10}$$

We see that if we are to copy the proof for Fourier series, we need:

i) Bounds for $\alpha_m = \frac{\gamma_{m-1}}{\gamma_m}$;

ii) Bounds for the orthonormal polynomials $\{p_m\}$.

Actually i) is trivial for $d\alpha$ having compact support, but harder for $d\alpha$ with non–compact support. However, the really hard part is the bounds for $\{p_n\}$, and in order to get suitable bounds on $\{p_n\}$, we shall need to restrict our $d\alpha$.

We also need

iii) Bounds for the Hilbert transform.

Here the analogue of Riesz' result for the circle is

$$\|H[g]\|_{L_p(\mathbb{R})} \leq C\|g\|_{L_p(\mathbb{R})}, \tag{2.11}$$

with C independent of g, provided $1 < p < \infty$.

We close this section with more notation: We denote by $\mathcal{P}_n$ the polynomials of degree $\leq n$. Also, $C,\ C_1,\ C_2, \ldots$ denote constants independent of n. The same symbol does not necessarily denote the same constant in different occurrences. We indicate that C is independent of n and f by writing $C \neq C(n, f)$.

In addition to the usual $L_p[a, b]$ norms, we set

$$\|f\|_{L_p(d\alpha)} := \begin{cases} \left(\int |f|^p d\alpha\right)^{1/p}, & p < \infty \\ \text{ess sup}\,|f|, & p = \infty \end{cases},$$

where the sup is taken over the support of $d\alpha$. We write $f \in L_p(d\alpha)$ if $\|f\|_{L_p(d\alpha)}$ is finite.

3 LEGENDRE SERIES

The most significant problems in handling orthogonal polynomial expansions already arise with the *Legendre weight.* This is

$$d\alpha(x) = dx \quad \text{on} \quad [-1, 1].$$

Of course everything is known about Legendre polynomials, far more than we need for our study. In fact, here [49]

$$\alpha_m = \frac{\gamma_{m-1}}{\gamma_m} = \frac{1}{2} + O(m^{-2}),\ m \to \infty,$$

and the correct bound on p_m is

$$|p_m(x)| \le C|1-x^2|^{-1/4}, \ x \in (-1,1), \ m \ge 0,$$

with $C \ne C(m,x)$. Applying this in (2.10) gives

$$|S_m[f](x)| \le C_1(1-x^2)^{-1/4} \sum_{j=m-1}^{m} |H[fp_j](x)|,$$

and then the boundedness of the Hilbert transform, and our bound on p_j, $j = m-1$, m, gives

$$\|S_m[f](x)(1-x^2)^{1/4}\|_{L_p(\mathbb{R})} \le C_2 \|f(x)(1-x^2)^{-1/4}\|_{L_p(\mathbb{R})}. \tag{3.1}$$

Here of course $C_2 \ne C_2(f,m)$. Once we have this, Weierstrass' theorem for ordinary polynomials gives, if $1 < p < \infty$,

$$\lim_{m\to\infty} \left\| (f(x) - S_m[f](x))(1-x^2)^{1/4} \right\|_{L_p(\mathbb{R})} = 0,$$

$\forall f$ such that $f(x)(1-x^2)^{-1/4} \in L_p[-1,1]$.

This path that we have just taken must have been the one followed by the American mathematician H. Pollard in the 1940's. Perhaps he is better known for his solution of Bernstein's approximation problem [18], but certainly all modern studies on mean convergence of orthogonal polynomial expansions refer back to his 1947, 1948, 1949 papers [40–42] for ideas, so his contributions to orthogonal polynomials were very respectable.

One can speculate that in examining whether the factors $(1-x^2)^{\pm 1/4}$ were necessary in (3.1), Pollard looked at asymptotics for Legendre polynomials. The essence of these [49, p.194] is that

$$p_m(\cos\theta) = (\sin\theta)^{-1/2} \frac{\sqrt{2}}{\sqrt{\pi}} \cos\left(m\theta + \frac{\theta}{2} - \frac{\pi}{4}\right) + o(1),$$

$\theta \in (0,\pi)$. Now Pollard must have recalled his high school trigonometry:

$$\begin{aligned} &p_m(\cos\theta) - p_{m-2}(\cos\theta) \\ &\quad = (\sin\theta)^{-1/2} \frac{\sqrt{2}}{\sqrt{\pi}} \left[\cos\left(m\theta + \frac{\theta}{2} - \frac{\pi}{4}\right) - \cos\left((m-2)\theta + \frac{\theta}{2} - \frac{\pi}{4}\right) \right] + o(1) \\ &\quad = -2\frac{\sqrt{2}}{\sqrt{\pi}} (\sin\theta)^{1/2} \sin\left((m-1)\theta + \frac{\theta}{2} - \frac{\pi}{4}\right). \end{aligned}$$

Pollard used special properties of Legendre polynomials to extend this to all of $(-1,1)$:

$$|p_m(x) - p_{m-2}(x)| \leq C\left[1 - x^2 + m^{-2}\right]^{1/4}, \; x \in (-1,1). \tag{3.2}$$

(If you refer to Pollard's paper, the formulation is a little different because of his different normalization of p_n). So instead of a negative power of $1 - x^2$, we now have virtually a positive power of $1-x^2$. The remarkable thing is that this phenomenon of $p_n - p_{n-2}$ being small near the endpoints of the interval of orthogonality holds for very general weights, not just the Legendre weight.

The factor $p_m - p_{m-2}$ does not appear in (2.10), so one is tempted to say so what? Well, now comes Pollard's main contribution to orthogonal expansions: He worked out how to rewrite the Christoffel–Darboux formula in such a way as to exploit (3.2). *Pollard's decomposition*, as it is commonly known, works for arbitrary orthogonal polynomials. I have not seen a paper proving mean convergence of orthogonal polynomial expansions that has avoided its use, except for a paper of Pollard's, where a closely related trick is used. (If you have a counterexample, please, please send me the reference by e-mail).

Theorem 3.1 Pollard's Decomposition. For an arbitrary $d\alpha$ and $p_n = p_n(d\alpha)$, etc., we can write

$$K_m(x,y) = h_1(x,y) + h_2(x,y) + h_3(x,y), \tag{3.3}$$

where

$$h_1(x,y) := \frac{\alpha_m}{\alpha_m + \alpha_{m-1}} p_{m-1}(x)p_{m-1}(y), \tag{3.4}$$

$$h_2(x,y) := \frac{\alpha_m \alpha_{m-1}}{\alpha_m + \alpha_{m-1}} \frac{p_{m-1}(y)\left[p_m(x) - p_{m-2}(x)\right]}{x - y}, \tag{3.5}$$

and

$$h_3(x,y) := h_2(y,x). \tag{3.6}$$

Proof.

We use the elegant proof of Mhaskar and Xu [31]: Now

$$\begin{aligned} 2K_m(x,y) &= K_m(x,y) + K_{m-1}(x,y) + p_{m-1}(x)p_{m-1}(y) \\ &= \alpha_{m-1}\left[\frac{K_m(x,y)}{\alpha_m} + \frac{K_{m-1}(x,y)}{\alpha_{m-1}}\right] + \frac{\alpha_m - \alpha_{m-1}}{\alpha_m} K_m(x,y) + p_{m-1}(x)p_{m-1}(y). \end{aligned}$$

Transferring the middle term on the right-hand side to the left and simplifying, gives

$$K_m(x,y) = \frac{\alpha_m\alpha_{m-1}}{\alpha_m+\alpha_{m-1}}\left[\frac{K_m(x,y)}{\alpha_m}+\frac{K_{m-1}(x,y)}{\alpha_{m-1}}\right]+\frac{\alpha_m}{\alpha_m+\alpha_{m-1}}p_{m-1}(x)p_{m-1}(y). \tag{3.7}$$

Finally, the Christoffel-Darboux formula (2.8) gives

$$\begin{aligned}\frac{K_m(x,y)}{\alpha_m}+\frac{K_{m-1}(x,y)}{\alpha_{m-1}} &= \frac{p_m(x)p_{m-1}(y)-p_{m-1}(x)p_m(y)+p_{m-1}(x)p_{m-2}(y)-p_{m-2}(x)p_{m-1}(y)}{x-y}\\ &= \frac{p_{m-1}(y)(p_m-p_{m-2})(x)-p_{m-1}(x)(p_m-p_{m-2})(y)}{x-y}.\end{aligned}$$

Substituting into (3.7) gives the result. ∎

We can now rewrite the partial sums of our orthogonal expansion as:

$$\begin{aligned}S_m[f](x) &= \frac{\alpha_m}{\alpha_m+\alpha_{m-1}}p_{m-1}(x)\int fp_{m-1}d\alpha\\ &\quad+\frac{\alpha_m\alpha_{m-1}}{\alpha_m+\alpha_{m-1}}(p_m-p_{m-2})(x)H[fp_{m-1}](x)\\ &\quad-\frac{\alpha_m\alpha_{m-1}}{\alpha_m+\alpha_{m-1}}p_{m-1}(x)H\left[f(p_m-p_{m-2})\right](x)\\ &=: S_{m,1}[f](x)+S_{m,2}[f](x)+S_{m,3}[f](x).\end{aligned} \tag{3.8}$$

Let us return to the Legendre weight. Clearly $S_{m,1}$ should be easy to handle, and in handling $S_{m,2}$ and $S_{m,3}$, we can hope to use the fact that $p_m - p_{m-2}$ is small near ± 1. Pollard defined

$$J(x,y) := \left|\frac{\left[(1-y^2)/(1-x^2)\right]^{\pm 1/4}-1}{x-y}\right|$$

and showed that the operator

$$f \to \int_{-1}^{1} f(y)J(x,y)dy$$

is bounded from $L_p[-1,1]$ to $L_p[-1,1]$ if $\frac{4}{3} < p < 4$ (perhaps one of the first use of these multiplication operators?) and in effect uses this to show that

$$\left\|(1-x^2)^{\pm 1/4} H[f](x)\right\|_{L_p[-1,1]} \leq C \left\|f(x)(1-x^2)^{\pm 1/4}\right\|_{L_p[-1,1]}, \tag{3.9}$$

with C independent of f, provided $\frac{4}{3} < p < 4$. Applying this in (3.8), and applying also our bounds for p_{m-1}, $p_m - p_{m-2}$ gives

Theorem 3.2 (Pollard, 1947) [40]. If $d\alpha(x) = dx$ on $[-1,1]$ is the Legendre weight on $[-1,1]$, and $\frac{4}{3} < p < 4$, then

$$\left\|S_m[f]\right\|_{L_p[-1,1]} \leq C\|f\|_{L_p[-1,1]},$$

with C independent of f and m.

We shall see that the restrictions on p are necessary in the next section.

4 NECESSARY CONDITIONS

Pollard proved in his 1947 paper that Theorem 3.2 failed for $1 < p < \frac{4}{3}$ and $p > 4$. He could not handle the case $p = \frac{4}{3}$ or $p = 4$. It was only in a 1952 paper that Newman and Rudin [39] showed that Theorem 3.2 failed for $p = \frac{4}{3}$ and $p = 4$ also. The author thanks Walter Rudin for this and the following remarks, given at the 2nd conference on function spaces. The 1952 paper was one of Rudin's first efforts, and he took a bus to New York to visit the Newman in question (J., not D.J. or M.), with whom he'd corresponded on this paper. Unfortunately, Newman, who had been suffering from a blinding eye disease, had already died in tragic circumstances. So he never got to meet his co-author.

The derivation of the necessary conditions is extremely general and extremely simple, so let us present the ideas in some generality. Suppose that for some absolutely continuous measure $d\alpha(x) = w(x)dx$ (recall that we call w a weight), and some $1 \leq p \leq \infty$, and some measurable non-negative functions u and v, we have

$$\left\|S_m[f]u\right\|_{L_p(\mathbb{R})} \leq C\|fv\|_{L_p(\mathbb{R})}, \tag{4.1}$$

with C independent of m and f. Of course the uniform boundedness principle, and the typical density of polynomials, show that this is typically equivalent to mean convergence of S_m. Recall our notation (2.5) for S_m. Then

$$c_m p_m = S_{m+1}[f] - S_m[f],$$

so (4.1) gives

$$|c_m| \|p_m u\|_{L_p(\mathbb{R})} = \left\| (S_{m+1}[f] - S_m[f])u \right\|_{L_p(\mathbb{R})} \leq 2C \|fv\|_{L_p(\mathbb{R})}.$$

Recalling our definition of c_m shows that $\forall f$, $\forall m$,

$$\frac{|\int f p_m w|}{\|fv\|_{L_p(\mathbb{R})}} \|p_m u\|_{L_p(\mathbb{R})} \leq 2C.$$

Taking sup's over f, duality gives:

Theorem 4.1 From (4.1) follows

$$\sup_m \|p_m w v^{-1}\|_{L_q(\mathbb{R})} \|p_m u\|_{L_p(\mathbb{R})} < \infty, \tag{4.2}$$

where $q := p/(p-1)$ is the conjugate parameter of p.

The remarkable thing about this simple necessary condition is that for a given weight w, one can use it to grind out necessary conditions for (4.1) to hold for given u and v, when one knows enough about p_n. The necessary conditions so obtained are usually virtually the sufficient ones, but to prove that they are sufficient involves considerable extra effort.

For the Legendre weight, for which $w \equiv 1$ on $[-1,1]$, we can take $u = v = 1$ in $[-1,1]$, and deduce that we need

$$\sup_m \|p_m\|_{L_q[-1,1]} \|p_m\|_{L_q[-1,1]} < \infty. \tag{4.3}$$

Now we know today that

$$\|p_m\|_{L_p[-1,1]} \sim \begin{cases} 1, & p < 4 \\ (\log m)^{1/4}, & p = 4 \\ m^{\frac{1}{2} - \frac{2}{p}}, & p > 4 \end{cases}.$$

Here and in the sequel, $\sim$ means that the ratio of the two sides is bounded above and below by positive constants independent of m. Using these in (4.3), it is easy to see that $\frac{4}{3} < p < 4$ is necessary for mean boundedness/convergence of Legendre expansions.

It is worth emphasizing that although the Legendre weight is very special, the range $\frac{4}{3} < p < 4$ occurs whenever we have a smooth positive weight on $(-1, 1)$. (Only singularities in the weight itself seem to override this range.) This is so, because we always get a factor something like $(1-x^2)^{-1/4}$ in the asymptotics for p_n, and then (4.2) gives the rest. The deeper reason for $(1-x^2)^{-1/4}$ appearing is that orthogonal polynomials on $[-1, 1]$ really "live on the unit circle", and the substitution $x = \cos\theta$ in any integral over $\theta \in [0, \pi]$ give a factor of $(1-x^2)^{-1/2}$ in the corresponding integral on $[-1, 1]$. For more orientation on this, see [12], [49], and the entertaining survey [37].

5 JACOBI, GENERALIZED JACOBI AND MORE GENERALIZED WEIGHTS ON $(-1, 1)$

The number of people that have worked directly or indirectly on mean convergence of orthogonal expansions for weights on $(-1, 1)$ is substantial. Amongst those that we are aware of, are H. Pollard (1947, 1948, 1949), J. Newman and W. Rudin (1952), G.M. Wing (1960), B. Muckenhoupt (1970), R. Askey (1972), V.M. Badkov (1973, 1974), P. Nevai (1979), A. Máté, P. Nevai and V. Totik (1986), Y. Xu (1993), (1994), Y.G. Shi (1993), H. König and N.J. Nielsen (1994). We insert the dates to give some idea of the development of the subject; they do not refer to our references.

Of course this list excludes those that have worked on transplantation, multiplier theorems and convolution structure, and on the more distantly related questions of summability, convergence a.e., and rates of convergence for smooth functions. I apologize in advance to those whose work has been excluded (please inform me by e-mail), and caution readers less familiar with the subject that I am an expert on weights on $(-\infty, \infty)$ rather than on weights on $(-1, 1)$. So the details here may well be incomplete.

Let us start by revisiting Pollard's work from 1948. Pollard recalled Riesz' result for Fourier series, and noted that if we consider the real version of these, more specifically, Fourier cosine series, then Riesz' result tells us everything about orthogonal expansions for the Chebyshev weight $1/\sqrt{1-x^2}$. To see this, recall that the orthonormal Chebyshev polynomials $p_n(x)$, satisfying

$$\int_{-1}^{1} p_n(x)p_m(x)\frac{dx}{\sqrt{1-x^2}} = \delta_{mn},$$

are given by

$$p_n(\cos\theta) = \left(\frac{2}{\pi}\right)^{1/2} \cos n\theta,\ n \geq 1.$$

So the substitution $x = \cos\theta$ transfers results on Fourier series to series of Chebyshev polynomials. In particular, we deduce that for all $1 < p < \infty$,

$$\int_{-1}^{1} |S_n[f]|(x)^p \frac{dx}{\sqrt{1-x^2}} \leq C \int_{-1}^{1} |f|(x)^p \frac{dx}{\sqrt{1-x^2}}, \tag{5.1}$$

with C independent of n and f, so

$$\lim_{n\to\infty} \int_{-1}^{1} |f(x) - S_n[f](x)|^p \frac{dx}{\sqrt{1-x^2}} = 0,$$

$\forall f$ for which the right–hand side of (5.1) is finite.

Of course the Legendre and Chebyshev weights are respectively the cases $\lambda = \frac{1}{2}$ and $\lambda = 0$ of the ultraspherical weight

$$U_\lambda(x) := \left(1 - x^2\right)^{\lambda - \frac{1}{2}}, \tag{5.2}$$

so it seemed natural to "interpolate": For which p is it true that when we take $d\alpha(x) = U_\lambda(x)dx$,

$$\int_{-1}^{1} |S_n[f]|(x)^p U_\lambda(x)dx \leq C \int_{-1}^{1} |f(x)|^p U_\lambda(x)dx,$$

with C independent of n and f? Pollard showed that the range of p is

$$2 - \frac{1}{\lambda+1} < p < 2 + \frac{1}{\lambda}, \tag{5.3}$$

for all $\lambda \geq 0$. (As with Legendre weight, he could not show the uniform boundedness holds at the boundary of the range. This can be resolved by the method of Newman and Rudin.)

In this paper, Pollard also considered more general weights, specified in terms of bounds on p_n, and also bounds on the polynomials q_n that are orthonormal with respect to $w(x)(1-x^2)$. One of the especially interesting features of this was that Pollard avoided hypotheses on $p_n - p_{n-2}$, replacing them with hypotheses on q_n, which are often easier to verify. He also considered Jacobi weights.

Perhaps the next important idea in the development of the subject was to try to increase the range of p, by replacing $d\alpha(x)$ in the integral measuring mean convergence by some other factor. Apparently a special case was due to G.M. Wing (1950), but mostly it was B. Muckenhoupt (1970). Let $1 < p < \infty$,

$$d\alpha(x) := (1-x)^{\alpha}(1+x)^{\beta}dx \quad \text{on} \quad [-1,1],$$

where $\alpha,\ \beta > -1$, and

$$u(x) := (1-x)^{a}(1+x)^{b}.$$

Muckenhoupt's result was that

$$\int_{-1}^{1} |S_n[f](x)u(x)|^p dx \leq C \int_{-1}^{1} |f(x)u(x)|^p dx, \tag{5.4}$$

with C independent of n and f, iff

$$\left| a - \frac{1}{2} - \frac{1}{2}\alpha + \frac{1}{p} \right| < \min\left\{ \frac{1}{4},\ \frac{1}{2} + \frac{1}{2}\alpha \right\} \tag{5.5}$$

and

$$\left| b - \frac{1}{2} - \frac{1}{2}\beta + \frac{1}{p} \right| < \min\left\{ \frac{1}{4},\ \frac{1}{2} + \frac{1}{2}\beta \right\}. \tag{5.6}$$

In particular, for a given $p,\ \alpha,\ \beta$ it is always possible to choose $a,\ b$ such that these are true.

One is tempted to guess that the next step would be to replace u on the right-hand side by a different Jacobi factor, and see what is possible. However, the next major leap forward is due to the Soviet mathematician Badkov, who in the early 1970's considered *generalized Jacobi weights*, that is

$$d\alpha(x) = w(x)dx \quad \text{on} \quad [-1,1],$$

where

$$w(x) := h(x)(1-x)^{\alpha} \left[\prod_{j=1}^{N} |x - t_j|^{A_j} \right] (1+x)^{\beta}, \tag{5.7}$$

where

$$-1 < t_1 < t_2 < \cdots < t_N < 1; \ \alpha,\ \beta,\ A_1,\ A_2, \ldots A_N > -1 \tag{5.8}$$

and h is a positive continuous function whose modulus of continuity

$$\omega(h;\delta) := \sup\Big\{\big|h(s) - h(t)\big| : |s-t| \le \delta\Big\}$$

satisfies

$$\int_0^1 \frac{\omega(h;t)}{t}dt < \infty. \tag{5.9}$$

Actually multiplication of a Jacobi weight by positive functions like h had already been considered by other authors starting with Pollard, under various conditions on h. Badkov's real contribution was the introduction of factors that could vanish, or become infinite, inside the interval of orthogonality. Badkov obtained the correct bounds on p_n, and all the other necessary estimates (a technical and non-trivial task), and proved a powerful extension of Muckenhoupt's Jacobi weight theorem: Let

$$u(x) := (1-x)^a \left[\prod_{j=1}^N |x - t_j|^{B_j}\right] (1+x)^b, \tag{5.10}$$

for some real a, b, B_j. Then for $d\alpha(x) = w(x)dx$, with w as above, we have (5.4), with C independent of n and f, iff (5.5) and (5.6) hold, and also

$$\left|B_j - \frac{1}{2} - \frac{1}{2}A_j + \frac{1}{p}\right| < \min\left\{\frac{1}{2}, \frac{1}{2} + \frac{1}{2}A_j\right\}, \ 1 \le j \le N. \tag{5.11}$$

The condition is slightly different for the exponents of factors with zeros inside $(-1,1)$ (namely $\frac{1}{2}$ rather than $\frac{1}{4}$ in the minimum on the right-hand side of (5.11)) because of the characteristic extra factor of $(1-x^2)^{-1/4}$ in the bounds for orthogonal polynomials near ± 1 (recall the Legendre weight).

Strangely enough, it is only recently (1993) that analogues of (5.4) with a different weighting factor have been investigated, by Yuan Xu [52] (an ex student of Donald Newman). Let us set

$$v(x) := (1-x)^c \left[\prod_{j=1}^N |x - t_j|^{D_j}\right] (1+x)^d, \tag{5.12}$$

where all c, d, D_j are real. Let $1 < p < \infty$, and $q := p/(p-1)$ be the conjugate parameter of p. Xu states his conditions more implicitly then did Badkov. Namely with $d\alpha(x) = w(x)dx$ as above, and u and v above, we have

$$\big\|S_n[f]u\big\|_{L_p(d\alpha)} \le C\big\|fv\big\|_{L_p(d\alpha)}, \tag{5.13}$$

with $C \neq C(n,f)$ iff

$$(5.14) \qquad u^p w \in L_1[-1,1]; \; v^{-q} w \in L_1[-1,1];$$

$$(5.15) \qquad u^p\left(w\sqrt{1-x^2}\right)^{-p/2} w \in L_1[-1,1]; \; v^{-q}\left(w\sqrt{1-x^2}\right)^{-q/2} w \in L_1[-1,1];$$

$$(5.16) \qquad u(x) \leq Cv(x), \; x \in (-1,1),$$

some $C > 0$. The energetic reader will check that these conditions reduce to Badkov's explicit ones in the case $v = u$!

One of Xu's motivations in considering $u \neq v$, was the subject of Lagrange interpolation at the zeros of orthogonal polynomials. Recall that if

$$-\infty < x_{nn} < x_{n-1,n} < x_{n-2,n} < \cdots < x_{1n} < \infty,$$

are the zeros of $p_n(d\alpha, x)$, then the *Lagrange interpolation polynomial* $L_n[f]$ to f at $\{x_{jn}\}$ is a polynomial of degree at most $n-1$, satisfying

$$L_n[f](x_{jn}) = f(x_{jn}), \; 1 \leq j \leq n.$$

There is an intimate connection between the convergence in L_p of orthonormal expansions $S_n[f]$, and their discrete cousins $L_n[f]$. We cannot hope to review this here, but see [24], [35], [37], [46], [52], [53].

In fact much of the work on mean convergence of orthonormal expansions since Badkov's has been inspired by problems of Turan, Askey, Nevai, and others, on mean convergence of Lagrange interpolation. In his landmark 1979 memoir [35], P. Nevai solved one of Turan's problems on Lagrange interpolation, but he also obtained powerful necessary conditions for mean convergence of quite general orthogonal expansions [35, p.154]. He seems the first to have considered inequalities involving an L_p norm for $S_n[f]$, and an L_r, $r \neq p$, norm of f. His results were subsequently generalized together with his collaborators A. Máté, and V. Totik. This trio of Hungarian born maestros (both Máté and Nevai now live in the U.S. and Totik spends half his time in Tampa with E.B. Saff), have collaborated on more than fifteen influential papers on orthogonal polynomials.

Whereas Nevai's memoir gave necessary conditions for mean convergence of $S_n[f]$ when $d\alpha$ is in *Szegö's class*, that is

$$(5.17) \qquad \int_{-1}^{1} \log \alpha'(x) \frac{dx}{\sqrt{1-x^2}} > -\infty,$$

the 1986 result of Máté, Nevai and Totik [26] dealt with $\alpha'(x)$ merely positive a.e. in $(-1,1)$ and is still state of the art: (For further orientation on the relevance of the behaviour of α' to the behaviour of $p_n(d\alpha,\cdot)$, see [37], [38], [47], ...)

Let $d\alpha$ have support in $[-1,1]$ and α' be positive a.e. in $[-1,1]$, and let $0 < p \leq \infty$, $1 \leq r \leq \infty$. Let u and v be non-negative and positive on a set of positive measure, with v finite on a set of positive measure. Let $r' := r/(r-1)$ and

$$\sigma(x) := \left[\alpha'(x)\sqrt{1-x^2}\right]^{1/2}. \tag{5.18}$$

From

$$\|S_n[f]u\|_{L_p(d\alpha)} \leq C\|fv\|_{L_r(d\alpha)}, \tag{5.19}$$

with $C \neq C(n,f)$, (where we take $(fv)(x) = 0$ if $f(x) = 0$ and $v(x) = \infty$), follows that

$$u \in L_p(d\alpha);\ v^{-1} \in L_{r'}(d\alpha) \tag{5.20}$$

$$\left(\int_{-1}^{1} \left(u(x)/\sigma(x)\right)^p \alpha'(x)dx\right)^{1/p} < \infty; \tag{5.21}$$

$$\left(\int_{-1}^{1} \left(u(x)\sigma(x)\right)^{-r'} \alpha'(x)dx\right)^{1/r'} < \infty. \tag{5.22}$$

Here we make obvious modifications when p or $r' = \infty$.

In the special case where $d\alpha$ is a *Pollaczek-type weight*, for example

$$d\alpha(x) = \exp\left(-(1-x^2)^{-\tau}\right)dx \ \text{ on } \ [-1,1],$$

where $0 < \tau < \frac{1}{2}$, P. Nevai had already noted in his memoir that mean convergence can occur in L_p only for $p = 2$.

Some other very interesting work where Lagrange interpolation and orthogonal expansions interact is due to König and Nielsen [16,17]. They study orthogonal expansions in Banach spaces which admit boundedness of the Hilbert transform (so called UMD spaces).

The most recent contribution to necessary conditions comes from a very talented young Chinese mathematician Ying Guang Shi [46]. While his result is less explicit than that of Máté, Nevai and Totik, and does not immediately imply theirs, it applies to very general weights. Effectively, Shi has shown that we can separate the product of terms in (4.2), and show that they must separately be bounded: (For more restricted weights, Nevai had this in his memoir.)

Let $0 < p < \infty$, $1 < r < \infty$, $r' := r/(r-1)$, u, v be non–negative functions with

$$\int_{-1}^{1} u(x)\alpha'(x)dx;\ \int_{-1}^{1} v(x)\alpha'(x)dx > 0; \tag{5.23}$$

and

$$u,\ v,\ v^{1/(1-q)} \in L_1(d\alpha). \tag{5.24}$$

Then from

$$\left[\int_{-1}^{1} |S_n[f](x)|^p u(x)d\alpha(x)\right]^{1/p} \leq C\left[\int_{-1}^{1} |f(x)|^r v(x)d\alpha(x)\right]^{1/r}, \tag{5.25}$$

with $C \neq C(n,f)$, follows

$$\sup_{n\geq 1}\left[\int_{-1}^{1} |p_n(x)|^p u(x)d\alpha(x)\right]^{1/p} < \infty; \tag{5.26}$$

$$\sup_{n\geq 1}\left[\int_{-1}^{1} |p_n(x)|^{r'} v(x)^{1-r'}d\alpha(x)\right]^{1/r'} < \infty. \tag{5.27}$$

Moreover, if $p > 2$,

$$\int_{-1}^{1} \left[\alpha'(x)(1-x^2)^{1/2}\right]^{-p/2} u(x)\alpha'(x)dx < \infty; \tag{5.28}$$

and if $1 < r \leq 2$, then

$$\int_{-1}^{1} \left[\alpha'(x)(1-x^2)^{1/2}\right]^{r/2(1-r)} v(x)^{1/(1-r)}\alpha'(x)dx < \infty. \tag{5.29}$$

The recent emphasis on necessary results for $d\alpha$ supported on $[-1,1]$, and the lack of matching sufficiency results, is due to the lack of suitable bounds for new classes of orthogonal polynomials. A forthcoming memoir of the author and A.L. Levin [21] will give all the required bounds and estimates for orthonormal polynomials for weights $d\alpha(x) = w(x)dx$ on $[-1, 1]$, where w vanishes rapidly at ± 1.

Archetypal examples are

$$w(x) := \exp\bigl(-(1 - x^2)^{-\tau}\bigr), \ \tau > 0,$$

or

$$w(x) := \exp\bigl(-\exp_k\left((1 - x^2)^{-\tau}\right)\bigr), \ \tau > 0, \ k \geq 1,$$

where

$$\exp_k := \exp\bigl(\exp(\exp\ldots)\bigr)$$

is the kth iterated exponential. So for large classes of weights violating Szegö's condition (5.17), it will be possible to establish results on mean convergence of $S_n[f]$.

However, we note that the type of possible results will have a very different flavour to the classical Jacobi/generalized Jacobi weights. They will have the form

$$\int_{-1}^{1} |S_n[f](x) w^{1/2}(x)|^p dx \leq C \int_{-1}^{1} |f(x) w^{1/2}(x)|^p dx,$$

$C \neq C(n, f)$ at least for $\frac{4}{3} < p < 4$, and the insertion of other factors will be necessary for other values of p. The inclusion of a power of the weight in the pth power is essential. The basic reason is that because the weight decays rapidly near ± 1, the orthogonal polynomials can become very large near ± 1. The reader will see more of this in our discussion of weights on $\mathbb{R}$ below.

6 THE HERMITE WEIGHT

The theory of orthogonal polynomials for weights with non–compact support is far less developed than the theory for weights with compact support. It is true that there is a beautiful general theory involving the moment problem going back to Stieltjes (1890's), Hamburger, Nevanlinna and others (1920's); L_2 density of polynomials à là Riesz; and Bernstein's L_∞ weighted approximation problem solved by Pollard, Mergelyyan and Akhiezer in the 1950's, but the sort of quantitative estimates needed for mean convergence of orthogonal expansions were available only for a few classical weights on $\mathbb{R}$ until very recently.

Perhaps the main reason for this is that for weights on $(-1,1)$, apart from the explicit formulae for the p_n of Jacobi weights, there was also the Bernstein–Szegö formula for the orthogonal polynomials $p_n(w,x)$ corresponding to the weight

$$w(x) := \sqrt{1-x^2}/\sqrt{P(x)},\ x \in (-1,1),$$

where P is a polynomial of degree at most $2n$, positive in $(-1,1)$. Via approximation, this and other formulae permitted the necessary analysis for quite general weights on $(-1,1)$. There is still no analogue of this formula for weights on $(-\infty,\infty)$.

The Hermite weight

$$w(x) := \exp(-x^2),\ x \in (-\infty,\infty),$$

and its Laguerre cousins

$$w(x) := x^\alpha e^{-x},\ x \in (0,\infty),\ \alpha > -1,$$

were about the only weights with non–compact support which permitted analysis.

We shall concentrate on even weights on the whole real line, as there is a simple relationship between such weights on $\mathbb{R}$, and their one–sided cousins: If w is an even weight on $\mathbb{R}$, and we set

$$w_1(x) := \frac{1}{\sqrt{x}} w(\sqrt{x});\ w_2(x) = \sqrt{x}\, w(\sqrt{x}),\ x \in (0,\infty),$$

then the substitution $x = \sqrt{t}$ in the orthogonality relationships for $p_n(w,x)$ shows that

$$p_n(w_1;t) = p_{2n}(w;\sqrt{t});\ p_n(w_2;t) = \frac{1}{\sqrt{t}} p_{2n+1}(w;\sqrt{t}).$$

In particular most properties of the Laguerre polynomials can be derived from corresponding properties of the orthonormal polynomials for the generalized Hermite weight $|x|^a \exp(-x^2)$ on $(-\infty,\infty)$.

At the same time as Pollard considered mean convergence for Jacobi weights, he also looked at the Hermite weight. But as he noted in his 1948 paper, he could only come up with a somewhat negative result: If $1 < p < \infty$, and for the Hermite weight $w(x) := \exp(-x^2)$, we have

$$\int_{\mathbb{R}} |S_n[f]|^p w \le C \int_{\mathbb{R}} |f|^p w,$$

with $C \neq C(n, f)$, then $p = 2$. With hindsight, we can say that this is the wrong formulation.

It had to wait until 1965 for Richard Askey and Steven Wainger [1] to give the correct formulation. They knew that as $n \to \infty$, we have for $x = \sqrt{2n}\cos\theta$

$$p_n(w, x)w^{1/2}(x)\left(1 - \left(x/\sqrt{2n}\right)^2\right)^{1/4} = \cos\left(nh(\theta) + g(\theta)\right) + o(1),$$

where $h(\theta)$ and $g(\theta)$ are explicitly given functions, and the asymptotic is uniform for $\theta \in [\epsilon, \pi - \epsilon]$, any fixed $\epsilon > 0$. This suggested that where w is small, $|p_n|$ can be large, and so we need to weight terms involving p_n with a factor of $w^{1/2}$.

They proved that

$$\left\|S_n[f]w^{1/2}\right\|_{L_p(\mathbb{R})} \leq C\left\|fw^{1/2}\right\|_{L_p(\mathbb{R})},$$

with $C \neq C(n, f)$ if $\frac{4}{3} < p < 4$, and this result failed if $1 \leq p \leq \frac{4}{3}$ or $p \geq 4$. As the reader might have guessed, this is the same $\frac{4}{3}$ and 4 as for the Legendre weight, and for the same reason: A factor of $(\ldots - x)^{-1/4}$ occurs in the asymptotic for p_n.

Just as B. Muckenhoupt extended the range of p for mean convergence for Jacobi weights by inserting factors on both sides of the integral, so in 1970 he did the same for Laguerre and Hermite weights. He looked for estimates of the type

$$\left\|S_n[f]w^{1/2}u_b\right\|_{L_p(\mathbb{R})} \leq C\left\|fw^{1/2}u_B\right\|_{L_p(\mathbb{R})}, \tag{6.1}$$

where

$$u_a(x) := \left(1 + |x|\right)^a, \; x \in \mathbb{R}, \; a \in \mathbb{R}, \tag{6.2}$$

and $C \neq C(n, f)$. In particular, he was able to prove that for a given $1 < p < \infty$, we can always find b and B for which (6.1) is valid. So by inserting mere powers of $1 + |x|$, which are negligible compared to any power of w, we could ensure mean convergence for all p.

As powerful as Muckenhoupt's results were, the real significance of his paper lay outside orthogonal expansions. In the course of his investigations, after applying Pollard's decomposition, he realized that he needed weighted inequalities for the Hilbert transform, such as [33, p.440]

$$\left\|H[f]u_b\right\|_{L_p(\mathbb{R})} \leq C\left\|fu_B\right\|_{L_p(\mathbb{R})}, \tag{6.3}$$

and he proved amongst other things that (6.3) is true if

$$B > -\frac{1}{p}; \; b < 1 - \frac{1}{p}; \; b \leq B. \tag{6.4}$$

(These conditions later turned out to be necessary and sufficient).

We already encountered the need for weighted inequalities such as (3.9) in Pollard's work, but Muckenhoupt was the first to take these to the limit. He began to investigate weighted inequalities of the form (6.3) for general factors replacing u_b, u_B culminating in his famous A_p condition, which appeared in the 1973 paper of Hunt, Muckenhoupt and Wheeden [13]: Let $u\colon \mathbb{R} \to [0, \infty)$ be measurable and $1 < p < \infty$. Then

$$\int_{\mathbb{R}} |H[f](x)|^p u(x)dx \leq C \int_{\mathbb{R}} |f(x)|^p u(x)dx, \tag{6.5}$$

with $C \neq C(f)$ iff $\exists\, C_1 > 0$ such that $\forall$ interval $[a, b]$

$$\left[\frac{1}{b-a}\int_a^b u\right]\left[\frac{1}{b-a}\int_a^b u^{-1/(p-1)}\right]^{p-1} \leq C_1. \tag{6.6}$$

A more elegant way to state this is to write $u := U^p$, and involves the conjugate parameter $q := p/(p-1)$ of p. Then (6.5) becomes

$$\|H[f]U\|_{L_p(\mathbb{R})} \leq C_2 \|fU\|_{L_p(\mathbb{R})}, \tag{6.7}$$

iff $\exists\, C_3$ such that $\forall$ interval $[a, b]$,

$$\frac{1}{b-a}\|U\|_{L_p[a,b]}\|U^{-1}\|_{L_q[a,b]} \leq C_3. \tag{6.8}$$

(The author thanks Michael Cwikel of the Technion–IIT for telling him of this nicer formulation by Yoram Sagher).

Of course weighted inequalities for maximal functions, Fourier transforms, and other convolution operators have all been explored along the same lines, not only for the same weight on each side, but also for a pair of weights (U, V) (so V replaces U in the right–hand side of (6.7)). Ironically, while the necessary and sufficient conditions for pairs of weights have been solved for most of these other operators [48], there is still a gap between the explicit necessary and sufficient conditions on (U, V) for

$$\|H[f]U\|_{L_p(\mathbb{R})} \leq C\|fV\|_{L_p(\mathbb{R})},$$

$C \neq C(f)$. For $p = 2$, Cotlar and Sadosky obtained an implicit necessary and sufficient condition on (U, V) [7], [8] (the author thanks R. Rochberg for these references) but as far as this author is aware, there are no identical necessary and sufficient conditions for $p \neq 2$, least of all explicit ones.

Perhaps the development of Muckenhoupt's A_p condition helps to emphasize some of the true value of investigating mean convergence of orthogonal expansions: On its own it is a difficult, somewhat technical, subject, but it has stimulated investigations into fundamental inequalities on operators, special functions, and orthogonal polynomials, that may be very widely applied.

There was another type of Hilbert inequality in Muckenhoupt's 1970 papers that has apparently not been extended: For simplicity, let us consider $b = B = \frac{1}{4}$ in (6.3), so that (6.4) and hence (6.3) fails for $p \leq \frac{4}{3}$. Muckenhoupt showed that nevertheless, we can get a result by restricting ourselves to finite intervals, and indicate the dependence on the size of the interval. For $T > 0$, let

$$\Lambda(T) := \begin{cases} 1, & p > 4/3 \\ \log T, & p = 4/3 \\ T^{1/p-3/4}, & 1 < p < 4/3 \end{cases}.$$

Then for $f \colon [-T, T] \to \mathbb{R}$,

$$\|H[f]u_{1/4}\|_{L_p[-T,T]} \leq C\Lambda(T)\|fu_{1/4}\|_{L_p[-T,T]},$$

with $C \neq C(f, T)$. Perhaps the extension of such results would be extremely difficult to formulate, but it would be nice to have results for smooth functions u, of the form

$$\|H[f]u\|_{L_p[a,b]} \leq \Lambda(u, a, b)\|f\|_{L_p(\mathbb{R})},$$

giving explicit dependence on u, a, and b. The theory around Koosis weights [8] does not give this.

Well, with all the buildup, what about Muckenhoupt's result for Hermite weights? Let $1 < p < \infty$, b, $B \in \mathbb{R}$. Then (6.1) holds with $C \neq C(n, f)$ provided

$$b \begin{cases} < 1 - 1/p, & p \leq 4 \\ \leq 2/3 + 1/(3p), & p > 4 \end{cases}; \tag{6.9}$$

$$B \begin{cases} \geq -1 + 1/(3p), & p < 4/3 \\ > -1/p, & p \geq 4/3 \end{cases}; \tag{6.10}$$

$$b \begin{cases} \leq B + 1 - 4/(3p), & p < 4/3 \\ \leq B, & 4/3 \leq p \leq 4 \\ \leq B - 1/3 + 4/(3p), & p > 4 \end{cases}, \tag{6.11}$$

and if equality holds in (6.11), then equality does not occur in (6.9) or (6.10). Moreover, these inequalities are sharp except possibly in cases of equality, where extra log factors need to be inserted. Some applications of these and related results to expansions in several problems appear in the monograph of Thangavelu [50].

7 FREUD WEIGHTS

The 1970's saw not only the development of Muckenhoupt's A_p theory, but also the beginning of a quantitative theory for weights on the whole real line. The initiator of this was the Hungarian mathematician G. Freud, together with his former student P. Nevai. One of Freud's biggest contributions was the introduction of *infinite–finite range inequalities*, which show that in studying $P(x)\exp(-|x|^\alpha)$, for $P \in \mathcal{P}_n$, it suffices to work on $[-Cn^{1/\alpha}, Cn^{1/\alpha}]$. Here $C \neq C(n,P)$. Thus we can work on a finite interval whose size depends on the degree of the polynomial, but not on the particular polynomial.

In honour of his work, weights of the form $w := W^2 := e^{-2Q}$, where Q is even and of smooth polynomial growth at infinity, are often called *Freud weights* [37]. The archetypal example is that just mentioned,

$$(7.1) \qquad w_\alpha(x) := W_\alpha(x)^2, \text{ where } W_\alpha(x) := \exp(-Q_\alpha(x)) = \exp\left(-\frac{1}{2}|x|^\alpha\right), \ \alpha > 1.$$

Expressing our weight as W^2 simplifies some formulations, for example, recall that we needed $w^{1/2}$ in the Askey–Wainger and Muckenhoupt results above. When $w := W^2$, where Q is even and of faster than polynomial growth at infinity, w is called an *Erdős weight*, for it was the maestro that first studied these weights, determining the distribution of zeros of p_n.

Some of the greatest advances in the theory of orthogonal polynomials for weights on $\mathbb{R}$ came in the 1980's with the use of potential theory, independently by E.A. Rahmanov [43], and H.N. Mhaskar and E.B. Saff [28], [29]. (There were already bits of potential theory in earlier work of P. Nevai, P. Erdös, G. Freud, and J.L. Ullmann).

Of particular interest for the approximation theoretic side of weights on $\mathbb{R}$, are the sharp form of the infinite–finite range inequalities, due to Mhaskar and Saff. This required the so–called *Mhaskar–Rahmanov–Saff* number a_u. Given Q that is even, and such that $xQ'(x)$ is positive and increasing in $(0,\infty)$, with limits 0 and ∞ at 0 and ∞ respectively,

a_u is the positive root of the equation

$$u = \frac{2}{\pi} \int_0^1 a_u t Q'(a_u t) \frac{dt}{\sqrt{1-t^2}}, \; u > 0. \tag{7.2}$$

For example, for $Q = Q_\alpha$,

$$a_u = C u^{1/\alpha}, \; \alpha > 0, \tag{7.3}$$

where C is an explicit constant involving gamma functions.

One of the significances of the number is the Mhaskar–Saff identity [28], [29], [30]

$$\|Pe^{-Q}\|_{L_\infty(\mathbb{R})} = \|Pe^{-Q}\|_{L_\infty[-a_n, a_n]}, \; P \in \mathcal{P}_n. \tag{7.4}$$

At least asymptotically as $n \to \infty$, a_n is the "smallest" number for which this can hold. Among the extensions of this to L_p, $p < \infty$ norms, we mention one due to A.L. Levin and the author [19], which is the sharpest for a large class of Q:

$$\|Pe^{-Q}\|_{L_p(\mathbb{R})} \le C\|Pe^{-Q}\|_{L_p[-a_n, a_n]}, \; P \in \mathcal{P}_n, \tag{7.5}$$

where $C \neq C(n, P)$. We refer the reader to other surveys and recent articles [23], [38] for further orientation on this topic.

Using the devices of potential theory and orthogonal polynomials, A.L. Levin and the author [19] proved most of the type of bounds that are needed for mean convergence of orthogonal expansions, and in so doing, resolved an old question of P. Nevai. Let $W := e^{-Q}$, where $Q\colon \mathbb{R} \to \mathbb{R}$ is even, Q'' is continuous in $(0, \infty)$, $Q' > 0$ in $(0, \infty)$, and for some A, $B > 1$,

$$A \le 1 + xQ''(x)/Q'(x) \le B, \; x \in (0, \infty). \tag{7.6}$$

(For example, Q_α satisfies this with $A = B = \alpha$, if $\alpha > 1$). Our 1992 paper established that if $w := W^2 = e^{-2Q}$ satisfies these hypotheses, then

$$\sup_{x \in \mathbb{R}} |p_n(W^2, x)| W(x) \sim a_n^{-1/2} n^{1/6}, \; n \ge 1, \tag{7.7}$$

and

$$\sup_{x \in \mathbb{R}} |p_n(W^2, x)| \left| 1 - \frac{|x|}{a_n} \right|^{1/4} W(x) \sim a_n^{-1/2}. \tag{7.8}$$

Here recall that $\sim$ means that the ratio of both sides is bounded above and below by positive constants independent of n.

A partial analogue of (7.7–7.8) was later obtained for w_α, $\alpha \leq 1$ [20], where new features arise that are in a sense associated with the indeterminacy, or close to indeterminacy, of the moment problem. In the sharper direction, powerful asymptotics for $p_n(w_\alpha, x)$, $\alpha > 1$, have been given by E.A. Rahmanov [44], and his techniques hold great promise for future developments. Still further refinements of asymptotics have been obtained for $p_n(w_\alpha, x)$, α a positive even integer, by S.W. Jha.

The bounds (7.7–8) were sufficient for applications to Lagrange interpolation, quadrature sums, and some other approximation theoretic applications, but recall that for orthogonal expansions, we needed Pollard's decomposition, and that required a bound on $p_n - p_{n-2}$. This proved to be a real sticking point when S.W. Jha and the author tried to attack the mean convergence of $S_n[f]$ for $W^2 = e^{-2Q}$, where Q satisfies (7.6). Earlier work of Mhaskar and Xu [31] had been unable to reproduce the sharpness of Muckenhoupt's results when applied to the Hermite weight for similar reasons: The bounds available at the time for p_n and $p_n - p_{n-2}$ were inadequate.

Jha and I did not try to use Pollard's alternative of obtaining bounds for q_n, the orthonormal polynomial for $(1-(x/a_n)^2)w(x)$ on $[-a_n, a_n]$, and that may be an option. Nor did we try a total alternative of avoiding Pollards' tricks, and using *commutators*, as suggested by Michael Cwikel. Both of these may be worth investigating for the future.

Our partially successful approach was to establish bounds for $p_n - p_{n-2}$ by placing hypotheses on the recurrence coefficients $\{\alpha_n\}$ in the three term recurrence relation, which were known to be true at least for w_α, α a positive integer. Under very general conditions, Mhaskar, Saff, and the author, had proved in (1988) that if $a_n = a_n(Q)$, $w = W^2 = e^{-2Q}$, and α_n is the recurrence coefficient for $p_n(W^2, x)$, then

$$\lim_{n\to\infty} \alpha_n/a_n = \frac{1}{2},$$

thereby resolving an old conjecture of Freud. (See [38]).

Shing Whu Jha and the author required somewhat more, namely

$$\alpha_n/a_n = \frac{1}{2}(1 + O(n^{-2/3})), \tag{7.9}$$

and

$$\alpha_n/\alpha_{n-1} = 1 + O(1/n), \tag{7.10}$$

and then using the Dombrowski–Fricke formula [11], [36], we were able to prove that

$$\sup_{x\in\mathbb{R}} |p_n - p_{n-2}|(x)W(x)\left[\left|1-\frac{|x|}{a_n}\right| + n^{-2/3}\right]^{-1/4} \leq C. \tag{7.11}$$

Steven Damelin, a Ph.D. student of the author, has observed that (7.10) essentially implies (7.9), given what else we know about zeros of $p_n(w,x)$.

Fortunately, these (7.9–7.10) are not vacuous hypotheses: Máté, Nevai and Zaslavsky [27] used a fascinating interplay between combinatorics and orthogonal polynomials to show that for $p_n(w_\alpha, x)$, $\alpha = 2, 4, 6, \ldots$, we have a complete asymptotic expansion for α_n/a_n: For each fixed k,

$$\alpha_n/a_n = \frac{1}{2}\left\{1 + \sum_{j=1}^{k} c_j n^{-2j} + o(n^{-2k})\right\}, \; n \to \infty.$$

The author is certain that the weaker (7.9-7.10) are true for $p_n(w_\alpha, x)$, all $\alpha > 1$, and that this will be proved in the next few years.

To state our result, we need the function

$$L_{\sigma,\tau}(n) := \begin{cases} (\log(n+1))^{|\sigma|}, & \text{if } \sigma = \tau \\ 1, & \text{if } \sigma \neq \tau \end{cases}.$$

Let b, $B \in \mathbb{R}$, $u_a(x) := (1+|x|)^a$ and $w := W^2 := e^{-2Q}$ satisfy all the hypotheses at (7.6). Then for

$$\|S_n[f]Wu_b\|_{L_p(\mathbb{R})} \leq C\|fWu_B\|_{L_p(\mathbb{R})}, \tag{7.12}$$

with $C \neq C(n,f)$, it is necessary that all the following hold:

(I)

$$b < 1 - \frac{1}{p}; \; B > -\frac{1}{p}; \; b \leq B. \tag{7.13}$$

(II) If $p < \frac{4}{3}$,

$$a_n^{\max\{b, -\frac{1}{p}\} - B} n^{\frac{1}{6}\left(\frac{4}{p} - 3\right)} = O\left(1/L_{b,-1/p}(n)\right). \tag{7.14}$$

If $p = \frac{4}{3}$ or 4, then $b < B$;
If $p > 4$,

$$a_n^{b-\min\{B,1-\frac{1}{p}\}} n^{\frac{1}{6}(1-\frac{4}{p})} = O(1/L_{B,1-1/p}(n)). \tag{7.15}$$

If in addition, we have the bound (7.11), then these conditions are necessary and sufficient for (7.12) to hold.

Thus our necessary conditions apply at least for w_α, $\alpha > 1$, and our necessary and sufficient conditions apply at least for $\alpha = 2, 4, 6, \ldots$. In this special case, where we have a simple formula for a_n (recall (7.3)), we may also reformulate these conditions more compactly: (7.14) and (7.15) become respectively

$$\max\left\{b, -\frac{1}{p}\right\} - B + \frac{\alpha}{6}\left(\frac{4}{p} - 3\right)\begin{cases} \leq 0, & b \neq -1/p \\ < 0, & b = -1/p \end{cases}$$

$$b - \min\left\{B, 1 - \frac{1}{p}\right\} + \frac{\alpha}{6}\left(1 - \frac{4}{p}\right)\begin{cases} \leq 0, & B \neq 1 - 1/p \\ < 0, & B = 1 - 1/p \end{cases}.$$

For the Hermite case $\alpha = 2$, these can be reformulated as Muckenhoupt's conditions, except that they close slight gaps in his cases of equality. We shall spare the reader our L_1 analogue.

We finish by noting that there is ample work to do: The author's Ph.D. student, Steven Damelin, is working on mean convergence of orthogonal expansions for Erdös weights, fresh from success there for Lagrange interpolation [9], [10], but analogues for non–even weights on $(-\infty, \infty)$ have not been explored at all, and are worthwhile. And as Ying Guang Shi and Máté, Nevai and Totik have shown, there are in 1994 still worthwhile problems in this topic even for weights on $(-1, 1)$.

REFERENCES

1. R. Askey and S. Wainger, Mean Convergence of Expansions in Laguerre and Hermite Series, *Amer. J. Math.* **87** *(1965), 695–708.*
2. V.M. Badkov, Convergence in Mean of Fourier Series in Orthogonal Polynomials, *Mathematical Notes* **14** *(1973), 651–658.*
3. V.M. Badkov, Convergence in Mean and almost everywhere of Fourier Series in Polynomials orthogonal on an Interval, *Math USSR. Sbornik* **24** *(1974), 223–256.*
4. V.M. Badkov, Approximation Properties of Fourier Series in Orthogonal Polynomials, *Russian Math Surveys* **33** *(1978), 53–117.*
5. W.C. Bauldry, A. Máté, P. Nevai, Asymptotics for Solutions of Systems of Smooth Recurrence Equations, *Pacific Journal of Mathematics* **133** *(1988) 209–227.*

6. T. Chihara, *An Introduction to Orthogonal Polynomials*, Gordon and Breach, New York, 1978.
7. M. Cotlar and C. Sadosky, *On the Helson–Szegö Theorem and a related class of modified Toeplitz kernels*, (in) Proc. Sympos. Pure and Applied Math., Vol. 35 (eds. G. Weiss and S. Wainger), Amer. Math. Soc., Providence, pp.383–407, 1979.
8. M. Cotlar and C. Sadosky, *On Some L_p Versions of the Helson–Szegö Theorem*, (in) Conference on Harmonic Analysis in Honour of Antoni Zygmund (eds. W. Beckner et al.), Wadsworth, Belmont, pp.306–317, 1983.
9. S.B. Damelin and D.S. Lubinsky, Necessary and Sufficient Conditions for Mean Convergence of Lagrange Interpolation for Erdös Weights I, *submitted.*
10. S.B. Damelin and D.S. Lubinsky, Necessary and Sufficient Conditions for Mean Convergence of Lagrange Interpolation for Erdös Weights II, *submitted.*
11. J.M. Dombrowski and P. Nevai, Orthogonal Polynomials, Measures and Recurrence Relations,*SIAM J. Math. Anal.* **17** *(1986), 752–759.*
12. G. Freud, *Orthogonal Polynomials.* Pergamon Press/Akademiai Kiado, Budapest/Oxford, 1971.
13. R.A. Hunt, B. Muckenhoupt, and R.L. Wheeden, Weighted Norm Inequalities for the Conjugate Function and Hilbert Transform, *Trans. Amer. Math. Soc.* **176** *(1973), 227–251.*
14. S.–W. Jha, A. Máté and P. Nevai, *Asymptotics for the Solutions of Systems of Smooth Recurrence Equations and their Application to Orthogonal Polynomials*, (in) Approximation Theory VI, Vol. II, (eds. C.K. Chui, L.L. Schumaker and J.D. Ward), Academic Press, San Diego, pp.345–348, 1989.
15. S.-W. Jha and D.S. Lubinsky, Necessary and Sufficient Conditions for Mean Convergence of Orthogonal Expansions for Freud Weights, *Constr. Approx. (1994), to appear.*
16. H. König, *Vector Valued Lagrange Interpolation and Mean Convergence of Hermite Series*, Proc. Essen Conference on Functional Analysis, North Holland, to appear.
17. H. König and N.J. Nielsen, Vector Valued L_p Convergence of Orthogonal Series and Lagrange Interpolation, *Forum Mathematicum* **6** *(1994), 183–207.*
18. P. Koosis, *The Logarithmic Integral I*, Cambridge Studies in Advanced Mathematics, Vol. 12, Cambridge University Press, Cambridge, 1988.
19. A.L. Levin and D.S. Lubinsky, Christoffel Functions, Orthogonal Polynomials and Nevai's Conjecture for Freud Weights, *Constr. Approx.* **8** *(1992), 463–535.*
20. A.L. Levin and D.S. Lubinsky, Orthogonal Polynomials and Christoffel Functions for $\exp(-|x|^\alpha)$, $\alpha \leq 1$, *J. Approx. Th., to appear.*
21. A.L. Levin and D.S. Lubinsky, Christoffel Functions and Orthogonal Polynomials for Exponential Weights on $(-1,1)$, *Memoirs of the Amer. Math. Soc.* **111** *No. 535 (1994), (146pp.).*
22. A.L. Levin, D.S. Lubinsky and T.Z. Mthembu, Christoffel Functions and Orthogonal Polynomials for Erdös Weights on $(-\infty,\infty)$, *Rendiconti di Mathematica di Roma, Serie VII* **14** *(1994), 199–290.*
23. D.S. Lubinsky, An Update on Orthogonal Polynomials and Weighted Approximation on the Real Line, *Acta Math Applic.* **33** *(1993), 121–164.*
24. D.S. Lubinsky and D.M. Matjila, Necessary and Sufficient Conditions for Mean Con-

vergence of Lagrange Interpolation for Freud Weights, *SIAM J. Math. Anal., to appear.*
25. D.S. Lubinsky and F. Moricz, The Weighted L_p Norms of Orthonormal Polynomials for Freud Weights, *J. Approx. Theory* **77** *(1994), 229-248.*
26. A. Máté, P. Nevai and V. Totik, Necessary Conditions for Weighted Mean Convergence of Fourier Series in Orthogonal Polynomials, *J. Approx. Th.* **46** *(1986), 314-322.*
27. A. Máté, P. Nevai and T. Zaslavsky, Asymptotic Expansions of Ratios of Coefficients of Orthogonal Polynomials with Exponential Weights, *Trans. Amer. Math. Soc.* **287** *(1985), 495-505.*
28. H.N. Mhaskar and E.B. Saff, Extremal Problems for Polynomials with Exponential Weights, *Trans. Amer. Math. Soc.* **285** *(1984), 203-234.*
29. H.N. Mhaskar and E.B. Saff, Where Does the Sup–Norm of a Weighted Polynomial Live?, *Constructive Approximation* **1** *(1985), 71-91.*
30. H.N. Mhaskar and E.B. Saff, Where Does the L_p–Norm of a Weighted Polynomial Live? *Trans. Amer. Math. Soc.* **303** *(1987), 109-124.*
31. H.N. Mhaskar and Y. Xu, Mean Convergence of Expansions in Freud–Type Orthogonal Polynomials, *SIAM J. Math. Anal.* **22** *(1991), 847-855.*
32. B. Muckenhoupt, Mean Convergence of Hermite and Laguerre Series. I, *Trans. Amer. Math. Soc.* **147** *(1970), 419-431.*
33. B. Muckenhoupt, Mean Convergence of Hermite and Laguerre Series. II, *Trans. Amer. Math. Soc.* **147** *(1970), 433-460.*
34. B. Muckenhoupt, Mean Convergence of Jacobi Series, *Proc. Amer. Math. Soc.* **23** *(1970), 306-310.*
35. P. Nevai, *Orthogonal Polynomials*, Memoirs of the American Mathematical Society, No. 213. Providence, R.I., American Mathematical Society 1979.
36. P. Nevai, Exact Bounds for Orthogonal Polynomials Associated with Exponential Weights, *J. Approx. Theory* **44** *(1985), 82-85.*
37. P. Nevai, Geza Freud, Orthogonal Polynomials and Christoffel Functions: A Case Study, *J. Approx. Theory* **48** *(1986), 3-167.*
38. P. Nevai (Editor), *Orthogonal Polynomials, Theory and Practice.* NATO ASI Series C, Vol. 294, Kluwer, Dordrecht, 1990.
39. J. Newman and W. Rudin, Mean Convergence of Orthogonal Series, *Proc. Amer. Math. Soc.* **3** *(1952), 219-222.*
40. H. Pollard, The Mean Convergence of Orthogonal Series I, *Trans. Amer. Math. Soc.* **62** *(1947), 387-403.*
41. H. Pollard, The Mean Convergence of Orthogonal Series II, *Trans. Amer. Math. Soc.* **63** *(1948), 355-367.*
42. H. Pollard, The Mean Convergence of Orthogonal Series III, *Duke Math. J.* **16** *(1949), 189-191.*
43. E.A. Rahmanov, On Asymptotic Properties of Polynomials Orthogonal on the Real Axis, *Math. USSR. Sbnornik* **47** *(1984), 155-193.*
44. E.A. Rahmanov, *Strong Asymptotics for Orthogonal Polynomials with Exponential Weights on* $\mathbb{R}$, (in) Methods of Approximation Theory in Complex Analysis and Mathematical Physics (A.A. Gonchar and E.B. Saff, eds.), Nauka, Moscow, pp.71–97, 1992.
45. M. Riesz, Sur les Fonctions Conjugees, *Mathematische Zeitschrift* **27** *(1927), 218-244.*

46. Ying Guang Shi, Bounds and Inequalities for General Orthogonal Polynomials on Finite Intervals, *J. Approx. Theory* **73** *(1993), 303–333.*
47. H. Stahl and V. Totik, *General Orthogonal Polynomials.* Encyclopaedia of Mathematics and its Applications, Vol. 43, Cambridge University Press, Cambridge, 1992.
48. E. Stein, *Harmonic Analysis: Real Variable Methods, Orthogonality and Oscillatory Integrals*, Princeton University Press, Princeton, 1993.
49. G. Szegö, *Orthogonal Polynomials*, American Mathematical Society Colloquium Publications, Vol. 23, 4th edn., American Mathematical Society, Providence, 1975.
50. S. Thangavelu, *Lectures on Hermite and Laguerre Expansions*, Mathematical Notes Vol. 42, Princeton University Press, Princeton, 1993.
51. G.M. Wing, The Mean Convergence of Orthogonal Series, *Amer. J. Math.* **72** *(1950), 792–807.*
52. Y. Xu, Mean Convergence of Generalized Jacobi Series and Interpolating Polynomials I, *J. Approx. Theory* **72** *(1993), 237–251.*
53. Y. Xu, Mean Convergence of Generalized Jacobi Series and Interpolating Polynomials II, *J. Approx. Theory* **76** *(1994), 77–92.*

Copies of Classical Sequence Spaces in Vector-Valued-Function Banach Spaces

JOSÉ MENDOZA* Departamento de Análisis Matemático, Universidad Complutense de Madrid, 28040 Madrid (Spain)

Dedicated to the memory of Prof. doña Fernanda Fuentes

1 INTRODUCTION

A very classical problem in Banach space theory is the study of the subspaces of a given Banach space, and particularly the search of conditions for a Banach space to contain a copy (or a complemented copy) of any of the classical sequence spaces. Let us recall for instance the famous theorems of Bessaga and Pelczynski characterizing the Banach spaces which have copies of c_0 or complemented copies of ℓ_1, and the Rosenthal characterization of Banach spaces which have a copy of ℓ_1 (see [18] or [38]). We are interested here in the search of copies and complemented copies of classical sequence spaces in some spaces of vector-valued functions.

Let X be a Banach space, let (Ω, Σ, μ) a finite measure space, and let $1 \leq p \leq \infty$. We denote by $L_p(\mu; X)$ the Banach space of all X-valued p-Bochner μ-integrable (μ-essentially bounded, when $p = \infty$) functions with its usual norm. If K is a compact Haussdorff space, we denote by $C(K; X)$ the Banach space of all continuous X-valued functions defined on K, endowed with the supremum norm.

In case X is the scalar field we simply denote $L_p(\mu)$ and $C(K)$.

Let X, Y be Banach spaces, we say that X has a (complemented) copy of Y if X has a (complemented) subspace which is isomorphic to Y. We simply denote $X \supset Y$ ($X \underset{(c)}{\supset} Y$).

---

* Supported in part by D.G.I.C.Y.T. grant PB91-0377

We use standard notation as in [18], [19] or [38].

Our aim is to survey the results related with the following

Question 1. When do the spaces $L_p(\mu; X)$ and $C(K; X)$ contain a copy (or a complemented copy) of c_0, ℓ_1 or ℓ_∞?

In the last twenty years many authors have been working in the different aspects of this question and thanks to them we have now an almost complete answer to it. We are mainly interested in giving a general idea of their contributions, pointing out some key ideas and results lying behind the main theorems. We will only outline a few non technical proofs to give some of the flavor of the subject.

We also consider several related questions and problems, and so in some way we are continuing and bringing up to date part of the work made by E. and P. Saab in [50].

In order to avoid trivial situations we will always suppose K infinite, X infinite-dimensional and $L_p(\mu)$ infinite-dimensional (this last condition simply means that the corresponding measure space (Ω, Σ, μ) is not trivial, i.e. it has infinitely many disjoint measurable sets of positive measure).

When one considers Question 1, the first thing to realize is the following easy and well known facts:

$$L_p(\mu; X) \underset{(c)}{\supset} L_p(\mu), X \qquad C(K, X) \underset{(c)}{\supset} C(K), X$$

It follows from this that if, let us say, $C(K)$ or X have a (complemented) copy of a Banach space Y, then also $C(K, X)$ has a (complemented) copy of Y (and the analogous result is true for $L_p(\mu; X)$). So very often the main problem is to know if the converse of this is true or not. So we have the following

Question 2. Let us suppose that $L_p(\mu; X)$ (or $C(K, X)$) has a (complemented) copy of c_0, ℓ_1 or ℓ_∞ . Does this imply that $L_p(\mu)$ (or $C(K)$) or X has a (complemented) copy of c_0, ℓ_1 or ℓ_∞ ?

Note that an affirmative answer to Question 2 is equivalent to the following complete and natural answer to Question 1 :

(A) $L_p(\mu; X)$ (or $C(K, X)$) has a (complemented) copy of Y if and only if $L_p(\mu)$ (or $C(K)$) or X has a (complemented) copy of Y (Y is any of the spaces we are considering: c_0, ℓ_1 or ℓ_∞).

We will see that in many cases, but not always, (A) is true.

2 THE RESULTS

The first answers to our questions were given by Kwapien and Pisier:

Theorem 1 (Kwapien 1974 [36]). *For $1 \le p < \infty$, $L_p(\mu; X) \supset c_0$ if and only if $X \supset c_0$.*

Theorem 2 (Pisier 1978 [42]). *For $1 < p < \infty$, $L_p(\mu; X) \supset \ell_1$ if and only if $X \supset \ell_1$.*

Since $L_p(\mu)$ has no copies of c_0 for $1 \leq p < \infty$ and $L_p(\mu)$ has no copies of ℓ_1 for $1 < p < \infty$, the preceding theorems are in fact affirmative answers to Question 2.

Kwapien actually proved the result for $p = 1$ because it has already been shown by Hoffmann-Jørgensen [32] that this was enough to get the general case. Kwapien's result as well as Hoffmann-Jørgensen's are very much involved in probability theory in Banach spaces.

Concerning theorem 2, the main tool Pisier used was a remark by Odell and Stegall [48, p. 377] on Rosenthal's theorem about copies of ℓ_1.

Soon after Kwapien and Pisier, Bourgain [8, 10] provided new proofs of Theorems 1 and 2, which have been very useful in subsequent works. Let us mention at least his result about c_0 from which Theorem 1 is just a corollary (using again Hoffmann-Jørgensen's results). It is crucial in the proof of Theorem 4.

Theorem 3 (Bourgain 1978 [8]**).** *Let c_{00} be the normed subspace of c_0 of all sequences of only finitely many non-zero terms. For each $\omega \in \Omega$ let $| \,.\, |_\omega$ be a seminorm on c_{00} such that for each $x \in c_{00}$, $| x |_\omega$ (as a function of ω) is integrable on (Ω, Σ, μ). Let us define the seminorm $\| \,.\, \|$ on c_{00} by*

$$\| x \| = \int_\Omega | x |_\omega \, d\mu(\omega) \quad \text{for each } x \in c_{00}$$

Let us suppose that a sequence $\{x_i\}$ of vectors of c_{00} is a c_0-basis for $\| \,.\, \|$. Then there exist $\omega_0 \in \Omega$ and a subsequence of $\{x_i\}$ which is a c_0-basis for $| \,.\, |_{\omega_0}$

Another affirmative answer to Question 2 was given by E. and P. Saab in the following

Theorem 4 (E. and P. Saab 1982 [49]**).** *$C(K, X) \underset{(c)}{\supset} \ell_1$ if and only if $X \underset{(c)}{\supset} \ell_1$.*

Let us stop for a while to think how to prove the preceding theorem. Let us begin thinking how to prove that $C(K)$ does not contain a *complemented* copy of ℓ_1 (remember that many $C(K)$ spaces *do contain* copies of ℓ_1, see Theorem 6). Perhaps the simplest proof is the following:

Let us suppose that $C(K)$ contains a complemented copy of ℓ_1. Then $C(K)^*$, the dual of $C(K)$, which we identify with the space $\mathcal{M}(K)$ of all regular measures of bounded variation on the Borel subsets of K, contains a copy of ℓ_∞, and therefore it contains a c_0-sequence $\{\mu_n\}$. Take $\lambda = \sum_{k=1}^{\infty} \frac{1}{2^k} \mid \mu_k \mid$ ($\mid \mu \mid$ denotes the variation of μ). Radon-Nikodym theorem allows us to consider $\{\mu_n\}$ as a sequence in $L_1(\lambda)$. But this is of course a contradiction (use for example that $L_1(\lambda)$ is weakly complete).

Observe now that the preceding proof generalizes easily to prove Theorem 4 for spaces X whose dual X^* enjoyes Radon-Nikodym property. It is enough to apply Theorem 1 and the classical Bessaga-Pelczynski theorem on complemented copies of ℓ_1. To overcome the restriction on X the main tools E. and P. Saab used were lifting theory and Theorem 3 instead of Theorem 1.

After the preceding results the following one was surprising because it was the first flatly negative answer to Question 2. It was obtained independently by Cembranos and Freniche (remember we are assuming K infinite and X infinite-dimensional):

Theorem 5 (Cembranos-Freniche 1984 [14] [30]**).** *$C(K, X)$ always contains a complemented copy of c_0.*

Sketch of Proof. The main ingredient in the proof is Josefson-Nissenzweig theorem. It provides a normalized *weak**-null sequence $\{x_n^*\}$ in X^*. Take a bounded sequence $\{x_n\}$ in X such that $x_n^*(x_n) = 1$, and a sequence $\{t_n\}$ in K of different points with no accumulation point at any of its terms (i.e. the relative topology to the sequence would be the discrete ropology). Using Urysohn's theorem we have a normalized sequence $\{f_n\}$ in $C(K)$ of disjointly supported functions such that $f_n(t_n) = 1$ for each n. It is not hard to show that $\{f_n(.)x_n\}$ is a complemented c_0-sequence, since

$$\begin{array}{rcl} C(K,X) & \longmapsto & C(K,X) \\ \phi & \longmapsto & \sum_{n=1}^{\infty} x_n^*(\phi(t_n))f_n(.)x_n \end{array}$$

is a projection. □

Let us see why the preceding theorem is a negative answer to Question 2. Of course, X may fail to contain a copy of c_0. On the other hand, it is well known that $C(K)$ always contains copies of c_0, but in general it does not contain *complemented* copies of it. Remember for example that ℓ_∞ is actually a $C(K)$ space (taking as K the Stone-Čech compactification of the natural numbers) and Phillips theorem guarantees that it does not contain complemented copies of c_0. In general, to get a $C(K)$ space with no complemented copies of c_0, it is enough to take any extremally disconnected compact space K, because the corresponding $C(K)$ is a Grothendieck space.

Remark 1. At this point we already have several answers to Question 1 and we should remark that there are also some obvious ones. They follow easily from the following well known chains of containments

$$C(K,X) \underset{(c)}{\supset} C(K) \supset c_0 \quad L_1(\mu;X) \underset{(c)}{\supset} L_1(\mu) \underset{(c)}{\supset} \ell_1$$

$$L_\infty(\mu;X) \underset{(c)}{\supset} L_\infty(\mu) \underset{(c)}{\supset} \ell_\infty \supset c_0, \ell_1$$

(in fact, it is well known that $L_\infty(\mu)$ and ℓ_∞ are isomorphic; see for instance [38, p. 111]).

In the next theorem copies of ℓ_1 in $C(K,X)$ are considered. The scalar case is quite a classical result due to Pełczynski and Semadeni [41]. The vector-valued case is perhaps a well known old result, but we have only found it in [13].

We need to recall that a compact Hausdorff space is said to be scattered (or dispersed) if each of its closed subsets have isolated points (see [51]). The simplest example of a compact scattered space is the Alexandroff compactification of the natural numbers, and a typical example of a non scattered compact is $[0,1]$.

Theorem 6.

(1) $C(K) \supset \ell_1$ *if and only if* K *is not scattered.*

(2) $C(K,X) \supset \ell_1$ *if and only if* $X \supset \ell_1$ *or* $C(K) \supset \ell_1$

Skecht of Proof of Part 2. Assume K scattered and let us suppose that $C(K, X)$ contains a copy of ℓ_1. An old technique (see for example the proof of [27, VI.7.6.] allows us to assume K is metrizable. But then K must be countable [51, 8.5.5.]. So, let $K = \{t_m\}$ and let $\{\phi_n\}$ be an ℓ_1-sequence in $C(K, X)$. If X does not contain ℓ_1 then Rosenthal's theorem says that, for each m, $\{\phi_n(t_m)\}$ has a weakly Cauchy subsequence. But then, by a standard diagonal argument, we can assure that $\{\phi_n\}$ has a subsequence, which we continue to denote in the same way, such that $\{\phi_n(t_m)\}$ is weakly Cauchy for each m. Therefore [20] $\{\phi_n\}$ is weakly Cauchy in $C(K, X)$. This contradicts the fact that it is an ℓ_1-sequence. □

Once the problems about copies of c_0 or ℓ_1 were solved, it was natural to consider ℓ_∞. This is done in the next two theorems.

Theorem 7 (Drewnowski 1990 [23]). *$C(K, X) \supset \ell_\infty$ if and only if $C(K) \supset \ell_\infty$ or $X \supset \ell_\infty$.*

Theorem 8 (Mendoza 1990 [39]). *For $1 \leq p < \infty$ then $L_p(\mu; X) \supset \ell_\infty$ if and only if $X \supset \ell_\infty$*

Of course, the two preceding theorems are affirmative answer to Question 2, because $L_p(\mu)$ has no copies of ℓ_∞ for $1 \leq p < \infty$. The second theorem was proved using ideas of the proof of the first one, and in both of them some ideas introduced by Rosenthal [46] and Kalton [34] are crucial.

The next step is to consider *complemented* copies of c_0 or ℓ_1 in $L_p(\mu; X)$ (remember that the problem is meaningless for ℓ_∞ because it is injective, i.e. it is complemented whenever it is a subspace of a Banach space). Let us begin by the c_0 case:

Theorem 9. *Let $1 \leq p \leq \infty$, then we have*

(1) *If the measure μ is purely atomic then $L_p(\mu; X) \underset{(c)}{\supset} c_0$ if and only if $X \underset{(c)}{\supset} c_0$.*
(2) *If the measure μ is not purely atomic then $L_p(\mu; X) \underset{(c)}{\supset} c_0$ if and only if $X \supset c_0$.*

Many people has contribued to the proof of the preceding theorem, and it was completed only very recently:

(1) The non trivial part is necessity. The case $p = \infty$ was proved by Leung and Räbiger in 1990 [37]. The case $1 \leq p < \infty$ was proved by Bombal in 1992 [4].
(2) In this case none of the implications is trivial. Sufficiency was shown by Emmanuele [28] (see also Bourgain [9]) . Necessity in the case $1 \leq p < \infty$ is Kwapien's Theorem 1; in the case $p = \infty$, it was shown by Díaz [15] in 1994.

It is worth to note that in the first case we actually have a vector sequence space: when the measure μ is purely atomic $L_p(\mu; X)$ is isomorphic to the sequence space $\ell_p(X)$, i.e. the ℓ_p-sum of the space X.

Of course, in the first case we have an affirmative answer to Question 2, that is, we have in some way a "good" behavior. The second case (for non purely atomic measures) however provides another negative answer.

Let us finally consider complemented copies of ℓ_1 in $L_p(\mu; X)$.
For $1 < p < \infty$ we have again a "good" behavior:

Theorem 10 (Mendoza 1992 [40]). *For $1 < p < \infty$, $L_p(\mu; X) \underset{(c)}{\supset} \ell_1$ if and only if $X \underset{(c)}{\supset} \ell_1$.*

The preceding theorem was first proved in a particular case by Bombal [1]. He used E. and P. Saab 's theorem 4 and we would like to remark again that it lays on Bourgain's theorem 3. Later the author was able to reduce the general case to the one studied by Bombal, using the following criterium to find complemented ℓ_1-sequences. We think that it is essentially due to Rosenthal [46, 47] (see also [11, Proposition 11 of Appendix I] for an improvement of the result) and well known, but we believe that it is not easy to find explicitely in the literature.

Theorem 11. *Let $\{x_n\}$ be a bounded sequence in X and let us suppose that there exists a weakly unconditionally Cauchy series $\sum x_n^*$ in X^* such that $x_n^*(x_n)$ does not converge to zero. Then $\{x_n\}$ has a complemented ℓ_1-subsequence.*

Skecht of Proof. Let $\{x_n\}$ and $\sum x_n^*$ be as in the statement. Taking subsequences, which we continue to denote in the same way, we can assume that there exists $\delta > 0$ such that

$$| x_n^*(x_n) | \geq \delta \text{ for all } n.$$

Now Rosenthal's disjointification lemma [18, p. 82] guarantees that there are subsequences, which we continue to denote again in the same way, such that

$$\sum_{n \neq m} | x_n^*(x_m) | < \delta/2 \text{ for all } m.$$

It is not hard to show that $\{x_n\}$ is an ℓ_1-sequence and also that $R : X \longmapsto \ell_1$, defined by $R(x) = \{x_n^*(x)\}$, $S : \ell_1 \longmapsto \ell_1$, defined by $S(\{t_n\}) = \{\sum_{n=1}^{\infty} t_n x_m^*(x_n)\}_m$, and $T : \ell_1 \longmapsto X$, defined by $T(\{t_n\}) = \sum_{n=1}^{\infty} t_n x_n$ are well defined bounded linear operators. Finally, it is straightforward to prove that S is an isomorphism and $TS^{-1}R$ is a projection. □

For $L_\infty(\mu; X)$ a first (negative) answer to Question 2 was given by Montgomery-Smith in [50]. Using ideas of Johnson [33], he provided an example of a Banach space X with no copies of ℓ_1 for which $L_\infty(\mu; X)$ has a complemented copy of ℓ_1. This made us think that it would be very difficult to find a reasonable characterization of when $L_\infty(\mu; X)$ has a complemented copy of ℓ_1. However Díaz [16] has found an interesting characterization which holds at least for Banach lattices X. To give it we should remember that X is said to contain uniformly complemented copies of ℓ_1^n if there are $M > 0$, a sequence F_n of finite dimensional subspaces of X and a sequence of projections P_n from X onto F_n, such that $d(F_n, \ell_1^n) \leq M$, where d is the Banach-Mazur distance, and $\| P_n \| \leq M$ for each natural number n.

Theorem 12 (Díaz 1994 [16]). *If X is a Banach lattice then $L_\infty(\mu; X) \underset{(c)}{\supset} \ell_1$ if and only if X contains uniformly complemented copies of ℓ_1^n.*

Actually it is shown in [16] that sufficiency in Theorem 10 holds for all Banach spaces X. The lattice assumption is only used in the proof of necessity.

We close the section with a table which summarizes our exposition.

	$C(K,X)$	$L_1(\mu;X)$	$L_p(\mu;X)$ $1<p<\infty$	$L_\infty(\mu;X)$
$\supset c_0$	always (Trivial)	$X \supset c_0$ (Theorem 1)	$X \supset c_0$ (Theorem 1)	always (Trivial)
$\underset{(c)}{\supset} c_0$	(*) always (Theorem 5)	(*) See Theorem 9	(*) See Therorem 9	(*) See Theorem 9
$\supset \ell_1$	$X \supset \ell_1$ or K is not scattered (Theorem 6)	always (Trivial)	$X \supset \ell_1$ (Theorem 2)	always (Trivial)
$\underset{(c)}{\supset} \ell_1$	$X \underset{(c)}{\supset} \ell_1$ (Theorem 4)	always (Trivial)	$X \underset{(c)}{\supset} \ell_1$ (Theorem 10)	(*) See Theorem 12
$\supset \ell_\infty$	$C(K) \supset \ell_\infty$ or $X \supset \ell_\infty$ (Theorem 7)	$X \supset \ell_\infty$ (Theorem 8)	$X \supset \ell_\infty$ (Theorem 8)	always (Trivial)

TABLE 1. In the first row we have the spaces we are studying, in the first column we see the different containments we are considering, and in each box we put the condition which must be satistied and the corresponding theorem if it is not trivial. "Trivial" means that it follows from Remark 1, and (*) means that (A) is not true in that case.

3 COMMENTS AND OPEN PROBLEMS

In this section we would like to mention some open problems and lines of work related with the preceding results.

I. Of course, the first problem we consider will be to complete our table. Only one question remains:

Problem 1. *When does $L_\infty(\mu;X)$ contain a complemented copy of ℓ_1?* And in view of Theorem 12 and the remark following it, the first question would be: *Is it true that if X is a Banach space for which $L_\infty(\mu;X)$ contains a complemented copy of ℓ_1 then X must contain uniformly complemented copies of ℓ_1^n?*

We should remark that Theorem 12 shows for the first time a connection between our problems an the local theory of Banach spaces, and this seems to deserve a further study.

II. In order to consider analogous problems to the one studied here, a natural thing is to add to our list of spaces (c_0, ℓ_1 and ℓ_∞) the ℓ_r spaces ($1<r<\infty$), i.e. we can ask:

Problem 2. *When does $C(K,X)$ (or $L_p(\mu;X)$) have a (complemented) copy of ℓ_r ($1<r<\infty$)?*

The general problem seems very difficult because no good general characterization of spaces containing ℓ_r $(1 < r < \infty)$ is known. However, some partial answers have already been given (see [5], [44], [45]).

And in the same way we can consider $L_1([0,1])$:

Problem 3. *If $L_p(\mu; X)$ $(1 < p < \infty)$ contains a copy of $L_1([0,1])$ does X contain a copy of $L_1([0,1])$?* or, in general, *when does $L_p(\mu; X)$ have a copy of $L_1([0,1])$?*

This question was posed to the author by J. Diestel and it can also be found in [50, Question 10]. The answer to the first part of Problem 3 is yes if X is a dual space [50], but we do not know a complete answer even for $p = 2$.

III. The results considered here have already inspired the search of copies of sequence spaces in several different situations. Let us mention some of them: tensor products (see [50]); function spaces more general than $L_p(\mu; X)$ spaces, like Orlicz vector valued function spaces $L_\Phi(X)$ (see, for instance [1], [2] or [3]), or even the more general $E(X)$ spaces, where E is an order continuous Banach lattice which has weak unit as considered in [52] (see for instance [29] or [40]); vector measure spaces or operator spaces (as in [23] or [24]); and vector-valued Hardy spaces (as in [22]). It seems however that many problems in these directions are still open.

IV. It is well known that vector-valued function spaces may be viewed as operator spaces. For example, the space $\mathcal{K}(X, C(K))$ of all compact operator from X into $C(K)$ is isometrically isomorphic to $C(K, X^*)$ [27, VI.7.1], and $L_\infty(\mu; X)$ is very useful in the representation of operators from $L_1(\mu)$ into X [19, Chapter III]. For this reason the results considered here have inspired and have found applications in the study of several operator spaces (see [6], [7], [21], [25], and see also [31] and [12] in connection with [25]). In particular in some complementation problems more or less related to the famous open problem of the uncomplementability of the space $\mathcal{K}(X, Y)$ of compact oparators from X to Y on the space $\mathcal{L}(X, Y)$ of all operators from X to Y. This should not be surprising because it was already shown in [34, Theorem 6] that such a problem is close to the one of containments of copies of c_0 or ℓ_∞ in operator spaces.

V. We have so far been interested in *isomorphic* copies of spaces. Another natural point of view is to consider *isometric* copies. Some results in this line have been recently obtained by Koldobsky [35] (see also [45]).

VI. Instead of asking about subspaces we could also ask about quotients. That is,

Problem 4. *When do the spaces considered here have quotients isomorphic to any of the classical sequence spaces?*

In fact Díaz [17] has already got an interesting answer to the preceding question. He has shown that: "*For $1 \leq p \leq \infty$, if the measure μ is not purely atomic and X^* contains a copy of ℓ_1, then $L_p(\mu; X)$ has a quotient isomorphic to c_0*". And this result has allowed him to conclude the following characterization: "*For $1 \leq p \leq \infty$, if the measure μ is not purely atomic, then $L_p(\mu; X)$ is a Grothendieck space if and only if X is reflexive*".

The result we have just mentioned recalls us that many properties considered in Banach space theory are connected with the presence of copies or quotients of certain spaces, for instance, property (V^*) of Pelczynski is connected with the existence of complemented ℓ_1-sequences [3], and Grothendieck property with the lack of complemented copies of c_0 [14]. Of course these connections have already been exploited ([3], [13], [14], [17], [30], [43]), but it seems natural to think that they could be used further.

VII. Let us finally consider the following problem: It is well known that ℓ_∞ and $L_\infty([0,1])$ are isomorphic [38, p. 111]. However, let us suppose that X has a copy but not a complemented copy of c_0. For instance X may be ℓ_∞ or ℓ_∞/c_0 (a well studied space, see for example [26]), or more generally X may be any Grothendieck $C(K)$ space or a quotient of it. It follows from Theorem 8 that $L_\infty([0,1],X)$ *does* contain a complemented copy of c_0, but $\ell_\infty(X)$ *does not*. Hence $L_\infty([0,1],X)$ and $\ell_\infty(X)$ are not isomorphic. Then the following natural question arises:

Problem 5. *In which conditions are* $L_\infty([0,1],X)$ *and* $\ell_\infty(X)$ *isomorphic?*

References

1. F. Bombal, *On ℓ^1 subspaces of Orlicz vector-valued function spaces,* Math. Proc. Camb. Phil. Soc. (1987), 101, 107-112.
2. F. Bombal, *On embedding ℓ^1 as a complemented subspace of Orlicz vector valued function spaces,* Revista Matemática de la Universidad Complutense de Madrid (1988), Vol. 1, 13-17.
3. F. Bombal, *On (V^*) sets and Pełczynski's property (V^*),* Glasgow Math. J. 32 (1990), 109-120.
4. F. Bombal, *Distinguished Subsets in Vector Sequence Spaces,* Progress in Functional Analysis, Proceedings of the Peñíscola Meeting on occasion of 60th birthday of M. Valdivia (Edited by J. Bonet et al.), Elsevier Science Publishers 1992, 293-306.
5. F. Bombal and B. Porras, *Strictly Singular and Strictly Cosingular Operators on $C(K,X)$,* Math. Nachr. 143 (1989), 355-364.
6. J. Bonet, P. Domański, M. Lindström and M.S. Ramanujan, *Operator spaces containing c_0 or ℓ_∞,* preprint.
7. J. Bonet, P. Domański and M. Lindström, *Cotype and complemented copies of c_0 in spaces of operators,* preprint.
8. J. Bourgain, *An averaging result for c_0-sequences,* Bull. Soc. Math. Belg. Ser. B 30 (1978), 83-87.
9. J. Bourgain, *A Note on the Lebesgue spaces of vector valued functions,* Bull. Soc. Math. Belg. Ser. B, 31 (1979), 45-47.
10. J. Bourgain, *An averaging result for ℓ_1-sequences and applications to weakly conditionally compact sets in $L_1(\mu)$,* Israel J. of Math., 32 (1979), 289-298.
11. J. Bourgain, *New classes of $\mathcal{L}^p$-spaces,* Lecture Notes in Math. 889. Springer Verlag 1981.
12. V. Caselles, *A Characterization of Weakly Sequentially Complete Banach Lattices,* Math. Z. 190 (1985), 379-385.
13. P. Cembranos, *Algunas propiedades del espacio de Banach $C(K,X)$,* Ph. D. Thesis, Universidad Complutense de Madrid, Madrid 1984.
14. P. Cembranos, *$C(K,X)$ contains a complemented copy of c_0,* Proc. Amer. Math. Soc. 91 (1984), 556-558.
15. S. Díaz, *Complemented copies of c_0 in $L_\infty(\mu;X)$,* Proc. Amer. Math. Soc. 120 (1994), 1167-1172.
16. S. Díaz, *Complemented ℓ_1-copies through local Banach theory* , preprint.
17. S. Díaz, *Grothendieck's property in $L_p(\mu;X)$,* to appear in Glasgow Math. J.
18. J. Diestel, *Sequences and series in Banach spaces,* Graduate texts in Math., n. 92. Springer 1984.
19. J. Diestel and J. J. Uhl Jr., *Vector Measures,* Math. Surveys n. 15 (American Mathematical Society, 1977).
20. I. Dobrakov, *On representation of linear operators on $C_0(T,X)$,* Czech. Math. J. 20 (1971), 13-30.
21. P. Domański and L. Drewnowski, *Uncomplementability of the spaces of norm continuous functions in some spaces of "weakly" continuous functions,* Studia Math. 97 (1991), 245-251.
22. P. Dowling, *On complemented copies of c_0 in vector-valued Hardy spaces,* Proc. Amer. Math. Soc. 107 (1989), 251-254.
23. L. Drewnowski, *Copies of ℓ_∞ in an operator space,* Math. Proc. Camb. Phil. Soc. 108 (1990), 523-526.

24. L. Drewnowski, *When does $ca(\Sigma, X)$ contain a copy of ℓ_∞ or c_0?*, Proc. Amer. Math. Soc. 109 (1990), 747-752.
25. L. Drewnowski and G. Emmanuele, *The problem of complementability for some spaces of vector measures of bounded variation with values in Banach spaces containing copies of c_0*, Studia Math. 104 (1993), 111-123.
26. L. Drewnowski and J. W. Roberts, *On the primariness of the Banach space ℓ_∞/c_0* , Proc. Amer. Math. Soc. 112 (1991), 949-957.
27. N. Dunford and J. T. Schwartz, *Linear Operators*, Part I, Interscience, New York 1958.
28. G. Emmanuele, *On complemented copies of c_0 in $L_p(\mu; X)$, $1 \leq p < \infty$* Proc. Amer. Math. Soc. 104 (1988), 785-786.
29. G. Emmanuele, *Copies of ℓ_∞ in Köthe spaces of vector valued functions,* Illinois J. of Math. 36 (1992), 293-296.
30. F. Freniche, *Barrelledness of the Space of Vector Valued and Simple Functions,* Math. Ann. 267 (1984), 479-486.
31. F. Freniche and L. Rodríguez-Piazza, *Linear projections from a space of measures onto its Bochner integrable functions subspace,* preprint.
32. J. Hoffmann-Jørgensen, *Sums of independent Banach space valued random variables,* Studia Math. 52 (1974), 159-186.
33. W.B. Johnson, *A complementary universal conjugate Banach space and its relation to the Approximation Problem,* Israel J. of Math. 13 (1972), 301-310.
34. N. J. Kalton, *Spaces of compact operators,* Math. Ann. 208 (1974), 267-278.
35. A. Koldobsky, *Isometric stability property of certain Banach spaces,* preprint.
36. S. Kwapien, *Sur les espaces de Banach contenant c_0,* Studia Math. 52 (1974), 187-188.
37. D. Leung and F. Räbiger, *Complemented copies of c_0 in ℓ_∞-sums of Banach spaces,* Illinois J. of Math. 34 (1990), 52-58.
38. J. Lindenstauss and L. Tzafriri, *Classical Banach spaces I*, Springer 1977.
39. J. Mendoza *Copies of ℓ_∞ in $L_p(\mu; X)$,* Proc. Amer. Math. Soc. 109 (1990), 125-127.
40. J. Mendoza, *Complemented copies of ℓ_1 in $L_p(\mu; X)$,* Math. Proc. Camb. Phil. Soc. (1992), 111, 531-534.
41. A. Pelczynski and Z. Semadeni, *Spaces of continuous functions. III. Spaces $C(\Omega)$ for Ω without perfect subsets,* Studia Math. 18 (1959), 211-222.
42. G. Pisier, *Une propriété de stabilité de la classe des espaces ne contenant pas ℓ_1*, C.R.Acad.Sci. Paris Sér. A 286 (1978),747-749.
43. N. Randrianantoanina, *Complemented copy of ℓ_1 and Pełczynski's property (V) and (V^*) in Banach spaces,* preprint.
44. Y. Raynaud, *Sous-espaces ℓ_r et geometrie des espaces $L_p(L_q)$ et L_Φ*, C. R. Acad. Sci. Paris, Ser. I, 301 (1985), 299-302.
45. Y. Raynaud, *Sur les sous-espaces de $L_p(L_q)$*, Seminaire d'Analyse Fonctionelle 1984/85, Publ. Math. Univ. Paris VII, 26, Univ. Paris VII, Paris 1986, 49-71.
46. H. P. Rosenthal, *On relatively disjoint families of measures with some applications to Banach space theory,* Studia Math. 37 (1970), 13-16.
47. H. P. Rosenthal, *On injective Banach spaces and the spaces $L_\infty(\mu)$ for finite measures μ,* Acta Math. 124 (1970), 205-248.
48. H. P. Rosenthal, *Point-wise Compact Subsets of the First Baire Class,* Amer. J. Math. 99 (1977), 362-378.
49. E. Saab and P. Saab, *A stability property of a class of Banach spaces not containing a complemented copy of ℓ_1,* Proc. Amer. Math. Soc. 84 (1982), 44-46.
50. E. Saab and P. Saab, *On Stability Problems of Some Properties in Banach Spaces,* Function Spaces, Lecture Notes in Pure and Applied Mathematics 136 (edited by K. Jarosz), Marcel Dekker 1992, 367-394.
51. Z. Semadeni, *Banach Spaces of Continuous Functions,* PWN-Polish Scientific Publishers, Warszawa 1971.
52. M. Talagrand, *Weak Cauchy sequences in $L_1(E)$* , Amer. J. of Math. 106 (1984), 703-724.

Boyd Indices of Orlicz–Lorentz Spaces

STEPHEN J. MONTGOMERY-SMITH* Department of Mathematics, University of Missouri, Columbia, Missouri 65211

ABSTRACT

Orlicz–Lorentz spaces provide a common generalization of Orlicz spaces and Lorentz spaces. In this paper, we investigate their Boyd indices. Bounds on the Boyd indices in terms of the Matuszewska–Orlicz indices of the defining functions are given. Also, we give an example to show that the Boyd indices and Zippin indices of an Orlicz–Lorentz space need not be equal, answering a question of Maligranda. Finally, we show how the Boyd indices are related to whether an Orlicz–Lorentz space is p-convex or q-concave.

1 INTRODUCTION

The Boyd indices of a rearrangement invariant space are of fundamental importance. They were originally introduced by Boyd (1969) for the purpose of showing certain interpolation results. Since then, they have played a major role in the theory of rearrangement invariant spaces (see, for example, Bennett and Sharpley (1988), Lindenstrauss and Tzafriri (1979) or Maligranda (1984)).

Orlicz–Lorentz spaces provide a common generalization of Orlicz spaces (see Orlicz(1932) or Luxembourg (1955)) and Lorentz spaces (see Lorentz (1950) or Hunt (1966)), and have been studied by many authors, including, for example, Maligranda (1984), Mastyło (1986)

*Research supported in part by N.S.F. Grants DMS 9001796 and DMS 9001357.

and Kamińska (1990a, 1990b, 1991). In particular, Maligranda posed a question about the Boyd indices of these spaces.

In this paper, we first give some fairly elementary estimates for the Boyd indices of Orlicz–Lorentz spaces. Then we give an example that show that these estimates cannot be improved, thus answering Maligranda's question. Finally we show how knowledge of the Boyd indices gives information about the p-convexity or q-concavity of the Orlicz–Lorentz space.

2 DEFINITIONS

In discussing Orlicz–Lorentz spaces, it will be convenient to talk about them in the more general framework of rearrangement invariant spaces. Unfortunately, the definitions in the literature usually require that the spaces be quasi-normed, which is not always the case with the Orlicz–Lorentz spaces. For this reason we introduce the following definition of rearrangement invariant spaces.

DEFINITION If $(\Omega, \mathcal{F}, \mu)$ is a measure space, we denote the measurable functions, modulo functions equal to zero almost everywhere, by $L_0(\mu)$. We say that a *Köthe functional* is a function $\|\cdot\| : L_0(\mu) \to [0, \infty]$ satisfying

i) if $f \in L_0(\mu)$, then $\|f\| = 0 \Leftrightarrow f = 0$;
ii) if $f \in L_0(\mu)$ and $\alpha \in \mathbf{C}$, then $\|\alpha f\| = |\alpha| \, \|f\|$;
iii) if $f, g \in L_0(\mu)$, then $|f| \le |g| \Rightarrow \|f\| \le \|g\|$;
iv) if $f_n, f \in L_0(\mu)$, then $|f_n| \nearrow |f| \Rightarrow \|f_n\| \to \|f\|$;
v) if $f_n \in L_0(\mu)$, then $\|f_n\| \to 0 \Rightarrow f_n \to 0$ in the measure topology.

A *Köthe space* is a pair $(X, \|\cdot\|)$, where $\|\cdot\|$ is a Köthe functional, and $X = \{f \in L_0(\mu) : \|f\| < \infty\}$. Usually, we will denote a space by a single letter, X, and denote its functional by $\|\cdot\|_X$.

DEFINITION If $f : \Omega \to \mathbf{C}$ is a measurable function, we define the *non-increasing rearrangement* of f to be

$$f^*(x) = \sup\{\, t : \mu(|f| \ge t) \ge x \,\}.$$

A *rearrangement invariant space* is a Köthe space such that if $f, g \in L_0(\mu)$, and $f^* \le g^*$, then $\|f\| \le \|g\|$.

Now we define the Orlicz–Lorentz spaces. We refer the reader to Montgomery-Smith (1992) for a motivation of the following definitions.

DEFINITION A *φ-function* is a function $F : [0, \infty) \to [0, \infty)$ such that

i) $F(0) = 0$;
ii) $\lim_{t \to \infty} F(t) = \infty$;
iii) F is strictly increasing;
iv) F is continuous;

We will say that a φ-function F is *dilatory* if for some $1 < c_1, c_2 < \infty$ we have $F(c_1 t) \ge c_2 F(t)$ for all $0 \le t < \infty$. We will say that F satisfies the *Δ_2-condition* if F^{-1} is dilatory.

If F is a φ-function, we will define the function $\tilde{F}(t)$ to be $1/F(1/t)$ if $t > 0$, and 0 if $t = 0$.

We say that two φ-functions F and G are *equivalent* (in symbols $F \asymp G$) if for some number $c < \infty$ we have that $F(c^{-1}t) \le G(t) \le F(ct)$ for all $0 \le t < \infty$.

We will denote the φ-function $F(t) = t^p$ by T^p.

DEFINITION (See Orlicz (1932) or Luxembourg (1955).) If $(\Omega, \mathcal{F}, \mu)$ is a measure space, and F is a φ-function, then we define the *Luxemburg functional* of a measurable function f by

$$\|f\|_F = \inf\left\{ c : \int_\Omega F(|f(\omega)|/c)\, d\mu(\omega) \le 1 \right\},$$

The *Orlicz space* is the associated Köthe space, and is denoted by $L_F(\Omega, \mathcal{F}, \mu)$ (or $L_F(\mu)$, $L_F(\Omega)$ or L_F for short).

DEFINITION If $(\Omega, \mathcal{F}, \mu)$ is a measure space, and F and G are φ-functions, then we define the *Orlicz–Lorentz functional* of a measurable function f by

$$\|f\|_{F,G} = \left\| f^* \circ \tilde{F} \circ \tilde{G}^{-1} \right\|_G .$$

The *Orlicz–Lorentz space* is the associated Köthe space, and is denoted by $L_{F,G}(\Omega, \mathcal{F}, \mu)$ (or $L_{F,G}(\mu)$, $L_{F,G}(\Omega)$ or $L_{F,G}$ for short).

We will write $L_{F,p}$, $L_{p,G}$ and $L_{p,q}$ for L_{F,T^p}, $L_{T^p,G}$ and L_{T^p,T^q} respectively.

It is an elementary matter to show that the Orlicz and Orlicz–Lorentz spaces are rearrangement invariant spaces. We note that $\|\cdot\|_{F,F} = \|\cdot\|_F$, and that $\|\chi_A\|_{F,G} = \tilde{F}^{-1}(\mu(A))$.

Now we define the various indices that we use throughout this paper. Obviously, the most important of these are the Boyd indices. These were first introduced in Boyd (1969). We will follow Maligranda (1984) for the names of the other indices, but will modify the definitions so as to be consistent with the notation used in Lindenstrauss and Tzafriri (1979). Thus other references to these indices will often reverse the words 'upper' and 'lower', and use the reciprocals of the indices used here. The Zippin indices were introduced in Zippin (1971), and the Matuszewska–Orlicz indices in Matuszewska and Orlicz (1960 and 1965). The Zippin indices are sometimes called *fundamental indices.*

DEFINITION For a rearrangement invariant space X, we let the *dilation operators* $d_a : X \to X$ be $d_a f(x) = f(ax)$ for $0 < a < \infty$. We define the *lower Boyd index* to be

$$p(X) = \sup\left\{ p : \text{for some } c < \infty \text{ we have } \|d_a\|_{X\to X} \le c a^{-1/p} \text{ for } a < 1 \right\}.$$

We define the *upper Boyd index* to be

$$q(X) = \inf\left\{ q : \text{for some } c < \infty \text{ we have } \|d_a\|_{X\to X} \le c a^{-1/q} \text{ for } a > 1 \right\}.$$

We define the *lower Zippin index* to be

$$p_z(X) = \sup\left\{ p : \begin{array}{c} \text{for some } c < \infty \text{ we have } \|d_a\chi_A\|_X \le c a^{-1/p} \|\chi_A\|_X \\ \text{for all } a < 1 \text{ and measurable } A \end{array} \right\}.$$

We define the *upper Zippin index* to be

$$q_z(X) = \inf\left\{ q : \begin{array}{c} \text{for some } c < \infty \text{ we have } \|d_a\chi_A\|_X \le c a^{-1/q} \|\chi_A\|_X \\ \text{for all } a > 1 \text{ and measurable } A \end{array} \right\}.$$

DEFINITION For a φ-function F, we define the *lower Matuszewska–Orlicz index* to be

$$p_m(F) = \sup \{ p : \text{for some } c > 0 \text{ we have } F(at) \geq c a^p F(t) \text{ for } 0 \leq t < \infty \text{ and } a > 1 \}.$$

We define the *upper Matuszewska–Orlicz index* to be

$$q_m(F) = \inf \{ q : \text{for some } c < \infty \text{ we have } F(at) \leq c a^q F(t) \text{ for } 0 \leq t < \infty \text{ and } a > 1 \}.$$

Thus, for example,

$$p(L_{p,q}) = q(L_{p,q}) = p_z(L_{p,q}) = q_z(L_{p,q}) = p_m(T^p) = q_m(T^p) = p.$$

We also note the following elementary proposition about the Matuszewska–Orlicz indices.

PROPOSITION 2.1 Let F be a φ-function.
i) F is dilatory if and only if $p_m(F) > 0$.
ii) F satisfies the Δ_2-condition if and only if $q_m(F) < \infty$.

It was conjectured, at one time, that the Boyd and Zippin indices coincide. This is a natural conjecture in view of the fact that these indices do coincide for almost all 'natural' rearrangement spaces, for example, the Orlicz spaces and the Lorentz spaces. However Shimogaki (1970) gave an example of a rearrangement invariant Banach space where these indices differ.

Maligranda (1984) posed a conjecture (Problem 6.1) that would imply that the Boyd indices and Zippin indices coincide for the Orlicz–Lorentz spaces. One of the main purposes of this paper is to show that this is not the case.

In the sequel, we will always suppose that the measure space is $[0, \infty)$ with the Lebsgue measure λ.

3 ESTIMATES FOR THE BOYD INDICES OF THE ORLICZ LORENTZ SPACES

The first results that we present give estimates for the Boyd indices. These estimates are not very sophisticated. However, as we will show in Section 4, they cannot be improved, at least in the form in which they are given. It would be nice to give better estimates at some point in the future, which would make use of more detailed structure information of the defining functions of the Orlicz–Lorentz space.

THEOREM 3.1 Let F and G be φ-functions. Then
i) $p_m(F) \geq p(L_{F,G}) \geq p_m(F \circ G^{-1}) p_m(G) \geq p_m(F) p_m(G)/q_m(G)$;
ii) $q_m(F) \leq q(L_{F,G}) \leq q_m(F \circ G^{-1}) q_m(G) \leq q_m(F) q_m(G)/p_m(G)$.

This will follow from the following propositions.

PROPOSITION 3.2 Let X be a rearrangement invariant space, and let F and G be φ-functions.
i) $p(X) \leq p_z(X)$ and $q(X) \geq q_z(X)$.
ii) $p(L_F) = p_z(L_F)$ and $q(L_F) = q_z(L_F)$.

iii) $p_z(L_{F,G}) = p_m(F)$ and $q_z(L_{F,G}) = q_m(F)$.

Proof: See Maligranda (1984) for part (i), and see Lindenstrauss and Tzafriri (1979) for part (ii). Part (iii) is clear.

PROPOSITION 3.3 Let F_1, F_2 and G be φ-functions.

i) $p(L_{F_1,G}) \geq p_m(F_1 \circ F_2^{-1}) p(L_{F_2,G})$.

ii) $q(L_{F_1,G}) \leq q_m(F_1 \circ F_2^{-1}) q(L_{F_2,G})$.

Proof: We will show (i). The proof of (ii) is similar.

We note that if $p_1 < p_m(F_1 \circ F_2^{-1})$, and if $p_2 < p(L_{F_2,G})$, then there is a constant $c_1 < \infty$ such that for any $t \geq 0$ and $0 < a < 1$ we have

$$a\tilde{F}_1 \circ \tilde{F}_2^{-1}(t) \leq \tilde{F}_1 \circ \tilde{F}_2^{-1}(c_1 \, a^{1/p_1} t),$$

and there is a constant $c_2 < \infty$ such that for any $f \in L_0$ and $0 < b < 1$ we have

$$\|d_{c_1 b} f\|_{F_2,G} \leq c_2 b^{-1/p_2} \|f\|_{F_2,G} .$$

Therefore,

$$\begin{aligned} \|d_a f\|_{F_1,G} &= \left\| x \mapsto f^* \left(a \tilde{F}_1 \circ \tilde{G}^{-1}(x) \right) \right\|_G \\ &\leq \left\| x \mapsto f^* \circ \tilde{F}_1 \circ \tilde{F}_2^{-1} \left(c_1 \, a^{1/p_1} \tilde{F}_2 \circ \tilde{G}^{-1}(x) \right) \right\|_G \\ &\leq c_2 a^{-1/p_1 p_2} \|f\|_{F_1,G} . \end{aligned}$$

Therefore $p(L_{F_1,G}) \geq p_1 p_2$, and the result follows.

Proof of Theorem 3.1: The first inequality follows from Proposition 3.2. The second inequality follows from Propositions (3.2) and (3.3). The third inequality follows because

$$p_m(F \circ G^{-1}) \geq p_m(F) p_m(G^{-1}) = p_m(F)/q_m(G).$$

4 BOYD INDICES CAN DIFFER FROM ZIPPIN INDICES

Now we show that Theorem 3.1 cannot be improved. In so doing, we answer Problem 6.1 posed by Maligranda (1984), by showing that the Boyd indices and Zippin indices do not necessarily coincide for the Orlicz–Lorentz spaces.

THEOREM 4.1 Given $0 < p < q < \infty$, there is a φ-function G such that $p_m(G) = p$, $q_m(G) = q$, $p(L_{1,G}) = p/q$, and $q(L_{1,G}) = q/p$.

We also have the following interesting example, that shows that an Orlicz–Lorentz space need not be quasi normed just because its defining functions are dilatory.

THEOREM 4.2 There is a dilatory φ-function G such that $L_{1,G}$ is not a quasi-Banach space.

At the heart of these results is the following lemma.

LEMMA 4.3 Suppose that $0 < p, q < \infty$, $a > 1$ and $n_0, n_1 \in \mathbf{N}$ are such that

$$(n_1 - n_0)a^{-p}\left(1 - a^{-(p+q)}\right) + a^{-2p-q} = 1.$$

Suppose that G is a φ-function such that for some $L, M > 0$ we have that

$$\begin{aligned}\tilde{G}(Ma^{2n}t) &= La^{(p+q)n}t^p \\ \tilde{G}(Ma^{2n+1}t) &= La^{(p+q)n+p}t^q\end{aligned}$$

for $1 \le t \le a$ and $n_0 \le n \le n_1 + 1$. Then for all $0 \le \theta \le \inf\{q/p, 1\}$, there are functions f and g such that we have

$$\|d_{a^{-\theta}} f\|_{1,G} = a^{(q/p)\theta}\|f\|_{1,G} \quad \text{and} \quad \|d_{a^{\theta}} g\|_{1,G} = a^{-(q/p)\theta}\|g\|_{1,G}.$$

Proof: We define the functions f and g by

$$f(Mx) = \begin{cases} M^{-1}a^{-2n_0-3} & \text{if } 0 \le x < a^{2n_0} \\ M^{-1}a^{-2n-3} & \text{if } a^{2n} \le x < a^{2n+2} \text{ and } n_0 \le n \le n_1 \\ 0 & \text{if } a^{2n_1+2} \le x \end{cases}$$

$$g(Mx) = \begin{cases} M^{-1}a^{-2n_0-3-(p/q)\theta} & \text{if } 0 \le x < a^{2n_0+\theta} \\ M^{-1}a^{-2n-3-(p/q)\theta} & \text{if } a^{2n+\theta} \le x < a^{2n+2+\theta} \text{ and } n_0 \le n \le n_1 \\ 0 & \text{if } a^{2n_1+2+\theta} \le x \end{cases}$$

so that $g = a^{-(p/q)\theta}d_{a^{-\theta}}f$. Then it is sufficient to show that $\|f\|_{1,G} = \|g\|_{1,G} = 1$. We will only show that $\|g\|_{1,G} = 1$, as setting $\theta = 0$ gives the other equality.

First, we note that if

$$La^{(p+q)n+p\theta} \le x < La^{(p+q)(n+1)+p\theta}$$

then

$$Ma^{2n+\theta} \le \tilde{G}^{-1}(x) < Ma^{2n+2+\theta}$$

and so

$$g^* \circ \tilde{G}^{-1}(x) = M^{-1}a^{-2n-3-(p/q)\theta}$$

implying that

$$G \circ g^* \circ \tilde{G}^{-1}(x) = 1/\tilde{G}(Ma^{2n+3+(p/q)\theta}) = 1/\left(La^{(p+q)n+2p+q+p\theta}\right) = L^{-1}a^{-(p+q)n-2p-q-p\theta}.$$

Similarly, if $0 \le x < La^{(p+q)n_0+p\theta}$, then $G \circ g^* \circ \tilde{G}^{-1}(x) = L^{-1}a^{-(p+q)n_0-2p-q-p\theta}$. Hence

$$\begin{aligned}&\int_0^\infty G \circ g^* \circ \tilde{G}^{-1}(x)\,dx \\ &= \sum_{n=n_0}^{n_1} \int_{La^{(p+q)n+p\theta}}^{La^{(p+q)(n+1)+p\theta}} G \circ g^* \circ \tilde{G}^{-1}(x)\,dx + \int_0^{La^{(p+q)n_0+p\theta}} G \circ g^* \circ \tilde{G}^{-1}(x)\,dx \\ &= \sum_{n=n_0}^{n_1} \left(La^{(p+q)(n+1)+p\theta} - La^{(p+q)n+p\theta}\right) L^{-1}a^{-n(p+q)-2p-q-p\theta} \\ &\quad + La^{(p+q)n_0+p\theta}L^{-1}a^{-n_0(p+q)-2p-q-p\theta} \\ &= (n_1 - n_0)a^{-p}\left(1 - a^{-(p+q)}\right) + a^{-2p-q} \\ &= 1,\end{aligned}$$

as required.

Proof of Theorem 4.1: Construct sequences of numbers a_k, b_k, M_k and N_k ($k \geq 0$) such that M_k and N_k are integers, $a_k, b_k > 0$,

$$\begin{aligned} M_k a_k^{-p}\left(1 - a_k^{-(p+q)}\right) + a_k^{-2p-q} &= 1, \\ N_k b_k^{-q}\left(1 - b_k^{-(p+q)}\right) + b_k^{-p-2q} &= 1, \end{aligned}$$

$a_k \to \infty$, and $b_k \to \infty$. Define sequences A_k and B_k inductively as follows: $A_0 = B_0 = 1$, $B_k = A_k a_k^{2M_k+2}$, and $A_{k+1} = B_k b_k^{2N_k+2}$ for $k \geq 0$. Define G by

$$\begin{aligned} G(1) &= 1, \\ G(A_k a_k^{2n} t) &= G(A_k) a_k^{(p+q)n} t^p \\ G(A_k a_k^{2n+1} t) &= G(A_k) a_k^{(p+q)n+p} t^q \end{aligned}$$

for $0 \leq n \leq M_k$ and $1 \leq t \leq a_k$,

$$\begin{aligned} G(B_k b_k^{2n} t) &= G(B_k) b_k^{(p+q)n} t^q \\ G(B_k b_k^{2n+1} t) &= G(B_k) b_k^{(p+q)n+q} t^p \end{aligned}$$

for $0 \leq n \leq N_k$ and $1 \leq t \leq b_k$, and

$$G(t) = \tilde{G}(t)$$

for $t < 1$. Clearly $p_m(G) = p$ and $q_m(G) = q$. From Lemma 4.3, we have that $p(L_{1,G}) = p/q$ and $q(L_{1,G}) = q/p$.

Proof of Theorem 4.2: Let $q = 1$, and construct sequences of numbers p_k, a_k and N_k ($k \geq 0$) such that N_k is an integer, $a_k > 0$,

$$N_k a_k^{-p_k}\left(1 - a_k^{-(p_k+q)}\right) + a_k^{-2p_k-q} = 1,$$

$p_k \to \infty$, and $a_k^{q/p_k} \to \infty$. Define a sequence A_k inductively as follows: $A_0 = 1$, and $A_{k+1} = A_k a_k^{2N_k+2}$ for $k \geq 0$. Define G by

$$\begin{aligned} G(1) &= 1, \\ G(A_k a_k^{2n} t) &= G(A_k) a_k^{(p_k+q)n} t_k^p \\ G(A_k a_k^{2n+1} t) &= G(A_k) a_k^{(p_k+q)n+p_k} t^q \end{aligned}$$

for $0 \leq n \leq M_k$ and $1 \leq t \leq a_k$, and

$$G(t) = \tilde{G}(t)$$

for $t > 1$. Then $p_m(G) = 1$. From Lemma 4.3, we have that $p(L_{1,G}) = 0$, and so by Theorem 5.3(ii) below, $L_{1,G}$ cannot be a quasi-Banach space.

5 CONVEXITY AND CONCAVITY OF ORLICZ–LORENTZ SPACES

An important property that one might like to know about Köthe spaces is whether it is p-convex or q-concave for some prescribed p or q. These questions have already been settled for Orlicz spaces and Lorentz spaces.

For Lorentz spaces, it is almost immediate from their definition (Bennett and Sharpley (1988) or Hunt (1966)) that $L_{p,q}$ is q-convex if $p \geq q$, and p-concave if $p \leq q$. However, outside of these ranges, it is more difficult. In general, it is only the case that $L_{p,q}$ is $q\wedge(p-\epsilon)$-convex and $p\vee(q+\epsilon)$-concave. These results are shown in many places, for example, in Bennett and Sharpley (1988) or Hunt (1966). For Orlicz–Lorentz spaces, the same methods of proof work, and we present these results here.

First we define the notions of p-convexity and q-concavity. These notions may also be found in, for example, Lindenstrauss and Tzafriri (1979).

DEFINITION If X is a Köthe space, we say that X is *p-convex*, respectively *q-concave*, if for some $C < \infty$ we have

$$\left\| \left(\sum_{i=1}^n |f_i|^p \right)^{1/p} \right\|_X \le C \left(\sum_{i=1}^n \|f_i\|_X^p \right)^{1/p},$$

respectively

$$\left\| \left(\sum_{i=1}^n |f_i|^q \right)^{1/q} \right\|_X \ge C^{-1} \left(\sum_{i=1}^n \|f_i\|_X^q \right)^{1/q},$$

for any $f_1, f_2, \dots, f_n \in X$.

The most elementary result about p-concavity and q-convexity is the following. This corresponds to the result that $L_{p,q}$ is q-convex if $p \ge q$, and p-concave if $p \le q$.

THEOREM 5.1 Let F and G be φ-functions.

i) If $G \circ T^{1/p}$ is equivalent to a convex function and $\tilde{G} \circ \tilde{F}^{-1}$ is concave, then $L_{F,G}$ is p-convex.

ii) If $G \circ T^{1/q}$ is equivalent to a concave function and $\tilde{G} \circ \tilde{F}^{-1}$ is convex, then $L_{F,G}$ is q-concave.

Proof: We will only prove (i), as the proof of (ii) is similar. We first use the identity

$$\|f\|_{F\circ T^p, G\circ T^p} = \| |f|^p \|_{F,G}^{1/p}$$

to notice that without loss of generality we may take $p = 1$.

From Hardy, Littlewood and Pólya (1952), Chapter X, it follows that

$$\|f\|_{F,G} = \sup \left\| f \circ \sigma \circ \tilde{F} \circ \tilde{G}^{-1} \right\|_G,$$

where the supremum is over all measure preserving maps $\sigma : [0,\infty) \to [0,\infty)$. Since G is convex, it follows from Krasnosel'skiĭ and Rutickiĭ (1961) that $\|\cdot\|_G$ is 1-convex. Now the result follows easily.

However, if we take the Boyd indices into account, we can also obtain the following results. These correspond to the result that says that $L_{p,q}$ is $q\wedge(p-\epsilon)$-convex and $p\vee(q+\epsilon)$-concave.

To state and prove these results, it is first necessary to recall notation and results from Montgomery-Smith (1992).

DEFINITION If F and G are φ-functions, then say that F is *equivalently less convex than* G (in symbols $F \prec G$) if $G \circ F^{-1}$ is equivalent to a convex function. We say that F is *equivalently more convex than* G (in symbols $F \succ G$) if G is equivalently less convex than F.

A φ-function F is said to be an *N-function* if it is equivalent to a φ-function F_0 such that $F_0(t)/t$ is strictly increasing, $F_0(t)/t \to \infty$ as $t \to \infty$, and $F_0(t)/t \to 0$ as $t \to 0$.

A φ-function F is said to be *complementary* to a φ-function G if for some $c < \infty$ we have

$$c^{-1}t \leq F^{-1}(t) \cdot G^{-1}(t) \leq ct \qquad (0 \leq t < \infty).$$

If F is an N-function, we will let F^* denote a function complementary to F.

An N-function H is said to satisfy *condition* (J) if

$$\left\| 1/\tilde{H}^{*-1} \right\|_{H^*} < \infty.$$

To give some intuitive feeling for N-functions that satisfy condition (J), we point out that these are functions that equivalent to slowly rising convex functions, for example,

$$F(t) = \begin{cases} t^{1+1/\log(1+t)} & \text{if } t \geq 1 \\ t^{1-1/\log(1+1/t)} & \text{if } t \leq 1. \end{cases}$$

THEOREM 5.2 (Montgomery-Smith, 1992) Let F, G_1 and G_2 be φ-functions such that one of G_1 and G_2 is dilatory, and one of G_1 or G_2 satisfies the Δ_2-condition. Then the following are equivalent.

i) For some $c < \infty$, we have that $\|f\|_{F,G_1} \leq c\,\|f\|_{F,G_2}$ for all measurable f.

ii) There is an N-function H satisfying condition (J) such that $G_1 \circ G_2^{-1} \succ H^{-1}$.

Now, we are ready to state the main results of this section.

THEOREM 5.3 Let F and G be φ-functions, and $0 < p < \infty$.

i) If the lower Boyd index $p(L_{F,G}) > p$, and if $G \succ H^{-1} \circ T^p$ for some N-function satisfying condition (J), then $L_{F,G}$ is p-convex.

ii) If $L_{F,G}$ is p-convex, then the lower Boyd index $p(L_{F,G}) \geq p$, and $G \succ H^{-1} \circ T^p$ for some N-function satisfying condition (J).

Note that in part (i), it is not sufficient to take $p(L_{F,G}) = 1$. This is shown by the example $L_{1,q}$ for $1 < q < \infty$, which is known to be not 1-convex (Hunt, 1966).

THEOREM 5.4 Let F and G be φ-functions such that G is dilatory and $p(L_{F,G}) > 0$, and let $0 < q < \infty$.

i) If the lower Boyd index $q(L_{F,G}) < q$, and if $T^q \circ G^{-1} \succ H^{-1}$ for some N-function satisfying condition (J), then $L_{F,G}$ is q-concave.

ii) If $L_{F,G}$ is q-convex, then the lower Boyd index $q(L_{F,G}) \leq p$, and $T^q \circ G^{-1} \succ H^{-1}$ for some N-function satisfying condition (J).

Proof of Theorem 5.3: As in the beginning of the proof of Theorem 5.1, we may suppose without loss of generality that $p = 1$.

The proof of (i) uses fairly standard techniques (Bennett and Sharpley, 1988). First, by Theorem 5.2, we may assume that G is equivalent to a convex function. Next, for any measurable function f, we define

$$f^{**}(x) = \frac{1}{x}\int_0^x f^*(\xi)\,d\xi = \int_0^1 d_a f^*(x)\,da.$$

Then we have the Hardy inequality holding, that is, for some $c < \infty$ we have that $\|f\|_{F,G} \leq \|f^{**}\|_{F,G} \leq c\,\|f\|_{F,G}$. The left hand inequality is obvious. For the right hand inequality, since

$p(L_{F,G}) > 1$, we know that for some $p > 1$ and some $c_1 < \infty$ we have that $\|d_a\|_{L_{F,G} \to L_{F,G}} \leq c_1 a^{-1/p}$ for all $a < 1$. Hence

$$\begin{aligned} \|f^{**}\|_{F,G} &= \left\| \int_0^1 d_a f^* \, da \right\|_{F,G} \\ &= \left\| \int_0^1 d_a f^* \circ \tilde{F} \circ \tilde{G}^{-1} \, da \right\|_G \\ &\leq c_2 \int_0^1 \left\| d_a f^* \circ \tilde{F} \circ \tilde{G}^{-1} \right\|_G \, da \end{aligned}$$

(as G is equivalent to a convex function)

$$\begin{aligned} &= c_2 \int_0^1 \|d_a f^*\|_{F,G} \, da \\ &\leq c_2 \int_0^1 c_1 a^{-1/p} \, da \, \|f^*\|_{F,G} \\ &\leq c_1 c_2 \frac{p}{p-1} \|f\|_{F,G} \,. \end{aligned}$$

But, the functional that takes f to $\|f^{**}\|_{F,G}$ is 1-convex. This is because for any $x_0 > 0$, we have that

$$f^{**}(x_0) = \sup_{\lambda(A) = x_0} \int_A |f(x)| \, dx.$$

(See Hardy, Littlewood and Pólya (1952), Chapter X, or Lindenstrauss and Tzafriri (1979).) Hence, $(f+g)^{**} \leq f^{**} + g^{**}$. Also, by Krasnosel'skiĭ and Rutickiĭ (1961), it follows that $\|\cdot\|_G$ is 1-convex. Therefore,

$$\begin{aligned} \left\| \sum_{i=1}^n |f_i| \right\|_{F,G} &\leq \left\| \left(\sum_{i=1}^n |f_i| \right)^{**} \right\|_{F,G} \\ &\leq \left\| \sum_{i=1}^n f_i^{**} \right\|_{F,G} \\ &= \left\| \sum_{i=1}^n f_i^{**} \circ \tilde{F} \circ \tilde{G}^{-1} \right\|_G \\ &\leq c_2 \sum_{i=1}^n \left\| f_i^{**} \circ \tilde{F} \circ \tilde{G}^{-1} \right\|_G \\ &= c_2 \sum_{i=1}^n \|f_i^{**}\|_{F,G} \\ &\leq c\, c_2 \sum_{i=1}^n \|f_i\|_{F,G} \,, \end{aligned}$$

as desired.

To show (ii), we note that if a is the reciprocal of an integer, then there are functions $g_1, g_2, \ldots, g_{a^{-1}}$, with disjoint supports, and each with the same distribution as f, so that $g_1 + g_2 + \ldots + g_{a^{-1}}$ has the same distribution as $d_a f$. Hence

$$\|d_a f\|_{F,G} \leq c \left(\|g_1\|_{F,G} + \|g_2\|_{F,G} + \ldots + \|g_{a^{-1}}\|_{F,G} \right) = c\, a^{-1} \|f\|_{F,G} \,.$$

Hence $p(L_{F,G}) \geq 1$.

To show that $G \succ H^{-1}$ for some N-function satisfying condition (J), we note the following inequalities.

$$\begin{aligned}\|f\|_{F,G} &= \left\| \int_0^\infty \chi_{|f|\geq t}\, dt \right\|_{F,G} \\ &\leq c \int_0^\infty \left\| \chi_{|f|\geq t} \right\|_{F,G} dt \\ &= c \int_0^\infty \tilde{F}^{-1}(\mu\{|f| \geq t\})\, dt \\ &= c\, \|f\|_{F,1}\,.\end{aligned}$$

Now the result follows immediately from Theorem 5.2.

Proof of Theorem 5.4: As in the proof of Theorem 5.1, we may assume that $q = 1$. To prove (i) we first note, by Theorem 5.2, we may assume that G^{-1} is equivalent to a convex function. Since G is dilatory, it follows that G is equivalent to a concave function (see Montgomery-Smith (1992), Lemma 5.5.2).

Next, for any measurable function f, we define

$$f_{**}(x) = f^*(x) + \frac{1}{x}\int_x^\infty f^*(\xi)\, d\xi = f^*(x) + \int_1^\infty d_a f^*(x)\, da.$$

Then, for some $c < \infty$ we have that $\|f\|_{F,G} \leq \|f_{**}\|_{F,G} \leq c\, \|f\|_{F,G}$. The left hand inequality is obvious.

For the right hand inequality, we argue as follows. Since $q(L_{F,G}) < 1$, we know that for some $q < 1$ and some $c_1 < \infty$ we have that $\|d_a\|_{L_{F,G}\to L_{F,G}} \leq c_1 a^{-1/q}$ for all $a > 1$. Since G is dilatory, it is easy to see that there is there some $p > 0$ such that $G \circ T^{1/p}$ is equivalent to a convex function. Let $q < r < 1$. Then there is a constant $c_2 < \infty$, depending upon r only, such that

$$\begin{aligned}f_{**}(x) &\leq c_2 \left((f^*(x))^p + \frac{1}{x^{p/r}} \int_x^\infty \xi^{p/r-1} (f^*(\xi))^p\, d\xi \right)^{1/p} \\ &= c_2 \left((f^*(x))^p + \int_1^\infty a^{p/r-1} (d_a f^*(x))^p\, da \right)^{1/p}.\end{aligned}$$

For if the right hand side is less than or equal to 1, then it is easily seen that

$$f^*(\xi) \leq 1 \wedge \left(\frac{x}{\xi - x} \right)^{1/r} \qquad (\xi > x),$$

and hence

$$f_{**}(x) \leq \int_1^\infty 1 \wedge (\theta - 1)^{-1/r}\, d\theta.$$

Thus we have the following inequalities.

$$\begin{aligned}\|f_{**}\|_{F,G} &\leq c_2 \left\| \left((f^*)^p + \int_1^\infty a^{p/r-1} (d_a f^*)^p\, da \right)^{1/p} \right\|_{F,G} \\ &= c_2 \left\| \left((f^* \circ \tilde{F} \circ \tilde{G}^{-1})^p + \int_1^\infty a^{p/r-1} (d_a f^* \circ \tilde{F} \circ \tilde{G}^{-1})^p\, da \right)^{1/p} \right\|_G \\ &\leq c_3 \left(\left\| f^* \circ \tilde{F} \circ \tilde{G}^{-1} \right\|_G^p + \int_1^\infty a^{p/r-1} \left\| d_a f^* \circ \tilde{F} \circ \tilde{G}^{-1} \right\|_G^p da \right)^{1/p}\end{aligned}$$

(as $G \circ T^{1/p}$ is equivalent to a convex function)

$$= c_3 \left(\|f^*\|_{F,G}^p + \int_1^\infty a^{p/r-1} \|d_a f^*\|_{F,G}^p \, da \right)^{1/p}$$
$$\leq c_1 c_3 \left(1 + \int_1^\infty a^{p/r-p/q-1} \, da \right)^{1/p} \|f\|_{F,G} .$$

But, the functional that takes f to $\|f_{**}\|_{F,G}$ is 1-concave. This is because for any $x_0 > 0$, we have that

$$f_{**}(x_0) = \frac{1}{x_0} \int_0^\infty f(\xi) \, d\xi - f^{**}(x_0).$$

Hence, $(f+g)_{**} \geq f_{**} + g_{**}$. Also, by an argument similar to that given in M.A. Krasnosel'skiĭ and Rutickiĭ (1961), it follows that $\|\cdot\|_G$ is 1-concave. Therefore,

$$\begin{aligned}
\sum_{i=1}^n \|f_i\|_{F,G} &\leq \sum_{i=1}^n \|f_{i**}\|_{F,G} \\
&= \sum_{i=1}^n \left\| f_{i**} \circ \tilde{F} \circ \tilde{G}^{-1} \right\|_G \\
&\leq c_3 \left\| \sum_{i=1}^n f_{i**} \circ \tilde{F} \circ \tilde{G}^{-1} \right\|_G \\
&= c_3 \left\| \sum_{i=1}^n f_{i**} \right\|_{F,G} \\
&\leq c_3 \left\| \left(\sum_{i=1}^n |f_i| \right)^{**} \right\|_{F,G} \\
&\leq c c_1 \left\| \sum_{i=1}^n |f_i| \right\|_{F,G} ,
\end{aligned}$$

as desired.

To show (ii), we note that if a is an integer, then there are functions $g_1, g_2, \ldots, g_a$, with disjoint supports, and each with the same distribution as $d_a f$, so that $g_1 + g_2 + \ldots + g_a$ has the same distribution as f. Hence

$$\|f\|_{F,G} \geq c^{-1} \left(\|g_1\|_{F,G} + \|g_2\|_{F,G} + \ldots + \|g_a\|_{F,G} \right) = c^{-1} a \, \|d_a f\|_{F,G} .$$

Hence $q(L_{F,G}) \leq 1$.

To show that $G \prec H$ for some N-function satisfying condition (J), we note the following inequalities.

$$\begin{aligned}
\|f\|_{F,G} &= \left\| \int_0^\infty \chi_{|f| \geq t} \, dt \right\|_{F,G} \\
&\geq c^{-1} \int_0^\infty \left\| \chi_{|f| \geq t} \right\|_{F,G} dt \\
&= c^{-1} \int_0^\infty \tilde{F}^{-1}(\mu\{|f| \geq t\}) \, dt \\
&= c^{-1} \|f\|_{F,1} .
\end{aligned}$$

Now the result follows immediately from Theorem 5.2.

6 ADDITIONAL COMMENTS

First, we remark that there is another definition of Orlicz–Lorentz spaces given by Torchinsky (1976) (see also Raynaud (1990)). If F and G are φ-functions, then we define

$$\|f\|_{F,G}^{T} = \inf\left\{ c : \int_0^\infty G(\tilde{F}^{-1}(x) f^*(x)/c)\, \frac{dx}{x} \le 1 \right\},$$

If F is dilatory and satisfy the Δ_2-condition, and if G is dilatory, then it is very easy to calculate the Boyd indices of these spaces — they are precisely the same as their corresponding Matuszewska–Orlicz indices. This follows from the fact that under these conditions, $\left\|\chi_{[0,t]}\right\|_{F,G}^{T} \approx \tilde{F}^{-1}(t)$ (See Raynaud (1990) for more details).

We also pose some questions.

i) What is the dual of an Orlicz–Lorentz space (when the space itself is 1-convex)? Is it another Orlicz–Lorentz space?

ii) Is it possible to find more precise estimates for the Boyd indices of Orlicz–Lorentz spaces?

An approach to the last problem (at least for giving necessary and sufficient conditions for $p(L_{1,G}) = q(L_{1,G}) = 1$ is suggested in Montgomery-Smith (1991).

ACKNOWLEDGEMENTS

This paper is an extension of work that I presented in my Ph.D. thesis (1988). I would like to express my thanks to D.J.H. Garling, my Ph.D. advisor, as well as the Science and Engineering Research Council who financed my studies at that time.

I would also like to express gratitude to A. Kamińska, W. Koslowski and N.J. Kalton for their keen interest and useful conversations.

REFERENCES

1. C. Bennett and R. Sharpley, *Interpolation of Operators,* Academic Press 1988.
2. D.W. Boyd, Indices of function spaces and their relationship to interpolation, *Canad. J. Math.* **21** *(1969), 1245–1254.*
3. D.W. Boyd, Indices for the Orlicz spaces, *Pacific J. Math.* **38** *(1971), 315–323.*
4. G.H. Hardy, J.E. Littlewood and G. Pólya, *Inequalities,* Cambridge University Press, 1952.
5. R.A. Hunt, On $L(p,q)$ spaces, *L'Enseignement Math. (2)* **12** *(1966), 249–275.*
6. A. Kamińska, Some remarks on Orlicz–Lorentz spaces, *Math. Nachr.* **147**, *(1990), 29–38.*
7. A. Kamińska, Extreme points in Orlicz–Lorentz spaces, *Arch. Math.* **55**, *(1990), 173–180.*
8. A. Kamińska, Uniform convexity of generalized Lorentz spaces, *Arch. Math.* **56**, *(1991), 181–188.*
9. M.A. Krasnosel'skiĭ and Ya.B. Rutickiĭ, *Convex Functions and Orlicz Spaces,* P. Noordhoof Ltd., 1961.

10. G.G. Lorentz, Some new function spaces, *Ann. Math.* **51** *(1950), 37–55.*
11. J. Lindenstrauss and L. Tzafriri, *Classical Banach Spaces I—Sequence Spaces,* Springer-Verlag 1977.
12. J. Lindenstrauss and L. Tzafriri, *Classical Banach Spaces II—Function Spaces,* Springer-Verlag 1979.
13. W.A.J. Luxemburg, *Banach Function Spaces,* Thesis, Delft Technical Univ. 1955.
14. L. Maligranda, Indices and interpolation, *Dissert. Math.* **234** *(1984), 1–49.*
15. M. Mastyło, Interpolation of linear operators in Calderon–Lozanovskii spaces, *Comment. Math.* **26,2** *(1986), 247–256.*
16. W. Matuszewska and W. Orlicz, On certain properties of φ-functions, *Bull. Acad. Polon. Sci., Sér. Sci. Math. Astronom. Phys.* **8** *(1960), 439–443.*
17. W. Matuszewska and W. Orlicz, On some classes of functions with regard to their orders of growth, *Studia Math.* **26** *(1965), 11–24.*
18. S.J. Montgomery-Smith, *The Cotype of Operators from $C(K)$,* Ph.D. thesis, Cambridge, August 1988.
19. S.J. Montgomery-Smith, Orlicz–Lorentz Spaces, *Proceedings of the Orlicz Memorial Conference*, (Ed. P. Kranz and I. Labuda), Oxford, Mississippi (1991).
20. S.J. Montgomery-Smith, Comparison of Orlicz–Lorentz spaces, *Studia Math.* **103,** *(1992), 161–189.*
21. W. Orlicz, Über eine gewisse Klasse von Räumen vom Typus B, *Bull. Intern. Acad. Pol.* **8** *(1932), 207–220.*
22. Y. Raynaud, On Lorentz–Sharpley spaces, *Proceedings of the Workshop "Interpolation Spaces and Related Topics", Haifa, June 1990.*
23. T. Shimogaki, A note on norms of compression operators on function spaces, *Proc. Japan Acad.* **46** *(1970), 239–242.*
24. A. Torchinsky, Interpolation of operators and Orlicz classes, *Studia Math.* **59** *(1976), 177–207.*
25. M. Zippin, Interpolation of operators of weak type between rearrangement invariant spaces, *J. Functional Analysis* **7** *(1971), 267–284.*

A Note on the Looman–Menchoff Theorem

N.V.RAO Department of mathematics, The University of Toledo, Toledo, Ohio 43606

Abstract

In this note we shall obtain a generalization of the classical Looman-Menchoff theorem: If f is a complex valued function of the complex variable $z = x + iy$ defined on a domain G where it is continuous and has first order partial derivatives everywhere and further satisfies the Cauchy-Riemann equations $f_x + if_y = 0$ almost everywhere, then f is holomorphic in G. We shall prove the same theorem without the hypothesis of continuity, instead, we shall assume $\int_G \log^+ |f| dxdy < \infty$. This condition is much weaker than that of Tolstov, which is: f is bounded, and that of Sindalovskii: f is in $L^1(G)$. This theorem is true in a slightly more general form — allowing some exceptional sets and assuming slightly more around those sets. These are explained in the main body of this note.

1.Introduction

A very detailed historical account of the Looman-Menchoff theorem can be found in Gray and Morris [**GM**]. For our purposes it is more convenient to introduce the following space of functions defined on a domain G in the complex plane.

A complex valued function defined on G is said to belong to $\mathcal{M}(G)$ if it satisfies the following conditions:

1.a) *f is continuous in x, y separately;*

1.b) *the first order partial derivatives f_x, f_y exist everywhere in G except on a set $E = \cup_{n=1}^{\infty} E_n$ where each E_n is closed and of finite Hausdorff length;*

1.c) *the Cauchy-Riemann equations $f_x + if_y = 0$ are satisfied almost everywhere.*

In 1913, Montel [**M**] announced that any bounded function satisfying 1.a), 1.b), and 1.c) with the exceptional set E empty is holomorphic in G. The Loman-Menchoff theorem was in response to Montel. It first appeared in the book of Saks [**Sa**] around 1930. But Looman and Menchoff had to assume the joint continuity of f instead of boundedness. Thus when in 1942, Tolstov [**T1**] vindicated Montel, the picture became complete or at least it was felt that way. But it was a complete surprise when Sindalovskii [**Si**] in 1985 proved that one need only assume that f is locally in L^1. Also he asked whether the result would remain valid if L^1 condition is replaced by L^p, $p > 0$. I answered this question affirmatively in [**R**]. But it was a corollary to a very complicated version of the Looman-Menchoff theorem in that paper. In this note we want to give a simpler, more direct answer.

In order to make sense out of the myriad generalizations of this theorem, it helps to keep in mind the following — I borrow this from Gray and Morris: A Green-type theorem and a Morera-type theorem give a Goursat-type theorem. Also as can be seen from the works of Tolsrov [**T1, T2**], Cohen [**C**], and Fesq [**F**], local boundedness cannot be avoided in the case of Green-type theorems. So in order to go from boundedness to summability, Sindalovskii uses a maximum principle — thus essentially saying, summability implies boundedness or a Phragmen-Lindelöf-type theorem and a Morera-type theorem give a Goursat-type theorem. In fact, Sindalovskii proves the following theorem (theorem 3 of his paper) restated in our notation:

If f belongs to $\mathcal{M}(G)$ and has the "Lindelöf property" in G for angles formed by vertical and horizontal lines, the f is analytic in G.

He does not have to assume that f belongs to L^1 locally or anything else. His definition of Lindelöf property is as follows:

A function f has the Lindelöf property with respect to angles formed by horizontal and vertical lines, if, for every circular sector, S, with $\overline{S} \subset G$, and with vertex A and sides parallel to the axes, inside which f is analytic and continuous on the closure except possibly at A, f is continuous at A also.

Here continuity is restricted to $\overline{S}$. Also, as was noted by Sindalovskii, Tolstov [**T3**] used the maximum principle in a similar way previously in deriving regularity for bounded functions satisfying the Laplace equation.

The most crucial idea leading to the surprising generalization of the Looman-Menchoff theorem by Sindalovskii was that

if f belongs to $L^{1,\text{loc}}(G)$ and $\mathcal{M}(G)$, then it has the Lindelöf property in G.

In view of this, I shall restrict myself to presenting a stronger form of the Phragmen-Lindelöf theorem and also a much simpler proof which allows us to prove the following theorem:

Theorem 1. *If f belongs to $\mathcal{M}(G)$ and also*

$$\int_G \log^+ |f| dx dy < \infty \tag{1.1}$$

locally, then f is holomorphic in G.

It is easy to deduce the sufficiency of the $L^{p,\text{loc}}$ for $p > 0$ from this.

Thus our theorem can be deduced from the theorem 3 of Sindalovskii quoted above and the following:

If f satisfies (1.1) *locally, then f has the Lindelöf property in G.*

And this would follow from the following version of the Pgragmen-Lindelöf:

Theorem 2. *Let S be the sector $\{|z| \leq R,\ x \geq 0,\ y \geq 0\}$ in the complex plane. Assume that f is continuous in $S \setminus \{0\}$, holomorphic in the interior S° of S, $f(z)$ has finite limits as z tends to 0 along the axes, and satisfies* (1.1) *on S. Then f is continuous on S.*

This would follow by a standard argument (Boas [**B, pp243–244**]) by proving

Theorem PL. *S stands for the same as in* theorem 2. *If f is holomorphic in the interior of S, satisfies* (1.1) *on S, and for every $a \in \partial S \setminus \{0\}$,*

$$\limsup_{z \to a, z \in S^\circ} |f(z)| \leq M,$$

then $|f(z)| \leq M$ in all of S.

2. Proof of PL.

Consider the function

$$v = (\log |f(z)| - \log M)^+$$

on S. Clearly v is a non-negative subharmonic function on S, continuous on S except possibly at 0, and $v = 0$ on $\partial S \setminus \{0\}$. Further $v \in L^1(S)$. Now we extend v by reflection below the x-axis and to the left of y-axis. The extension v^* satisfies for any z in S,

$$v^*(z) = v^*(\overline{z}) = v^*(-\overline{z}).$$

If we denote the domain of v^* by $3S$, we notice that v^* is subharmonic in $3S$ and belongs to $L^1(3S)$. Hence for any $a \in S$ such that $0 < |a| < R/2$, the disc D with center a and radius $|a|$ would be entirely contained in $3S$ and

$$\pi |a|^2 v(a) \leq \int_D v^* dx dy. \tag{2.1}$$

Therefore for any given $0 < r < R/2$ and $M(r) = \sup_{|a|=r} v(a)$,

$$\pi r^2 M(r) \leq \int_{3D} v^*(z) dx dy \tag{2.2}$$

where $3D$ represents $\{z \in 3S,\ 0 < |z| < 2r\}$. Combining the fact that $v^* \in L^1(3S)$, we obtain

$$M(r) = o(r^{-2}) \quad \text{as} \quad r \to 0.$$

Now we can apply the usual trick [**B, p244**] to obtain $v \leq 0$ and so $v = 0$. This proves our assertion PL.

Remark. We cannot improve our PL as can be seen by the example

$$f(z) = e^{\frac{i}{z^2}}.$$

References

[**B**] R.P.Boas, *Invitation to complex analysis*, Random House, New York, 1986.

[**C**] Paul J.Cohen, *On Green's theorem*, Proc. Amer. Math. Soc., 10(1959), 109–112.

[**F**] Robert M.Fesq, *Green's formula, linear continuity and Hausdorff measure*, Trans. Amer. Math. Soc. 118(1965), 105–112.

[**GM**] J.D.Gray and S.A.Morris, *When is a function that satisfies Cauchy-Riemann equations analytic?*, Amer. Math. Monthly, 85(1978), 24 6–256.

[**M**] P.Montel, *Sue les différentiales totales et les fonctions mo nogènes*, C.R.Acad.Sci.Paris 156(1913),1820–1822.

[**R**] N.V.Rao, *A generalization of the Looman-Menchoff theorem*, Israel J. Math. Vol.70,No.1(1990), 93–103.

[**Sa**] S.Saks, *Théorie de l'intégrale*, Monografje Matematyczne II, Varsovie, 1933.

[**Si**] G.Kh.Sindalovskii, *On the Cauchy-Riemann conditions in the class of functions with summable modulus and some boundary properties of holomorphic functions*, Mat. Sb. 128(170)No.3(1985), 364–382 (Russian).

[**T1**] G.P.Tolstov, *Sur la différentiale totale*, Mat. Sb. 9(51) (1942), 461–468.

[**T2**] G.P.Tolstov, *Sur les fonctions bornées verifiant les conditions Cauchy-Riemann*, Mat. Sb. 10(52)(1942), 79–86.

[**T3**] G.P.Tolstov, *On bounded functions satisfying a Laplace equation*, Mat. Sb. 29(71)(1951), 559–563.

On the Extreme Point Intersection Property

T.S.S.R.K. Rao Indian Statistical Institute, R.V. College P.O, Bangalore - 560 059, INDIA

Introduction

In this paper we study an extreme point intersection property introduced by Lindenstrauss [7], for function spaces and spaces of operators. For a Banach space X, let X_1 denote the unit ball and $\partial e X_1$ the set of extreme points of X_1. Taking an equivalent form of the original definition (see Theorem 4.7 of [7] and Theorem 2.2 of [5]) we say that a Banach space X has the E.P.I.P if

$$|x^*(x)| = 1 \quad \forall x \in \partial e X_1, \quad \forall x^* \in \partial e X_1^*.$$

Here we are interested in studying this property for the space of operators $\mathcal{L}(X,Y)$ and for certain function spaces.

It is apparent from the work of Lima [5] that questions of this nature are intimately related to the question of 'extreme operators' being 'nice', ie., $T \in \partial e\mathcal{L}(X,Y)_1 \Rightarrow T^*(\partial e Y_1^*) \subset \partial e X_1^*$. This latter question has been well studied when the the domain or the range is a $C(K)$ space, instead of listing all the relevant literature here we refer the interested reader to the survey article [12] and [1,8,7].

Main results. We will be using several times the following result from [6].

Theorem 3.7 ([6]) : Let $\lambda \in \partial_e X_1^{**}$ and $y^* \in \partial_e Y_1^*$. Suppose either λ or y^* is a w^*-denting point. Then the functional $\lambda \otimes y^*$ has unique norm preserving extension from $\mathcal{K}(X,Y)$ to $\mathcal{L}(X,Y)$.

We first give a simple proof of a result of Lima [5] and a partial converse of it.

Let X and Y be Banach spaces.

Theorem 1 : a) Suppose X^* has t he E.P.I.P and that $T \in \partial e\mathcal{L}(X,Y)_1 \Rightarrow T^*(\partial eY_1^*) \subset \partial eX_1^*$. Then $\mathcal{L}(X,Y)$ has the E.P.I.P.

b) Suppose X is such that X_1^{**} is the w^*-closed convex hull of its w^*-denting points. If $\mathcal{L}(X,Y)$, has the E.P.I.P then, $T \in \partial e\mathcal{L}(X,Y)_1 \Rightarrow T^*(\partial eY_1^*) \subset \partial eX_1^*$.

The hypothesis on X in b) is satisfied when X has the R.N.P (see [2]).

Proof : a) For $x^{**} \in X^{**}, y^* \in Y^*$ denote by $x^{**} \otimes y^*$ the functional defined on a $T : X \to Y$ by

$$(x^{**} \otimes y^*)(T) = x^{**}(T^*(y^*)).$$

With this notation it is clear that the set

$$\mathcal{A} = \{x^{**} \otimes y^* : x^{**} \in \partial eX_1^{**} \text{ and } y^* \in \partial eY_1^*\}$$

is a norm determining set for $\mathcal{L}(X,Y)$. Hence by an application of the Hahn-Banach separation theorem and the Milman's converse of the Krein-Milman Theorem [4], we have

$$\partial e\mathcal{L}(X,Y)_1^* \subset \mathcal{A}^{-w^*}.$$

Now let $T \in \partial e\mathcal{L}(X,Y)_1, x^{**} \in \partial eX_1^{**}, y^* \in \partial eY_1^*$. Since $T^*(\partial eY_1^*) \subset \partial eX_1^*$ and since X^* has the E.P.I.P.

$$|(x^{**} \otimes y^*)(T)| = |x^{**}(T^*(y^*))| = 1.$$

By taking w^*-limits we have, for any

$$J \in \partial e\mathcal{L}(X,Y)_1^*, \qquad |J(T)| = 1$$

Therefore $\mathcal{L}(X,Y)$ has the E.P.I.P.

b) Note that if

$$\mathcal{A} = \{x^{**} \in \partial e X_1^{**} : x^{**} \text{ is a } w^*\text{-denting point } \}$$

then $\partial e X_1^{**} \subset \mathcal{A}^{-w^*}$.

Let $x^{**} \in \mathcal{A}$. It follows from Theorem 3.7 in [6] that $\forall y^* \in \partial e Y_1^*$,

$$x^{**} \otimes y^* \in \partial e \mathcal{L}(X, Y)_1^*.$$

Now let $T \in \partial e \mathcal{L}(X, Y)_1$, fix $y^* \in \partial e Y_1^*$. Since $\mathcal{L}(X, Y)$ has the E.P.I.P

$$|x^{**}(T^*(y^*))| = 1$$

Taking w^*-limits we have

$$|x^{**}(T^*(y^*))| = 1 \qquad \forall x^{**} \in \partial e X_1^{**}.$$

Therefore $T^*(y^*) \in \partial e X_1^*$. Hence $T^*(\partial e Y_1^*) \subset \partial e X_1^*$.

Corollary 1 : If Y is such that every point of $\partial e Y_1^*$ is a w^*-denting point and $\mathcal{L}(X, Y)$ has the E.P.I.P then $T \in \partial e \mathcal{L}(X, Y)_1 \Rightarrow T^*(\partial e Y_1^*) \subset \partial e X_1^*$.

Proof : It again follows from Theorem 3.7 in [6] that $\forall y^* \in \partial e Y_1^*, x^{**} \in \partial e X_1^{**}, x^{**} \otimes y^*$ is in $\mathcal{L}(X, Y)_1^*$.

Remark 1 : If one merely assume that Y_1^* is the w^*-closed convex hull of its w^*-denting points, then the same conclusion can be drawn under the extra assumption that T is a compact operator.

Specializing to the situation when $Y = C(K)$ and identifying $\mathcal{L}(X, C(K))$ with the space $C(K, (X^*, w^*))$ (functions that are continuous when X^* has the w^*-topology) we have the following:

Theorem 2 : Suppose

a) X_1^{**} is the w^*-closed convex hull of w^*-denting points.

or

b) K is dispersed and w^*-extreme points (i.e., extreme points that are w^*-continuous) form a w^*-dense set in $\partial e X_1^{**}$.

holds

If $C(K,(X^*,w^*))$ has the E.P.I.P, then every extreme point of $C(K,(X^*,w^*))_1$ takes extreme values and X^* has the E.P.I.P.

Proof : Assume (a).

We only need to show that X^* has the E.P.I.P. Let $x^* \epsilon \partial e X_1^*, x^{**} \in \partial e X_1^{**}$ be a w^*-denting point. Since $C(K,(X^*,w^*))$ has the E.P.I.P and since $\delta(k) \otimes x^{**} \in \partial e C(K,(X^*,w^*))_1^*$ we have

$$|x^{**}(x^*)| = 1$$

and once more using the hypothesis we have $|x^{**}(x^*)| = 1 \quad \forall x^{**} \in \partial e X_1^{**}$.

Assume (b).

Let $\Gamma =$ set of isolated points of K. Then Γ is dense in K.

Let $f \in \partial e C(K,(X^*,w^*))_1$.

For $k \in \Gamma$, it follows from the remarks made in [1] that $f(k) \in \partial e X_1^*$.

Now fix a $k_0 \in K$ and let $k_\alpha \to k_0, k_\alpha \in \Gamma$ and let x_0 be a w^*-extreme point of $\partial e X_1^{**}$. Since $\delta(k_\alpha) \otimes x_0 \in \partial e C(K,(X^*,w^*))_1^* \quad \forall \alpha, |f(k_\alpha)(x_0)| = 1$ and hence $|f(k_0)(x_0)| = 1$. Now use again the w^*-density of such points to conclude, $f(k_0) \epsilon \partial e X_1^*$. That X^* has the E.P.I.P is also deduced similarly.

Remark 2 : The condition in b) is met when X_1 is the closed convex hull of strong extreme points. This theorem is the full converse of Lima's Theorem from [5] since it gives E.P.I.P on X^* and "niceness" of extreme operators.

Corollary 2 : Let K be a compact extremally disconnected space and suppose X_1 is the closed convex hull of its extreme points. Then if X has the E.P.I.P, $C(K,(X^*,w^*))$ has the E.P.I.P. and every extreme point of $C(K,(X^*,w^*))_1$ takes extreme values.

Proof : It is easy to see that the hypothesis on X implies that $\partial e X_1^*$ is w^*-closed and X^* has the E.P.I.P. Therefore using Theorem 4 of [8] it follows that every extreme point of $C(K,(X^*,w^*))_1$ takes extreme values and that the space has the E.P.I.P follows from Theorem 1.

In the next proposition we consider a situation where E.P.I.P of $\mathcal{L}(X,Y)$ induces the E.P.I.P on Y (Note that in the case of $\mathcal{L}(X,C(K)), C(K)$ space already has the E.P.I.P)

Proposition 1 : Let X be such that for some $x_0^* \epsilon \partial e X_1^*$, span $\{x_0^*\}$ is a semi-L-summand and Y is such that Y_1^* is the w^*-closed convex hull of w^*-denting points. If $\mathcal{L}(X,Y)$ has the E.P.I.P then Y has the E.P.I.P.

Proof : Write $F = \{x \in X : \|x\| = 1 = x_0^*(x)\}$. It follows from Theorem 3.5 of [5] that $X_1 = \bar{C}0(\Pi F)$ where Π denotes the unit circle.

Fix $x_0^{**} \in \partial e X_1^{**}$ such that $x_0^{**}(x_0^*) = 1$. Let $e \in \partial e Y_1$. We claim that the operator $x_0^* \otimes e$ is an extreme operator.

For, if $x_0^* \otimes e = \frac{T_1+T_2}{2}, \|T_i\| = 1, T_i \in \mathcal{L}(X,Y)$. Then for $x \in \Pi F$

$$te = \frac{T_1x + T_2x}{2} \qquad \text{where} |t| = 1.$$

Hence $T_1x = T_2x$ and therefore $T_1 = T_2 = x_0^* \otimes e$. Again if y^* is a w^*-denting point,

$$|(x_0^{**} \otimes y^*)(x_0^* \otimes e)| = |y^*(e)| = 1.$$

Hence Y has the E.P.I.P.

Function spaces

In this section we consider the E.P.I.P for various function spaces. Like before we repeatedly use the fact that if $\mathcal{A} \subset \partial e X_1^*$ is a norm determining set and $|x^*(x)| = 1 \quad \forall x^* \in \mathcal{A}, x \epsilon \partial e X_1$ then X has the E.P.I.P.

Our first result here gives a simple proof of a Theorem 2.10 of Sharir [9].

Proposition 2 : Let $(\Omega, \mathcal{B}, \mu)$ be a positive measure space. If X has the E.P.I.P then so does $L^1(\mu, X)$.

Proof : Let $\mathcal{A} = \{f \in L^\infty(\mu, X^*)_1 : f$ is a.e., $\partial e X_1^*$ valued $\}$. It is easy to see that $\mathcal{A}$determines the norm of $L^1(\mu, X)$.

If $f \in \partial e L^1(\mu, X)_1$ then $f = t\frac{\chi_A}{\mu(A)}x$ where $x \in \partial e X_1, |t| = 1$ and A is a μ-atom (see [10]).

Now for $g \in \mathcal{A}$

$$|\int g(w)(f(w))d\mu(w)| = 1$$

since X has the E.P.I.P and g is constant on the atom.

In case of the space $M(K, X)$, the space of X-valued regular Borel measures we have the following.

Corollary 3 : If K is dispersed or X has the R.N.P then $M(K, X)$ has the E.P.I.P when ever X has the E.P.I.P.

Proof : Use the results of [3] to conclude that in either case $M(K, X) = L^1(\mu, X)$ where $M(K) = L^1(\mu)$.

Proposition 3 : If $\{X_i\}$ is a family of Banach spaces having the E.P.I.P then the ℓ^∞-direct sum $\oplus_\infty X_i$ has the E.P.I.P.

Proof : It is well known that

$$(\oplus_\infty X_i)^* = \oplus_1 X_i^* \oplus_1 (\oplus_{c_0} X_i)^\perp$$

Where $\oplus_1$ denotes the ℓ^1-direct sum and $\oplus_{c_0}$ the c_0-direct sum.

Now it $\mathcal{A} = \{(0, \cdots, 0, x_i^*, 0 \cdots) : x_i^* \in \partial e X_i^*\}$ then $\mathcal{A} \subset \partial e(\oplus_\infty X_i)_1^*$ and determines the norm on $\oplus_\infty X_i$. Since each X_i has the E.P.I.P we get the desired conclusion.

In the case of continuous function space $C(K, X)$, we have the following.

Theorem 3 : $C(K, X)$ has the E.P.I.P iff X has the E.P.I.P and

Proof : Recall that $\partial e C(K, X)_1^* = \{\delta(k) \otimes x^* : k \in K, x^* \in \partial e X_1^*\}$ where $\delta(k)$ denotes the evaluation functional at k.

Suppose $C(K, X)$ has the E.P.I.P. Let $f \in \partial e C(K, X), k \in K$. Since $\forall x^* \in \partial e X_1^*$,
$|x^*(f(k))| = 1$ we get that $f(k) \in \partial e X_1^*$. It is clear that X has the E.P.I.P. The converse is similarly proved.

Corollary 4 : Let X be a real Banach space with the 3.2.I.P, then every extreme point $C(K, X)_1$ takes extreme values.

Proof : Let us recall that X has the 3.2.I.P, if any collection of 3 pairwise intersecting closed balls has non-empty intersection. It follows from Corollary 2 on Page 43 of [7] that $C(K, X)$ has the 3.2.I.P. From the original definition of the E.P.I.P, it is clear that 3.2.I.P $\Rightarrow$ E.P.I.P. Hence the conclusion follows.

Using a theorem of Lindenstrauss that $L^1(\mu)$ has the 3.2.I.P. (see[7]), we get another proof of a theorem of Werner, [11].

Corollary 5 : Every extreme point of $C(K, L^1(\mu))$ takes extreme values.

Corollary 6 : Let X be a real Banach space of finite dimension. If X has the E.P.I.P, then $C(K, X)$ has the E.P.I.P.

Proof : Suppose X has the E.P.I.P. Since the distance between any two distinct extreme points is 2, we see that X_1 is a polyhedron.

Let $f \in \partial eC(K, X)_1$. If $\Gamma = \{k \in K : f(k) \in \partial eX_1\}$ then it follows from Theorem 3 [1] the Γ is dense in K. Now since X has the E.P.I.P, arguments given similar to the ones given before show that $\Gamma = K$.

Therefore $C(K, X)$ has the E.P.I.P.

Using again the remark from [1] that if $f \in \partial eC(K, X)_1$ and k_0 is an isolated point, then $f(k_0) \in K$ we see that :

Corollary 7 : If K is dispersed and X has the E.P.I.P then $C(K, X)$ has the E.P.I.P.

Remark : These ideas can be used to give a simple proof of theorem 5 in [8].

References

1. R.M. Blumenthal, J. Lindenstrauss and R.R. Phelps, Extreme operators into $C(K)$, Pacific, J. Math. 15 (1956), 747-756.

2. R.D. Bourgin : Geometric aspects of convex sets with the Radon Nikodym property, Springer LNM # 993, 1983.

3. M. Cambern and P. Greim : The bidual of $C(X, E)$, Proc. Amer. Math. Soc. 85 (1982), 53-58.

4. J. Diestel : Sequences and Series in Banach Spaces, GTM # 92, Springer 1984.

5. Å. Lima : Intersection properties of balls in spaces of compact operators, Ann. Inst. Fourier. Grenoble 28 (1978), 35-65.

6. Å. Lima, E. Oja, T.S.S.R.K. Rao and D. Werner : Geometry of operator spaces, Michigan Math.J.,to appear.

7. J. Lindenstrauss : Extension of compact operators, Memoir, Amer. Math. Soc. 48 (1964).

8. M. Sharir : Characterization and properties of extreme operators into $C(Y)$, Israel J. Math. 12 (1972), 174-183.

9. M. Sharir : Extremal structure in operator spaces, Trans. Amer. Math. Soc. 186 (1973), 91-111.

10. K. Sundaresan : Extreme points of the unit cell in Lebesgue-Bochner function spaces, Colloq. Math. 22 (1970), 111-119.

11. D. Werner : Extreme points in function spaces, Proc. Amer. Math. Soc. 89 (1983) 598-600.

12. D. Werner : Extreme points in spaces of operators and vector-valued measures, Proc. 12th Winter School on Abstract. Analysis, Srni (Bohemian Weald) 5, 1984, 135-143 (Supplemento ai Rendicont. Circolo. Math. di Palermo).

A Singular Integral Operator with a Large Eigenvector

Richard Rochberg Department of Mathematics, Box 1146, Washington University, St. Louis MO 63130-4899, rr@math.wustl.edu

1 Background

The results in [R] give upper bounds on the size of eigenvectors for ceratin classes of of singular integral operators acting on $L^2(\mathbf{R}^d)$. In this paper we construct an explicit family of examples which show that those upper bounds are sharp.

The examples we will work with are operator whose integral kernel is a pointwise majorant for a paraproduct. In fact these operators are also the moduli of the paraproducts in the operator theoretic sense. Actually the term "paraproduct" is used for a large and informally defined class of operators. The simplest example is the dyadic paraproduct acting on $L^2(\mathbf{R})$ which we now define. Let $\mathbf{Q}$ be the collection of dyadic cubes in $\mathbf{R}$. For Q in $\mathbf{Q}$ let φ_Q be the normalized characteristic function, $\varphi_Q(x) = |Q|^{-1/2}\chi_Q(x)$ and let h_Q be the Haar function, $h_Q(x) = -|Q|^{-1/2}$ on the left half of Q and $|Q|^{-1/2}$ on the right half. Recall that $\{h_Q\}$ is an orthonormal basis of $L^2(\mathbf{R})$. Given a sequence of scalars $\Lambda = \{\lambda_Q\}_{Q\in\mathbf{Q}}$ we define the associated dyadic paraproduct, S, by

$$Sf = \sum \lambda_Q < f, \varphi_Q > h_Q.$$

S need not be bounded even if $\{\lambda_Q\}$ is bounded. However, if

$$\Lambda \in l^p \tag{1.1}$$

for some finite p then S is bounded and in fact is compact and in the Schatten-von Neumann ideal S_p.

This work supported in part by a grant from the NSF.

The analysis of this operator in [R] actually doesn't make use of the oscillation in the h_Q. Rather the analysis is done on the majorant T given by

$$Tf = \sum |\lambda_Q| < f, \varphi_Q > \varphi_Q. \tag{1.2}$$

T is as a sum of expectation operators. For Q in $\mathbf{Q}$, let E_Q be the associated expectation operator

$$E_Q(f)(x) = \left(\frac{1}{|Q|}\int_Q f(t)dt\right)\chi_Q(x).$$

Thus

$$T = \sum |\lambda_Q| E_Q.$$

Note that if Λ is bounded then, formally, T has an integral kernel which is $O(|x-y|^{-1})$. Furthermore, and of particular interest to us, the integral kernel for T is a pointwise majorant for the integral kernels of S and of S^*.

(We also note in passing that there is some operator theoretic relation between operators such as S as operators such as T. Using the fact that the Haar system is orthonormal we find that

$$S^*Sf = \sum \lambda_Q^2 < f, \varphi_Q > \varphi_Q.)$$

If (1.1) holds for some p, $1 < p < \infty$, then T will be a positive compact operator and will have a sequence of eigenvalues which is in l^p. In that case Theorem A of [R] gives information about the size of the eigenvectors. Let p' be the conjugate index, $1/p + 1/p' = 1$. Roughly the eigenvectors can have singularities no worse than the singularity at the origin of $\exp((\log(1/|x|))^{1/p'})$. The precise result is

Theorem 1: *Suppose T is given by (1.2), that (1.1) holds for some finite p, and $Tf = \lambda f$, $\|f\|_2 = 1$, $\lambda > 0$. There are positive constants c_i which depend only on Λ so that for every Q in $\mathbf{Q}$ and for μ with $\lambda |Q|^{1/2}\mu > c_1$*

$$|Q \cap \{x : |f(x)| \geq \mu\}| \leq |Q| \exp(-c_2(\log\ c_3\lambda|Q|^{1/2}\mu)^{p'}).$$

There is a similar but slightly more complicated result if T is only assumed to be compact.

The proof of Theorem 1 doesn't actually require (1.2). It only requires that Λ be in weak l^p; $\Lambda \in l^{p,\infty}$. That hypothesis puts T in the slightly larger ideal $S_{p,\infty}$.

To help put this result in perspective we look at an example where generalized eigenvectors can be computed explicitly. As we said, many operators are called paraproducts; one is the following operator on the Hardy space $H^2(\mathbf{R})$. Given f in H^2 we can regard f as an analytic function in the upper half-plane. Pick and fix a function b analytic in the upper half-plane which tends to 0 at ∞. We define $S_b f$, the paraproduct with symbol b applied to f, as (the boundary value function on $\mathbf{R}$ of) the analytic function in the upper half-plane given by

$$S_b f(z) = \int_{i\infty}^{z} b'(\zeta) f(\zeta) d\zeta. \tag{1.3}$$

We can solve the eigenvalue equation explicitly and obtain a standard of comparison. If $Sf = \lambda f$ then $b'f = \lambda f'$. Hence $f = c \exp(b/\lambda)$. The necessary and sufficient condition for S to be in S_p is that b be in the Besov space B_p. In that case b will decay at infinity and hence f will not. Thus, although f is in L^2_{loc} it fails to be in L^2 and hence is a generalized eigenvector but not an actual eigenvector. However, in terms of local behavior, we are in perfect agreement with Theorem A — exponentials of functions in B_p satisfy the conclusion of Theorem A. A related context in which exponentials of functions in B_p occur is in the very nice result of Peller [Pe] who shows that Schmidt vectors of Hankel operators in S_p (and hence also the eigenvectors of the absolute values of Hankel operators) are dominated by exponentials of B_p functions.

2 The Example

We want to show that Theorem 1 is relatively sharp.

For $n = 1, 2, \ldots$ let $Q_n = (0, 2^{-n})$ and write λ_n for λ_{Q_n} and E_n for the associated expectation operator; $E_n = E_{Q_n}$.

Consider

$$T = \sum \lambda_n E_n \tag{2.1}$$

where Λ consists of positive terms and is in $l^{p,\infty}$. T is a positive compact operator and is given by integration against a positive kernel. Hence there will be a positive function F which is an eigenvector for the largest eigenvalue (and that eigenvalue will be the operator norm of T). We now estimate that function.

Write $\Delta_n = \chi_{Q_n} - \chi_{Q_{n+1}}$. Then

$$F = \sum c_n \Delta_n \tag{2.2}$$

for some nonnegative numbers $\{c_n\}$. Hence $Tf = \sum d_n \Delta_n$ where, using (2.1),

$$d_n = \sum_{j \le n} \lambda_j E_j(F).$$

Using (2.2) we see

$$E_j(F) = 2^j \sum_{k=j}^{\infty} 2^{-k-1} c_k.$$

Thus

$$d_n = \sum_{j \le n} \sum_{k=j}^{\infty} \lambda_j 2^{j-k-1} c_k$$

$$= \sum_{k \le n} \sum_{j=0}^{k} \lambda_j 2^{j-k-1} c_k + \sum_{k > n} \sum_{j=0}^{n} \lambda_j 2^{j-k-1} c_k$$

$$= \sum_{k\leq n}(\sum_{j=0}^{k} 2^j\lambda_j)2^{-k-1}c_k + \sum_{k>n}(\sum_{j=0}^{n} 2^j\lambda_j)2^{-k-1}c_k.$$

Set

$$A_k = \sum_{j=0}^{k} 2^j\lambda_j.$$

So

$$d_n = \sum_{k\leq n} A_k 2^{-k-1}c_k + A_n\sum_{k>n} 2^{-k-1}c_k.$$

So

$$d_{n+1} - d_n = (A_{n+1} - A_n)2^{-n-2}c_{n+1} + (A_{n+1} - A_n)\sum_{k>n+2} 2^{-k-1}c_k.$$

Now we use the fact that all the c_j are positive.

$$d_{n+1} - d_n = 2^{n+1}\lambda_{n+1}2^{-n-2}c_{n+1} + (\text{positive})$$

$$= 2^{-1}\lambda_{n+1}c_{n+1} + (\text{positive}).$$

Now we assume that F is an eigenvector and for notational convenience we suppose the eigenvalue is $2\ \lambda$. Hence $c_k = 2\ d_k$ and

$$d_{n+1} - d_n = \lambda\ \lambda_{n+1}d_{n+1} + (\text{positive}).$$

So

$$1 - \frac{d_n}{d_{n+1}} \geq \lambda\ \lambda_{n+1}$$

and hence

$$\frac{d_n}{d_{n+1}} \leq 1 - \lambda\ \lambda_{n+1} \leq \frac{1}{1+\lambda\lambda_{n+1}}\ .$$

We now multiply these estimates for $n = 1, 2, ..., N$ and then take reciprocals. We normalize by $d_1 = 1$.

$$d_{N+1} \geq \prod_{n=1}^{N}(1 + \lambda\ \lambda_{n+1}) = \exp\sum_{1}^{N}\log(1 + \lambda\ \lambda_{n+1}).$$

If

$$\lambda_1 \leq 1 \text{ and } \lambda_n \searrow 0$$

then we have $\log(1 + \lambda\ \lambda_n) \sim \lambda_n$. In this case

$$d_{N+1} \geq \exp\ c\sum_{1}^{N}\lambda_{n+1}.$$

For the particular choice $\lambda_n = n^{-1/p}$ this gives

$$d_N \geq \exp\ c\ N^{1/p'}.$$

and hence, for small positive x

$$f(x) \geq c_1 \exp(c_2(\log(1/x))^{1/p'}).$$

This shows that the form of estimate in Theorem 1 is appropriate and that we can't get by with Λ in a larger $l^{p,q}$ space.

3 Questions

- Can S and/or S^* have eigenvectors that are this large ?
- These results are for an operator T which is of a form similar to the form of the operator S^*S. Are there analogous results for operators which have a form similar to the operator SS^*?

REFERENCES

1. Peller, V. V. (1988). Smoothness of Schmidt functions of smooth Hankel operators, Function Sapces and applications (Lund, 1986), 337-364, *Lecture Notes in Mathematics 1302*, Springer, New York.
2. Rochberg, R. Size estimates for eigenvalues of singular integral operators and Schrödinger operators and for derivatives of quasiconformal mappings, *Amer. J. Math., to appear.*

Some Very Dense Subspaces of $C(X)$

S. J. SIDNEY Department of Mathematics, The University of Connecticut, Storrs, Connecticut 06269-3009, U.S.A.

ABSTRACT

It is shown that the space $E_0(X)$ of continuous functions on a compact metric space X that are locally constant on a (varying) dense open subset of X, endowed with the uniform norm, is barreled, even though it is not complete unless X contains a discrete dense open subset. This provides not only previously unrecognized examples of barreled spaces, but also new methods for demonstrating that function spaces are barreled. Related results are obtained.

1 INTRODUCTION

In an earlier paper dealing with functions operating on real function spaces, Alain Bernard proved [**B**, Lemme 12] a slightly stronger version of the following assertion: If E is a real *Banach function space* on X (a point-separating subspace of $C(X)$ that contains the constant functions and is a Banach space in a norm that dominates the uniform norm) and if φ is a nonconstant continuous real-valued function on an interval I that *operates from* $C(X)$ *to* E (that is, whenever $u \in C(X)$ and $u(X) \subseteq I$, it follows that $\varphi \circ u \in E$), then $E = C(X)$. This lemma is crucial for proving the following localization result [**B**, Théorème 3], which is deeper than it appears: If E is a real Banach function space on X, if E is *locally trivial* (that is, every point of X has a compact neighborhood V such that the restriction of E to V is $C(V)$), and if some nonaffine continuous function on an interval operates on E, then $E = C(X)$. The localization result, in turn, allows one to pass from theorems concluding that E is locally trivial under appropriate hypotheses to theorems concluding that $E = C(X)$.

Bernard's paper in this conference proceedings gives, in effect, a new treatment of [**B**, Lemme 12], and of its application to certain spaces $E_0(X)$. If X is a compact Hausdorff space, $E_0(X)$ denotes the subspace of $C(X)$ consisting of those functions that are locally constant on a (varying with the function) dense open subset of X. If $E_0 = E_0([0,1])$ (so the standard Cantor function is a typical member of E_0) then every function in E_0 operates from $C(X)$ to $E_0(X)$ provided X has a dense locally connected open subset; consequently, in this case $E_0(X)$ is not "Banachizable" unless $E_0(X) = C(X)$.

Bernard's work actually shows that for X having a dense locally connected open subset, $E_0(X)$, endowed with the uniform norm, possesses a number of functional-analytic properties which imply, among other things, this non-Banachizability when $E_0(X) \neq C(X)$, even though in many cases (for instance, E_0 itself) $E_0(X)$ is of the first category; most of these properties are equivalent to one another, and mean in some sense that the given normed space is dense in its completion in a very strong way. At the conference it was pointed out that these equivalent properties are also equivalent to the space being barreled. (Recall that a *barrel* in a normed linear space E is a closed absorbent absolutely convex subset of E, and E is said to be *barreled* if every barrel in E is a neighborhood of 0.) Thus Bernard's results show that $E_0(X)$ is barreled if X has a dense locally connected open subset.

It turns out that $E_0(X)$ in the uniform norm is *always* barreled, as are a number of closely related spaces. In this paper we shall content ourselves with treating the case when X is metrizable. More elaborate versions of these results will appear elsewhere.

If X is metrizable or has a dense locally connected open subset, one can check that $E_0(X)$ must separate the points of X; see the lemma in section 3 below for the metrizable case, and note that the other case follows from Bernard's work. Since the conference, Mary Ellen Rudin and Walter Rudin [**R**] have constructed a nontrivial X such that $E_0(X)$ consists of only the constant functions.

The plan of the rest of the paper is as follows. In the next section we shall recall only those relations between being barreled and having various other properties that we will use in our investigations in this paper. In section 3 we shall prove that if X is a compact metric space then $E_0(X)$ with the uniform norm is barreled; along the way we can observe (as do the Rudins) that in this case $E_0(X)$ separates the points of X. Finally, in the last section we shall see that if X and Y are infinite compact Hausdorff spaces, then the *algebraic* tensor product of $C(X)$ and $C(Y)$ is *not* barreled; this will show also that an implication given in section 2 cannot be reversed.

2 BARRELED SPACES

Consider a normed linear space E; it is immaterial whether the scalars are real or complex. The completion and topological dual of E will be denoted, respectively, by $\hat{E}$ and $E^\star$. The properties of interest to us are listed below. *Caution:* The names we are assigning them are not standard, and are being given just for convenience.

We will say that E has the *weak sequential property* if whenever (λ_n) is a sequence in $E^\star = \hat{E}^\star$ which has the property that $\lambda_n(x) \longrightarrow 0$ for every $x \in E$, it follows that also $\lambda_n(x) \longrightarrow 0$ for every $x \in \hat{E}$. This amounts to saying that the *sequential* weak-$\star$ topologies on $E^\star = \hat{E}^\star$ induced by E and $\hat{E}$ are identical, even though the corresponding weak-$\star$ topologies are different unless $E = \hat{E}$, that is, unless E is complete. An equivalent (by the usual Banach-Steinhaus theorem applied to $\hat{E}$, and a standard density argument) form of the property which is the one we will use is the following: If a sequence (λ_n) in $E^\star$ is weak-$\star$ null (in the duality with E), then the set $\{\lambda_n\}$ is norm-bounded in $E^\star$.

We will say that E has the *Banach dichotomy* if either it is already complete in its given norm, or no norm on E that dominates the given norm can be complete.

Last, E has the *Baire nested subspace property* if it is *not* the union of an increasing sequence of proper closed linear subspaces.

The following theorem summarizes the facts about barreled spaces that we shall need. Characterization (a) is standard.

Theorem 1. Let E be a normed linear space. Then each of the conditions (a), (b) and (c) listed below is necessary and sufficient that E be barreled. If E is barreled then E has the Baire nested subspace property.

- (a) Every weak-$\star$ bounded subset of $E^\star$ is norm-bounded.
- (b) E has the weak sequential property.
- (c) Every subspace of $\hat{E}$ that contains E has, in the norm inherited from $\hat{E}$, the Banach dichotomy.

Proof. Suppose E is barreled and S is a weak-$\star$ bounded subset of $E^\star$. $\{x \in E : \sup_S |\lambda(x)| \leq 1\}$ is a barrel, so is a neighborhood of 0, hence S is norm-bounded in $E^\star$. Thus (a) holds.

It is immediate that (a) implies (b).

Conversely, suppose E contains a barrel Q which is not a neighborhood of 0. Take $x_n \in E$ not in Q but of norm smaller than n^{-2}. By separation theorems there is $\lambda_n \in E^\star$ of norm n such that $|\lambda_n(x)| \leq |\lambda_n(x_n)| \leq n^{-1}$ for all $x \in Q$. It follows that $\lambda_n(x) \longrightarrow 0$ for all $x \in Q$, and so for all $x \in E$. Thus if E is not barreled then (b) does not hold.

Note that conditions (a) and (b) are "upward hereditary" in that if E enjoys it then so does any other subspace of $\hat{E}$ containing E; thus the same is true for the property of being barreled.

Now suppose E is barreled in its original norm, F is a subspace of $\hat{E}$ that contains E, and N is a complete norm on F that dominates the norm inherited from $\hat{E}$. If $Q = \{x \in F : N(x) \leq 1\}$ and B is the closure of Q in F with respect to the norm inherited from $\hat{E}$, then B is a barrel in F with respect to the inherited norm, so by the preceding remark contains $U \equiv \{x \in F : \|x\|_{\hat{E}} \leq r\}$ for some $r > 0$. The usual geometric series/successive approximations argument used to prove the open mapping theorem in Banach spaces now works here. Specifically, given $x \in U$, inductively select vectors $x_n \in Q \quad (n \geq 0)$ so that $\|\sum_{k=0}^{n} 2^{-k} x_k - x\|_{\hat{E}} < r 2^{-(n+1)}$. If $y_n = \sum_{k=0}^{n} 2^{-k} x_k$ then the sequence (y_n) is N-Cauchy, so converges (N) to a vector $z \in F$, $N(z) \leq 2$. Then also $y_n \longrightarrow z \quad (\|\cdot\|_{\hat{E}})$, so $z = x$. Thus if $x \in U$ then $N(x) \leq 2$, that is, $N \leq 2r^{-1}\|\cdot\|_{\hat{E}}$ on F, and the two norms on F are equivalent, making $\|\cdot\|_{\hat{E}}$ complete on F. Thus if E is barreled then (c) holds.

Suppose now that (b) fails, so there is a sequence (λ_n) in $E^\star = \hat{E}^\star$ that is weak-$\star$ null on E, but not norm-bounded in $E^\star$. By passing to a subsequence if necessary, we may assume that $\|\lambda_n\| \longrightarrow \infty$. For $x \in \hat{E}$ define $N_1(x) = \sup_n |\lambda_n(x)|$, and let $E_1 = \{x \in \hat{E} : N_1(x) < \infty\}$, a proper (by the Banach-Steinhaus theorem) subspace of $\hat{E}$ that contains E. Then $N(x) = \max\{\|x\|_{\hat{E}}, N_1(x)\}$ is a complete norm on E_1 that dominates the norm inherited from $\hat{E}$, that is, (c) fails.

Finally, suppose E_n is an increasing sequence of proper closed linear subspaces of E. Take $\lambda_n \in E^\star$ of norm n that annihilates E_n. Trivially $\lambda_n \longrightarrow 0$ pointwise on the union of the E_n, so if that union is all of E then (b) fails. Thus if E is barreled then it has the Baire nested subspace property. The proof is complete. ○

3 $E_0(X)$ IS BARRELED

It happens that $E_0(X)$ is barreled for *every* compact Hausdorff space X, whether or not $E_0(X)$ separates the points of X. In this paper, however, we shall restrict our attention to the case of metrizable X, for which point separation always takes place, and leave the general situation to be treated elsewhere. The proof of the following lemma is somewhat different (and less general) than that given in [**R**].

Lemma. [**R**] If X is a metrizable compact space then $E_0(X)$ separates the points of X.

Proof. Let p and q be distinct points of X and let h be a continuous function from X into $[0,1]$ such that $h(p) = 0$ and $h(q) = 1$. If $\{G_n\}$ is a countable base for the topology of X, a set S obtained by choosing one point from each nonempty $h(G_n)$ is an at most countable subset of $[0,1]$, so it is not hard to construct in $[0,1]$ a Cantor set K that does not meet S. If φ is a Cantor function mapping $[0,1]$ onto itself in nondecreasing fashion, constant on the intervals comprising the complement of K in $[0,1]$, then $f \equiv \varphi \circ h$ belongs to $E_0(X)$ and maps p to 0 and q to 1. ○

Theorem 2. If X is a metrizable compact space then $E_0(X)$, endowed with the uniform norm, is barreled.

Proof. We shall show that $E_0(X)$ has the weak sequential property, whence the result will follow from Theorem 1.

By the lemma together with the Stone-Weierstrass theorem, $E_0(X)$ is (uniformly) dense in $C(X)$; in particular, then, the dual space of $E_0(X)$ is canonically identified with $M(X)$, the space of Baire measures on X. Also, the space of real functions in $E_0(X)$ is closed under truncation.

Suppose, to obtain a contradiction, that $M(X)$ contains a sequence (λ_n) such that $\|\lambda_n\| \longrightarrow \infty$ but $\int u d\lambda_n \longrightarrow 0$ for every $u \in E_0(X)$. Passing to a subsequence if necessary, there is $x_0 \in X$ such that $|\lambda_n|(V) \longrightarrow \infty$ for every neighborhood V of x_0, where $|\lambda|$ denotes the total variation measure associated to the measure λ. Let $\gamma_n = \lambda_n(\{x_0\})$ and $\lambda'_n = \lambda_n - \gamma_n\delta$ where δ is point-mass at x_0. Passing to a further subsequence if necessary, we may arrange that $|\lambda'_n|(V) \longrightarrow \infty$ for every neighborhood V of x_0.

Build a decreasing sequence (V_k) of neighborhoods of x_0, a subsequence (λ_{n_k}) of (λ_n), and a sequence (u_k) of functions in $E_0(X)$ that vanish at x_0 to satisfy

(1) $\operatorname{diam}(V_k) \longrightarrow 0$,

$$j < k \Longrightarrow V_k \cap \operatorname{supp}(u_j) = \emptyset, \ \ |\lambda'_{n_j}|(V_k) \le 1;$$

$$(2) \quad |\lambda'_{n_k}|(V_k) > k2^k,$$

$$j < k \Longrightarrow |\int u_j d\lambda'_{n_k}| \le 1;$$

$$(3) \quad \|u_k\| \le 1, \ \ \operatorname{supp}(u_k) \subseteq V_k \setminus \{x_o\}, \ \ \int u_k d\lambda'_k \ge k2^k.$$

Here "diam" is diameter and "supp" is closed support, that is, the closure of the set on which the given function is nonzero.

Let $u = \sum_1^\infty 2^{-k} u_k \in C(X)$. This sum is actually finite off any neighborhood of x_0, so u belongs to $E_0(X)$ locally except at x_0, hence $u \in E_0(X)$. But (using the fact that $u(x_0) = 0$)

$$\begin{aligned} &|\int u d\lambda_{n_k}| = |\int u d\lambda'_{n_k}| \\ &\ge 2^{-k}|\int u_k d\lambda'_{n_k}| - \sum_{j<k} 2^{-j}|\int u_j d\lambda'_{n_k}| - \sum_{j>k} 2^{-j}|\int u_j d\lambda'_{n_k}| \\ &\ge 2^{-k}(k2^k) - \sum_{j<k} 2^{-j} - \sum_{j>k} 2^{-j}|\lambda'_{n_k}|(V_j) > k-1 \longrightarrow \infty, \end{aligned}$$

giving the sought contradiction. ○

Theorem 2 is of interest only when $E_0(X)$ is not complete, or what is the same (in view of the lemma), when $E_0(X) \neq C(X)$. The following proposition shows that this is the usual situation.

Proposition. If X is a metrizable compact space, $E_0(X)$ is complete if and only if X contains a discrete dense open subset (that is, a dense subset consisting entirely of isolated points of X).

Proof. Clearly if X contains such a subset then $E_0(X) = C(X)$. Suppose that X does not contain such a subset. Let $(V_n)_{n\in\mathbf{N}}$ be a sequence (perhaps with repetitions) consisting of those members of a countable base for the topology of X that are not singletons. We shall inductively select sequences $(u_n)_{n=1}^\infty$ and $(w_n)_{n=0}^\infty$ in $C(X)$ so that w_n is nonconstant on V_n. First set $w_0 = 0$. If $n > 0$ and u_k $(1 \le k \le n-1)$ and w_k $(0 \le k \le n-1)$ have been chosen in $C(X)$ so that w_k is nonconstant on V_k $(1 \le k \le n-1)$, let $u_n = 0$ if w_{n-1} is nonconstant on V_n, otherwise let u_n be 0 off V_n, nonconstant on V_n, and such that $|u_n| < 2^{-n}\operatorname{diam}(w_k(V_k))$ on X for $1 \le k \le n-1$; let $w_n = w_{n-1} + u_n$. Then $\sum u_n = \lim w_n$ converges uniformly to some $w \in C(X)$, and w is not constant on any V_n. Consequently the (open) set of points of X at which w is locally constant is discrete, so by assumption is not dense in X, and w does not belong to $E_0(X)$. ○

4 TENSOR PRODUCTS

Let X and Y be compact Hausdorff spaces. If $f \in C(X)$ and $g \in C(Y)$, the *simple tensor* $f \otimes g$ is the member of $C(X \times Y)$ defined by $(f \otimes g)(x, y) = f(x)g(y)$. The *algebraic tensor product* of $C(X)$ and $C(Y)$, denoted $C(X) \otimes C(Y)$, is the subspace of $C(X \times Y)$ consisting of the finite linear combinations of simple tensors. As usual, $C(X) \otimes C(Y)$ is endowed with the uniform norm, and (by the Stone-Weierstrass theorem) is a dense subalgebra of $C(X \times Y)$. We shall close the paper with the following result.

Theorem 3. Let X and Y be compact Hausdorff spaces. Then $C(X) \otimes C(Y)$, endowed with the uniform norm, has the Baire nested subspace property. Furthermore, if X and Y are both infinite, then $C(X) \otimes C(Y)$ is *not* barreled.

In particular this shows that the implication "barreled implies Baire nested subspace property" in Theorem 1 is not reversible. That $C(X) \otimes C(Y)$ is not barreled when X and Y are infinite actually follows from Theorem 1 and the fact that the projective tensor product of $C(X)$ and $C(Y)$ is not all of $C(X \times Y)$, since the latter implies that $C(X) \otimes C(Y)$ does not have the Banach dichotomy; instead we give a direct proof that can be adapted to give a proof of the projective tensor product result, and is a variant of a proof given in a related context in [**G**, proof of Theorem 2.4]. Of course, if X or Y is finite then $C(X) \otimes C(Y) = C(X \times Y)$ is complete, and certainly barreled.

Proof of Theorem 3. Let $E = C(X) \otimes C(Y)$.

Suppose E is the union of an increasing sequence (E_n) of closed subspaces of E. For $g \in C(Y)$ let $E_n(g) = \{f \in C(X) : f \otimes g \in E_n\}$. The $E_n(g)$ form an increasing sequence of closed subspaces of $C(X)$ whose union is $C(X)$, so $E_n(g) = C(X)$ for some $n \in \mathbf{N}$ (depending on g). In a word: for each $g \in C(Y)$, $C(X) \otimes g \equiv \{f \otimes g : f \in C(X)\} \subseteq E_n$ for some $n \in \mathbf{N}$. Let $F_n = \{g \in C(Y) : C(X) \otimes g \subseteq E_n\}$. The F_n form an increasing sequence of closed subspaces of $C(Y)$ whose union is $C(Y)$, so $F_n = C(Y)$ for some $n \in \mathbf{N}$. This says that for this n, E_n contains all the simple tensors, so $E_n = E$, proving that E has the Baire nested subspace property.

Now suppose X and Y are both infinite. The proof that E is not barreled will be given in the complex case. For each $n \in \mathbf{N}$ chose n distinct points $x_{n,1}, \ldots, x_{n,n}$ in X and n distinct points $y_{n,1}, \ldots, y_{n,n}$ in Y and let $\delta_{n,j,k}$ denote point mass at $(x_{n,j}, y_{n,k}) \in X \times Y$. Let $\omega = e^{2\pi i/n}$ and $\lambda_n = n^{-7/4} \sum_{j=1}^{n} \sum_{k=1}^{n} \omega^{jk} \delta_{n,j,k} \in E^\star = \hat{E}^\star$ where $\hat{E} = C(X \times Y)$, so $\|\lambda_n\| = n^{1/4}$. Suppose $f \otimes g$ is a simple tensor. Set $f(x_{n,j}) = f_j$ and $g(y_{n,k}) = g_k$. Then, using the Schwarz inequality,

$$|\lambda_n(f \otimes g)|^2 = n^{-7/2} |\sum_{j=1}^{n} f_j [\sum_{k=1}^{n} g_k \omega^{jk}]|^2 \leq n^{-7/2} [\sum_{j=1}^{n} |f_j|^2] [\sum_{j=1}^{n} | \sum_{k=1}^{n} g_k \omega^{jk}|^2]$$

$$\leq n^{-5/2} \|f\|^2 \sum_{j=1}^{n} \sum_{k=1}^{n} \sum_{r=1}^{n} g_k \overline{g_r} \omega^{jk} \overline{\omega}^{jr} = n^{-5/2} \|f\|^2 \sum_{k=1}^{n} \sum_{r=1}^{n} g_k \overline{g_r} \sum_{j=1}^{n} \omega^{(k-r)j}$$

$$= n^{-5/2} \|f\|^2 \sum_{k=1}^{n} n|g_k|^2 \leq n^{-1/2} \|f\|^2 \|g\|^2,$$

so $|\lambda_n(f \otimes g)| \leq n^{-1/4} \|f\| \|g\|$. Thus $\lambda_n(f \otimes g) \longrightarrow 0$ for every simple tensor $f \otimes g$, so $\lambda_n \longrightarrow 0$ pointwise on E, yet $\|\lambda_n\| \longrightarrow \infty$. We see that E does not have the weak sequential property, so by Theorem 1 it is not barreled. The theorem is now proven. ○

ACKNOWLEDGEMENTS

Thanks are due to Alain Bernard and Denise Muraz for the form of the proof of Theorem 2 given here, to the several *conférenciers* who brought the "barreled" connection to Bernard's and my attention, and to the referee for suggesting a number of useful modifications to my original manuscript. A large debt is owed to Krzysztof Jarosz and his associates for organizing a superb conference.

REFERENCES

B. A. Bernard, Une fonction non Lipschitzienne peut-elle opérer sur un espace de Banach de fonctions non trivial?, *J. Functional Analysis* **122** *(1994), 451-477.*

G. I. Glicksberg, *Recent Results on Function Algebras*, CBMS Regional Conference Series in Mathematics number 11, Amer. Math. Soc. 1972.

R. M. E. Rudin and W. Rudin, Continuous functions that are locally constant on dense sets, preprint.

Antosik's Interchange Theorem

CHARLES SWARTZ Department of Mathematical Sciences, New Mexico State University, Las Cruces, NM 88003

A problem frequently encountered in analysis is the existence and equality of the iterated series $\sum_i \sum_j x_{ij}$ and $\sum_j \sum_i x_{ij}$. There are a number of sufficient conditions which guarantee the existence and equality of these iterated series, but they are often difficult to verify in practice, particularly in infinite dimensional spaces. On the other hand, Antosik has given an interesting sufficient condition which guarantees the existence and equality of iterated series and which is often easily verified in practice. In this note we describe Antosik's result, indicate two applications of the result and give a generalization of Antosik's result.

Throughout this note let G be a Hausdorff Abelian topological group. If $\{x_j\}$ is a sequence in G and σ is an infinite subset of $\mathbb{N}$, we denote by $\sum_{j \in \sigma} x_j$ the sum of the series $\sum_{j=1}^{\infty} x_{n_j}$ where the elements of σ are arranged in a subsequence $\sigma = \{n_1 < n_2 < \cdots\}$ and the subseries converges in G; if $\sigma \subset \mathbb{N}$ is finite, the meaning of $\sum_{j \in \sigma} x_j$ is clear. A series $\sum x_j$ in G is said to be subseries convergent if the subseries $\sum x_{n_j}$ converges in G for every subsequence $\left\{x_{n_j}\right\}$ of $\{x_j\}$.

For the proof of Antosik's result we use a vector version of a result from summability theory due to Hahn and Schur. For the convenience of the reader, we state the version which we use.

Theorem 1. *Let $x_{ij} \in G$ be such that $\sum_j x_{ij}$ is subseries convergent for every $i \in \mathbb{N}$.*

If $\lim_i \sum_{j\in\sigma} x_{ij}$ exists for every $\sigma \subset \mathbb{N}$ and if $x_j = \lim_i x_{ij}$, then

(i) *$\sum x_j$ is subseries convergent and*

(ii) *$\lim_i \sum_{j\in\sigma} x_{ij} = \sum_{j\in\sigma} x_j$ uniformly for $\sigma \subset \mathbb{N}$.*

For a proof of Theorem 1 see [AS1] or [AS2]. As observed in [AS1] §9, Theorem 1 gives a generalization of a well-known scalar result of Hahn and Schur which, in particular, asserts that a sequence in ℓ^1 is weakly convergent if and only if it is norm convergent ([Sw1] 16.14).

We now state and prove Antosik's Theorem.

Theorem 2 (Antosik's Interchange Theorem). *Let $x_{ij} \in G$ for $i,\ j \in \mathbb{N}$.*

Suppose $\sum_{i=1}^{\infty} \sum_{j\in\sigma} x_{ij}$ converges for every $\sigma \subset \mathbb{N}$. Then the double series $\sum_{i,j} x_{ij}$ converges and

$$\sum_{i,j} x_{ij} = \sum_{i=1}^{\infty}\sum_{j=1}^{\infty} x_{ij} = \sum_{j=1}^{\infty}\sum_{i=1}^{\infty} x_{ij}.$$

Proof: Note that the hypothesis implies that the series $\sum_{i=1}^{\infty} x_{ij}$ converges for every j [take $\sigma = \{j\}$]. Set $z_{mj} = \sum_{i=1}^{m} x_{ij}$. Then for $\sigma \subset \mathbb{N}$, $\sum_{j\in\sigma} z_{mj} = \sum_{i=1}^{m} \sum_{j\in\sigma} x_{ij}$ converges to $\sum_{i=1}^{\infty} \sum_{j\in\sigma} x_{ij}$ as $m \to \infty$. By Theorem 1 the series $\sum_{j=1}^{\infty} \left(\sum_{i=1}^{\infty} x_{ij} \right)$ is subseries convergent and

$$\lim_m \sum_{i=1}^{m} \sum_{j\in\sigma} x_{ij} = \sum_{j\in\sigma} \sum_{i=1}^{\infty} x_{ij}$$

uniformly for $\sigma \subset \mathbb{N}$. Hence, the double limit

$$\lim_{m,n} \sum_{i=1}^{m} \sum_{j=1}^{n} x_{ij} = \sum_{i,j} x_{ij}$$

exists and equals

$$\sum_{i=1}^{\infty}\sum_{j=1}^{\infty} x_{ij} = \sum_{j=1}^{\infty}\sum_{i=1}^{\infty} x_{ij}$$

Remark 3. Antosik established a version of Theorem 2 for spaces with a (sequential) convergence structure ([A]). Antosik required that the iterated series $\sum_{i=1}^{\infty}\sum_{j=1}^{\infty} x_{in_j}$ converge for every increasing sequence $\{n_j\}$. However, this condition is equivalent to the hypothesis in Theorem 2 since it implies that the series $\sum_{i=1}^{\infty} x_{ij}$ converges for every $j \in \mathbb{N}$ [consider the difference between the two series $\sum_{i=1}^{\infty}\sum_{k=1}^{\infty} x_{in_k}$ and $\sum_{i=1}^{\infty}\sum_{k=1}^{\infty} x_{im_k}$, where $n_k = k$ for every k and $m_k = \{1,\cdots,j-1,j+1,\cdots\}$]. The proof above for series in a topological group was given in [Sw2].

We now give an application of Antosik's Interchange Theorem by deriving a version of the Orlicz-Pettis Theorem for series in a topological vector space with a Schauder basis. The first Orlicz-Pettis Theorem of this type was obtained by Stiles for series in an F-space with a Schauder basis; Stiles' result seems to be the first version of the Orlicz-Pettis Theorem for series in a non-locally convex space [see the remarks in [K]].

Let E be a Hausdorff topological vector space with a Schauder basis $\{b_i\}$ and associated coordinate functionals $\{f_i\}$. We do not assume that the $\{f_i\}$ are continuous although this is the case if E is a complete metric linear space. Let $F = \{f_i : i \in \mathbb{N}\}$. We show that the Antosik Interchange Theorem can be used to establish an Orlicz-Pettis Theorem for the weak topology $\sigma(E,F)$ and the original topology of E.

Theorem 4. *If $\sum x_i$ is $\sigma(E,F)$ subseries convergent in E, then $\sum x_i$ is subseries convergent in the original topology of E.*

Proof: Let $\{n_j\}$ be an increasing sequence of positive integers and let $x = \sum_{j=1}^{\infty} x_{n_j}$ be the $\sigma(E,F)$ sum of the series. For each i, $\sum_{j=1}^{\infty} \left\langle f_i, x_{n_j}\right\rangle b_i$ converges in E to $\langle f_i, x\rangle b_i$ since

$\sum_{j=1}^{\infty} \left\langle f_i, x_{n_j}\right\rangle = \langle f_i, x\rangle$. Then

$$\sum_{i=1}^{\infty}\sum_{j=1}^{\infty} \left\langle f_i, x_{n_j}\right\rangle b_i = \sum_{i=1}^{\infty} \langle f_i, x\rangle b_i = x.$$

By Antosik's Theorem 2,

$$\sum_{i=1}^{\infty}\sum_{j=1}^{\infty} \langle f_i, x_j\rangle b_i = \sum_{j=1}^{\infty}\sum_{i=1}^{\infty} \langle f_i, x_j\rangle b_i = \sum_{j=1}^{\infty} x_j = x$$

where the series converge in the original topology of E. Since the same argument can be applied to any subsequence of $\{x_j\}$, the result follows.

Example 5. Let E be a vector space of scalar-valued sequences. If E has a vector topology such that the coordinate maps $x = \{x_j\} \to x_j$ are continuous from E into the scalar field for each j, then E is called a K-space. Let e^k be the sequence with a 1 in the k^{th}coordinate and 0 in the other coordinates. If each $x = \{x_k\} \in E$ has a series expansion $x = \sum_{k=1}^{\infty} x_k e^k$(convergence in E), then E is called an AK-space, i.e., $\{e^k\}$ is a Schauder basis for E ([Z], [Wi2]). If E is an AK-space and $F = \{e^k : k \in E\}$, then $\sigma(E, F)$ is just the topology of coordinatewise convergence on E. It follows from Theorem 4 that a series in E is subseries convergent in the topology of coordinatewise convergence if and only if the series is subseries convergent in the original topology of E.

The proof of Theorem 4 is applicable to more general situations than that of a topological vector space with a Schauder basis. Let (G, τ) be an Abelian Hausdorff topological group and assume there exists a sequence of homomorphisms $P_i : G \to G$ such that $\sum_{i=1}^{\infty} P_i x = x$ (convergence in τ) for each $x \in G$. In Theorem 4 above, we may take $P_i x = \langle f_i, x \rangle b_i$. If G is a Hausdorff topological vector space and each P_i is a projection, then a sequence $\{P_i\}$ satisfying the condition above is called a Schauder decomposition ([LT]).

Theorem 6. *Let σ be a Hausdorff group topology on G and assume that each P_i is $\sigma - \tau$ continuous. If $\sum x_i$ is σ subseries convergent, then $\sum x_i$ is τ subseries convergent.*

Proof: Let $\{n_j\}$ be an increasing sequence of positive integers and let $\sum_{j=1}^{\infty} x_{n_j}$ be the σ sum of this series. For each i, $\sum_{j=1}^{\infty} P_i(x_{n_j}) = P_i\left(\sum_{j=1}^{\infty} x_{n_j}\right)$, where the convergence is in τ by the continuity of P_i. Hence,

$$\sum_{i=1}^{\infty}\sum_{j=1}^{\infty} P_i(x_{n_j}) = \sum_{i=1}^{\infty} P_i\left(\sum_{j=1}^{\infty} x_{n_j}\right) = \sum_{j=1}^{\infty} x_{n_j},$$

where the convergence is in τ. By Antosik's Theorem 2,

$$\sum_{j=1}^{\infty}\sum_{i=1}^{\infty} P_i(x_j) = \sum_{j=1}^{\infty} x_j$$

converges in τ. Since the same argument can be applied to any subsequence of $\{x_j\}$, the result follows.

As noted above Theorem 4 is a special case of Theorem 6 where we take $P_i x = \langle f_i, x \rangle b_i$, and Theorem 6 is applicable to spaces having a Schauder decomposition. Many spaces of vector-valued sequences have Schauder decompositions so Theorem 6 is applicable to such spaces; for example, $\ell^p(X)$, with the topology of coordinatewise convergence, where X is a Banach space and $0 < p < \infty$.

Let E and F be scalar sequence spaces and $A = [a_{ij}]$ an infinite scalar matrix. If $x \in E$, we denote the kth coordinate of x by x_k so that $x = \{x_k\}$; we denote by e^k the sequence with 1 in the kth coordinate and 0 in the other coordinates. We say that A maps E into F if the series $\sum_{j=1}^{\infty} a_{ij}x_j$ converge for every i and $\left\{\sum_{j=1}^{\infty} a_{ij}x_j\right\} = Ax \in F$; if A maps E into F, we write $A\colon E \to F$. The classical Hellinger-Toeplitz Theorem asserts that if $A : \ell^2 \to \ell^2$, then A is (norm) continuous ([HT]). We show how Antosik's Theorem can be used to obtain a generalization of this result.

If x and y are scalar sequences, we write xy for the pointwise product of x and y; $xy = \{x_i y_i\}$. If $\sigma \subset \mathbb{N}$, let C_σ be the characteristic function of σ. The space E is said to be monotone if $C_\sigma x \in E$ when $\sigma \subset \mathbb{N}$, $x \in E$.

The α-dual (β-dual) of E is $E^\alpha = \{\{y_i\} : \sum_{i=1}^{\infty} |x_i y_i| < \infty$ for every $x \in E\}$ $(E^\beta = \{\{y_i\} : \sum_{i=1}^{\infty} x_i y_i$ converges for every $x \in E\})$. Of course, if E is monotone, $E^\alpha = E^\beta$. If $x \in E$ and $y \in E^\beta$, we write $y \cdot x = \sum_{i=1}^{\infty} y_i x_i$; this defines a duality between E, E^β and E, E^α. The weak topologies induced by these dualities are denoted by $\sigma(E, E^\beta)$, etc.

Let $A^T = [a_{ji}]$ be the transpose of A. Note that if $A : E \to F$, then the series $\sum_{i=1}^{\infty} a_{ij} y_i$ converges for every $j \in \mathbb{N}$ and $y \in F^\beta$ so $A^T : F^\beta \to s$, the vector space of all sequences. We consider conditions guaranteeing that $A^T : F^\beta \to E^\beta$.

Theorem 7. *Assume that E is monotone and $A : E \to F$. Then*

(i) $y \cdot Ax = A^T y \cdot x$ *for every* $x \in E, y \in F^\beta$.

(ii) $A^T : F^\beta \to E^\beta$ *and* A^T *is* $\sigma(F^\beta, F) - \sigma(E^\beta, E)$ *continuous.*

(iii) A is $\sigma(E, E^\beta) - \sigma(F, F^\beta)$ *continuous.*

Proof: Let $x \in E$, $y \in F^\beta$. Then $y \cdot Ax = \sum_{i=1}^{\infty} \sum_{j=1}^{\infty} y_i a_{ij} x_j$. Since x can be replaced in this sum by $C_\sigma x$ for any $\sigma \subset \mathbb{N}$, Antosik's Theorem is applicable and yields

$$y \cdot Ax = \sum_{j=1}^{\infty} x_j \sum_{i=1}^{\infty} a_{ij} y_i = x \cdot A^T y$$

so **(i)** holds. **(ii)** and **(iii)** follow immediately from **(i)**.

It follows from **(iii)** and standard results, that A is also continuous with respect to the Mackey (strong) topologies on E and F from the duality between E, E^β and F, F^β ([Wil], [Sw1]). In particular, if $A : \ell^2 \to \ell^2$, then A is norm continuous since $(\ell^2)^\beta = \ell^2$; this is the classical Hellinger-Toeplitz Theorem.

The proof of Theorem 7 is easily adapted to the case of vector-valued sequence spaces and operator-valued measures (see [Sw2]). It should also be pointed out there are stronger results than Theorem 7 which involve the weak sequential completeness of E^β, a property guaranteed by the monotonicity of E ([Swt1],[Swt2], [St1]).

Antosik's Theorem gives a sufficient condition for the interchange of the orders of summation in an iterated series, but the condition is by no means necessary. For example, let $\sum a_i$ and $\sum b_j$ be convergent real-valued series and set $x_{ij} = a_i b_j$. Then

$$\sum_{i=1}^{\infty}\sum_{j=1}^{\infty} x_{ij} = \sum_{j=1}^{\infty}\sum_{i=1}^{\infty} x_{ij},$$

but if the series $\sum b_j$ is conditionally convergent, the hypothesis of Theorem 2 is certainly not satisfied. We give a generalization of Antosik's Theorem which includes the above example.

The basic idea in the generalization of Antosik's Theorem is to replace the family of all subsets of $\mathbb{N}$ by a smaller family, for this, we require a variant of the Hahn-Schur Theorem 1 with the family of all subsets of $\mathbb{N}$ replaced by a smaller family. The type of family which we consider was introduced by Samaratunga and Sember ([SaSe]). By an interval in $\mathbb{N}$ we mean a set of the form $[m,n] = \{k \in \mathbb{N} : m \le k \le n\}$, where $m < n$. A family $\mathcal{F}$ of subsets of $\mathbb{N}$ is called an $IQ\sigma$-family if $\mathcal{F}$ contains the finite subsets and whenever $\{I_j\}$ is a pairwise disjoint sequence of intervals, there is a subsequence $\{I_{n_j}\}$ such that $\bigcup_{j=1}^{\infty} I_{n_j} \in \mathcal{F}$ ([SaSe]).

Example 8. Let $\sum b_j$ be a convergent series in $\mathbb{R}$. Then $\mathcal{F} = \left\{\sigma \subset \mathbb{N} : \sum_{j\in\sigma} b_j \text{ converges}\right\}$ is an $IQ\sigma$-family. For let $\{I_j\}$ be a pairwise disjoint sequence of intervals which we can assume to be increasing, i.e., $\max I_j < \min I_{j+1}$. Since $\sum b_j$ converges, there exists a subsequence $\{I_{n_j}\}$ such that $\left|\sum_{j\in\tau} b_j\right| \le 1/2^j$ for every $\tau \subset I_{n_j}$. Then $\sigma = \bigcup_{j=1}^{\infty} I_{n_j} \in \mathcal{F}$. Of course, if $\sum b_j$ is conditionally convergent, $\mathcal{F}$ is a proper subset of the power set of $\mathbb{N}$.

If $\mathcal{F}$ is a family of subsets of $\mathbb{N}$ and $x_j \in G$, we say the series $\sum x_j$ is $\mathcal{F}$ convergent if $\sum_{j\in\sigma} x_j$ converges for every $\sigma \in \mathcal{F}$.

We now establish a version of Theorem 1 for $IQ\sigma$-families.

Theorem 9. *Let $\mathcal{F}$ be an $IQ\sigma$-family containing* $\mathbb{N}$ *and* $\sum_j x_{ij}$ be *$\mathcal{F}$-convergent for every $i \in \mathbb{N}$. If $\lim_i \sum_{j\in E} x_{ij}$ exists for every $E \in \mathcal{F}$ and $x_j = \lim_i x_{ij}$, then $\sum_{j=1}^{\infty} x_j$* converges and $\lim_i \sum_{j=1}^{\infty} x_{ij} = \sum_{j=1}^{\infty} x_j$ *[note $\lim_i \sum_{j=1}^{\infty} x_{ij}$ exists by hypothesis].*

Proof: Set $z = \lim_i \sum_{j=1}^{\infty} x_{ij}$. If the desired conclusion fails to hold, there exist an increasing sequence $\{n_k\}$ and a closed, symmetric neighborhood of 0, U, such that $z - \sum_{j=1}^{n_k} x_j \notin U$ for all k. Thus,

$$\begin{aligned} z-\sum_{j=1}^{n_k} x_j &= \lim_i \left(\sum_{j=1}^{n_k} x_{ij} + \sum_{j=n_k+1}^{\infty} x_{ij} \right) - \sum_{j=1}^{n_k} x_j \\ &= \lim_i \left(\sum_{j=1}^{n_k} (x_{ij}-x_j) + \sum_{j=n_k+1}^{\infty} x_{ij} \right) \\ &= \lim_i \sum_{j=n_k+1}^{\infty} x_{ij} \end{aligned}$$

so $\lim_i \sum_{j=n_k+1}^{\infty} x_{ij} \notin U$ for all k.

Choose k_1 such that $\sum_{j=n_1+1}^{\infty} x_{k_1 j} \notin U$ and set $n_1 = \ell_1$. Pick a closed, symmetric neighborhood of 0, V, such that $V + V \subset U$. There exists $m_1 > \ell_1 + 1$ such that $\sum_{j=m_1+1}^{\infty} x_{k_1 j} \in V$. Hence, $\sum_{j=\ell_1+1}^{m_1} x_{k_1 j} \notin V$. Let $I_1 = \{\ell_1+1, \cdots, m_1\}$. Now choose $k_2 > k_1$ and $\ell_2 > m_1$ such that $\sum_{j=\ell_2+1}^{\infty} x_{k_2 j} \notin U$. As above, there exists $m_2 > \ell_2 + 1$ such that $\sum_{j=\ell_2+1}^{m_2} x_{k_2 j} \notin V$. Let $I_2 = \{\ell_2 + 1, \cdots, m_2\}$. Continuing this construction produces a sequence of disjoint, finite intervals $\{I_j\}$ with

$$\sum_{j\in I_i} x_{k_i j} \notin V \quad \text{for all} \quad i. \qquad (*)$$

Consider the matrix $M = \left[\sum_{k\in I_j} x_{k_i k} \right]$. The columns of M converge to $\sum_{k\in I_j} x_k$. By the $IQ\sigma$-property, for any subsequence of $\{I_j\}$, there is a further subsequence $\left\{I_{p_j}\right\}$ such that $A = \cup I_{p_j} \in \mathcal{F}$ and $\lim_i \sum_{j=1}^{\infty} \sum_{k\in I_{p_j}} x_{k_i k} = \lim_i \sum_{k\in A} x_{k_i k}$ exists. Hence, M is a $\mathcal{K}$-matrix so by the Antosik-Mikusinski Theorem ([AS] 2.2, [Sw1] 9.2), the diagonal of M converges to 0. But, this contradicts (*).

We use Theorem 9 to obtain a variant of Antosik's Theorem.

Theorem 10 ([St2]). *Let $\mathcal{F}$ be an $IQ\sigma$-family containing $\mathbb{N}$. If $\sum_{i=1}^{\infty} \sum_{j\in\sigma} x_{ij}$ converges for every $\sigma \in \mathcal{F}$, then*

$$\sum_{i=1}^{\infty}\sum_{j=1}^{\infty} x_{ij} = \sum_{j=1}^{\infty}\sum_{i=1}^{\infty} x_{ij}.$$

Proof: As in the proof of Theorem 2, if $z_{mj} = \sum_{i=1}^{m} x_{ij}$, then for any $\sigma \in \mathcal{F}$,

$\lim_m \sum_{j\in\sigma} z_{mj}$ exists and equals $\sum_{i=1}^{\infty}\sum_{j\in\sigma} x_{ij}$. From Theorem 9 $\sum_{j=1}^{\infty}\left(\sum_{i=1}^{\infty} x_{ij}\right)$ converges and equals $\sum_{i=1}^{\infty}\sum_{j=1}^{\infty} x_{ij}$.

Now consider the example above where $\sum a_i, \sum b_j$ converge and $x_{ij} = a_i b_j$. By Example 8 the set of all $\sigma \subset \mathbb{N}$ such that

$$\sum_{i=1}^{\infty}\sum_{j\in\sigma} x_{ij} = \sum_{i=1}^{\infty} a_i \sum_{j\in\sigma} b_j$$

converges is an $IQ\sigma$-family so Theorem 10 is applicable.

REFERENCES

[**A**] P. Antosik, *On Interchange of Limits, Generalized Functions, Convergence Structures and their Applications,* Plenum, N.Y., 1988, 367-374.

[**AS1**] P. Antosik and C. Swartz, Matrix Methods in Analysis, *Springer Lecture Notes in Mathematics,* 1113, Heidelberg, 1985.

[**AS2**] P. Antosik and C. Swartz, Multiplier Convergent Series and the Schur Theorem, *Southeast Asia Bull. Math.,* 15 (1991), 173-182.

[**HT**] E. Hellinger and O. Toeplitz, Gründlagen für eine Theorie den unendlichen Matrizen *Math. Ann.,* 69 (1910), 289-330.

[**K**] N. Kalton, The Orlicz-Pettis Theorem, Contemporary Math., *Amer. Math. Soc.,* Providence, 1980, 91-100.

[**LT**] J. Lindenstrauss and L. Tzafriri, *Classical Banach Spaces I,* Springer-Verlag, Berlin, 1977.

[**SaSe**] R. Samaratunga and J. Sember, Summability and Substructures of $2^{\mathbb{N}}$, *Southeast Asia Math. Bull.*, 66 (1990), 237-252.

[**S**] W. Stiles, On Subseries Convergence in F-spaces, *Israel J. Math.,* 8 (1970), 53-56.

[**St1**] C. Stuart, *Weak Sequential Completeness in Sequence Spaces,* Ph.D. dissertation, New Mexico State University, 1992.

[**St2**] C. Stuart, Interchanging the limits in a double series, preprint.

[**Sw1**] C. Swartz, *An Introduction to Functional Analysis,* Marcel Dekker, N.Y., 1992.

[Sw2] C. Swartz, Iterated Series and the Hellinger-Toeplitz Theorem, *Public, Mat.,* 36 (1992), 167-173.

[Sw3] C. Swartz, Subseries Convergence in Spaces with Schauder Basis, *Proc. Amer. Math. Soc.,* to appear.

[Swt1] J. Swetits, A Characterization of a Class of Barrelled Sequence Spaces, *Glasgow Math. J.,* 19 (1978), 27-31.

[Swt2] J. Swetits, On the Relationship between a Summability Matrix and its Transpose, J. Australian M.S., (1980), 362-368.

[Wi1] A. Wilansky, *Modern Methods in Topological Vector Spaces,* McGraw-Hill, N.Y., 1978.

[Wi2] A. Wilansky, *Summability through Functional Analysis,* North-Holland, Amsterdam, 1984.

[Z] K. Zeller and W. Beekman, *Theorie der Limitierungsverfahren,* Springer-Verlag, Berlin, 1970.

On the Continuity of Random Derivations

M. V. VELASCO Departamento de Análisis Matemático, Universidad de Granada, 18071 Granada, SPAIN. E-mail address: MVVELASCO@UGR.ES

A. R. VILLENA Departamento de Análisis Matemático, Universidad de Granada, 18071 Granada, SPAIN. E-mail address: AVILLENA@UGR.ES

Abstract. In this note, we discuss a randomized version of the classical notion of a deriv ation and the corresponding extension of the Johnson-Sinclair theorem about the continuity of derivations on semisimple Banach algebras to the case of random derivations.

A classical topic in the theory of automatic continuity is the investigation of algebraic conditions on a Banach algebra that imply an analytic property, namely the continuity of every derivation acting on such an algebra.

We recall that a *derivation* on a Banach algebra A is a linear mapping $D : A \to A$, which satisfies the condition that

$$D(ab) = D(a)b + aD(b), \ \forall a, b \in A, \tag{1}$$

Since in 1953 Kaplansky [10] conjectured the *continuity of every derivation acting on a C^*-algebra*, a result which was proved by Sakai [14] in 1960, an extensive literature on the automatic continuity of derivations arose (see for example [1-3],[5-10],[12-14]). Perhaps the most outstanding result in this context is a theorem by Johnson and Sinclair [9], which asserts *the continuity of all derivations acting on semisimple Banach algebras.*

Derivations have also been studied systematically in a somewhat more general context, namely in the case of certain module-valued mappings. Recall that a *module derivation* on a Banach algebra A is a linear mapping D from A into a Banach A-bimodule X, which satisfies condition (1), with respect to module multiplication. The question of continuity of such derivations gave a new perspective to the theory of automatic continuity of derivations.

In 1972, Ringrose [12] generalized the Sakai theorem by showing that *all module derivations acting on C^* -algebras are continuous.* A generalized version of the Ringrose theorem was obtained by Bade and Curtis in [2, Corollary 2.6].

However it is known that the Johnson-Sinclair theorem cannot be established for arbitrary module derivations, because non-continuous derivations of this kind have been constructed in [2].

Derivations have been investigated also from the nonassociative point of view. In this context, we recall that a *Jordan derivation* on a Banach algebra A is a linear operator D on A such that

$$D(a^2) = aD(a) + D(a)a, \ \forall a \in A,$$

or equivalently such that

$$D(a \circ b) = D(a) \circ b + a \circ D(b), \ \forall a, b \in A,$$

where, as usual, $a \circ b = \frac{1}{2}(ab + ba)$.

In 1970, Sinclair [13] conjectured that *every Jordan derivation on a semisimple Banach algebra is continuous* and proved that *every continuous Jordan derivation on a Banach algebra is a derivation.* Therefore, it follows from the Johnson-Sinclair theorem that a Jordan derivation on a semisimple Banach algebra is continuous precisely when it is a derivation.

In 1975, Cusack [6] established that *every Jordan derivation acting on a semiprime Banach algebra is a derivation.* Thanks to the Johnson-Sinclair theorem, this result allows to prove, as Brešar observed [5], that *Jordan derivations acting on semisimple Banach algebras are continuous.*

Our purpose is now to discuss a "randomization" of the Johnson-Sinclair theorem and its consequences for Jordan derivations.

For a Banach space Y and a fixed probability space $(\Omega, \Sigma, \mathbb{P})$, we will denote by $\mathcal{L}_0(\mathbb{P}, Y)$ the linear space of all Y-valued Bochner random variables on $(\Omega, \Sigma, \mathbb{P})$, that is the almost surely limits of simple measurable Y-valued functions on Ω. We endow $\mathcal{L}_0(\mathbb{P}, Y)$ with its usual topology, namely the topology of the convergence in probability [11; Section II]. With the usual almost surely identification on $\mathcal{L}_0(\mathbb{P}, Y)$ we obtain a complete metrizable topological linear space.

When Y is replaced by a Banach algebra A, it can be shown [4; Section 1.3.F] that $\mathcal{L}_0(\mathbb{P}, A)$ is in fact an algebra whose multiplication, the natural one, is continuous with respect to convergence in probability.

Given two Banach spaces X and Y, a *random operator* from X to Y is a mapping $T : X \to \mathcal{L}_0(\mathbb{P}, Y)$. The random operator T will be called *linear* if $T(\alpha x + \beta y) = \alpha T(x) + \beta T(y)$, almost surely, $\forall x, y \in X$ and α, β constants.

In the formulation of an equation used to model a given physical phenomenon, the coefficients and parameters are often determined experimentally. Hence, when dealing with certain parameters in physics, it is, in many instances, more appropriate to consider random variables instead of constants. The same thing can be said about the coefficients and forcing functions of many equations. A more realistic formulation of the equations arising in applied mathematics would therefore typically involve the study of random equations. Linear random operators arise extensively in the theory of random equations, which is presently a very active area of mathematical research. We refer to [4] and [15], where many further references can be found.

DEFINITION 1. A *random derivation* on a Banach algebra A is a linear random operator D on A such that

$$D(ab) = D(a)b + aD(b), \text{ almost surely } \forall a, b \in A.$$

Obviously, every derivation may be regarded, in a trivial way, as a random derivation by considering the trivial probability.

Given a Banach space Y and $r > 0$, let $\mathcal{L}_r(\mathbb{P}, Y)$ denote the space of all Y-valued random variables with *finite* r-th *moment*, that is

$$\mathcal{L}_r(\mathbb{P}, Y) = \{\mathbf{y} \in \mathcal{L}_0(\mathbb{P}, Y) : \int_\Omega \|\mathbf{y}\|^r d\mathbb{P} < \infty\}.$$

Note that the subspace $\mathcal{L}_r(\mathbb{P}, Y)$ has its own topology, given by the convergence in r-mean. This topology, which is stronger than the probability topology, is induced by the paranorm

$$\|\mathbf{x}\|_r = \int_\Omega \|\mathbf{x}\|^r d\mathbb{P}, \ \mathbf{x} \in \mathcal{L}_r(\mathbb{P}, Y),$$

when $0 < r < 1$, while for $r \geq 1$, it is induced by the seminorm

$$\|\mathbf{x}\|_r = \left(\int_\Omega \|\mathbf{x}\|^r d\mathbb{P}\right)^{\frac{1}{r}}, \ \mathbf{x} \in \mathcal{L}_r(\mathbb{P}, Y).$$

In either case, under the usual almost surely identification, $\mathcal{L}_r(\mathbb{P}, Y)$ is a complete metrizable topological linear space.

If a random operator T from a Banach space X to Y takes its values in $\mathcal{L}_r(\mathbb{P}, Y)$, then T is said to have r-th *moment.* For a random operator with r-th moment, there are two natural notions of continuity: the continuity in r-mean and the continuity in probability. However, it is an easy consequence of the closed graph theorem that, for linear random operators, these two notions of continuity are equivalent [19; Proposition 5]. Therefore, it is not ambiguous to say that such an operator T is *continuous.* This applies, in particular, to the case of random derivations. Thus, when a linear random operator T has a moment of order r, it suffices to verify the continuity in probability to establish that T is continuous in both probability and r-mean. Since, in this sense, $\mathcal{L}_r(\mathbb{P}, Y)$ and $\mathcal{L}_0(\mathbb{P}, Y)$ play the same role, our attention will be centred on $\mathcal{L}_0(\mathbb{P}, Y)$ from now on.

The main automatic continuity result in this context is the following theorem, whose proof was the goal of [16].

THEOREM 2 (Random Johnson-Sinclair theorem). *Every random derivation acting on a semisimple Banach algebra is continuous.*

This theorem is a generalization of the Johnson-Sinclair theorem. We note that random derivations can be considered as module derivations taking their values in a complete metrizable topological algebra, namely $\mathcal{L}_r(\mathbb{P}, A)$, $r \geq 0$. However, Theorem 2 is not contained in the theorem of Ringrose [12, Theorem 2] and the Bade-Curtis results [2].

The pattern of the proof of Theorem 2 is somewhat similar to the proof of the Johnson-Sinclair theorem. The proof is divided into two parts. In one of them, we work with the primitive ideals of A of infinite codimension, using ideas that Johnson and Sinclair give in the corresponding part of their work, but making essential refinements of some of them. In the other part, we deal with the primitive ideals of finite codimension of the algebra A. The key of the proof is to measure the degree of continuity that a not necessarily continuous linear random operator has. This is possible thanks to the random closed graph theorem, which will be discussed below.

If $T : X \to \mathcal{L}_0(\Omega, Y)$ is a not necessarily continuous linear random operator and if Ω_0 is a measurable subset of Ω with non-zero probability, then, in order to study the behaviour of T on Ω_0, we consider Ω_0 as a new probability space with the structure inherited from Ω and define the linear random operator $T_{\Omega_0} : X \to \mathcal{L}_0(\Omega_0, Y)$ by $T_{\Omega_0}(x) = T(x)_{|\Omega_0}$. The number

$$\beta(T) = \sup\{\mathbb{P}[\Omega_0] : T_{\Omega_0} \text{ is continuous}\}$$

may be interpreted as *the probability with which T is continuous.* Since the classical closed graph theorem shows that the continuity of a linear operator between two complete metrizable linear spaces is equivalent to the fact that the separating subspace of the operator equals zero (*i.e.*, the operator has closed graph), it seems reasonable to consider the number

$$\alpha(T) = \inf\{\mathbb{P}[\mathbf{y} = 0] : \mathbf{y} \in \mathcal{S}(T)\}$$

where $\mathcal{S}(T) = \{y \in Y : \exists x_n \to 0 \text{ with } T(x_n) \to y \text{ in probability}\}$ is the *separating subspace* of T. The value $\alpha(T)$ may be interpreted as *the probability of T having closed graph* (we observe that the graph of T is closed if, and only if, $\alpha(T) = 1$).

THEOREM 3 (Random closed-graph theorem) [18, Theorem 7]. *Let T be a linear random operator from a Banach space X to a Banach space Y. Then the probability with which T is continuous coincides with the probability with which T has closed graph. More precisely, $\beta(T)$ is attained as a maximum, $\alpha(T)$ is attained as a minimum, and the two numbers coincide.*

As a consequence of this result, other quantifications of the probability with which a linear random operator T is continuous may be obtained. For instance, it has been shown in [18, Corollary 8] that this probability equals the value of the following limit

$$\lim_{\varepsilon \to 0} \underline{\lim}_{x \to 0} \mathbb{P}[\|T(x)\| \leq \varepsilon].$$

In a similar way, given a linear random operator $T : A \to \mathcal{L}_0(\Omega, A)$ on a Banach algebra A, we consider the number

$$\delta(T) = \inf\{\mathbb{P}[T(ab) = T(a)b + aT(b)],\ a, b \in A\}$$

as *the probability with which T is a derivation.* The following generalization of Theorem 2 has been proved in [16; Corollary 4.3].

THEOREM 4. *The probability with which a linear random operator T on a semisimple Banach algebra A is continuous is at least the probability with which T is a derivation. More precisely, $\delta(T)$ is attained as a maximum and $\alpha(T) = \beta(T) \geq \delta(T)$.*

Finally, we define a *random Jordan derivation* on a Banach algebra A as a linear random operator D on A such that

$$D(a^2) = D(a)a + aD(a), \text{ almost surely } \forall a \in A.$$

We can improve the result proved by Cusack [6] and Brešar [5] as follows [17; Corollary 4.3].

THEOREM 5. *A linear random operator on a semiprime Banach algebra is a random Jordan derivation if, and only if, it is a random derivation.*

Therefore, the next result on automatic continuity is a consequence of the random Johnson- Sinclair theorem.

THEOREM 6. *Every random Jordan derivation acting on a semisimple Banach algebra is continuous.*

Since Jordan derivations can be regarded as random Jordan derivations, the above theorem is a generalization of the corresponding classical result.

For every linear random operator T on a Banach algebra A, we can consider *the probability with which T is a Jordan derivation* as the number

$$\delta_J(T) = \inf\{\mathbb{P}[T(a^2) = T(a)a + aT(a)],\ a \in A\}$$

The following more general result has been obtained in [17; Corollary 4.7].

THEOREM 7. *The probability with which a linear random operator T on a semisimple Banach algebra A is continuous is at least the probability with which T is a Jordan derivation. More precisely, $\delta_J(T)$ is attained as a maximum and $\alpha(T) = \beta(T) \geq \delta_J(T)$.*

REFERENCES

1. Bade, W. G. & Curtis P. C. , Jr., Prime ideals and automatic continuity problems for Banach algebras. *J. Functional Anal. 29*, (1978), 88-103.
2. Bade, W. G. & Curtis P. C. , Jr., Continuity of derivations of Banach algebras. *J. Functional Anal. 16*, (1974), 372-387.
3. Bade, W. G. & Dales H. G., Continuity of derivations from radical convolution algebras. *Studia Mathematica, 95* (1989), 59-91.
4. Bharucha-Reid A.T., *Random Integral Equations*, Academic Press, New York, 1972.
5. Brešar M., Jordan derivations on semiprime rings, *Proc. Amer. Math. Soc., 104*, (1988), 1003-1006.
6. Cusack, J., Jordan derivations on rings, *Proc. Amer. Math. Soc., 53*, (1975), 321-324.
7. Cusack, J., Automatic continuity and topologically simple radical Banach algebras, *London Math. Soc., 12*, (1977), 493-500.
8. Garimella R. V., Continuity of derivations on some semiprime Banach algebra, *Proc. Amer. Math. Soc., 99*, (1987), 289-292.
9. Johnson, B. E. & Sinclair, A. M., Continuity of derivations and a problem of Kaplansky, *Amer. J. Math., 90*, (1968), 1067-1073.
10. Kaplansky I., Derivations on Banach algebras, *Seminars on analytic functions, 2*, Princeton, 1958.
11. Ledoux, M. & Talagrand, M., *Probability in Banach Spaces*, Springer-Verlag, Berlin, 1991.
12. Ringrose, J. R., Automatic continuity of derivations of operator algebras, *London Math. Soc., 2*, (1972), 432-438.

13. Sinclair A. M., Jordan homomorphisms and derivations on semisimple Banach algebras, *Proc. Amer. Math. Soc, 24* (1970), 209-214.
14. Sakai, S., On a conjecture of Kaplansky. *Tōhoku Math. J. 12* (1960), 31-33.
15. Skorohod, A. V., *Random Linear Operators*, D. Reidel Publishing Company, Holland, 1984.
16. Velasco, M. V. & Villena, A. R., Continuity of random derivations, *Proc. Amer. Math. Soc.* (to appear).
17. Velasco, M. V. & Villena, A. R., Random Jordan derivations, *Proc. III International Conference on Non Associative Algebra and its Applications.* Kluwer Academic Publisher (to appear).
18. Velasco, M. V. & Villena, A. R., A random closed graph theorem, *Submitted for publication.*
19. Velasco, M. V. & Villena, A. R., On the continuity of random operators, *Submitted for publication.*

Subdifferentiability and the Noncommutative Banach–Stone Theorem

W. WERNER Universität-GH, Fachbereich 17, 33095 Paderborn, Germany

1 Introduction

When Banach proved his version of what nowadays is known as the Banach-Stone theorem [4, Théorème IV.3], he used smooth points of the unit sphere of the underlying Banach spaces as a geometric invariant. (Here is one possible statement of this result: The linear isometric isomorphism between two spaces of continuous functions $C(K_1)$ and $C(K_2)$ are precisely the maps $f \mapsto u\Phi(f)$, where u denotes a unitary element of $C(K_2)$ and Φ is a *-algebraic isomorphism.) Our chief concern here is the extension by Kadison to the non-commutative case, i.e. to isometries that operate between two C*-algebras $\mathfrak{A}$ and $\mathfrak{B}$. It seems impossible to directly use one of the more classical concepts of differentiability for the proof of Kadison's result (see [32], [33] for a characterization of the points at which the norm has one of these properties). Nevertheless, a certain type of subdifferentiability is well suited even for the present situation: We will show that results of [9] allow to prove the non-commutative case as well.

There are nowadays numerous results that share some similarity with the Banach-Stone theorem, and it certainly cannot be the objective of this note to review them all. (Results that are close to the original statement are surveyed in [5], [17] as well as [28], for some more recent developments see e.g. [6], [7] or [18].) We prove Kadison's result in §3 and devote §2 to explain how the non-commutative case fits into the theory of infinite dimensional holomorphy.

Though we never will explicitly use or prove a result on the facial structure of any of the balls $B_{\mathfrak{A}} = \{a \in \mathfrak{A} \mid \|a\| \leq 1\}$ or $B_{\mathfrak{A}'}$, our proof comes closest to the one in [11], where exposed faces of $\mathfrak{A}'$ are in use. We try to clarify this point in §4.

For all unexplained notation of C*-algebra theory we refer the reader to [22].

2 Bounded symmetric domains and the Kaup-Upmeier theorem

We begin this section with a precise statement of Kadison's result:

THEOREM 1 The linear isometric isomorphism between two unital C*-algebras $\mathfrak{A}$ and $\mathfrak{B}$ are precisely the maps

$$f \mapsto u\Psi(f),$$

where u denotes a unitary element of $\mathfrak{B}$ and Ψ is a unital JB*-isomorphism.

'JB*' is short-hand for 'Jordan-Banach*'. Jordan algebras first appear in a paper of Jordan [19] in which he addresses the problem of finding an algebraic structure on a set of quantum mechanical observables, e.g. the space of all self-adjoint operators on a Hilbert space. Whereas the ordinary product of two such operators in general fails to be self-adjoint, squares and sums do share this property, as in consequence does

$$\frac{1}{2}\left[(A+B)^2 - A^2 - B^2\right] = \frac{AB+BA}{2} =: A \circ B,$$

the so called Jordan product of the two operators A and B.

For our purposes it is sufficient to know that, roughly and up to some exotic cases, it is a good idea to think of a JB*-algebra as a closed, $*$-invariant Jordan subalgebra of a C*-algebra. (At least for factors this fits the picture pretty well — see [14] for the whole story and [29] for a more recent survey.) Finally, a JB*-homomorphism is a complex linear and bounded mapping Ψ respecting the Jordan product as well as the $*$-operation. It is furthermore called unital, whenever $\Psi(1) = 1$.

The formal resemblance the Jordan product for C*-algebras has with the Lie bracket certainly does not suffice to explain the somewhat unexpected relation Theorem 1 has with holomorphy.

The first observation in this direction one can make is the fact that the structure of a (unital) C*-algebra makes it possible to define what is meant by a 'Möbius transformation' of the domain $D_{\mathfrak{A}}$, the interior of $B_{\mathfrak{A}}$: Given $a \in D_{\mathfrak{A}}$, we put

$$M_a(z) := (1-aa^*)^{-1/2}(a-z)(1-a^*z)^{-1}(1-a^*a)^{1/2}.$$

(Note that it is non-commutativity that makes this transformation look somewhat unfamiliar — exchange factors and you'll end up with the standard form of a Möbius transform.) As in the one-dimensional case, M_a transposes a and 0 (note that $a(1-a^*a)^{1/2} = (1-aa^*)^{1/2}a)$ and is furthermore holomorphic: There are n-linear forms $\widehat{p}_n$ on $\mathfrak{A}$ so that for the corresponding homogeneous monomials $p_n(z) := \widehat{p}_n(z, \cdots, z)$ we have for each $z \in D_{\mathfrak{A}}$ a representation $M_a(z) = \sum_{n=1}^{\infty} p_n(z)$. Furthermore, M_a is bijective on $D_{\mathfrak{A}}$: its inverse actually is given by M_{-a}, and so $M_a \in \operatorname{Aut} D_{\mathfrak{A}}$, the group of all biholomorphic automorphism of $D_{\mathfrak{A}}$. Now let $z_0 \in D_{\mathfrak{A}}$ be given and denote by R the map $z \to -z$. Then $\Phi := M_{z_0}^{-1} R M_{z_0}$ is a reflection (i.e. $\Phi^2 = \Phi$), and z_0 is an isolated (in fact the only) fixed point of Φ. Traditionally, a domain D in a Banach space X with these properties is called a (bounded) symmetric domain.

The vast majority of the domains one encounters as the open unit balls of C*-algebras, of course, is of infinite dimension. The theory of such domains often follows closely the pattern of their finite dimensional counterparts; there are equally often cases, however, where this is far from being true. One cannot hope, for example, to completely classify these domains as it is possible in the finite dimensional case (such a theory, among other things, would have to contain, as we have seen, a complete list of all C*-algebras). Another problem concerns the appropriate choice of topologies: Whereas in the finite dimensional case uniform convergence on compact sets is the natural candidate for holomorphic function spaces, there is no substitute for this topology that would be satisfactory in all cases. So one of the seemingly natural topologies (the topology of uniform convergence on balls $B \subseteq D$ with $d(B, \partial D) > 0$) converts $\operatorname{Aut} D$ into a topological group, which in the case of D_{ℓ^∞} turns out to be completely disconnected and not discrete [36, Théorème 2.7.]. It hence follows that $\operatorname{Aut} D$ furnished with this topology cannot be a Banach-Lie group — remember that a chart of a Banach manifold M is a *homeomorphism* from an open set $U \subseteq M$ onto an open set of a – fixed – Banach space X. That $\operatorname{Aut} D$ nevertheless carries a topology that permits to define the structure of a real analytic Banach-Lie group was finally shown in [34] and [35].

Though no Lie group structure on $\operatorname{Aut} D$ was known for some time, everybody seems to have been convinced that should there exist such a thing, the Lie algebra $\mathfrak{aut}\, D$ of $\operatorname{Aut} D$ would have be what it always (i.e. in the finite dimensional case) was: the Lie algebra of complete holomorphic vector fields defined on D. (Let us recall that a holomorphic vector field $X : D \to \mathfrak{A}$ is said to be complete iff for all $z_0 \in D$ the initial value problem $\phi'(t) = X(\phi(t))$, $\phi(0) = z_0$ has a solution ϕ defined on the whole of $\mathbf{R}$.) This idea turned out to be correct, and showed even quite helpful in the construction of the Lie group topology on $\operatorname{Aut} D$.

In the case of a C*-algebra $\mathfrak{A}$ complete holomorphic vector fields on $D_{\mathfrak{A}}$ are the mappings

$$(*) \qquad z \mapsto a + i(hz + d(z)) - za^*z,$$

where $a \in \mathfrak{A}$, $h \in \mathfrak{A}_{sa}$, and $d : \mathfrak{A} \to \mathfrak{A}$ is a *-derivation, i.e. $d(xy) = d(x)y + xd(y)$ and $d(x^*) = -d(x)^*$ for all $x, y \in \mathfrak{A}$. (Here we made tacitly use of Sinclair's characterization of hermitian operators *on* C*-algebras [31].) The Lie bracket of two vector fields $X_k(z) = a_k + i(h_k z + d_k(z)) - za_k^* z$, $k = 1, 2$, is given by

$$\begin{aligned}
\{X_1, X_2\}(z) &= X_1'(z)(X_2(z)) - X_2'(z)(X_1(z)) \\
&= i(H_1(a_2) - H_2(a_1)) \\
&\quad + [H_2, H_1](z) - (a_2 a_1^* - a_1 a_2^*)z - z(a_1 a_2^* - a_2 a_1^*) \\
&\quad + iz\left((H_1(a_2) - H_2(a_1))^* z,\right.
\end{aligned}$$

where we have set $H_k(z) = h_k z + d_k(z)$. Inspection of $(*)$ yields that we have a decomposition $\mathfrak{aut}\, D_{\mathfrak{A}} = \mathfrak{k} \oplus \mathfrak{p}$: $\mathfrak{k}$ consists of polynomials P that satisfy $P(-z) = -P(z)$, whereas $\mathfrak{p}$ contains those P with $P(z) = P(-z)$. Moreover, for each $a \in \mathfrak{A}$ there is exactly one $P_a \in \mathfrak{p}$ with $P_a(0) = a$. If $\widehat{Q}_a$ denotes the symmetrized bilinear form such that $P_a(z) = a - \widehat{Q}_a(z, z)$, then it is customary to write

$$\{x, y, z\} := \widehat{Q}_y(x, z),$$

and to call the expression $\{\cdot, \cdot, \cdot\}$ the (Jordan) triple product. This 'product' is symmetric bilinear in the two outer variables and antilinear in the middle variable. (It has some other important properties which we omit.) In the case of a C*-algebra $\mathfrak{A}$ we have by the above

$$\{x, y, z\} = \frac{1}{2}(xy^*z + zy^*x).$$

The connection to Theorem 1 is displayed by the formula

$$(**) \qquad \{x, y, z\} = (x \circ y^*) \circ z + (z \circ y^*) \circ x - (x \circ z) \circ y^*.$$

For, if Φ is an isometric isomorphism between unital C*-algebras $\mathfrak{A}$ and $\mathfrak{B}$, then, by Theorem 1 and $(**)$,

$$\Phi\{x, y, z\} = u\{\Psi(x), \Psi(y), \Psi(z)\} = \{\Phi(x), \Phi(y), \Phi(z)\}.$$

Conversely, each such triple isomorphism between C*-algebras is as in Theorem 1 (this again is a purely algebraic statement and follows from letting $\Psi(a) = \Phi(1)^*\Phi(a)$ — see [15, p. 25] for details) and consequently, we have the following concise reformulation of Theorem 1:

THEOREM 1* A linear map Φ is an isometric isomorphism between two unital C*-algebras $\mathfrak{A}$ and $\mathfrak{B}$ iff for all $x, y, z \in \mathfrak{A}$

$$\Phi\{x, y, z\} = \{\Phi(x), \Phi(y), \Phi(z)\}.$$

Let us finally mention two results that might seem more natural than Theorem 1* to those who think that triple products look extraneous to C*-algebras: One may in fact define an abstract notion of a JB*-triple system. It consists of a vector space on which there is a suitably defined triple product $\{\cdot,\cdot,\cdot\}$, a *-operation and a (complete) norm, all of which are required to collaborate in a reasonable way. This abstract (and quite cumbersome) notion becomes interesting through a result of Kaup (which extends previous work of Koecher [25] to the infinite dimensional setting) which states that all bounded symmetric domains are, up to biholomorphic equivalence, exactly the open unit balls of these JB*-triple systems. Let us add that with the exception of the somewhat more explicit formulae, all structural properties of the open unit balls $D_{\mathfrak{A}}$ that appear above — the fact that elements of $\mathfrak{aut}\, D_{\mathfrak{A}}$ are polynomials of degree at most two, the Cartan decomposition $\mathfrak{aut}\, D_{\mathfrak{A}} = \mathfrak{k} \oplus \mathfrak{p}$, the behavior of this decomposition under the Lie bracket as well as the definition of the triple product — turn out to be common features of bounded symmetric domains (and hence of JB*-triple systems). The reader interested in details is referred to [2], [12], [15], [16], [36] and the references given there.

Theorem 1* now extends naturally to JB*-triple systems [23]:

THEOREM 2 Two JB*-triple systems are isometrically isomorphic as Banach spaces if and only if they are isomorphic as JB*-triple systems.

As was said, JB*-triple systems owe their existence to bounded symmetric domains. Consequently, 'isomorphism' of JB*-triple could as well be read 'biholomorphic equivalence of open unit balls'. The last link in this chain of results hence might be the following one (which is taken from [24]):

THEOREM 3 The open unit balls of two Banach spaces X and Y are biholomorphically equivalent iff X and Y are isometrically isomorphic. Furthermore, each biholomorphic mapping $\Phi : D_X \to D_Y$ has the form $\Phi = \phi \circ T$, where T is a linear isometry from X onto Y and $\phi \in \operatorname{Aut} D_Y$.

3 A proof of Kadison's result

LEMMA 1 If $\mathfrak{M}$ and $\mathfrak{N}$ are von Neumann algebras equipped with their natural JB*-structure, then $\Psi : \mathfrak{M} \to \mathfrak{N}$ is a JB*-homomorphism if and only if $\Psi(\operatorname{Proj}\mathfrak{M}) \subseteq \operatorname{Proj}\mathfrak{N}$. It is injective iff $\Psi(p) \neq 0$ for all non-zero projections p.

PROOF: Suppose that Ψ maps projections onto projections. Let us first observe that this implies $\Psi(p)\Psi(q) = 0$ whenever the two projections p and q are orthogonal: In fact, since

$p+q$ in this case is a projection, we must have $\Psi(p)\Psi(q) = -\Psi(q)\Psi(p)$. But then $\Psi(p)\Psi(q) = -\Psi(p)\Psi(q)\Psi(p) = \Psi(q)\Psi(p)$ since Ψ preserves projections, and so $\Psi(p)\Psi(q) = 0$. Next, any self-adjoint element in $\mathfrak{M}$ is the norm-limit of a sequence of elements of the form $\sum_{i=1}^{n} \alpha_i p_i$ for some real scalars α_i and some orthogonal projections p_i. It therefore follows that $\Psi(a^2) = \Psi(a)^2$ for every self-adjoint element $a \in \mathfrak{M}$. If we write for an arbitrary element $x = h_1 + ih_2$ with $h_{1,2}$ self-adjoint, then it follows from the equation $x^2 = h_1^2 + i\left[(h_1+h_2)^2 - h_1^2 - h_2^2\right] - h_2^2$ that also $\Psi(x^2) = \Psi(x)^2$. But this implies $\Psi(a \circ b) = \frac{1}{2}\left(\Psi((a+b)^2) - \Psi(a^2) - \Psi(b^2)\right) = \Psi(a) \circ \Psi(b)$. In a similar way, $\Psi(a)$ must be self-adjoint for self-adjoint elements a, and, if x is an arbitrary element in $\mathfrak{M}$, $\Psi(x^*) = \Psi(h_1) - i\Psi(h_2) = \Psi(x)^*$.

Should $\Psi(p)$ be different from zero as soon as p is, then, as above, Ψ restricted to self-adjoint elements of the form $\sum_{i=1}^{n} \alpha_i p_i$ with $p_i \perp p_j$ for $i \neq j$ is injective. Since $\|\sum_{i=1}^{n} \alpha_i p_i\| = \max|\alpha_i|$ the same holds true for the restriction of Ψ to self-adjoint elements. Since Ψ preserves the *-operation, it must be injective on the whole of $\mathfrak{M}$.

For the reverse direction, note that by its very definition any JB*-homomorphism Ψ must satisfy $\Psi(a^2) = \Psi(a \circ a) = \Psi(a)^2$, preserve self-adjointness and hence map projections onto projections. □

The differentiability condition we would like to apply in the proof of Theorem 1 is the subject of the following

DEFINITION A real-valued function ψ defined on a Banach space X is said to be strongly subdifferentiable at a point $x \in X$ whenever the limit

$$\lim_{t \to 0+} \frac{\psi(x+tu) - \psi(x)}{t}$$

exists uniformly for $u \in S_X$, the unit sphere of X.

Note that for fixed $u \in S_X$ the above limit always exists in case that ψ is the norm function. When $\mathfrak{A}$ is a C*-algebra, the points of strong subdifferentiability for the norm have been identified in [9, Theorem 1]:

THEOREM 4 Let $\mathfrak{A}$ be a C*-algebra and suppose $a \in S_{\mathfrak{A}}$. Then the following conditions are equivalent.

(i) The norm of $\mathfrak{A}$ is strongly subdifferentiable at a

(ii) There is a partial isometry $v \in \mathfrak{A}$ such that

$$a \in F_{v,0} = \{x \in S_{\mathfrak{A}} \mid xv^* = vv^* \quad \text{and} \quad \|a - v\| < 1\}.$$

(iii) 1 is an isolated point in $\sigma(|a|)$.

Let x be an element of a Banach space X and put $\partial(x) = \{x' \in S_{X'} \mid x'(x) = \|x\|\}$. The next lemma mainly is a consequence of [9, Theorem 1] — rather to its proof than to the theorem itself.

LEMMA 2 Let $a \in S_{\mathfrak{A}}$ with polar decomposition $a = u|a|$ and suppose that v is a partial isometry in $\mathfrak{A}$ such that

$$av^* = vv^* \qquad \text{and} \qquad \|a - v\| < 1.$$

If p is a projection in $\mathfrak{A}$ with

$$|a|p = p \qquad \text{and} \qquad \| |a| - p\| < 1,$$

then $v^*v = p$ and $v = up$. Furthermore, we have $\partial(|a|) = \partial(p)$ and $\partial(a) = \partial(v)$.

PROOF: By assumption and Theorem 4(ii), any projection p with $|a|p = p$ is dominated by the spectral projection p_1 pertaining to $1 \in \sigma(|a|)$. If, in addition, $\|p - |a|\| < 1$, then $\|p - p_1\| = \|(p - |a|)p_1\| < 1$ and so $p = p_1$. We infer from the proof of (iii)⇒(i) of [9, Theorem 1] that necessarily $p_1 = p = v^*v$. In the same vein, $v = vv^*v = av^*v = ap_1$ and there can only be one partial isometry with $av^* = vv^*$ and $\|a - v\| < 1$. It follows from the proof of (ii)⇒(iii) of [9, Theorem 1] that $v = up$.

Recall from the same part of this proof that $\partial(|a|) = \partial(p)$. But if $\phi(v) = 1$, then $\phi u(p) = \phi(v) = 1$ and so $\phi(a) = \phi u(|a|) = 1$. In a similar way, $\phi(v) = 1$ whenever $\phi(a) = 1$. So $\partial(a) = \partial(v)$. □

LEMMA 3 Let v be a partial isometry, $a \in S_{\mathfrak{A}}$. Then

(i) $a \in F_{v,0}$ iff $\partial(a) = \partial(v)$.

(ii) If $x \in F_{v,0}$ satisfies $\|x - a\| < 1$ for all $a \in F_{v,0}$, then $x = v$.

PROOF: That $a \in F_{v,0}$ implies equality between $\partial(a)$ and $\partial(v)$ was observed in Lemma 2, and thus the first half of (i) already has been proven. Suppose conversely that $\partial(a) = \partial(v)$. Since the norm of a Banach space X is strongly subdifferentiable at a point x in a Banach space iff $\partial(x)$ is strongly exposed by x [13] this property then passes in light of Theorem 4 from the partial isometry v to a. It follows that there is a partial isometry w so that $aw^* = ww^*$ and $\|a - w\| < 1$. The first part of this proof permits to conclude that $\partial(w) = \partial(a) = \partial(v)$, which implies $v = w$ (and thus the proof has ended). In fact, suppose that $\mathfrak{A}$ is isomorphically represented as a C*-algebra of bounded operators on some Hilbert space $\mathcal{H}$. Then, in case v differs from w, it is easy to see that for one of them, v say, there is $\xi \in S_{\mathcal{H}}$, for which $\langle v(\xi), v(\xi)\rangle = 1$ but $\langle w(\xi), v(\xi)\rangle \neq 1$. So $a \mapsto \langle a(\xi), v(\xi)\rangle$ is in $\partial(v) \setminus \partial(w)$.

For the proof of (ii) suppose that $x \neq v$ and write $x = u|x|$ for the polar decomposition of x. Then, since by Lemma 2 $uv^*v = v$, it follows that $|x| \neq v^*v$, and we may invoke the continuous functional calculus to find $\xi \in C^*(|x|)$, the C*-algebra generated by $|x|$, with

$$v^*v\xi = v^*v, \qquad \|v^*v - \xi\| < 1 \qquad \text{and} \qquad \| |x| - \xi\| \geq 1$$

(In fact, if $C^*(|x|) = C_0L$ for some locally compact Hausdorff space L, and $v^*v = \chi_F$ with F a clopen subset of L, then the set $U = \{l \in L \setminus F \mid |x|(l) > 0\}$ is not empty. Fix $l_0 \in U$ and $\xi \in C_0L$ so that

$$-1 + \frac{|x|(l_0)}{2} \leq \xi|_{L\setminus F} \leq 0, \qquad \xi(l_0) = -1 + \frac{|x|(l_0)}{2}, \qquad \text{and} \qquad \xi|_F = 1.$$

Then $v^*v\xi = v^*v$, $\|v^*v - \xi\| = \|\xi|_{L\setminus F}\| < 1$ and $\| |x| - \xi\| \geq |x|(l_0) - \xi(l_0) > 1$.)

Since $v = uv^*v$ we have $uv^* = vv^*$ and, by part (i), $v^*u = v^*v$. Hence

$$v^*(u\xi) = v^*v, \qquad \|v - u\xi\| = \|uv^*v - u\xi\| < 1,$$

as well as $\|x - u\xi\| \geq \|u^*x - u^*u\xi\| = \| |x| - \xi\| \geq 1$. This contradiction shows (ii). □

PROOF OF THEOREM 1 We will first settle the case of two von Neumann algebras $\mathfrak{M}$ and $\mathfrak{N}$ and an isometric isomorphism $\Phi : \mathfrak{M} \to \mathfrak{N}$. Fix a partial isometry $v \in \mathfrak{M}$. Since by Theorem 4, v must be a point of strong subdifferentiability, $\Phi(v)$ must, and there is a partial isometry $w \in \mathfrak{N}$ with $\Phi(v)w^* = ww^*$ and $\|w - \Phi(v)\| < 1$. Denote by $\Phi' : \mathfrak{N}' \to \mathfrak{M}'$ the adjoint of Φ as linear operator between Banach spaces. By Lemma 3(i) we have, for any $x \in F_{v,0}$, $\partial(\Phi(x)) = \Phi'^{-1}\partial(x) = \Phi'^{-1}\partial(v) = \partial(\Phi(v)) = \partial(w)$ so that $\Phi(F_{v,0}) \subseteq F_{w,0}$. Similarly, when $y \in F_{w,0}$, we have $\partial(\Phi^{-1}(y)) = \Phi'\partial(y) = \Phi'\partial(w) = \Phi'\partial(\Phi(v)) = \partial(v)$, and so $\Phi(F_{v,0}) = F_{w,0}$. But $\|\Phi(v) - \Phi(x)\| < 1$ for all $x \in F_{v,0}$, hence Lemma 3(ii) implies that for all partial isometries $v \in \mathfrak{M}$ we must have

$$\Phi(F_{v,0}) = F_{\Phi(v),0}.$$

Notice that also $\Phi(F_v) = F_{\Phi(v)}$, where $F_v = \{a \in S_{\mathfrak{A}} \mid av^* = vv^*\}$: For, if $x \in F_v$, then, by convexity, the whole line segment $[x, v]$ belongs to F_v and therefore, $x = v + tx_0$, for some $t \geq 1$ and $v + x_0 \in F_{v,0}$. Since $\Phi(v + x_0) \in F_{\Phi(v),0}$, we have $\Phi(x_0)\Phi(v)^* = 0$ and so $\Phi(x)\Phi(v)^* = \Phi(v)\Phi(v)^*$, hence $\Phi(x) \in F_{\Phi(v)}$. The reverse inclusion follows in quite the same way.

Let $\Psi : \mathfrak{M} \to \mathfrak{N}$ be defined by

$$\Psi(a) := \Phi(a)\Phi(1)^*.$$

For any projection $p \in \mathfrak{M}$ we have $1 \in F_p$, hence

$$\Psi(p) = \Phi(p)\Phi(1)^* = \Phi(p)\Phi(p)^*.$$

Since $\Phi(p)$ by the above must be a partial isometry, $\Phi(p)\Phi(p)^*$ is a projection, which by assumptions made on Φ is different from zero. In consequence, by Lemma 1, Ψ is an injective JB*-homomorphism. We are thus left with showing that $\Phi(1)$ is unitary — it is then automatic that Ψ is a unital JB*-isomorphism. To this end, pick $a_0 \in \mathfrak{M}$ with $\Phi(a_0) = 1 - \Phi(1)^*\Phi(1)$. Then, since $\Phi(1)$ is a partial isometry,

$$\Psi(a_0) = (1 - \Phi(1)^*\Phi(1))\Phi(1)^* = 0$$

and therefore — Ψ was injective — $\Phi(1)^*\Phi(1) = 1$. We apply quite the same reasoning to $\Phi^* : a \mapsto \Phi(a^*)^*$ to show that $\Phi(1)\Phi(1)^* = 1$ and so, $\Phi(1)$ is unitary.

If, finally, $\mathfrak{A}$ and $\mathfrak{B}$ are arbitrary unital C*-algebras, then the above yields a unitary element $U \in \mathfrak{B}''$ and a JB*-isomorphism $\Psi_0 : \mathfrak{A}'' \to \mathfrak{B}''$ with $\Phi''(a) = U\Psi_0(a)$ for all $a \in \mathfrak{A}''$. Since $1 \in \mathfrak{A}$ and $\Psi_0(1) = 1$, it results that $U \in \mathfrak{B}$ and thus $\Psi_0(\mathfrak{A}) \subseteq \mathfrak{A}$. The conclusion now follows with U as before and $\Psi := \Psi_0|_{\mathfrak{A}}$. □

REMARKS

1. The case of Theorem 1 in which $\mathfrak{A}$ and $\mathfrak{B}$ are not supposed to be unital was settled in [27]. The difference to Theorem 1 consists in the fact that, in general, the unitary element u no longer belongs to $\mathfrak{B}$ but to the multiplier algebra of $\mathfrak{B}$. (Theorem 1*, by the way, remains unaltered when no unit is assumed to exist.)

2. For some extensions of Theorem 1 the reader is referred to [3], [8], [10], [21], [26] or [30].

3. It is possible to extend the method we apply here to the more general setting of BJ*-triple systems, i.e. one can prove a characterization similar to Theorem 4 for triple systems and use this result to obtain Theorem 2.

4 An afterthought: C*-algebras and faces

The sets F_u appearing in the above are easily seen to be (norm closed) faces of the unit ball of the C*-algebra $\mathfrak{A}$. By a result of Akeman and Pedersen [1, Theorem 4.10.] any face of $B_{\mathfrak{A}}$ is of the form $\{x \in S_{\mathfrak{A}} \mid xu^* = uu^*\}$, where, however, the partial isometry u in general belongs to $\mathfrak{A}''$. It turns out that there are two classes of faces. One (and hence the other) might be characterized as follows:

OBSERVATION For a norm-closed face F_u of the unit ball of a C*-algebra $\mathfrak{A}$ the following conditions are equivalent:

(i) There is a point x in F_u with $\|x - u\| < 1$

(ii) u belongs to $\mathfrak{A}$

In fact, if there is any x in F_u with $\|x-u\| < 1$ then x is a point of strong subdifferentiability for the norm of $\mathfrak{A}''$ and hence in $\mathfrak{A}$. So there is a partial isometry $w \in \mathfrak{A}$ such that $x \in F_{w,0}$. But $\partial(u) = \partial(x) = \partial(w)$ (in $\mathfrak{A}'''$) and so $u = w$. (Note that there are faces that contain points of strong subdifferentiability whose defining partial isometry does not belong to $\mathfrak{A}$.)

Also the norm closed faces of the dual unit ball of $\mathfrak{A}$ are explicitly known: They are of the form $F^u = \{\phi \in B_{\mathfrak{A}'} \mid \phi(u) = 1\}$, where u is a partial isometry in $\mathfrak{A}''$ ([1, Theorem 4.11.]). This result reveals a special feature of faces in the dual of a C*-algebra: By their very definition, they are exposed, (take $\ell^\infty(\Gamma)$, Γ an uncountable set, in duality with $\ell^1(\Gamma)$ to see that there are weak*-closed faces which are not exposed) and a moment's reflection shows that they are even strongly exposed by u, i.e. $u(x'_n) \to 1$ for a sequence $(x'_n) \subseteq B_{\mathfrak{A}'}$ entrains $d(x'_n, F^u) \to 0$): In fact, since u is a point of strong subdifferentiability of the norm in $\mathfrak{A}''$ it follows that $u(x'_n) \to 1$ implies $\|i_{X'}(x'_n) - \xi_n\| \to 0$ for some $\xi_n \in B_{\mathfrak{A}'''}$ norming u. By the principle of local reflexivity [37, II.E.14] this yields $\overline{\xi}_n \in F^u$ with $\|x'_n - \overline{\xi}_n\| \to 0$. Finally, the face F^u is weak*-closed, iff $F_u \neq \emptyset$, and it follows from the observation we made before that F^u is strongly exposed by an element $x \in \mathfrak{A}$ iff $u \in \mathfrak{A}$.

In [8] the authors use (and, independently, characterize) the faces F^u as the decisive geometrical invariant for their proof of Theorem 1 (as well as Theorem 2). So one might think of the present proof as one that is obtained by applying the 'facear' operation in the sense of [1]. The reader should note, however, that an argument which is based exclusively on the faces F_u still would lack a method that singles out the defining partial isometry u among the many others that F_u might contain.

REFERENCES

1. C. A. Akemann and G. K. Pedersen, *Facial structure in operator algebra theory.* Proc. London Math. Soc. **64** (1992) 418–448.
2. J. Arazy, *An application of infinite dimensional holomorphy to the geometry of Banach spaces.* In: *Geometrical Aspects of Functional Analysis*, Lect. Notes Math. 1267, Springer, 1987, 122–150.
3. J. Arazy, *Isometries of Banach algebras satisfying the von Neumann inequality.* Math. Scand (to appear).

4. S. Banach, *Théorie des opérations linéaires.* Monografie matematyczne, Warszawa, 1932.
5. E. Behrends, *M-structure and the Banach-Stone theorem.* Lect. Notes Math. 736, Springer, 1979.
6. E. Behrends, *Isomorphic Banach-Stone theorems and isomorphisms which are close to isometries.* Pac. J. Math. **133** (1988) 229–250.
7. M. Cambern and K. Jarosz, *Isometries of spaces of weak* continuous functions* Proc. Am. Math. Soc. **106** (1989) 707–712.
8. C. Chu, T. Dang, B. Russo and B. Ventura, *Surjective isometries of real C^*-algebras.* J. London Math. Soc. **47** (1993) 97–118.
9. M. Contreras, R. Payá and W. Werner, *C^*-algebras that are I-rings.* To appear.
10. D. P. O'Donovan and K. R. Davidson, *Isometric images of C^* algebras.* Can. Math. Bull. **27** (1984) 286–294.
11. T. Dang, Y. Friedman and B. Russo, *Affine geometric proofs of the Banach Stone theorems of Kadison and Kaup.* Rocky Mountain J. Math. **20** (1990) 409–428.
12. S. Dineen, *The Schwarz Lemma.* Oxford University Pres, 1989.
13. D. Gregory, *Upper semi-continuity of subdifferential mappings.* Canad. Math. Bull. **23** (1980) 11–19.
14. H. Hanche-Olsen and E. Størmer, *Jordan operator algebras.* Pitman Monographs and Studies in Mathematics 21, 1984.
15. L. A. Harris, *Bounded symmetric homogeneous domains in infinite dimensional spaces.* In: *Proceedings on Ininite Dimensional Holomorphy,* Lect. Notes Math. 364, Springer, 1974, 13–40.
16. L. M. Isidro and L. L. Stachó, *Holomorphic Automorphism Groups in Banach Spaces.* Math. Studies 105, North Holland, 1984.
17. K. Jarosz, *Perturbations of Banach Algebras.* Lect. Notes Math. 1120, Springer, 1985.
18. K. Jarosz, *Small isomorphisms of $C(X, E)$ spaces.* Pac. J. Math. **138** (1989) 295–315.
19. P. Jordan, *Über Verallgemeinerungsmöglichkeiten des Formalismus der Quantenmechanik.* Nachr. Akad. Wiss. Göttingen **41** (1933) 209–217.
20. R. V. Kadison, *Isometries of operator algebras.* Ann. of Math. **54** (1951) 325–338.
21. R. V. Kadison, *Transformation of states in operator theory and dynamics.* Topology **3** (1965) 177–198.
22. R. V. Kadison and J. R. Ringrose, *Fundamentals of the theory of operator algebras I,II.* Academic Press, New York-London, 1983/86.
23. W. Kaup, *A Riemann mapping theorem for bounded symmetric domains in complex Banach spaces.* Math. Z. **183** (1983) 503–529.
24. W. Kaup and H. Upmeier, *Banach spaces with biholomorphically equivalent unit balls are isomorphic.* Proc. Amer. Math. Soc. **58** (1978) 129–133.
25. M. Koecher, *An elementary approach to bounded symmetric domains.* Rice University, Houston, 1969.
26. R. L. Moore and T. T. Trent, *Isometries of nest algebras.* J. Funct. Anal. **86** (1989) 180–209.
27. A. Paterson and A. Sinclair, *Characterization of isometries between C^*-algebras.* J. London Math. Soc. **2** (1972) 755–761.
28. R. Rochberg, *Deformation theory for uniform algebras: An introduction.* In: *Banach algebras and several complex variables, Proc. Conf., New Haven* Contemp. Math. **32** (1984) 209–216.
29. A. Rodríguez Palacios, *Jordan structures in Analysis.* In W. Kaup, K. McCrimmon, and H. P. Petersson (ed.) *Jordan Algebras: Proc. Conf. Oberwolfach 1992* Walter de Gruyter, Berlin, 1994, 97–186.

30. F. W. Shultz, *Dual maps of Jordan homomorphisms and *-homomorphisms between C^*-algebras.* Pac. J. Math. **93** (1981) 435–441.
31. A. Sinclair, *Jordan homomorphisms and derivations on semi-simple Banach algebras.* Proc. Amer. Math. Soc. **24** (1970) 209–214.
32. K. F. Taylor and W. Werner, *Differentiability of the norm in von Neumann algebras.* Proc. Amer. Math. Soc. **119** (1993) 475–480.
33. K. F. Taylor and W. Werner, *Differentiability of the norm in C^*-algebras.* In: K. Bierstedt, A. Pietsch, W. Ruess, and D. Vogt (ed.) *Proc. Conf. Functional Analysis*, Lecture Notes in Pure and Applied Mathematics. Marcel Dekker, 1994, 329–344.
34. H. Upmeier, *Über die Automorphismengruppen von Banach-Manigfaltigkeiten mit invarianter Metrik.* Math. Ann. **223** (1976) 279–288.
35. J.-P. Vigué, *Sur le groupe des automorphismes analytiques d'un domaine borné d'un espace de Banach complexe.* CRAS **282** (1976) 111–114.
36. J.-P. Vigué, *Domaine bornés symétriques.* In: *Geometry Seminar Luigi Bianchi* Lect. Notes Math. 1022, Springer, 1983, 123–177.
37. P. Wojtaszczyk, *Banach spaces for analysts* Cambridge studies in advanced mathematics **25**, Cambridge University Press, 1991.

Index

Akopian's Theorem, 1-10
algebra of smooth functions, 167
amenability, 59
analytic functional, 246
Antosik's Interchange Theorem, 361-369
Apostol algebra, 224
AR-property, 85
Archimedean field, 29-30
automatic continuity, 223, 371

B_0-algebra, 12
Banach function algebra, 117-121, 139-144
Banach lattice, 265
Banach-Mazur distance, 51
Banach-Stone Theorem, 51, 377
band projection, 265
Bargman space, 123
barreled, 353-360
Bishop-Phelps Theorem, 19
Blaschke product, 147
bounded approximate identity, 117-121
bounded relative unit, 117-120
Boyd index, 323

C^*-algebra 377
Carleman class, 244
Carleson measure, 157
Carleson squares, 156
character, 139-144
Choquet boundary, 52
Christoffel-Darboux formula, 284
close homomorphism property, 65
conditional expectation operator, 102, 105
contractive projections, 101-108
Corona Theorem, 148
cyclic basis, 16

denting point, 114, 206, 339
δ-point, 113
Δ-point, 113
differential operator, 251
dilation, 124, 323
Ditkin, 117-121
Dungundji extension, 86

extended spectral radius 13
extended spectrum 13
extreme point, 114, 205, 339
Euclidean group, 126

f-algebra 29
fixed point property, 77-78
Fourier-Borel transform, 250
Fourier series, 282
Freud weights, 303

generalized Jacobi weights, 293
Gleason part, 118, 120-121, 151
Grothendieck property, 318
Grothendieck space, 314, 318-319

Haar measure, 124
Hadamard factorization theorem, 133
Hardy space 1-10, 318, 348
Hartogs' Theorem, 132
Hermite weight, 298
Hilbert transform, 282
$Hom(A)$, 139

interpolating Blaschke product, 148
interpolating sequence, 148
iterated series, 361

Jacobi weights, 291
JB^*-algebra, 378
Johnson-Sinclair Theorem, 371
Jordan derivation, 372

Kadec-Klee property, 72-74
Kadison Theorem, 378
Kaup-Upmeier Theorem, 378
Köthe space, 322
Krein-Milman Theorem, 86

L-projection, 265
L-space, 265
$L(l_p, l_q)$, 71-82
Laplace operator, 252
Laplace transformation, 247
Lebesgue-Bochner spaces, 101-108, 224, 311
Legendre polynomials, 285
Lindelöf property, 336
Liouville's Theorem, 133
Looman-Menchoff Theorem, 335
Lorentz space, 321
Luxembourg functional, 323

m-convex algebra 13
M-projection, 265
M-space, 265
Matuszewska-Orlicz index, 324
measurable cardinal, 140
modular ring, 30-31
Muckenhoupt's A_p condition, 301
multiplier, 184, 224

nearly uniformly convex, 76
nearly uniformly smooth, 76
normal structure, 78
norm-attaining 19-20
nuclear mapping, 128

operating function, 35-40, 43-48, 183
Orlicz space, 323
Orlicz-Lorentz space, 323
Orlicz-Pettis Theorem, 363
Orlicz vector-valued function space, 318
$Orth(A)$, 30
orthogonal polynomials, 283-284

paraproduct, 347
peak point, 52
Pełczyński property (V^*), 318
plurisubharmonicity, 125
Pollard's Decomposition, 287
polynomial, 140
potential theory, 303
primary ideals, 167
projective space, 126
projective representation, 126

radical Banach algebra, 64
random derivations, 371
Radon-Nikodym property, 22, 221
rearrangement invariant space, 322
retractions, 85-99
Riesz projection, 128
Riesz space, 29
Roberts sets, 86

Schauder decomposition, 104
Schatten-von Neumann ideal, 347
Schur's Lemma, 129
semisimple, 31
singular operators, 347
small-bound isomorphisms, 51-56
Sobolev algebra, 168
spectral approximation, 168
spectral synthesis, 167-180
strongly extreme point, 114, 206
subdifferentiability, 377
superdensity, 35-40

Szegö's class, 295

tensor product, 355, 359-360

ultraseparating space, 43-48
unconditional basis, 131, 134
uniform algebra, 117-121
uniformly complete, 29
uniformly convexifiable, 71, 81-82
uniformly Kadec-Klee property, 72, 74-81

weak amenability, 62
weakly amenable radical algebra, 67
weak peak point, 52

Zippin index, 323